Oceanography: Wave Dynamics

Oceanography: Wave Dynamics

Edited by Jeremy Harper

SYRAWOOD
PUBLISHING HOUSE

New York

Published by Syrawood Publishing House,
750 Third Avenue, 9th Floor,
New York, NY 10017, USA
www.syrawoodpublishinghouse.com

Oceanography: Wave Dynamics
Edited by Jeremy Harper

International Standard Book Number: 978-1-68286-661-0 (Hardback)

Cataloging-in-Publication Data

Oceanography : wave dynamics / edited by Jeremy Harper.
    p. cm.
Includes bibliographical references and index.
ISBN 978-1-68286-661-0
1. Oceanography. 2. Ocean waves. 3. Wave mechanics. 4. Fluid dynamics. I. Harper, Jeremy.
GC11.2 .O24 2019
551.46--dc23

TABLE OF CONTENTS

Permissions

List of Contributors

Index

PREFACE

The world is advancing at a fast pace like never before. Therefore, the need is to keep up with the latest developments. This book was an idea that came to fruition when the specialists in the area realized the need to coordinate together and document essential themes in the subject. That's when I was requested to be the editor. Editing this book has been an honour as it brings together diverse authors researching on different streams of the field. The book collates essential materials contributed by veterans in the area which can be utilized by students and researchers alike.

Oceanography is the study of oceans along with their physical and chemical processes. This field has relevance across a number of scientific fields such as hydrology, hydrography, meteorology, limnology, etc. Oceanography covers a range of topics including ocean acidification, ocean currents, wave dynamics, etc. From theories to research to practical applications, chapters related to all contemporary topics of relevance to this field have been included in this book. It covers some of the vital pieces of work being conducted across the world, on various topics related to oceanography. Those in search of information to further their knowledge will be greatly assisted by this book.

Each chapter is a sole-standing publication that reflects each author´s interpretation. Thus, the book displays a multi-facetted picture of our current understanding of application, resources and aspects of the field. I would like to thank the contributors of this book and my family for their endless support.

Editor

Observability of fine-scale ocean dynamics in the northwestern Mediterranean Sea

Rosemary Morrow[1], **Alice Carret**[1], **Florence Birol**[1], **Fernando Nino**[1], **Guillaume Valladeau**[2], **Francois Boy**[3], **Celine Bachelier**[4], and **Bruno Zakardjian**[5,6]

[1]LEGOS, IRD, CNRS, Université de Toulouse, Toulouse, 31400, France
[2]CLS Ramonville, St.-Agne, 31520, France
[3]CNES, Toulouse, 31400, France
[4]IRD, Brest, 29280, France
[5]Université de Toulon, CNRS, IRD, Mediterranean Institute of Oceanography (MIO), UM 110, 83957 La Garde, France
[6]Aix Marseille Université, CNRS, IRD, Mediterranean Institute of Oceanography (MIO), UM 110, 13288 Marseille, France

Correspondence to: Rosemary Morrow (rosemary.morrow@legos.obs-mip.fr)

Abstract. Technological advances in the recent satellite altimeter missions of Jason-2, SARAL/AltiKa and CryoSat-2 have improved their signal-to-noise ratio, allowing us to observe finer-scale ocean processes with along-track data. Here, we analyse the noise levels and observable ocean scales in the northwestern Mediterranean Sea, using spectral analyses of along-track sea surface height from the three missions. Jason-2 has a higher mean noise level with strong seasonal variations, with higher noise in winter due to the rougher sea state. SARAL/AltiKa has the lowest noise, again with strong seasonal variations. CryoSat-2 is in synthetic aperture radar (SAR) mode in the Mediterranean Sea but with lower-resolution ocean corrections; its statistical noise level is moderate with little seasonal variation. These noise levels impact on the ocean scales we can observe. In winter, when the mixed layers are deepest and the submesoscale is energetic, all of the altimeter missions can observe wavelengths down to 40–50 km (individual feature diameters of 20–25 km). In summer when the submesoscales are weaker, SARAL can detect ocean scales down to 35 km wavelength, whereas the higher noise from Jason-2 and CryoSat-2 blocks the observation of scales less than 50–55 km wavelength.

This statistical analysis is completed by individual case studies, where filtered along-track altimeter data are compared with co-located glider and high-frequency (HF) radar data. The glider comparisons work well for larger ocean structures, but observations of the smaller, rapidly moving dynamics are difficult to co-locate in space and time (gliders cover 200 km in a few days, altimetry in 30 s). HF radar surface currents at Toulon measure the meandering Northern Current, and their good temporal sampling shows promising results in comparison to co-located SARAL altimetric currents. Techniques to separate the geostrophic component from the wind-driven ageostrophic flow need further development in this coastal band.

1 Introduction

The ocean circulation in the northwestern Mediterranean Sea exhibits widespread mesoscale dynamics, with strongest values along the Northern Current which flows westwards along the French coast following the continental slope (Millot, 1999; Guihou et al., 2013). Observing the mesoscale variability is critical in this region since it plays a key role in the coupled ocean–atmospheric system that can lead to extreme precipitation events (Lebeaupin Brossier et al., 2015). Horizontal currents stirred by the mesoscales are important in the dispersion of pollutants and the monitoring of marine ecosystems. The vertical transport of heat, salt and nutrients is strongly driven by the smaller-scale dynamics, in the fronts and filaments surrounding these mesoscale eddies, and within the deep convection cells that form in the Gulf of Lyons in winter–spring (Herrmann et al., 2008).

Compared to other current systems at similar latitudes such as the Gulf Stream, the mesoscale variability in the northwestern Mediterranean Sea has a small Rossby radius of 5–15 km, varying seasonally with the stratification (Grilli and Pinardi, 1998). This makes the ocean dynamics of this region particularly difficult to observe and monitor. The surface mesoscale characteristics have been studied with satellite sea surface temperature (SST) and ocean colour data in clear-sky conditions (Robinson, 2010), but the mesoscale variability is often hidden in winter by clouds and in summer under the more homogenous warm surface layer. Numerical modelling studies are improving in resolution and in their internal physics to allow a better representation of the mesoscale variability (e.g. Herrmann et al., 2008), although these models need to be validated against observations.

In the global ocean, mapped satellite altimeter products have allowed unprecedented advances in understanding the mesoscale eddy variability and characteristics (Chelton et al., 2011). Altimetry measures sea surface height (SSH) that responds to mass and density changes over the entire water column, and as such, altimetry is the only satellite observation that can detect deep ocean changes. Deep-reaching mesoscale eddies can be tracked over many seasons or years (e.g. Morrow et al., 2004; Chelton et al., 2011), even if their surface signature disappears through air–sea interactions so that they become undetectable in satellite imagery. Although regional altimeter maps have been constructed with improved resolution and spatial scales adapted for the Mediterranean Sea (e.g. Pujol and Larnicol, 2005), the spacing between ground tracks still limits our ability to monitor scales less than 150 km wavelength (or 75 km diameter features) (Pascual et al., 2006). Thus we can only detect the larger mesoscale structures, missing most of the typical Rossby radius dynamics in the Mediterranean Sea.

Along-track altimeter data are able to detect finer scales than the mapped altimeter data, but the spatial scales we can resolve are still limited by the altimeter noise, the accuracy of the corrections and the processing methodology. However, over the last 5 years, there has been great progress in improving the quality of along-track satellite altimeter data for ocean studies. Of the three missions currently flying in the altimeter constellation, Jason-2 in Ku-band (launched in 2008) has benefitted from continually refined algorithms and corrections, and new waveform retrackers that allow more data points to be collected close to the coast and islands, and more stable performance with lower noise over the oceans (Dibarboure et al., 2011). SARAL/AltiKa (launched in 2013) was designed to have a smaller footprint and lower noise over all surfaces, due to the choice of antenna pattern, Ka-band frequency and its lower altitude (Verron et al., 2015). CryoSat-2 (launched in 2010) is primarily a cryosphere mission and not planned for ocean observations. Yet over the last years, considerable efforts have been made by the ESA SAMOSA project (Ray et al., 2015) and the CNES Cryosat-2 Processing Prototype (CPP) project (Boy et al., 2017) in

collaboration with oceanographers to improve the waveform retracking over the ocean and provide adequate corrections for ocean observations. CryoSat-2 is in low-resolution mode over most of the global ocean but has synthetic aperture radar (SAR) mode observations available over a few regions, including the Mediterranean Sea, with improved along-track sampling down to 300 m and reduced noise. However, certain ocean corrections are less accurate than on Jason-2 or SARAL, including the radiometer correction and the mean sea surface estimate, since CryoSat-2 is on a geodetic orbit. These three altimeter missions with different technologies and data processing will provide an ideal data set to test the improved observational capabilities in the NW Mediterranean Sea.

Previous studies have analysed the altimetric capabilities in the NW Mediterranean Sea from conventional along-track data (Bouffard et al., 2008, 2011; Birol and Delebecque, 2014; Birol and Nino, 2015), including using seasonal averaging to reduce the noise for Jason but maintaining along-track resolution (Birol et al., 2010). Here we will take a different approach, in order to measure the altimetric signal-to-noise ratio statistically in the different seasons. We will calculate along-track sea level anomaly (SLA) spectra (e.g. Fu, 1983), which allows us to observe the SLA spectral energy at different wavelengths, and also the time-averaged spectral noise at small wavelengths. In terms of signal, the spectral energy of SLA is higher at longer wavelengths, and lower at small wavelengths, and geostrophic turbulence theory involves a cascade of energy from the larger to smaller scales, leading to a steep spectral slope in wavenumber space. When spectra are averaged (over different ground tracks in a region and/or over time along the same ground track), the random altimeter noise averages out to create a flat spectral noise floor in the 1 Hz data. This spectral noise level then defines our altimeter noise. The intersection of this noise floor with the spectral slope will define the limit of the observable wavelengths, where the signal-to-noise ratio is statistically greater than 1.

Following Xu and Fu (2012) we will remove the spectral noise from the spectra before calculating the spectral slope, to improve the slope estimate and have more precise observational limits. This technique has been applied to the global altimeter data sets, for Jason-1 by Xu and Fu (2012) and for Jason-2, SARAL and CryoSat-2 by Dufau et al. (2016). Their results showed considerable geographical variations in the spectral slope, noise levels and mesoscale resolution (Xu and Fu, 2012), and strong seasonal variations in the noise level and the mesoscale observing capabilities (Dufau et al., 2016). Neither study included the smaller Mediterranean Sea region, due to the limited spatial coverage in this regional sea. In our analysis, we will concentrate on tracks having at least 200 km length.

These studies calculated their spectral slopes over a fixed "mesoscale" band from 70 to 250 km wavelength. The Mediterranean Sea, which is dominated by smaller dynam-

Table 1. Altimetric data used in this study.

Altimetric mission	Frequency band	High-frequency rate (average 1 Hz)[1]	Time period used	No. sections used in spectral averaged mean (seasonal)[2]
Jason-2	Ku	20 Hz – LRM	Jul 2008–Feb 2015	246 (summer: 65, winter: 58, spring: 71, autumn: 52)
SARAL	Ka	40 Hz – LRM	Mar 2013–Jan 2015	292 (summer: 66, winter: 66, spring: 96, autumn: 64)
CryoSat-2	Ku	20 hZ – SAR	Apr 2013–Apr 2014	276 (summer: 77, winter: 69, spring: 75, autumn: 55)

LRM: conventional low-resolution mode; SAR: synthetic aperture radar mode.
First number corresponds to the total number of 200 km sections used in the regionally averaged spectra (Fig. 3); numbers in brackets correspond to the number of sections used in each seasonal average (Fig. 4).

ical structures, may have different spectral energy and spectral slopes in this band compared to open-ocean regions. The surface sea-state conditions are also dominated by short wind waves and less by long swell, which may impact on the radar altimeter's noise level. Both of these features will be considered in the first section of this paper. We aim to investigate the noise levels for the most recent altimeter missions, estimated from their spectral noise level in the Mediterranean Sea. We will revisit the appropriate filtering to be applied to remove the noise in different seasons. We will then consider what scales of ocean dynamics can be observed today in the Mediterranean Sea with along-track altimetry and investigate how much of the seasonal dynamical signal is observable above the seasonal noise.

In the second part of this paper, we will use a complementary approach and focus on the observation of individual features using a combination of altimetry and a limited number of glider sections and 2 years of high-frequency (HF) radar observations filtered at similar scales. We will examine whether the ocean scales observable with altimetry are also captured by the co-located in situ data. Glider–altimetry comparisons have been used for previous altimetry missions in the NW Mediterranean Sea (e.g. Bouffard et al., 2010) but not for the three most recent missions. For the glider comparison, we only have a limited number of historical co-located sections, and so gliders were deployed specifically along altimetric tracks for each of the three missions, under different mesoscale conditions. For the HF radar, we will use a HF radar site near Toulon, as part of the MOOSE observational array (Quentin et al., 2013), with an offshore extent of 25–75 km from the coast. We will discuss the strengths and limits of the different measurement systems' observation in the coastal band.

2 Data sets used

2.1 Altimeter data

Along-track SSH observations from the most recent altimetry missions (Jason-2, CryoSat-2 and SARAL/AltiKa) are analysed over the NW Mediterranean Sea (Fig. 1) and over different periods (Table 1). The data are made available from AVISO/CNES. Jason-2 is a conventional pulse-width limited

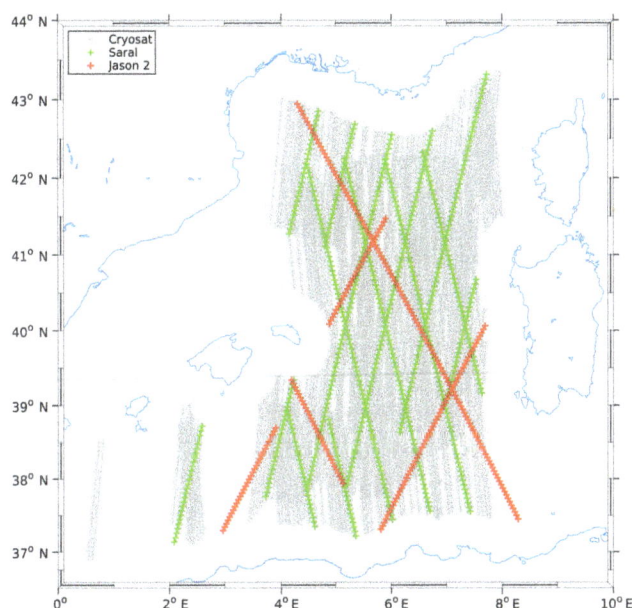

Figure 1. Distribution of altimeter tracks in the NW Mediterranean Sea showing the different missions: the 10-day repeat Jason-2 mission in red, 35-day repeat SARAL/AltiKa in green, and the 380-day repeat CryoSat-2 in grey. Only sections greater than 200 km are included in the spectral analysis, and only data more than 50 km from the coast are analysed to remove the increased errors in the coastal zone. The distance from the coast is calculated using the Stumpf database (http://oceancolor.gsfc.nasa.gov/DOCS/DistFromCoast).

altimeter operating in Ku-band (Lambin et al., 2010) and provides the longest time series: we use data over the 6.8-year period from July 2008 to February 2015. SARAL/AltiKa, with its 40 Hz Ka-band emitting frequency, its wider bandwidth, lower orbit, increased pulse repetitivity frequency and reduced antenna beamwidth, provides a smaller footprint and lower noise than the Ku-band altimeters (Verron et al., 2015). We use data from the nearly 2-year period from March 2013 to January 2015. CryoSat-2 is a synthetic interferometric altimeter (SIRAL) Ku-band instrument operating in three modes (low-resolution mode (LRM), synthetic aperture radar mode (SARM) and SAR interferometric mode). Only the SARM data are available over the Mediterranean Sea, and we use data from the CNES CryoSat-2 processing prototype (version 14) from CNES (Boy et al., 2017) over the 1-year period April 2013 to April 2014. For all three missions we will analyse the 1 Hz data only, which have a flat noise floor. Higher-frequency data (20 or 40 Hz) show a spectral bump at wavelengths less than 70 km, which does not allow us to estimate a stable noise floor (Dibarboure et al., 2011).

The choice to analyse different periods was dictated by the data availability and our desire to have longest possible time periods available for the seasonal analyses. The limited quantity of altimeter cycles considered during this period is compensated by the spatial averaging of available tracks in the NW Mediterranean Sea, which improves the statistical significance of our analysis.

Along-track SSH observations are maintained at their original observational position and corrected for all instrumental, environmental and geophysical corrections. Only the time variable part of the SSH is considered following Stammer (1997), Le Traon et al. (2008) and Xu and Fu (2011, 2012). SLAs are calculated for all missions relative to their precise along-track mean sea surface for Jason-2 and SARAL, both on a long-term repeat track. CryoSat is on a geodetic orbit, and its SLAs are calculated relative to a gridded mean sea surface (MSS_CLS2011, http://www.aviso.altimetry.fr/en/data/products/auxiliary-products/mss.html), which can introduce slightly higher errors over scales of 40–80 km wavelength (Dibarboure et al., 2011; Dufau et al., 2016). In the following analyses of spectra and geostrophic current anomalies, we will use the time-varying SLAs.

2.2 Glider data

A large number of gliders have been deployed in the NW Mediterranean Sea as part of the MOOSE project (http://www.moose-network.fr/gliders), with more than a hundred glider sections available in the region during the 6.5 years of our study. However, since our objective was to validate the smaller-scale structures that move rapidly, it was important that the glider and altimeter observations were co-located in space and time. Two glider sections were available along a Jason-2 track in September–October 2012. MOOSE and CNES also co-funded the deployment of gliders along three SARAL tracks as part of the Comsom campaign in October–November 2014, and along two CryoSat-2 tracks and three SARAL tracks in April–May 2015 (see Fig. 5a and Table 2).

Slocum gliders were used, diving at a 26° inclination with an average horizontal speed of around $0.35 \, \mathrm{m \, s^{-1}}$. They reach a maximum depth of 1000 m, and the distance between two surface positions is around 2–3 km. The deployments are made away from the coast to be in deep water, although an onboard captor can detect whether they approach the bottom before 980 m. The gliders were deployed a few days before the passage of the satellite in order to be sampling along the track when the altimeter passed. The altimeter passes every

10 days for Jason, and every 35 days for SARAL and in a given region every month for CryoSat-2. So with this type of precise-date deployment, there is no guarantee that the glider and altimeter pass will cross an energetic structure at the time and position that the altimeter passes.

For comparison with the altimeter data, we need to obtain steric heights from the glider relative to 1000 m. For this, we calculate a single vertical profile at the central position for each of the diagonal dives (descending or ascending) and calculate steric heights from the density anomalies. Geostrophic velocities are also calculated relative to the 1000 m depth.

There is an additional "drift" speed that can be added to this geostrophic velocity, associated with the lateral heading correction used to keep the glider on track against a strong current. This drift correction represents the total current over the upper 1000 m and will include the barotropic currents close to the continental slope, some ageostrophic surface currents and a correction for the upper baroclinic flow. This correction was generally small in our region except near the continental slope, and we will clearly identify when this correction is used in the following study.

2.3 HF radar data

As part of the MOOSE observing system, a HF radar system has been installed near Toulon (http://hfradar.univ-tln.fr/HFRADAR) to monitor the Northern Current, with gridded data available since 2012. HF radars measure the reflected radar signal from the ocean surface at a given lateral incidence angle. The surface currents are obtained after subtracting the surface wave speed, which is estimated from the measured frequency of the wave energy peak and the known frequency of the emitted radar signal. Two radars orientated with different angles allow the determination of the current direction.

The Toulon HF radar system uses two WERA radars that provide surface current vectors over a region extending 80–100 km offshore, with a spatial resolution of 3 km and an angular resolution of 2°. They operate at 16–17 Mhz. Observations are collected every 20 min and data have been edited and averaged daily over the period May 2012–September 2014. The surface current vectors represent the total current averaged over the upper 1 m of the ocean and include a significant ageostrophic component, not present in the altimetric currents.

3   Spectral analysis of along-track altimeter data

Spectral analyses are performed on each of the three altimeter missions, with their tracks shown in Fig. 1. Only data more than 50 km from the coast are analysed to avoid the increased errors in the coastal zone. Each track and cycle is then selected along a common segment of 200 km. This segment length was chosen to allow a large number of altimeter

Table 2. Characteristics of the co-located glider and altimeter track sections.

Altimeter track	Along-track filtering[1]	Glider name	Start date of section	End date of section	Section length (km)	No. glider profiles[2]
Jason 146	50	Campe	23 Sep 2012	8 Oct 2012	292	111
Jason 146	50	Campe	8 Oct 2012	23 Oct 2012	327	80
SARAL 846	**35**	Eudoxus	23 Oct 2014	29 Oct 2014	125	54
SARAL 57	**35**	Milou	27 Oct 2014	3 Nov 2014	164	92
SARAL 388	**30**	Milou	9 Nov 2014	13 Nov 2014	77	55
SARAL 973	35	Bonplan	13 Apr 2015	22 Apr 2015	180	101
SARAL 973	35	Tintin	17 Apr 2015	23 Apr 2015	115	58
SARAL 973	35	Tintin	8 May 2015	13 May 2015	99	56
CryoSat 493	35	Bonplan	24 Apr 2015	1 May 2015	166	101
CryoSat 493	35	Tintin	25 Apr 2015	4 May 2015	188	101

[1] Altimetric data are filtered with a Loess filter at different wavelength cutoffs depending on the mission and season (see text).

[2] All glider data are filtered with a two-step Butterworth filter which removes high-frequency signals < 30 km wavelength.

segments in different regions in between the numerous islands and to be more than 50 km from the coast, to avoid the increased errors in the coastal altimeter data. This segment length is also long enough to well resolve the dominant scales (Rossby radius of 5–15 km). Missing data are a problem for a stable spectral analysis. If fewer than three consecutive 1 Hz points are missing (20 km), the data are linearly interpolated; if a larger gap is present the cycle is eliminated from the analysis. Tracks passing over large islands are thus eliminated (see Fig. 1). Wavenumber spectral analysis is then performed by Fourier transform on the ensemble of the remaining segments for each mission (see Table 1). The cycles are averaged in wavenumber space for the entire period and for each season.

An example of the power spectral density (PSD) of SLA averaged for all of the Jason-2 data in the NW Mediterranean Sea over the period 2008–2015 is shown as the black curve in Fig. 2. The PSD is high at longer wavelengths (> 300 km). There is a cascade of energy over the mesoscale range from 50 to 300 km, but the spectra become whiter at small wavelengths (i.e. less than 50 km), where the weaker ocean energy is hidden by the stronger instrument and geophysical noise.

In the following seasonal analyses, the *noise level* will be calculated as a constant PSD value estimated between 12 and 25 km wavelength, as in Dufau et al. (2016) (e.g. black horizontal dashed line, Fig. 2).

Following the global studies made by Xu and Fu (2012) and Dufau et al. (2016), we then subtract this statistically stable noise level from the mean spectral curve, to obtain an unbiased spectral estimate corrected for the noise (red solid line curve, Fig. 2). The *spectral slope of this unbiased estimate* is steeper over the mesoscale range and corresponds to a $k^{-2.5}$ slope and the SLA PSD cascade continues more smoothly down to smaller wavelengths.

We define the *mesoscale observability limit* as the wavelength corresponding to the intersection of the spectral slope and the noise level, where the signal-to-noise ratio is greater

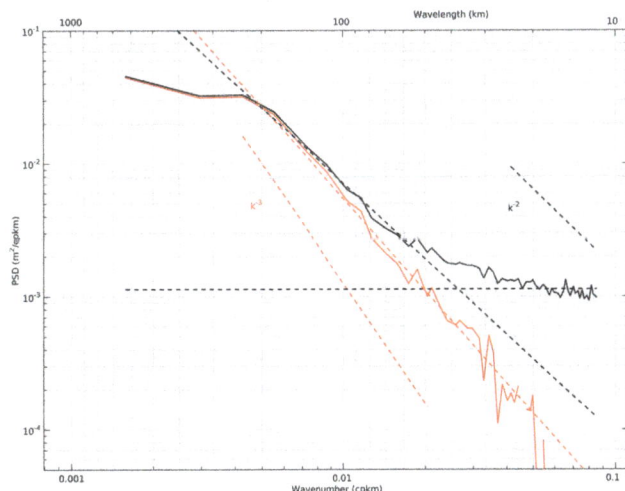

Figure 2. Mean wavenumber spectra (power spectral density) for Jason-2 sea level anomalies, averaged over all tracks in the NW Mediterranean Sea > 50 km from the coast (black curve) for the period 2008 to 2015. The estimated noise level is shown as the horizontal black dashed line. The unbiased spectra (red curve) are obtained by subtracting this constant noise from the original spectra. The spectral slope (red dashed line) is calculated between 50 and 200 km wavelength. The intersection between these two curves occurs around 50 km wavelength for this case, which represents the mesoscale observational limit, above which the mean signal-to-noise ratio is > 1.

than 1. This is a statistical representation of the average ocean and noise conditions over the entire period and over the entire region analysed. In some local cases, smaller energetic structures may still be observable above the altimetric noise. However in the following results, we will discuss this regional statistical approach.

The *mean spectra* for the three altimeter missions over the NW Mediterranean Sea are shown in Fig. 3a for the 200 km segment tracks in Fig. 1 and over the 13-month common

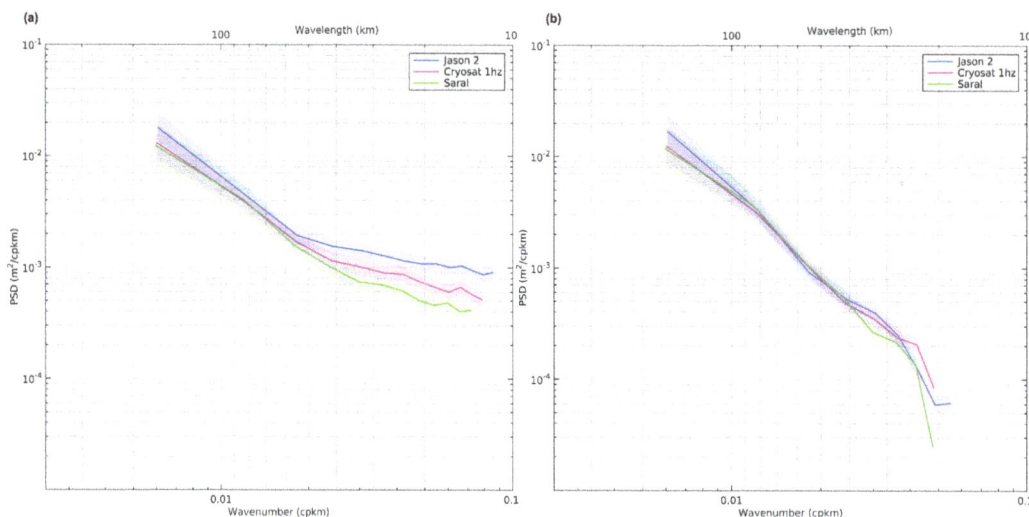

Figure 3. (a) Mean wavenumber spectra (power spectral density) for the three altimeter missions, averaged over the 200 km track segments in the NW Mediterranean Sea, > 50 km from the coast, and for the common period 1 April 2013–30 April 2014. Jason-2 is in blue, SARAL in green, CryoSat-2 SAR 1 Hz data in pink. **(b)** The unbiased spectra with a constant noise level removed, resulting in a mean $k^{-2.5}$ spectral slope. Shading represents the error bars, based on a chi-squared test with the number of degrees of freedom being wavenumber dependent. Table 1 gives the number of sections used.

data period from 1 April 2013 to 30 April 2014. The unbiased estimate with the noise removed is in Fig. 3b. Recall that the space–time samplings of the three missions are different, and as such they may capture different dynamics at different regions. So we do not expect the spectra to be perfectly aligned. More distinctive are the different noise levels between 15 and 100 km wavelength. Jason-2 has the highest noise level in this region, followed by CryoSat-2 in SAR mode. SARAL/AltiKa in Ka-band exhibits the lowest noise of all.

When a constant noise level is removed from each spectral PSD, the spectral slopes line up surprisingly well, given the different space–time sampling of the three missions over this 13-month period. The spectral slope is again around $k^{-2.5}$ from a fit to the unbiased spectra over the wavelength range from 50 to 200 km. These spectral slopes in the offshore regions of the Mediterranean Sea are quite shallow compared to the k^{-5} slopes expected for quasi-geostrophic theory (Stammer, 1997). The reason for this needs further investigation, but smaller slopes are also characteristic of open-ocean low-eddy-energy regions (Xu and Fu, 2012). For the Mediterranean Sea, the dominant mesoscale energy at small Rossby radius scales tends to flatten the spectra, but internal waves or mean sea surface errors in the CryoSat-2 data could also contribute to higher SSH energy at small scales and flatter spectra (Dufau et al., 2016).

The fact that the CryoSat-2 1 Hz data in SAR mode had a higher noise level than SARAL/AltiKa was unexpected. We verified that the CryoSat-2 20 Hz data were consistent with the 1 Hz averages, so this is not an averaging problem. The CryoSat-2 20 Hz SAR mode does exhibit a spectral hump for this region and time period that was not present in other regions with SAR data (Agulhas or tropical Pacific; S. Labroue, personal communication, 2016). This warrants further analysis of the particular surface roughness conditions occurring in the NW Mediterranean during this year, and further expertise in SAR processing for the Mediterranean conditions is needed. These results reinforce the very low noise level associated with the 40 Hz Ka-band SARAL data, averaged here to 1 Hz.

Seasonal spectra were also calculated from the longest time series possible, i.e. over 6.5 years for Jason-2 data, over 22 months for SARAL/AltiKa, and for the shorter 13-month period for CryoSat-2 (see Table 1). The spectral noise floor levels for the seasonal analyses are shown in Fig. 4a. Note the spectral units are in $m^2 \, cpkm^{-1}$, where cpkm refers to cycles per km. Jason-2 and SARAL/AltiKa show a large seasonal variability in their noise levels, with highest noise levels in winter ($1.2 \times 10^{-3} \, m^2 \, cpkm^{-1}$) and then autumn, due to the high sea-state roughness in these months from the stronger wind-wave conditions which increases the spectral SLA "hump" at wavelengths from 30 to 70 km (Dibarboure et al., 2014). In summer, the Jason-2 noise level is only $0.8 \times 10^{-3} \, m^2 \, cpkm^{-1}$, but this is still higher than the noise floor in any season for the SARAL or CryoSat-2 missions. SARAL with its small footprint has the lowest noise levels but has strong seasonal variability, with values ranging from a low $0.3 \times 10^{-3} \, m^2 \, cpkm^{-1}$ in summer to $0.7 \times 10^{-3} \, m^2 \, cpkm^{-1}$ in winter. The CryoSat-2 SAR mode shows very stable background noise levels over this 1-year record, varying between 0.6 and $0.8 \times 10^{-3} \, m^2 \, cpkm^{-1}$. The reasons for this stable seasonal noise level are not yet known.

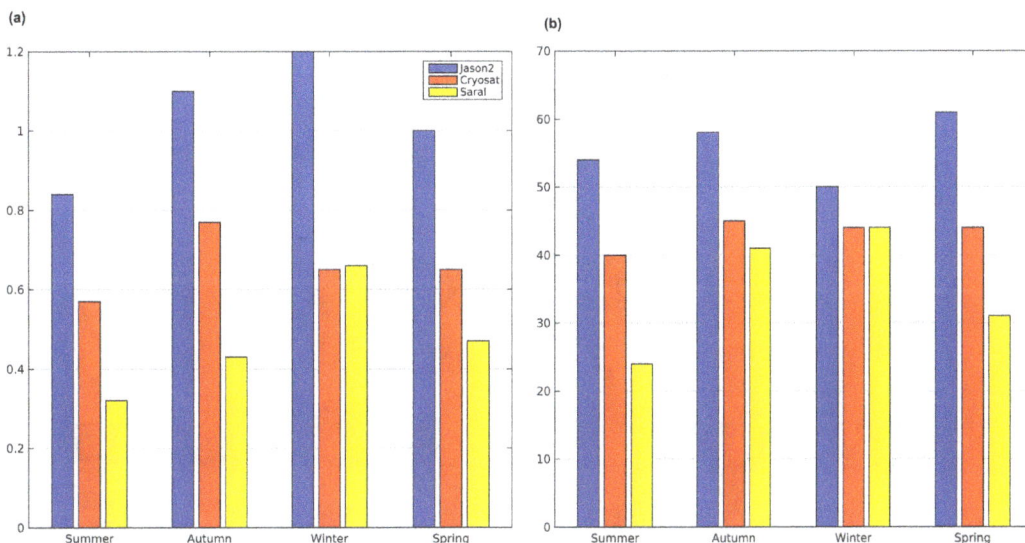

Figure 4. (a) Seasonal noise levels (in $10^{-3}\,\mathrm{m}^2\,\mathrm{cpkm}^{-1}$) for Jason-2 (blue), CryoSat-2 SAR mode (orange) and SARAL/AltiKa (yellow) derived from along-track wavenumber spectra. **(b)** Seasonal observational limits in terms of wavelength (in km) where the signal-to-noise ratio is > 1 for each altimeter mission. Table 1 gives the number of sections used.

However CryoSat-2 has a long repeat cycle (369 days), so different geographical regions are sampled in different seasons; there may be strong interannual variations in the wind-wave conditions that merit more detailed investigation. The additional mean SSH errors introduced due to the non-repeating track will also impact the CryoSat-2 spectra over all seasons.

Figure 4b shows the observational limits for each altimeter mission by season. Clearly, the background noise is not the only limiting factor on the scales of mesoscale energy that we can observe. The SLA energy at low wavelengths also varies from one season to another. In winter, when the mixed layers are deepest and energetic deep convection cells occur in the NW Mediterranean Sea (e.g. Herrmann et al., 2008), all of the altimeter missions can observe wavelengths down to 40–50 km (individual features of 20–25 km). In summer when the submesoscales are weaker, SARAL can detect ocean scales down to 35 km wavelength, whereas the higher noise from Jason-2 and CryoSat-2 blocks the observation of scales less than 50–55 km. This characteristic was also noted in the global analysis of Dufau et al. (2016). Unfortunately in winter, when we would like to observe the smaller energetic submesoscales, all of the radar altimeters observe higher noise levels associated with the higher wind-wave field.

4 Co-located altimeter and glider observations

The previous section highlighted that the altimetric noise was effectively masking the smaller-scale SLA signals in the along-track data. The smallest scales observable with a signal-to-noise ratio greater than 1 will vary from one altime-

ter mission to another and seasonally. Statistically, we cannot observe structures less than 35–45 km wavelength with SARAL, or 50–60 km wavelength with the higher noise of Jason-2. However, individual energetic features may be revealed above the statistical noise. We will explore this with a series of co-located along-track altimeter–glider sections and compare the vertical structure observed by the gliders with their steric height and geostrophic velocities.

In this section, the filtering of the along-track altimetry data is based on the standard Loess filtering applied to the CTOH coastal processed data (Birol et al., 2010; Birol and Nino, 2015). For each glider–altimeter comparison, the first estimate of the along-track altimeter filtering scales was based on the seasonal spectral analysis results for each altimeter mission (see Sect. 3). Other cut-off frequencies around this seasonal statistical value were also tested. The filter which gave the best results in terms of glider–altimeter correlation coefficient and which had the lowest cut-off wavelength was then chosen. The altimeter filter values are given in Table 2.

One should bear in mind that the glider steric height and geostrophic velocities (with or without their surface drift adjustment) will observe different dynamics from the altimetric sea level and geostrophic velocity anomalies. The steric height calculated from gliders represents the upper ocean baroclinic component due to the density anomalies above 1000 m depth. Altimetric SLAs include the full-depth baroclinic motions and the barotropic component, and the barotropic flow may be quite active in the NW Mediterranean Sea, in particular near the shelf break and slope (F. Lyard, personal communication, 2016). When the glider "surface drift" is added to the glider geostrophic currents relative to

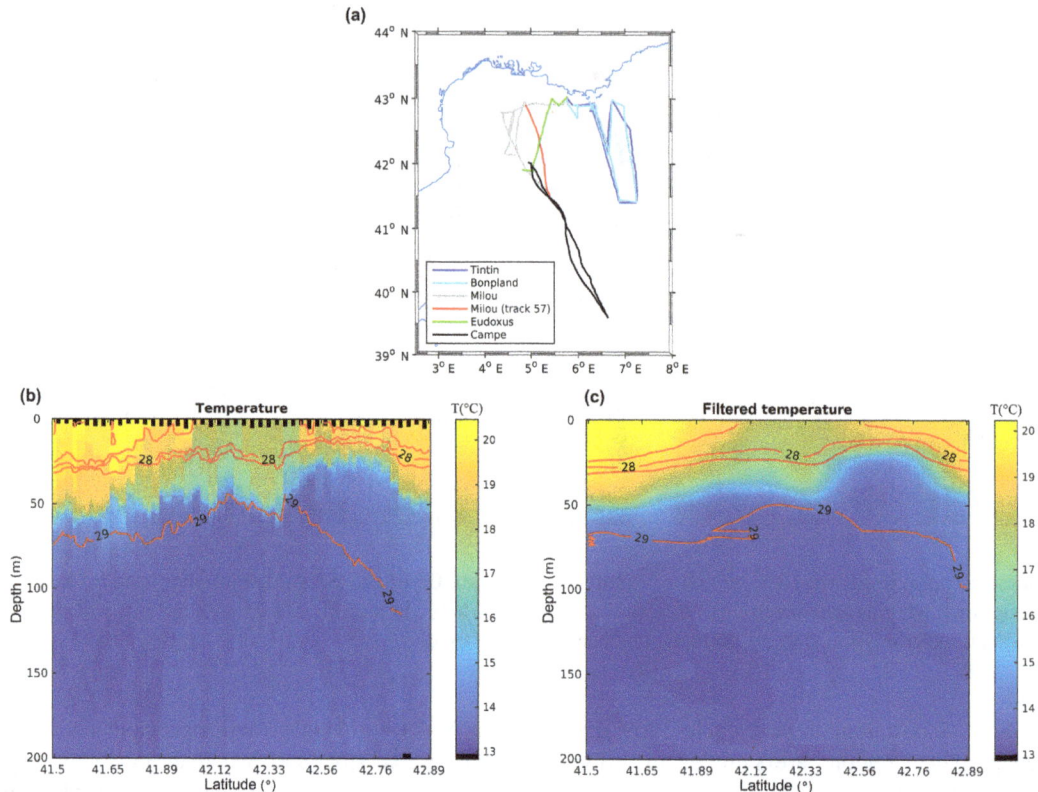

Figure 5. (a) Location of the different gliders used in this analysis. In red, the glider Milou section (155 km long) along the SARAL altimeter track 57 from 27 October to 3 November 2014. **(b)** Vertical temperature section from the Milou glider over the upper 200 m. **(c)** Filtered temperature section with cutoff at 30 km wavelength.

1000 m, this may partially correct for the missing barotropic component. Altimetry may also include other SLA signals, such as from internal tides or internal waves, which contribute as errors in the geostrophic velocity calculation (although tides are small in the Mediterranean Sea). In addition, the altimetric SLAs have the mean ocean circulation removed, whereas the gliders provide the total upper ocean baroclinic flow. For consistency, the mean dynamic topography and mean geostrophic velocities derived from Rio et al. (2014) are added to the altimetric data for this comparison. The third main difference is the time taken to make a section over 100 to 300 km. The altimeter makes a "snapshot" of the section as it passes at 7 km s^{-1} (200 km in 30 s) whereas the glider moves at 0.35 m s^{-1} (200 km in 6.5 days). We will see that slow-moving structures may be well-sampled by both; rapidly evolving smaller-scale structures are harder to co-locate.

One crucial point is that the gliders have their own noise and also measure HF ageostrophic ocean structures that will not be observable with altimetry. Figure 5 shows a vertical temperature section over the upper 200 m from the glider Milou along the SARAL altimeter track 57 from 27 October to 3 November 2014. Figure 5b shows the very small-scale signals in the upper ocean temperature structure along this

164 km long section. These may be associated with noise in the glider heading or from the processing steps, or from internal waves or rapid submesoscale structures. To remove these scales, we have applied a recursive Butterworth second-order along-track filter to the density data, before calculating the steric height or geostrophic anomalies, with a filter cut-off at 30 km wavelength, designed to retain the typical Rossby radius scales of 10–15 km in the NW Mediterranean Sea. This filtering step was recommended from previous glider studies (e.g. Durand et al., 2016). An example of the filter applied to the same temperature section is shown in Fig. 5c. Similar filtering is applied to the different glider sections presented below.

Ten glider sections are available, co-located with altimeter tracks (details given in Table 2). Here we present three glider track sections along different altimeter mission tracks.

4.1 Jason-2–glider comparison over a large slow eddy

The glider Campe followed a Jason-2 track 146 over a 300 km section from 42 to 39.5° N over a 1-month period 23 September–23 October 2012. During this period, Jason-2 passed three times over the same track. Jason-2 data were filtered using a Loess filter with a 50 km cutoff for this summer–autumn section (Table 2). Figure 6a shows the

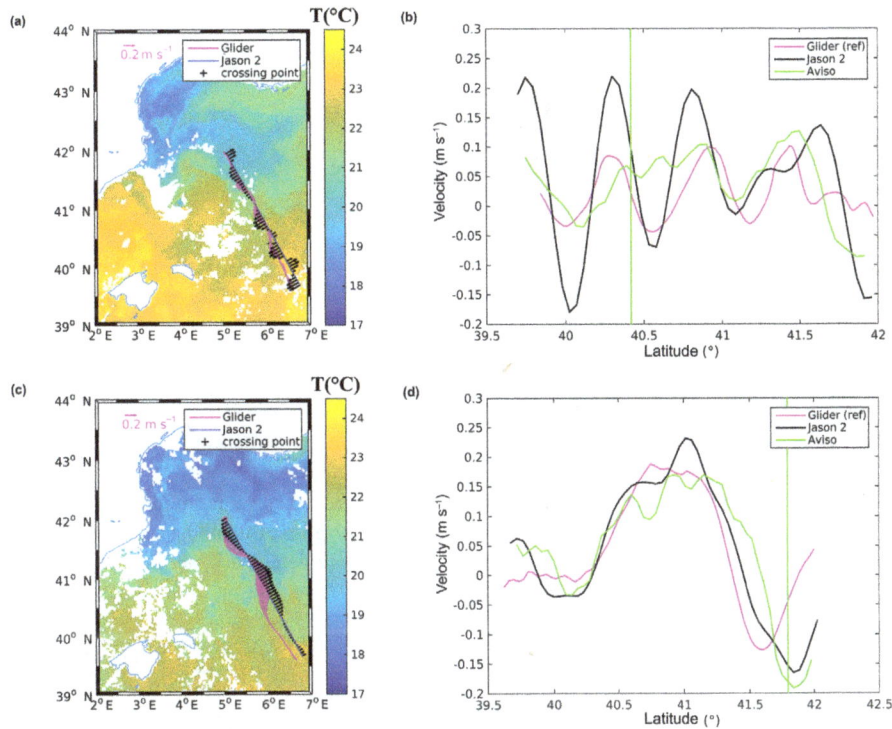

Figure 6. (a) Co-located Jason-2 track and currents (in black) and glider track and currents (in pink) for the southbound leg, overlaid with a satellite SST plot on 1 October 2012. **(b)** Along-track comparison of geostrophic velocities for the glider (including the drift velocities) in pink, and filtered along-track Jason-2 data in black. Mapped AVISO altimeter data, interpolated back onto the Jason-2 track, are in green. Green vertical line shows the position when the Jason-2 data and gliders are co-located in time. **(c)–(d)** Same but for the northbound section with SST fields from 21 October 2012; 48h SST fields at 0.02° resolution from CLS.

glider cross-track geostrophic currents (in pink) with the Jason-2 cross-track currents superimposed (black) for the southward passage on 1 October 2012, overlaid on the satellite SST for the same date. The northward passage centred on 21 October 2012 is in Fig. 6c. The southbound section in late September has weak currents and is located slightly to the west; the northbound section crosses a strong mesoscale structure with an eastward current from 40.3 to 41.3° N, then a westward return current from 41.3 to 42° N at the northern end, when the third Jason pass is co-located. The filtered glider data and the filtered Jason data are also shown for the southbound section (Fig. 6b) and the northbound section (Fig. 6d). The instant of the Jason-2 passage is marked by a vertical line – identifying the latitude where the glider and the Jason observations coincide exactly in time. The geostrophic currents from the AVISO 2-D maps are also shown for reference.

The southbound section crosses a series of small reversing currents around small SST structures of 30–50 km (Fig. 6a). The glider and along-track Jason-2 data show cross-track currents in phase, although the Jason-2 amplitudes are stronger (correlation, $r = 0.5$; RMSE $= 0.06\,\mathrm{m\,s^{-1}}$). This may be real (due to deeper baroclinic or barotropic structures not observed by the glider's upper 1 km observations) or in-

duced by the effects of filtering higher noise. The mapped AVISO data have similar amplitude to the glider data but are not in phase, which reduced their statistical correlation ($r = 0.4$; RMSE $= 0.06\,\mathrm{m\,s^{-1}}$). Adding the glider "drift" reference currents introduces little change to these results.

Three weeks later, the northbound section crosses a strong mesoscale eddy. The three data sets present similar eastward currents across the mesoscale eddy, and although the amplitude of the westward current near 42° N is similar, along-track altimetry positions the return flow 30 km further north than is detected by the glider. For this larger eddy, 100 km in diameter, the AVISO 2-D maps and the 50 km filtered along-track data both provide a good estimate of the glider's geostrophic currents ($r = 0.9$) with similar RMSE ($\sim 0.07\,\mathrm{m\,s^{-1}}$ for both data sets).

4.2 SARAL–glider comparison over a small rapid meander

Although a number of satellite underpasses were planned for SARAL, different deployment problems limited the number of successful intercomparisons (bad weather, gliders leaking, errors in estimating the satellite position, etc.). The longer sections did not necessarily cross any energetic features, and we eliminated sections where the currents were very weak.

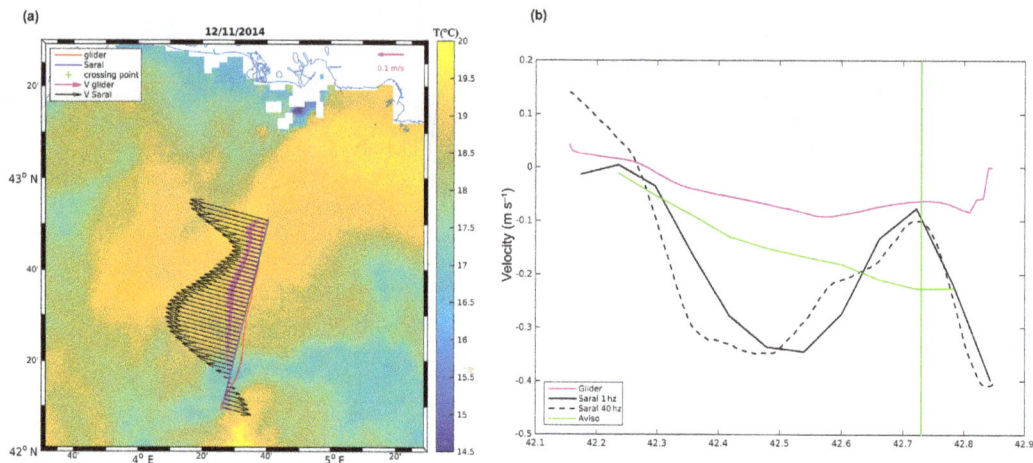

Figure 7. (a) Co-located SARAL track 388 and currents (in black) and glider track and currents (in pink), overlaid with a satellite SST plot on 12 November 2014. **(b)** Along-track comparison of geostrophic velocities for the glider (including the drift velocities) in pink, and filtered SARAL data in black. Mapped AVISO altimeter data, interpolated back onto the altimeter track, are in green. Green vertical line shows the position when the altimeter data and gliders are co-located. Daily SST fields at 0.02° resolution from CLS.

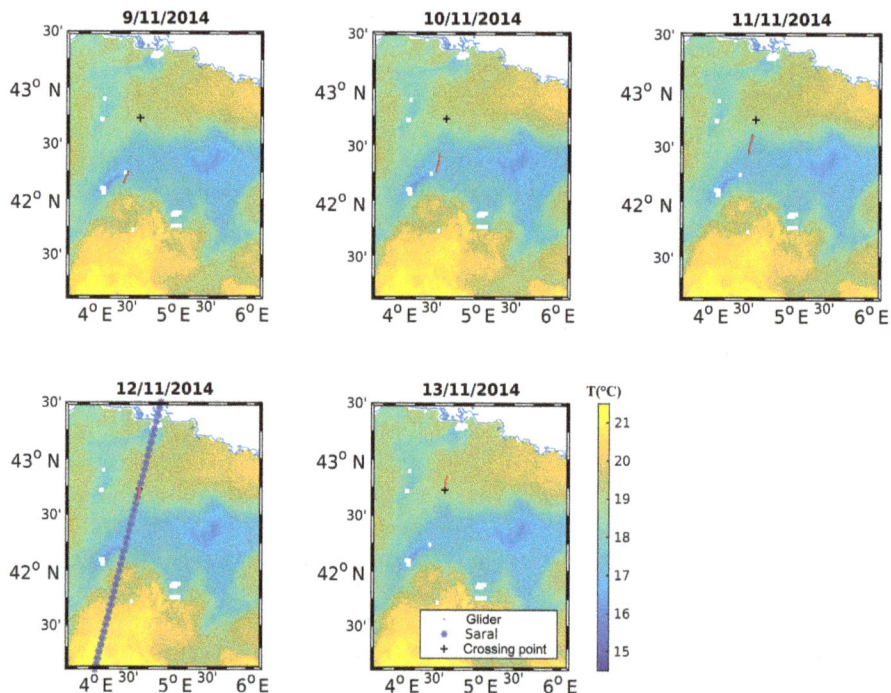

Figure 8. Five-day series of satellite SST maps for the period 9–13 November. The glider position is shown each day (in red), the SARAL–glider crossing position on 12 November (in black), and the SARAL track passing on 12 November 2014. Daily SST fields at 0.02° resolution from CLS.

The short section presented here highlights another difficulty – comparing small-scale structures in a rapidly evolving field.

Figure 7a shows an example of the SARAL–glider comparison for the SARAL track 388 and the glider Milou, which crossed a narrow, intense, westward current around 42.75° N, a broad, weak, westward current further south, and then touched an eastward return flow around 42.25° N. These nar-

row currents are the limit of the observability with the gliders, given the filtering cutoff at 30 km wavelength. In comparison, the altimeter data show a broad intense westward flow over the entire section, except for the return eastward flow in the south. The along-track comparison of their amplitudes (Fig. 7b) shows that the two systems measure similar currents at the exact time of the SARAL passage (vertical line), but otherwise the broad, intense westward flow captured by altimetry is not observed by the gliders. The mapped AVISO data are halfway between.

If the glider and altimeter observations are overlaid on a daily time series of satellite SST maps, the differences between these two observations becomes clearer. Figure 8 shows the 5 days needed by the glider to complete this 77 km section to 1000 m depth and the evolving SST conditions during this period. On the 9 November 2014, the glider was in the south and crossed a cold eastward-moving filament. On 10 November, the glider is in weaker conditions. On 11 November, the warmer westward-flowing current starts to shift southward and on 12 November, when Jason-2 passed over, the warm branch has extended south to 42.3° N.

This example highlights the difficulty in comparing sections constructed from 5 days of glider data with the near-instantaneous coverage from the along-track altimetry data. These small-scale structures less than 50 km evolve quickly, and having observations that are not exactly co-located in space and time leads to large differences.

4.3 CryoSat-2–glider comparisons

The third example concerns two gliders deployed at 1-day intervals along the CryoSat-2 track 493, which passed on 27 April 2015. CryoSat-2 SAR data are filtered at 35 km (see Sect. 3). Figure 9 shows that the two gliders and the CryoSat-2 data detect well the westward-flowing Northern Current near 42.5° N as well as an eastward return flow around 41.5° N. In contrast, the CryoSat-2 data overlay a weak cyclonic eddy centred on 42° N, which is also apparent in the mapped AVISO data but is not detected by the gliders. The CryoSat-2 data are included in the AVISO maps, so the two products show consistent results, though AVISO is smoother.

The along-track geostrophic currents (Fig. 9b) show that the two gliders, separated by 1 day, observe the same features. However, the peaks in westward flow, detected by the gliders at 42.6 and 42.1° N, are slightly more intense with the CryoSat-2 observations and had shifted southward when the altimeter observed them a few days later. Tintin is 1 day in advance of Bonplan-d as they move southward, and the southward shift in the westward flow is also observed between Tintin and Bonplan-d at 42° N. There is a good alignment of the eastward currents between the three observing systems around 41.7° N.

In summary, the glider–altimeter comparisons reveal the difficulty in validating the along-track altimetry data with observations that are not exactly co-located in time and space. The relatively slow gliders are able to capture the slower-moving larger eddies, as seen in our example with Jason-2 and highlighted by previous studies (Bouffard et al., 2010). However, the real improvement in altimetric signal-to-noise levels expected with SARAL and CryoSat-2 are not revealed in these glider comparisons, mainly because at the time of these altimeter observations, rather weak signals were de-

tected or the small-scale meanders were moving rapidly. In these cases, our observations approach the error levels of the two systems. Small offsets in the structure of the Northern Current could also be introduced by the removal of a mean sea surface from the CryoSat-2 data sets, which could induce errors on these small space scales (up to 80 km wavelength, Dufau et al., 2016). Although gliders can observe energetic small-scale structures in dedicated campaigns in the Mediterranean Sea (e.g. Bosse et al., 2015), the chance is small that these occur at the precise position and time when the gliders and altimeter tracks coincide. This comparison highlights the difficulty in setting up a validation campaign for altimetric observations of small-scale rapidly moving dynamics.

5   Co-located HF radar and SARAL altimeter

HF radar data provide an additional observation of the oceanic surface currents. In comparison to the geostrophic component of the flow obtained with altimetry and gliders, HF radars measure the total surface current, due to balanced geostrophic and unbalanced ageostrophic currents (wind-driven, inertial, tidal currents, etc.). The daily data set we used has been processed to remove the HF tides and inertial currents, retaining the geostrophic and wind-driven currents. Figure 10 shows an example of the HF radar total currents for one date, 20 October 2013 near Toulon, with the two coastal radar locations marked. The presence of the strong Northern Current is clearly visible in the 2-D HF radar current vectors, with a central jet only 10 km wide, the current spanning 20 km to its edges. This is clearly below the statistical observability limits from the spectral analysis of the three altimeter missions. The offshore extent of the HF radar data is from 25 to 75 km from the coast, which extends into the coastal band that was excluded from our spectral analysis, as it has frequently "noisy" altimeter data and corrections. The small spatial coverage of the HF radar means that no Jason-2 data cross this region, although we have one SARAL track passing through the centre (Fig. 10) and a number of non-repeating CryoSat-2 tracks. The angle of the SARAL track shown in Fig. 10 is such that the cross-track geostrophic currents are mainly orientated in the principal direction of the Northern Current. For this date (20 October 2013), the amplitude of the HF radar currents, projected in the altimetric cross-track direction (in red), is similar to the SARAL cross-track currents (in black), reaching 0.7–0.8 m s^{-1} within the Northern Current. Further offshore, the HF radar currents decrease gradually whereas the geostrophic altimetric currents are much weaker outside of the jet. The presence of ageostrophic currents in the HF radar data could contribute to this difference. Our statistical estimate of the spatial observability of SARAL observations in autumn is around 35 km wavelength (Sect. 3), representing feature structures across the current of around 17 km. Clearly at these scales,

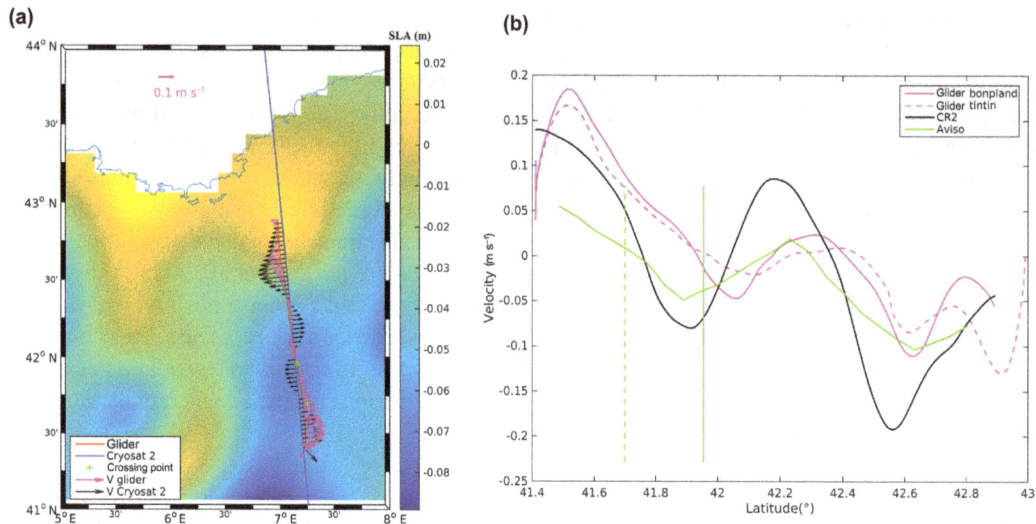

Figure 9. (a) Co-located CryoSat-2 track 493 currents (in black) and glider currents (in pink), overlaid with a satellite SST plot on 27 April 2015. Two gliders, Bonplan-d and Tintin, follow at 1-day intervals. **(b)** Along-track comparison of geostrophic velocities for the Bonplan-b glider (pink solid), and Tintin (pink dashed) with the filtered CryoSat-2 SAR data in black. Mapped AVISO altimeter data, interpolated back onto the altimeter track, are in green. Green vertical line shows the position when the altimeter data and gliders are co-located. (solid for Bonplan-d; dashed for Tintin). Daily SST fields at 0.02° resolution from CLS.

the 20 km wide Northern Current can be observed by the SARAL altimeter.

The advantage of the HF radar data set is its daily 2-D coverage at fine resolution, so we should not have the space–time offsets in the sampling of small-scale features that plagued the glider–altimeter comparisons. The disadvantage is that altimeter data in the last 10–50 km from the coast are noisy, and the ageostrophic wind-driven component of the HF radar surface currents can be strong here, in the region with strong mistral winds.

We have compared the observability of these near-shore currents with the finer-resolution SARAL altimeter time series, filtered at 35 km (see Sect. 3). SARAL data are available along this track every 35 days, and Fig. 11 shows the 18-month time series of cross-track surface velocities from the HF radar. The upper panel shows the full time series of HF radar currents projected perpendicular to the altimeter track; the middle panel shows the HF radar currents sampled at the same dates as the SARAL altimeter passes, and spatially sampled at 7 km as for the 1 Hz altimeter data. The bottom panel shows the SARAL 1 Hz geostrophic currents (mean and anomalies), filtered at 35 km. SARAL clearly detects more of the offshore return flow than the HF radar can but covers a similar data range as the HF radar to the coast. Along-track correlations of the HF radar and altimetric currents for this cross-track velocity component are between 0.7 and 0.9 for these 16 tracks, except for four dates, where the correlations drop below 0.5. The RMSE between the cross-track HF radar current amplitudes and the SARAL current amplitudes is shown in Fig. 12. Dates with low correlations (<0.5) are marked with the vertical dashed line, and these

have a higher RMSE. The RMSE is generally lower in the summer months when the wind is lower and increases in winter.

Wind forcing of the ageostrophic currents may explain part of the difference. If we consider the daily time series of HF radar data (Fig. 11a) and extract the outliers in cross-track velocity having $>1\sigma$ standard deviation from the mean, we find that these outliers are correlated at 0.84 with the cross-track wind at the same date (not shown). For the dates with weak correlations, wind may play a role for one date (December 2013), but the other dates have relatively low wind. The differences with SARAL are often associated with 10 km wide structures and close to the coast. This could be due to errors in either measurement system (e.g. for SARAL: the nearshore wave height bias, wet tropospheric corrections, mean sea surface errors) but also from rapid events that are detected by the altimeter 8 s "snapshot" but viewed differently with the HF radar 1-day averages (rapid meander, internal waves, etc.). Planned future analysis of the higher-frequency radar data and the 40 Hz altimeter data with appropriate filtering may help elucidate some of these differences.

6 Discussion

The along-track altimeter spectral analysis allows us to estimate the mean dynamical scales that can be observed today with different altimeter technology and associated processing, and in different seasons. In winter, when the mixed layers are deepest and the submesoscale is energetic, all of the altimeter missions can observe wavelengths down to 40–

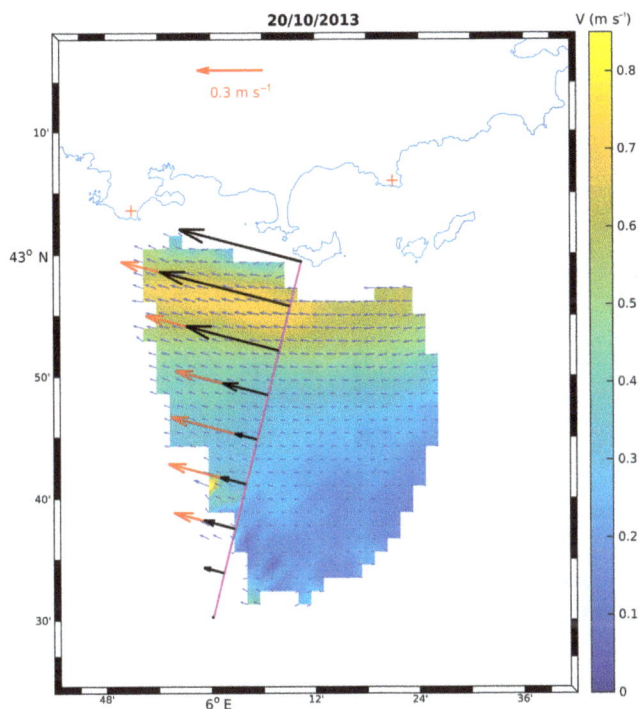

Figure 10. HF radar surface currents near Toulon for one date (20 October 2013); direction with small arrows, current speed is in colour. SARAL track 302 is marked in pink; 1 Hz cross-track geostrophic currents from SARAL altimetry are in black; the HF radar total currents projected in the altimetric cross-track direction are in red. The current scale of $0.3\,\mathrm{m\,s^{-1}}$ is associated with the projected currents. Positions of the two HF radar sites are marked with the red crosses on land.

50 km (individual feature diameters of 20–25 km). In summer when the submesoscales are weaker, SARAL can detect ocean scales down to 35 km wavelength, whereas the higher noise from Jason-2 and CryoSat-2 blocks the observation of scales less than 50–55 km wavelength.

This is a statistical view. There are limits in applying this too assiduously, especially as these statistics are calculated from relatively short records for SARAL, and only 13 months of reprocessed SAR data for CryoSat-2. We chose to analyse the longest time series possible for the seasonal calculations since the records are relatively short. However, entire years should be analysed to remove any sampling biases in these statistics. Given the long repeat time for CryoSat-2, we also measure different geographical regions in each season, which can introduce biases in our basin-scale averages. Interannual variations also occur in the dynamics in response to interannual atmospheric changes, which can lead to different deep convection events from one season to another (Adloff et al., 2015). Analysing a longer time series of SARAL and CryoSat data should improve the significance of these early results.

One application of this type of analysis is to improve the altimetric data post-processing to be adapted to the regional conditions. Today, along-track filtering is applied in a similar way to all altimeter missions to reduce the instrument and geophysical noise. Since consecutive altimeter points are laid down spatially, data are filtered spatially along the track to reduce this noise. Standard filtering in the AVISO along-track products DT2010 ranges from 55 km wavelength at high latitudes to around 250 km in the tropics (Dibarboure et al., 2011). The new AVISO products DT2014 apply lower along-track smoothing at 65 km wavelength, globally and for all missions (Pujol et al., 2016). This study suggests that the along-track filtering may be tuned in a regional study to be better adapted to the local dynamics and noise conditions. Thus in the NW Mediterranean Sea, filtering of Jason-2 data could vary seasonally from 50 km in winter to 60 km in autumn and spring (or a conservative 60 km year-round). SARAL could have a finer-scale along-track filtering applied, to retain wavelengths greater than 35 km in summer–autumn and 45 km in winter. A filter cutoff of 50 km year-round could be suitable for CryoSat-2. Knowing how this statistical signal-to-noise ratio varies from one mission to another, and seasonally, is very useful for regional applications, for local process studies or for data assimilation.

The in situ validation remains very limited in space and time and did not allow us to confirm whether these smaller scales are realistic ocean features. For the glider comparison with SARAL, small-scale structures were detected by both systems, but their rapid movement prevented us from giving a precise along-track co-location except for the short scales close to the temporal crossing point. Indeed, for advective dynamics to be resolved correctly, they should conform to the Friedrichs–Lewy condition, i.e. $U\,\Delta t\,/\,\Delta x < 1$. If we follow small structures with typical advection speeds of $U = 0.3\,\mathrm{m\,s^{-1}}$ (typical of the Northern Current), then we need time differences, Δt, of less than 1.35 days to resolve the smaller SARAL wavelengths at 35 km, and within 2 days for the Jason-2 and CryoSat-2 data to resolve 50 km wavelength structures. With the slow-moving gliders, we can only cover 30 km per day, and so our along-track intercomparisons should be limited to the ±30 km around the altimeter–glider crossing point. This places a very strong constraint on our in situ validation.

The SARAL intercomparison with the Toulon HF radar data was quite promising. Despite the apparent nearshore errors in the SARAL data, and the periods with strong wind-driven currents, the correlation between the SARAL geostrophic currents and HF radar total currents remained high. The position of the Toulon HF radar helps, as the observations are centred on the Northern Current, in a region where the current is strongly steered by bathymetry, and the geostrophic component is dominant. This example indicates that a strong coastal current, with a high signal-to-noise ratio, can be detected by satellite altimetry, even at 20 km from the coast. Improvements are still needed to reduce the altimetric errors in the nearshore region, and to compare the CryoSat-2 SAR current observations with the HF radar data.

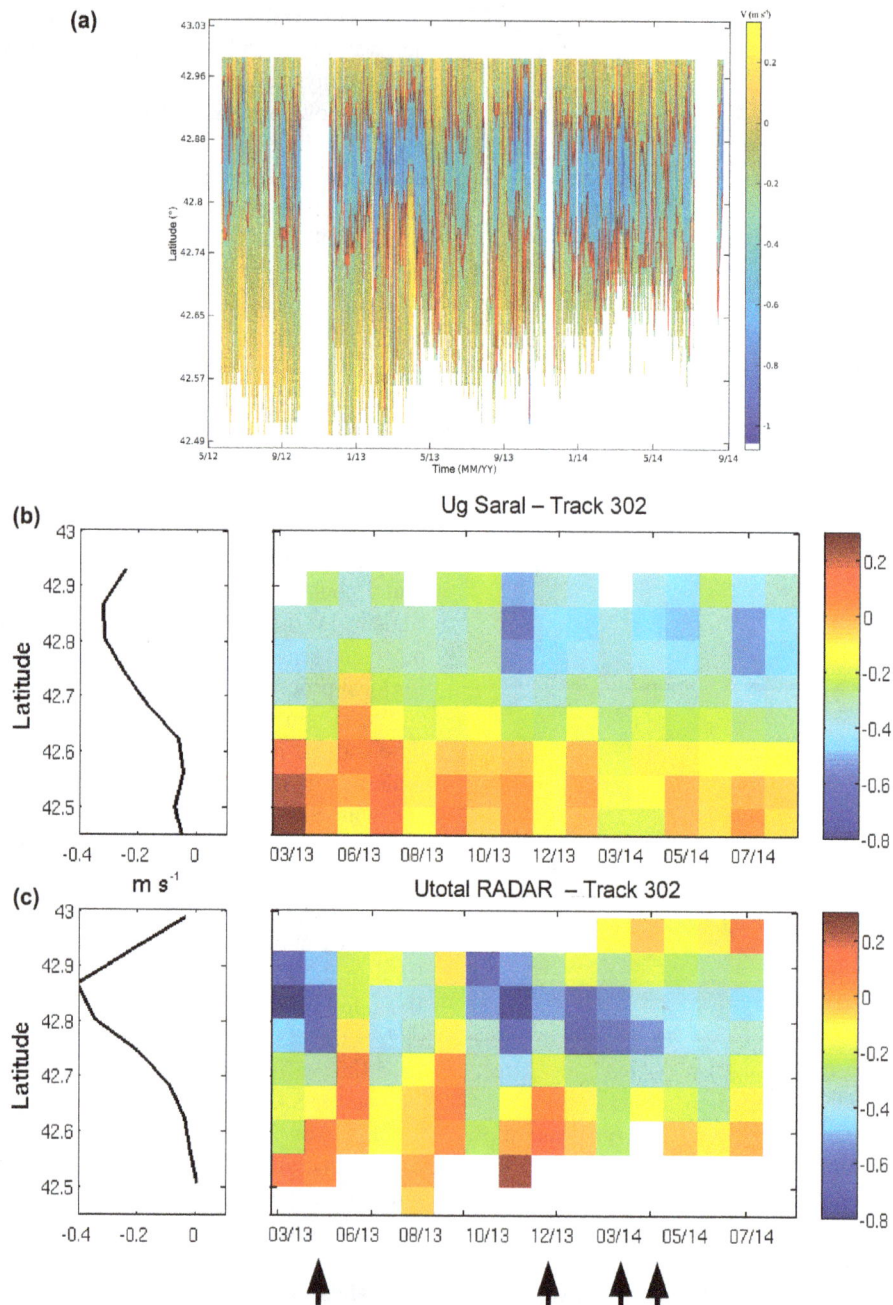

Figure 11. (a) Upper panel: 18-month time series of daily HF radar surface currents projected in the cross-track direction of the SARAL ground track. Red contours at $-0.3\,\mathrm{m\,s^{-1}}$ aid to delimit the westward Northern Current position. **(b)** Middle panel: extraction of these daily HF radar currents at the day of the SARAL observations. The temporal mean value is shown on the left. **(c)** Bottom panel: cross-track geostrophic currents from the SARAL altimeter data, filtered at 35 km wavelength. Arrows mark the dates with low correlations < 0.5.

This good intercomparison suggests that HF radar data may be combined with altimetry to extend the observations (duration and offshore extent) of the Northern Current and its recirculation near Toulon.

Another potential way to cross-validate the feature scales observed by the different altimeter missions is to use the crossover points between different missions. Figure 1 shows that there are many crossover points during this analysis period, especially from CryoSat-2 on its long-repeat 369-day orbit and even from Jason-1, which moved into a long-repeat 406-day geodetic orbit from April 2012 to 1 July 2013. Our analyses of the small, fast-moving features in this paper indicate that we really need crossover measurements overlapping within 1–2 days to capture these fine-scale features. These

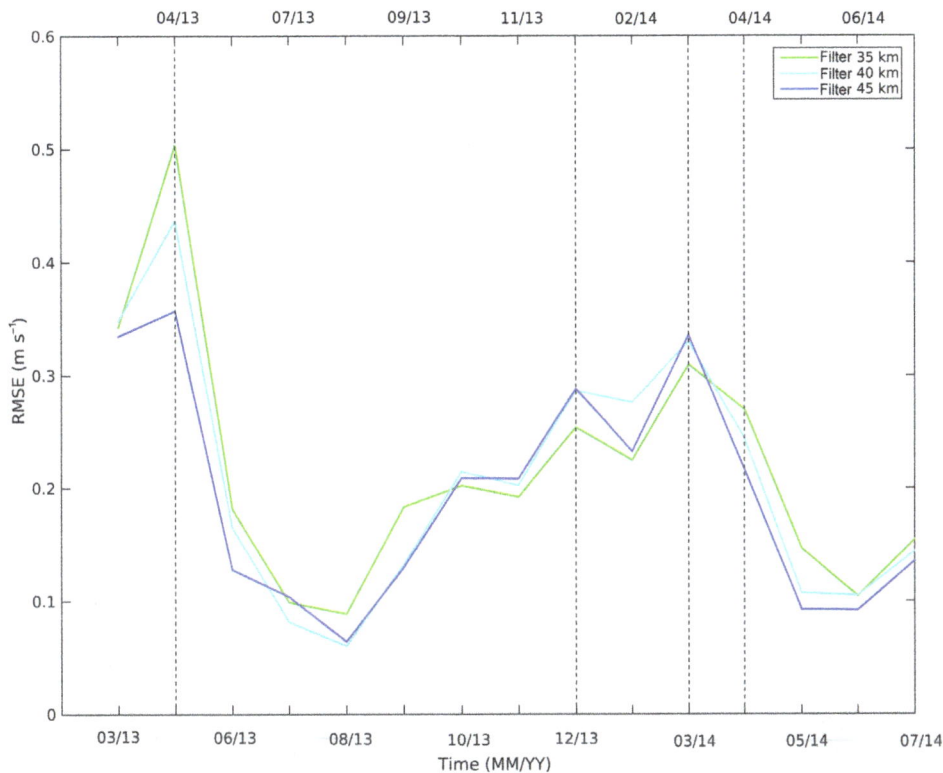

Figure 12. RMSE between the cross-track HF radar current amplitudes and the SARAL current amplitudes. Dates with low correlations (< 0.5) are marked with the vertical dashed line.

multi-altimeter overlapping passes are also interesting for the missions on a similar inclination, since their overlapping sections can be quite long. For example, SARAL and CryoSat may have long overlapping sections with a time difference of less than 2 days (see Fig. 1). Similar long sections may be available from the Jason-1 geodetic mission and Jason-2. At present, we are developing the code to calculate the crossovers from multi-satellite passes and select the passes based on their time differences. This analysis will be performed as part of our ongoing work in this region.

For the future altimetric missions, finer spatial sampling and lower noise levels should continue, with Sentinel-3 in global SAR mode launched in early 2016, and SWOT providing 2-D interferometric SAR heights and images and an order of magnitude lower noise in 2021. Similar wavenumber spectral analysis techniques could be applied to estimate the noise levels and observable spatial scales with these new missions. This study illustrates that the difficulties in setting up an adequate in situ validation for the small-scale, rapidly evolving dynamics will remain a challenge to resolve in the future.

Author contributions. This work was carried out by Alice Carret as part of her master's programme. Rosemary Morrow supervised the work and prepared the manuscript with contributions from all co-authors. Guillaume Valladeau and Francois Boy provided co-supervision. Florence Birol and Fernando Nino provided support with the analysis. Celine Bachelier processed the glider data and Bruno Zakardjian the HF radar data.

Competing interests. The authors declare that they have no conflict of interest.

Acknowledgement. This work was funded by an OSTST CNES TOSCA grant. The glider and HF radar data were funded as part of the French MOOSE Mediterranean observing system programme, with additional financial support from CNES as part of the Comsom glider campaign. We gratefully acknowledge the constructive comments by two reviewers and the editor, which helped to improve the manuscript.

Edited by: J. M. Huthnance

References

Adloff, A., Somot, S., Sevault, F., Jorda, G., Aznar, R., Déqué, M., Herrmann, M., Marcos, M., Dubois, C., Padorno, E., Alvarez-Fanjul, E., and Gomis, D.: Mediterranean Sea response to climate change in an ensemble of 21st century scenarios, Clim. Dynam., 45, 2775, doi:10.1007/s00382-015-2507-3, 2015.

Birol, F. and Delebecque, C.: Using High Sampling Rate (10/20 Hz) Altimeter Data for the Observation of Coastal Surface Currents: A Case Study over the Northwestern Mediterranean Sea, J. Marine Syst., 129, 318–333, doi:10.1016/j.jmarsys.2013.07.009, 2014.

Birol, F. and Niño, F.: Ku and Ka-Band Altimeter Data in the Northwestern Mediterranean Sea, Mar. Geod., 38, 313–327, doi:10.1080/01490419.2015.1034814, 2015.

Birol, F., Cancet, M., and Estournel, C.: Aspects of the seasonal variability of the Northern Current (NW Mediterranean Sea) observed by altimetry, J. Marine Syst., 81, 297–311, 2010.

Bosse, A., Testor, P., Mortier, L., Prieur, L., Taillandier, V., D'Ortenzio, F., and Coppola, L.: Spreading of Levantine Intermediate Waters by submesoscale coherent vortices in the northwestern Mediterranean Sea as observed with gliders, J. Geophys. Res.-Oceans, 120, 1599–1622, doi:10.1002/2014JC010263, 2015.

Bouffard, J., Vignudelli, S., Cipollini, P., and Ménard, Y.: Exploiting the potential of an improved multimission altimetric data set over the coastal ocean, Geophys. Res. Lett., 35, L10601, doi:10.1029/2008GL033488, 2008.

Bouffard, J., Pascual, A., Ruiz, S., Faugère, Y., and Tintoré, J.: Coastal and mesoscale dynamics characterization using altimetry and gliders: A case study in the Balearic Sea, J. Geophys. Res., 115, C10029, doi:10.1029/2009JC006087, 2010.

Bouffard, J., Roblou, L., Birol, F., Pascual, A., Fenoglio-Marc, L., Cancet, M., Morrow, R., and Ménard, Y.: Introduction and assessment of improved coastal altimetry strategies: case study over the North Western Mediterranean Sea, Chapter 12, in: Coastal Altimetry, edited by: Vignudelli, S., Kostianoy, A. G., Cipollini, P., and Benveniste, J., Springer-Verlag Berlin Heidelberg, 578 pp., doi:10.1007/978-3-642-12796-0_12, 2011.

Boy, F., Desjonquères, J.-D., Picot, N., Moreau, T., and Raynal, M.: CryoSat-2 SAR-Mode Over Oceans: Processing Methods, Global Assessment, and Benefits, IEEE Trans. Geosci. Remote Sens., 55, 148–158, doi:10.1109/TGRS.2016.2601958, 2017.

Chelton, D. B., Schlax, M. G., and Samelson, R. M.: Global observations of nonlinear mesoscale eddies, Prog. Oceanogr., 91, 167–216, 2011.

Dibarboure, G., Pujol, M.-I., Briol, F., Le Traon, P. Y., Larnicol, G., Picot, N., Mertz, F., and Ablain, M.: Jason-2 in DUACS: Updated System Description, First Tandem Results and Impact on Processing and Products, Mar. Geod., 34, 214–241, doi:10.1080/01490419.2011.584826, 2011.

Dibarboure, G., Boy, F., Desjonqueres, J. D., Labroue, S., Lasne, Y., Picot, N., Poisson, J. C., and Thibaut, P.: Investigating Short-Wavelength Correlated Errors on Low-Resolution Mode Altimetry, J. Atmos. Ocean. Tech., 31, 1337–1362, doi:10.1175/JTECH-D-13-00081.1, 2014.

Dufau, C., Orsztynowicz, M., Dibarboure, G., Morrow, R., and Le Traon, P.-Y.: Mesoscale resolution capability of altimetry: Present and future, J. Geophys. Res.-Oceans, 121, 4910–4927, doi:10.1002/2015JC010904, 2016.

Durand, F., Marin, F., Fuda, J. L., and Terre, T.: The East Caledonian Current : A Case Example for the Intercomparison between AltiKa and In Situ Measurements in a Boundary Current, Mar. Geod., 1–22, 2016.

Fu, L.-L.: On the wavenumber spectrum of oceanic mesoscale variability observed by the Seasat altimeter, J. Geophys. Res.-Oceans, 88, 331–334, 1983.

Grilli, F. and Pinardi, N.: The computation of Rossby radii between external forcing mechanisms and internal of deformation for the Mediterranean Sea, MTP News, 6, 4–5, 1998.

Guihou, K., Marmain, J., Ourmières, Y., Molcard, A., and Zakardjian, B.: Forget P., A case study of the mesoscale dynamics in the North-Western Mediterranean Sea: a combined data-model approach, Ocean Dynam., 63, 793–808, doi:10.1007/s10236-013-0619-z, 2013.

Herrmann, M., Somot, S., Sevault, F., Estournel, C., and Déqué, M.: Modeling deep convection in the Northwestern Mediterranean Sea using an eddy-permitting and an eddy-resolving model: case study of winter 1986–87, J. Geophys. Res., 113, C04011, doi:10.1029/2006JC003991, 2008.

Lambin, J., Morrow, R., Fu, L.-L., Willis, J. K., Bonekamp, H., Lillibridge, J., Perbos, J., Zaouche, G., Vaze, P., Bannoura, W., Parisot, F., Thouvenot, E., Coutin-Faye, S., Lindstrom, E., and Mignogno, M.: The OSTM/Jason-2 Mission, Mar. Geod., 33, 4–25, 2010.

Lebeaupin Brossier, C., Bastin, S., Béranger, K., and Drobinski, P. Regional mesoscale air–sea coupling impacts and extreme meteorological events role on the Mediterranean Sea water budget, Clim. Dynam., 44, 1029–1051, 2015.

Le Traon, P. Y., Klein, P., Hua, B. L., and Dibarboure, G.: Do Altimeter Wavenumber Spectra Agree with the Interior or Surface Quasigeostrophic Theory?, J. Phys. Oceanogr., 38, 1137–1142, doi:10.1175/2007JPO3806.1, 2008.

Millot, C.: Circulation in the Western Mediterranean Sea, J. Mar. Syst., 20, 423–442, 1999.

Morrow, R., Birol, F., Griffin, D., and Sudre, J.: Divergent pathways of cyclonic and anti-cyclonic ocean eddies, Geophys. Res. Lett., 31, L24311, doi:10.1029/2004GL020974, 2004.

Pascual, A., Pujol, M.-I., Larnicol, G., Le Traon, P.-Y., and Rio, M.-H.: Mesoscale mapping capabilities of multisatellite altimeter missions: First results with real data in the Mediterranean Sea, J. Mar. Syst., 65, 190–211, doi:10.1016/j.jmarsys.2004.12.004, 2006.

Pujol, M.-I. and Larnicol, G.: Mediterranean sea eddy kinetic energy variability from 11 years of altimetric data, J. Mar. Syst., 58, 121–142, 2005.

Pujol, M.-I., Faugère, Y., Taburet, G., Dupuy, S., Pelloquin, C., Ablain, M., and Picot, N.: DUACS DT2014: the new multi-mission altimeter data set reprocessed over 20 years, Ocean Sci., 12, 1067–1090, doi:10.5194/os-12-1067-2016, 2016.

Quentin, C., Barbin, Y., Bellomo, L., Forget, P., Gagelli, J., Grosdidier, S., Guerin, C.-A., Guihou, K., Marmain, J., Molcard, A., Zakardjian, B., Guterman, P., and Bernardet, K.: HF radar in French Mediterranean Sea: an element of MOOSE Mediterranean Ocean Observing System on Environment, OCOSS'2013 Proceedings, 25–30, 2013.

Ray, C., Martin-Puig, C., Clarizia, M. P., Ruffini, G., Dinardo, S., Gommenginger, C., and Benveniste, J.: SAR altimeter backscattered waveform model, Trans. Geosci. Remote Sens., 53, 911–919, doi:10.1109/TGRS.2014.2330423, 2015.

Rio, M.-H., Pascual, A., Poulain, P.-M., Menna, M., Barceló, B., and Tintoré, J.: Computation of a new mean dynamic topography for the Mediterranean Sea from model outputs, altimeter measurements and oceanographic in situ data, Ocean Sci., 10, 731–744, doi:10.5194/os-10-731-2014, 2014.

Robinson, I. S.: Discovering the ocean from Space. The unique applications of satellite oceanography, Springer, 638 pp., doi:10.1007/978-3-540-68322-3, 2010.

Stammer, D.: Global characteristics of ocean variability estimated from regional TOPEX/POSEIDON altimeter measurements, J. Phys. Oceanogr., 27, 1743–1769, 1997.

Verron, J., Sengenes, P., Lambin, J., Noubel, J., Steunou, N., Guillot, A., Picot, N., Coutin-Faye, S., Sharma, R., Gairola, R. M., Raghava Murthy, D. V. A., Richman, J. G., Griffin, D., Pascual, A., Rémy, F., and Gupta, P. K.: The SARAL/AltiKa altimetry satellite mission, Mar. Geod., 38, 2–21, doi:10.1080/01490419.2014.1000471, 2015.

Xu, Y. and Fu, L.-L.: Global Variability of the Wavenumber Spectrum of Oceanic Mesoscale Turbulence, J. Phys. Oceanogr., 41, 802–809, doi:10.1175/2010JPO4558.1, 2011.

Xu, Y. and Fu, L.-L.: The Effects of Altimeter Instrument Noise on the Estimation of the Wavenumber Spectrum of Sea Surface Height, J. Phys. Oceanogr., 42, 2229–2233, doi:10.1175/JPO-D-12-0106.1, 2012.

The "shallow-waterness" of the wave climate in European coastal regions

Kai Håkon Christensen[1,3], Ana Carrasco[1], Jean-Raymond Bidlot[2], and Øyvind Breivik[1,4]

[1]Norwegian Meteorological Institute, Henrik Mohns plass 1, 0313 Oslo, Norway
[2]European Centre for Medium-Range Weather Forecasts, Shinfield Park, Reading, RG2 9AX, UK
[3]Department of Geosciences, University of Oslo, Sem Sælands vei 1, 0316, Oslo, Norway
[4]Geophysical Institute, University of Bergen, Allégaten 70, 5007, Bergen, Norway

Correspondence to: Kai Håkon Christensen (kaihc@met.no)

Abstract. In contrast to deep water waves, shallow water waves are influenced by bottom topography, which has consequences for the propagation of wave energy as well as for the energy and momentum exchange between the waves and the mean flow. The ERA-Interim reanalysis is used to assess the fraction of wave energy associated with shallow water waves in coastal regions in Europe. We show maps of the distribution of this fraction as well as time series statistics from eight selected stations. There is a strong seasonal dependence and high values are typically associated with winter storms, indicating that shallow water wave effects can occasionally be important even in the deeper parts of the shelf seas otherwise dominated by deep water waves.

1 Introduction

The purpose of this brief note is to present some aspects of ocean surface waves related to bottom topography. If the wavelength is small compared to the local water depth, the waves are unaffected by the presence of the sea floor and the wave energy balance is dominated by input from wind, dissipation by wave breaking and white capping, and nonlinear wave–wave interactions. If the wavelength is large compared to the local water depth, the situation is quite different and the wave energy propagation will directly depend on the bottom topography, with implications for dissipation and sediment transport in the bottom boundary layer, wave–mean flow interactions through wave radiation stresses, modifica-

tion to the nonlinear wave–wave interactions, and so on (e.g., Komen et al., 1994; Smith, 2006).

The main aim of this study is to identify in which coastal regions in Europe shallow water wave effects may be important and to quantify the fraction of wave energy associated with ocean waves that can "feel" the bottom. As such, this note differs from previous studies that focus on the wave climate, employing either hindcasts (e.g., Gorman et al., 2003; Dodet et al., 2010; Reistad et al., 2011; Aarnes et al., 2012), reanalyses (e.g., Dee et al., 2011; Reguero et al., 2012), or climate projections (e.g., Wang et al., 2004; Hemer et al., 2013) to assess average and/or extreme values of typical wave parameters on regional or global scales. Typical wave conditions can be classified according to the shape of the two-dimensional wave spectrum (e.g., Boukhanovsky et al., 2007), utilizing the fact that the waves will often be a combination of remotely forced swell and locally generated wind waves. In coastal regions, a significant proportion of the wave energy may be associated with waves on intermediate depth, and at any specific location this proportion will vary in time due to variations in the local and remote forcing of the waves. It should be emphasized that we do not make a clear distinction here between intermediate and shallow water waves, for which the wavelength is much larger than the local depth.

2 Concept and methods

We divide the wave spectrum into high- and low-frequency parts, using prescribed values of the ratio n between the wave group and phase velocities to identify the frequency that sep-

Table 1. Station names, positions, and depths, in addition to verification statistics for significant wave height (H_s): scatter index (SI, standard deviation of error divided by the observation average) and bias. The rightmost column shows the number of collocated measurements used in deriving the statistics. The depths referred to here and in subsequent plots are the model depths.

Name	Latitude	Longitude	Depth (m)	H_s SI (%)	H_s bias (m)	Collocation numbers
LF3J	61.20	2.30	181	16.95	0.07	19 395
62023	51.40	−7.90	103	19.27	0.35	19 400
AUK	56.39	2.05	79	15.03	0.02	2572
62069	48.29	−4.97	63	19.53	0.22	6380
LF5U	56.50	3.21	60	14.49	−0.07	27 684
K13	53.20	3.22	29	15.94	−0.06	12 910
EURO	51.99	3.27	28	17.77	−0.09	12 303
BSH03	54.00	8.12	20	29.79	−0.28	9113

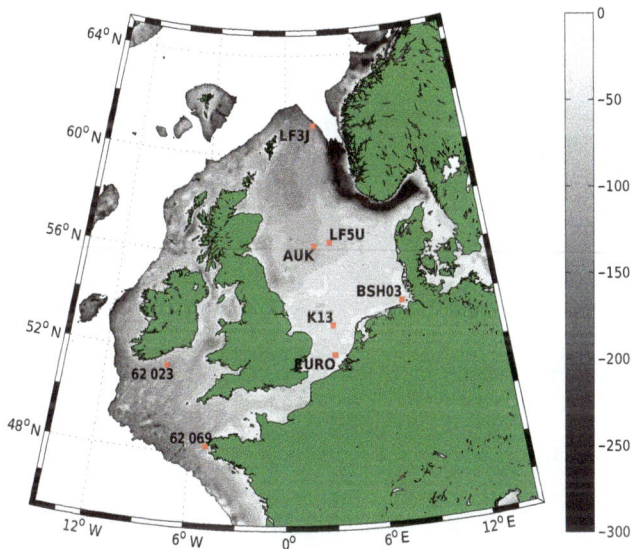

Figure 1. Map of station positions. Depths less than 300 m are indicated in gray.

arates the two parts. The wave energy in the low-frequency part is divided by the total wave energy, and maps and time series statistics of this ratio are presented. Since wave dispersion depends on the local water depth in shallow waters, the frequency limit for any given n will vary in space. The data are obtained from the wave model component of the ERA-Interim reanalysis (Dee et al., 2011).

2.1 Wave dispersion

The dispersion relation for surface gravity waves is

$$\omega^2 = gk \tanh(kh). \tag{1}$$

Here ω is the wave angular frequency, g is the acceleration due to gravity, k is the wave number, and h is the water depth. The phase velocity c in the direction of wave propagation is $c = \omega/k$. The group velocity is given by $c_g = d\omega/dk$, and

using Eq. (1) we have

$$n \equiv \frac{c_g}{c} = \frac{1}{2} + \frac{kh}{\sinh(2kh)}. \tag{2}$$

The ratio n between the group and the phase velocity is thus a function of the local water depth and the wave number. The limiting cases are for deep water ($kh \to \infty$), when $n = 1/2$, and for shallow water ($kh \to 0$), when $n = 1$ and the waves are non-dispersive. If $n > 1/2$, the waves are thus to some extent influenced by the bottom. In the present study we will consider n-values of 0.55, 0.65, 0.75, and 0.85. We will classify the waves according to their frequency $f = \omega/2\pi$, and for any given value of n the corresponding frequency f_n can be obtained from Eqs. (1) and (2). To investigate the "shallow-waterness" of a certain location we compute the ratio of energy E_{sw} of the waves that feel the bottom to the total energy E_{tot}:

$$r_n = \frac{E_{sw}}{E_{tot}} = \frac{\int_0^{2\pi} \int_0^{f_n} F \, df \, d\theta}{\int_0^{2\pi} \int_0^{\infty} F \, df \, d\theta}, \tag{3}$$

where $F(f, \theta)$ is the directional wave spectrum obtained from the reanalysis data.

There are several options for the choice of parameter for the frequency cutoff. The ratio n between the group and phase velocities occurs naturally in radiation stress theory, which is the main reason why we use it here. A simple example of how Eqs. (2) and (3) can be used is as follows: for monochromatic waves with energy E, the sum of the contribution to the radiation stress in the propagation direction from horizontal advection of momentum and the dynamical pressure below the mean (Eulerian) surface level is given by $2E(n - 1/2)$, which is zero for irrotational deep water waves (see Longuet-Higgins and Stewart, 1964, and also Whitham, 1962). The contribution from the divergence effect (e.g., Mcintyre, 1988) depends on the surface variance and yields an additional $E/2$. For any given n, the expression

$$\hat{S}_{xx} = r_n E_{tot}(2n - 1/2) \tag{4}$$

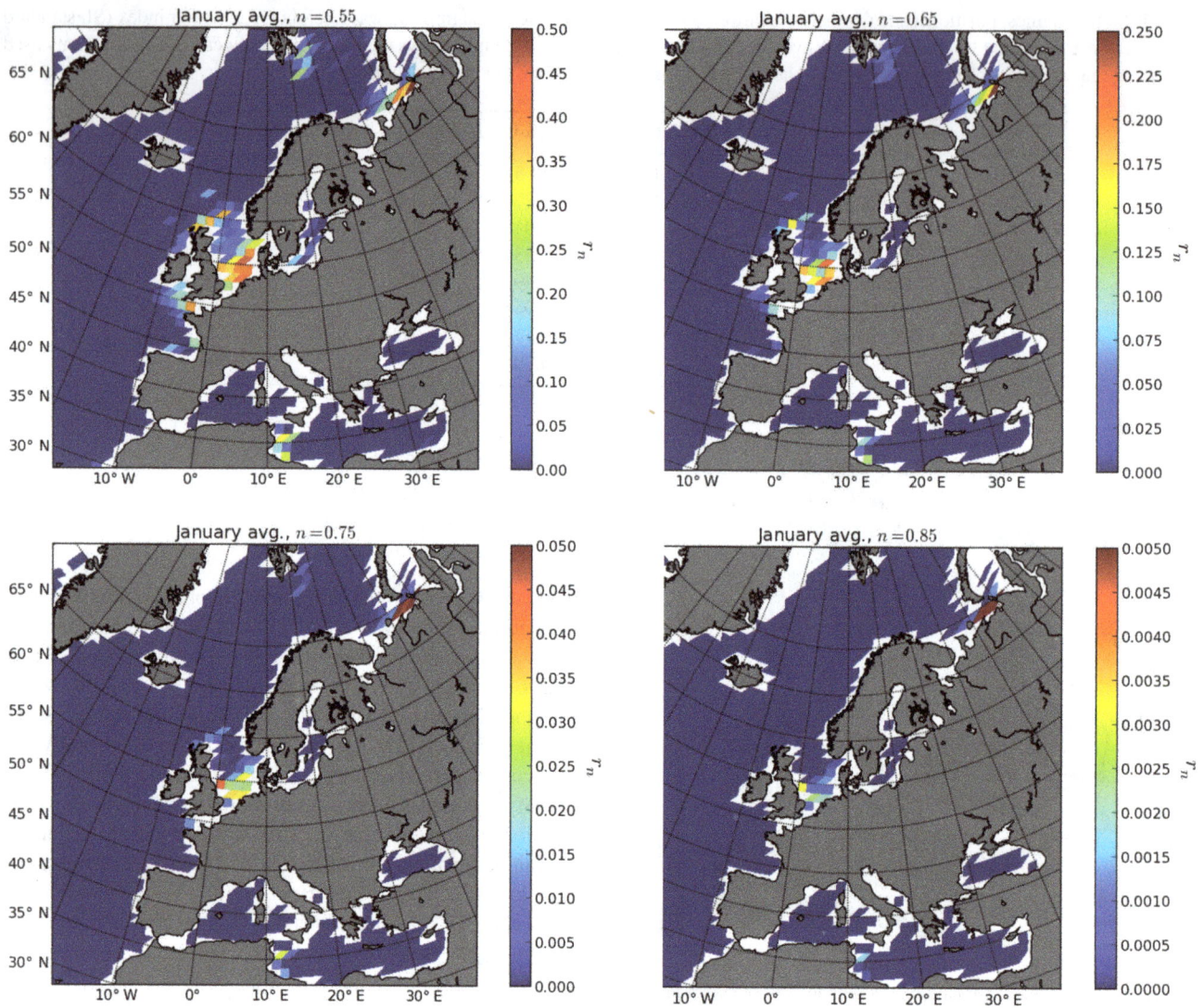

Figure 2. Average values of $r_{0.55}$, $r_{0.65}$, $r_{0.75}$, and $r_{0.85}$ in January for the period 1979–2012.

thus provides a lower bound (since n increases with wavelength) for the radiation stress $\hat{S}_{xx}$ in the mean wave direction and should be suitable for assessing an order of magnitude estimate. A similar expression for the transverse radiation stress component can easily be derived. The net effect on, e.g., the mean surface elevation will of course depend on the gradients in the radiation stresses and will vary from case to case.

2.2 ERA-Interim wave spectra

ERA-Interim (ERA-I) is a global coupled atmosphere–wave reanalysis starting in 1979 (Dee et al., 2011). An irregular latitude–longitude grid ensures relative constancy in atmospheric grid resolution towards the poles. T255 is the Gaussian grid with a spacing of the order 80 km, but atmospheric parameters are also made available (following bi-linear in-

terpolation) on a $0.75 \times 0.75°$ regular latitude–longitude grid. The model and data assimilation scheme of the reanalysis are based on Cycle 31r2 of the Integrated Forecast System (IFS). The WAM wave model is coupled to the atmospheric part of the IFS through the exchange of the Charnock parameter. See Janssen (1989, 1991, 2004) for details of the coupling and Dee et al. (2011) for an overview of the ERA-Interim reanalysis. The resolution of the wave model component is 1.0° on the Equator, but the resolution is kept approximately constant globally through the use of a quasi-regular latitude–longitude grid where grid points are progressively removed toward the poles (Janssen, 2004). The spectral range from 0.035 to 0.55 Hz is spanned with 30 logarithmically spaced frequency bands. The angular resolution is 15° (24 bins). Full two-dimensional spectra are archived every 6 h on the native grid. The ERA-I WAM implementation incorporates

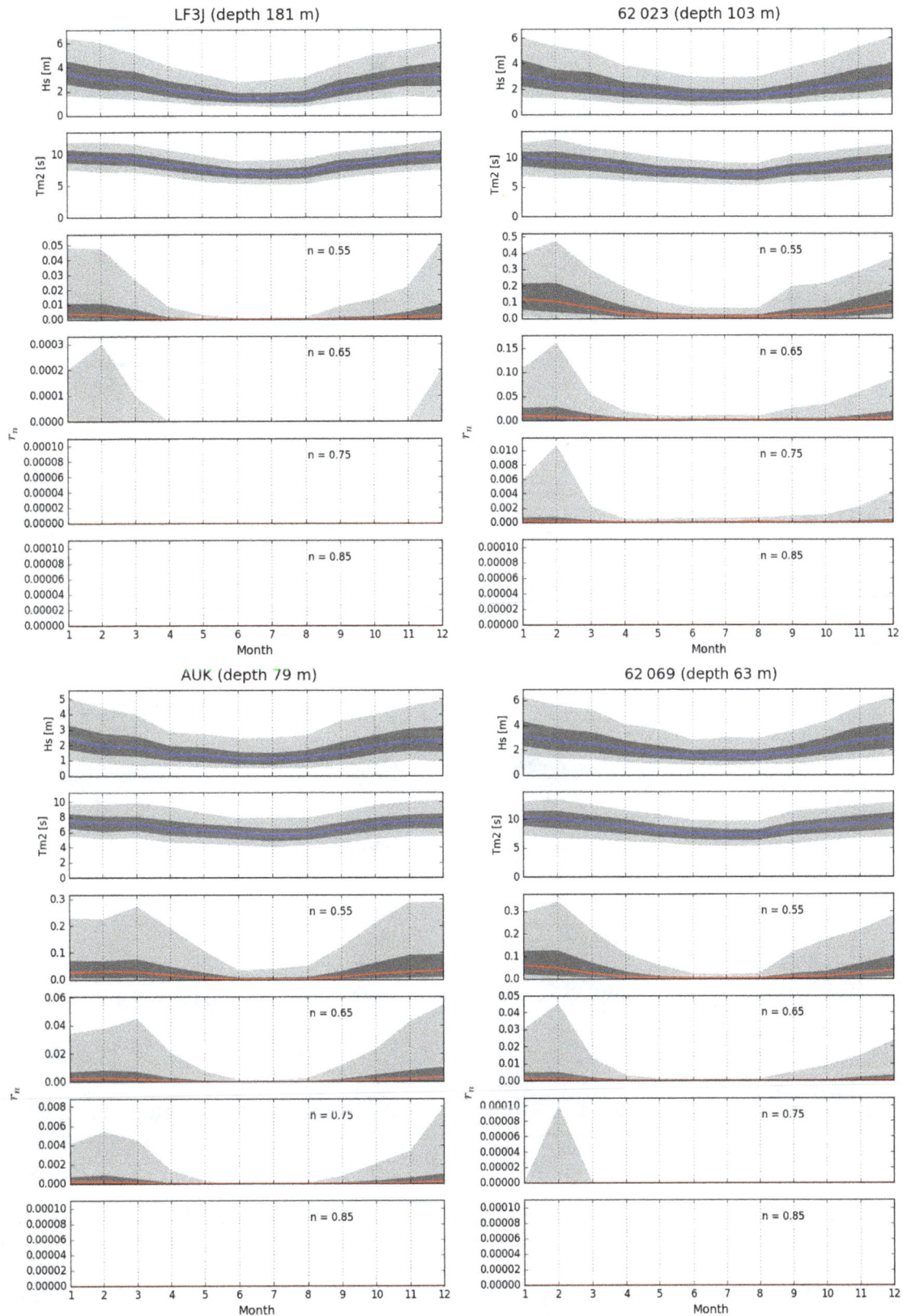

Figure 3. Monthly values of significant wave height, mean period, and r_n values for $n = 0.55, 0.65, 0.75,$ and 0.85 at stations LF3J, 62023, AUK, and 62069. Median values are given by red and blue lines; 25th to 75th percentiles are shown as dark gray; 5th to 95th percentiles are shown as light gray.

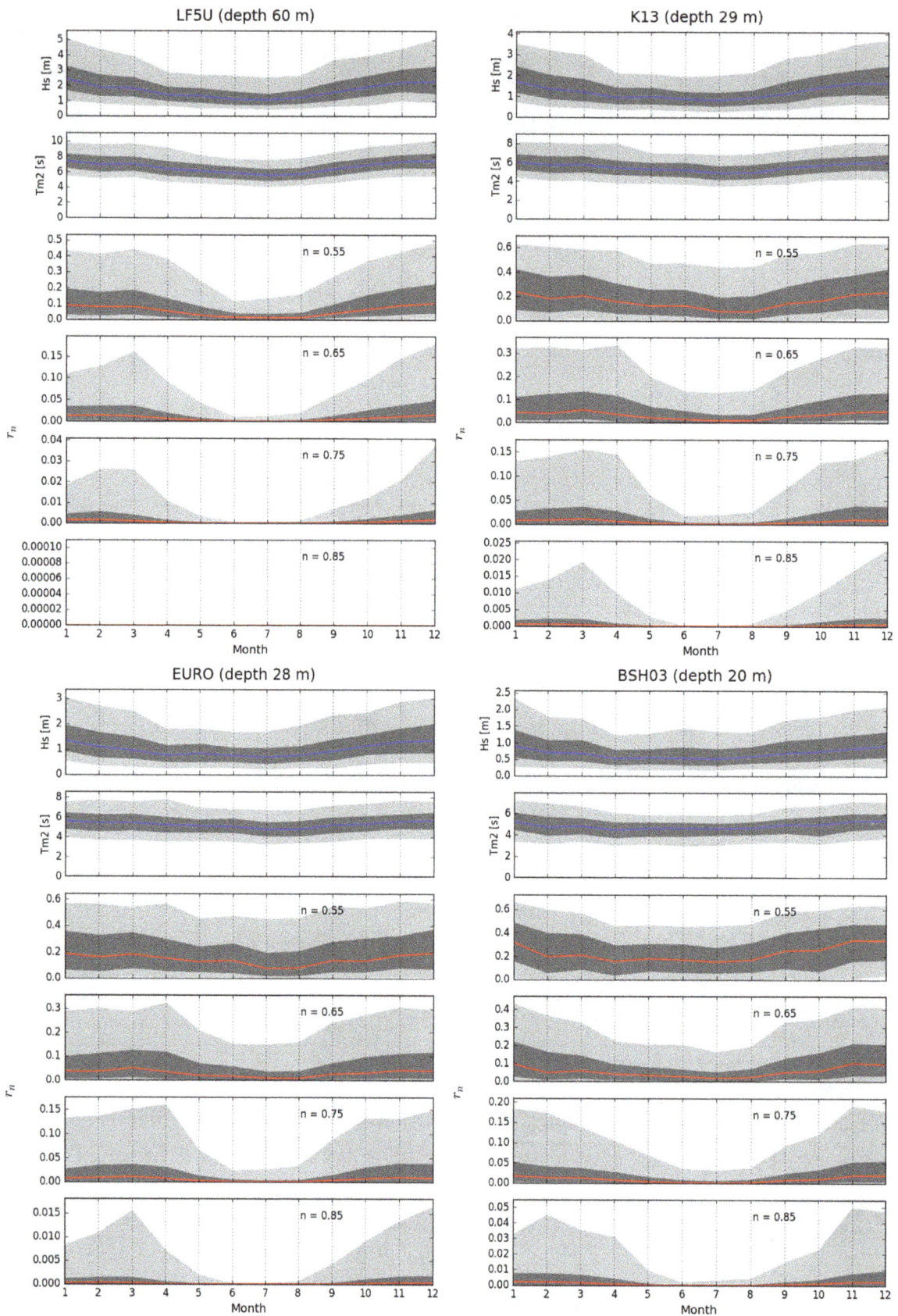

Figure 4. Same as Fig. 3 but for stations LF5U, K13, EURO, and BSH03.

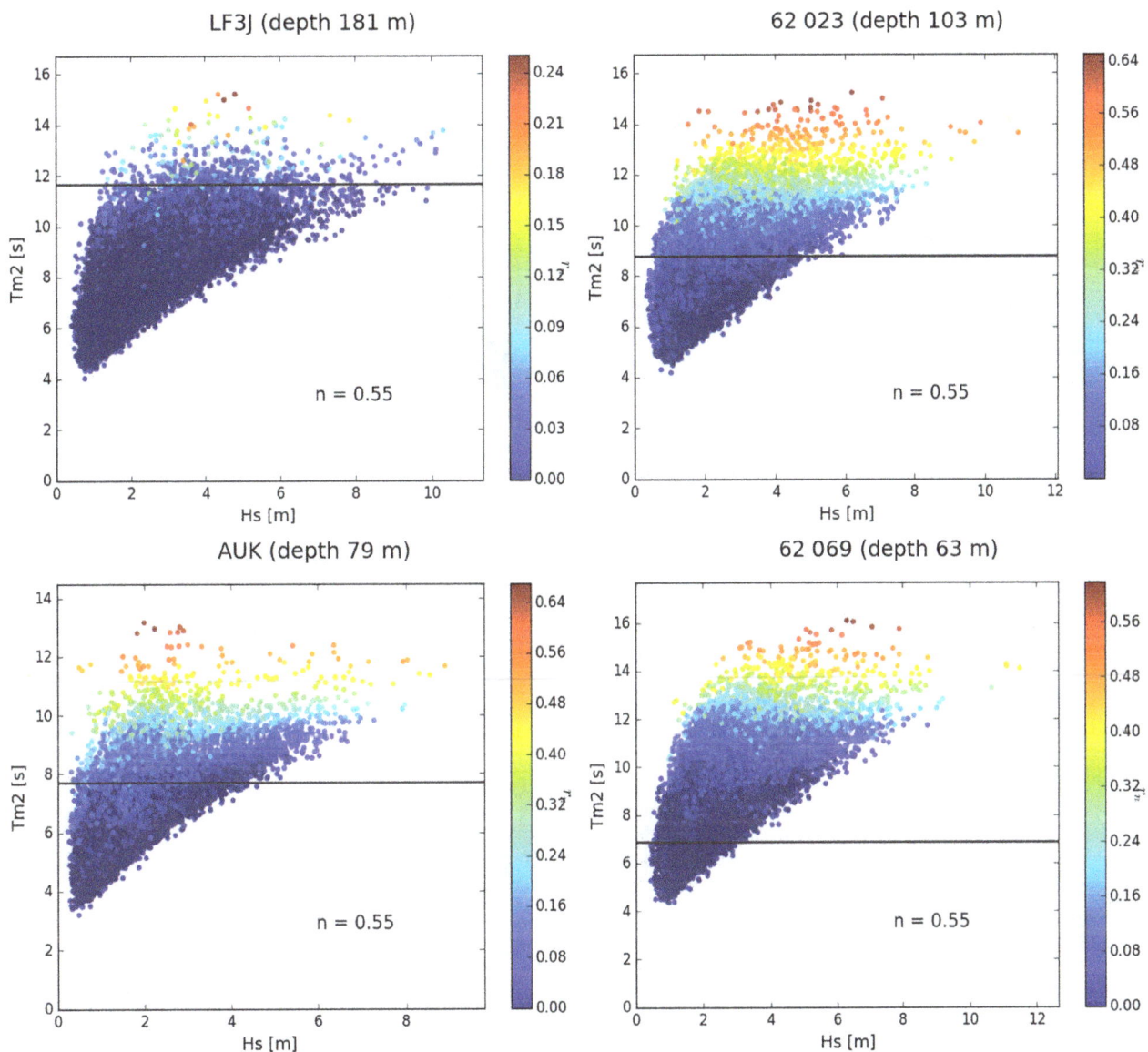

Figure 5. Scatter plot of all the data points for stations LF3J, 62023, AUK, and 62069, with colors indicating $r_{0.55}$ values. The gray line indicates the period corresponding to $n = 0.55$ for each station.

shallow-water effects important in areas like the southern North Sea (Komen et al., 1994).

2.3 Stations

In addition to presenting maps of the ratio r_n, we analyse eight stations in some detail using the 6-hourly time series from ERA-I. These stations correspond to locations with wave observations, and we have focused on the European Northwest Shelf Sea where shallow water waves are most prominent. The station names, positions, depths and some verification statistics are listed in Table 1, and the positions are also shown in Fig. 1. Three stations southwest of Ireland and the UK, and in the northern North Sea, are exposed

to long swell from the North Atlantic (62069, 62023, and LF3J), and all these stations are in intermediate to deep water (63, 103, and 181 m, respectively). Two stations are in intermediate depths in the middle of the North Sea (AUK and LF5U), while the rest are in the shallow southern part of the North Sea.

3 Results

We first investigate the spatial distribution of n. For this purpose we use monthly averages of the wave spectra. We then investigate the temporal variation of n at the eight stations defined in Sect. 2.3, presenting monthly median values as well as the 5th, 25th, 75th, and 95th percentiles. Finally, we plot

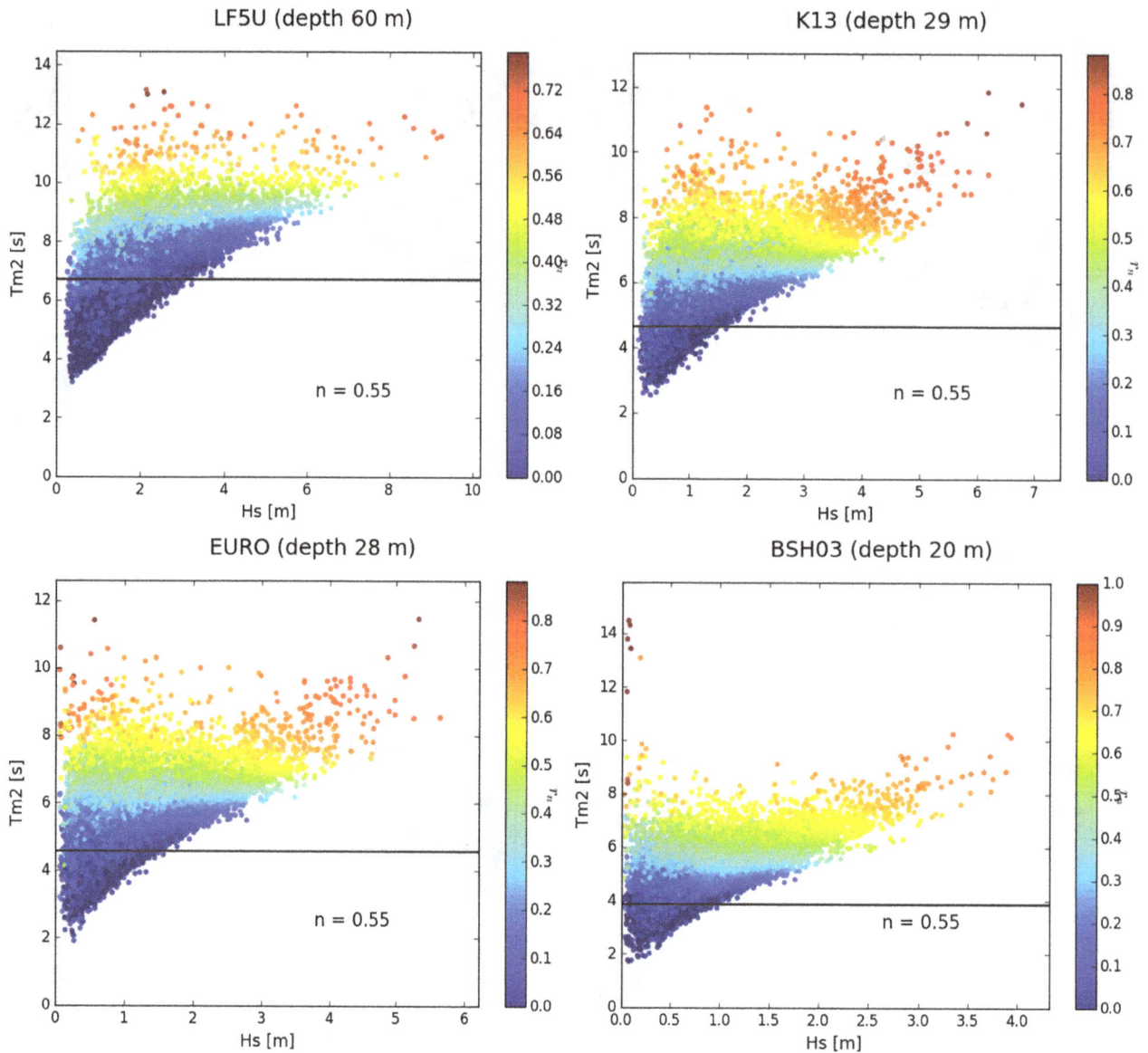

Figure 6. As Fig. 5 but for stations LF5U, K13, EURO, and BSH03.

n-values against mean period and significant wave height to investigate the variation of n with wave steepness.

3.1 Spatial distribution

Values of r_n are typically highest in the period December–March and Fig. 2 shows maps of the average values of r_n for January for the period 1979–2012. Unsurprisingly, the highest values are found in shallow waters, including the North Sea, southwest of Ireland and the UK, south of Spitsbergen, in the eastern part of the Barents Sea, and in the central Mediterranean. The monthly average r_n ratios become, by necessity, smaller for increasing n, and are for $n = 0.85$ vanishingly small everywhere.

3.2 Seasonal dependence

Figures 3–4 show monthly values of significant wave height (H_s), mean period (T_{m02}), and r_n for the eight stations listed in Table 1. The data are presented as median values and the 5th, 25th, 75th, and 95th percentiles for the period 2003–2013. Significant wave height and mean periods are highest in the winter months, and the spread is also larger. The values of r_n are quite small for the three stations with the largest depth, but we also see, e.g., values of $r_{0.65}$ reaching 15 % at station 62023 (103 m depth). Notably, the $r_{0.65}$ values are lower for the shallower AUK station (79 m depth), which is explained by this station being sheltered from the long swell originating in the North Atlantic. The r_n values are consistently lower in the summer months.

3.3 Dependence on wave steepness

Finally we investigate whether high r_n values are associated with a particular sea state, and Figs. 5–6 show scatter plots of all the data points in a H_s/T_{m02} diagram. We only consider $n = 0.55$. For the deepest station LF3J there are only a few cases with relatively high $r_{0.55}$ values (to put this in context: there are over 16 000 data points altogether). For the rest of the stations it is clear that $r_{0.55}$ is primarily correlated with the mean period, and not with the significant wave height, and high values can be found for both high and low waves. There is a lower limit to the mean period that increases with the wave height, however; hence, the average value of $r_{0.55}$ in general increases with H_s.

4 Conclusions

Data from the wave model component of the ERA-Interim reanalysis have been used to quantify the "shallow-waterness" of the wave climate in coastal regions in Europe. The "shallow-waterness" is here defined as the ratio r_n of wave energy of the components that are influenced by the bottom compared to the total wave energy. As can be expected, the ratios are largest during winter and on the European Northwest Shelf. Eight stations over that area have therefore been investigated in more detail.

This work has a bearing on coupled wave–ocean modeling systems; for example, shallow water wave-induced radiation stresses give rise to barotropic forcing terms that can play a role in storm surge modeling. The resolution of the ERA-Interim reanalysis is admittedly too coarse to provide much detail in several regions such as the Baltic Sea and the Mediterranean subbasins. The point is, however, that a straightforward analysis of standard two-dimensional wave spectra from any wave model can provide some guidance on whether or not certain dynamical processes related to the "shallow-waterness" are important. All the necessary information to evaluate Eq. (4) can essentially be shown in scatter plots like Figs. 5 and 6.

Similar methods to those we present here could also be used to investigate other processes, for example depth refraction, although the cutoff criterion should in this case be based on the ratio between deep water and local wave number values (e.g., Holthuijsen, 2007). In addition, the influence of currents on the waves could be included using relative instead of absolute frequencies, which is likely to play a role in places with strong tidal flows such as at station 62069 (Ardhuin et al., 2012).

With the exception of the shallowest parts of the shelf seas, the "shallow-waterness" is on average quite small, but occasional high values of r_n can be found at intermediate water depths ($\sim$ 100 m). Destructive storm surge events are typically caused by intense winter storms with high waves, and our results suggest that in such situations shallow water effects can be important even at great distances from the coast. The "shallow-waterness" is primarily correlated with the mean period and can be found for both high and low waves, but shallow water effects become increasingly important for higher waves since these are associated with longer mean periods.

Competing interests. The authors declare that they have no conflict of interest.

Acknowledgements. This work was funded by the European Union Seventh Framework Program FP7/2007–2013 under grant no. 283367 (MyOcean2). Øyvind Breivik was partly supported by European Union FP7 project MyWave (grant no. 284455) and in part by the ExWaMar project (grant no. 256466) funded by the Research Council of Norway. The authors also want to thank Fabrice Ardhuin for providing corrections and constructive comments on an earlier version of this paper.

Edited by: Mario Hoppema

References

Aarnes, O. J., Breivik, Ø., and Reistad, M.: Wave extremes in the Northeast Atlantic, J. Clim., 25, 1529–1543, https://doi.org/10.1175/JCLI-D-11-00132.1, 2012.

Ardhuin, F., Roland, A., Dumas, F., Bennis, A.-C., Sentchev, A., Forget, P., Wolf, J., Girard, F., Osuna, P., and Benoit, M.: Numerical wave modeling in conditions with strong currents: dissipation, refraction, and relative wind, J. Phys. Oceanogr., 42, 2101–2120, https://doi.org/10.1175/JPO-D-11-0220.1, 2012.

Bidlot, J.-R.: Grib and ASCII data, subset ERA-I for shallow water waves Ocean Science study [Data set], Zenodo, https://doi.org/10.5281/zenodo.831329, 2017.

Boukhanovsky, A. V., Lopatoukhin, L. J., and Guedes Soares, C.: Spectral wave climate of the North Sea, Appl. Ocean Res., 29, 146–154, https://doi.org/10.1016/j.apor.2007.08.004, 2007.

Dee, D., Uppala, S., Simmons, A. et al.: The ERA-Interim reanalysis: configuration and performance of the data assimilation system, Q. J. Roy. Meteor. Soc., 137, 553–597, https://doi.org/10.1002/qj.828, 2011.

Dodet, G., Bertin, X., and Taborda, R.: Wave climate variability in the North-East Atlantic Ocean over the last six decades, Ocean Model., 31, 120–131, https://doi.org/10.1016/j.ocemod.2009.10.010, 2010.

Gorman, R. M., Bryan, K. R., and Laing, A. K.: Wave hindcast for the New Zealand region: nearshore validation and coastal wave climate, New Zeal. J. Mar. Fresh., 37, 567–588, https://doi.org/10.1080/00288330.2003.9517190, 2003.

Hemer, M. A., Fan, Y., Mori, N., Semedo, A., and Wang, X. L.: Projected changes in wave climate from a multi-model ensemble, Nat. Clim. Change, 3, 471–476, https://doi.org/10.1038/nclimate1791, 2013.

Holthuijsen, L. H.: Waves in Oceanic and Coastal Waters, Cambridge University Press, Cambridge, 2007.

Janssen, P.: Wave-induced stress and the drag of air flow over sea waves, J. Phys. Oceangr., 19, 745–754, https://doi.org/10.1175/1520-0485(1989)019<0745:WISATD>2.0.CO;2, 1989.

Janssen, P.: Quasi-linear theory of wind-wave generation applied to wave forecasting, J. Phys. Oceangr., 21, 1631–1642, https://doi.org/10.1175/1520-0485(1991)021<1631:QLTOWW>2.0.CO;2, 1991.

Janssen, P.: The Interaction of Ocean Waves and Wind, Cambridge University Press, Cambridge, UK, 2004.

Komen, G. J., Cavaleri, L., Donelan, M., Hasselmann, K., Hasselmann, S., and Janssen, P. A. E. M.: Dynamics and Modelling of Ocean Waves, Cambridge University Press, Cambridge, 1994.

Longuet-Higgins, M. S. and Stewart, R.: Radiation stresses in water waves; a physical discussion, with applications, Deep-Sea Res., 11, 529–562, https://doi.org/10.1016/0011-7471(64)90001-4, 1964.

Mcintyre, M. E.: A note on the divergence effect and the Lagrangian-mean surface elevation in periodic water waves, J. Fluid Mech., 189, 235–242, https://doi.org/10.1017/S0022112088000989, 1988.

Reguero, B. G., Méndez, M., Méndez, F. J., Ménguez, R., and Losada, I. J.: A Global Ocean Wave (GOW) calibrated reanalysis from 1948 onwards, Coast. Eng., 65, 38–55, https://doi.org/10.1016/j.coastaleng.2012.03.003, 2012.

Reistad, M., Breivik, Ø., Haakenstad, H., Aarnes, O. J., Furevik, B. R., and Bidlot, J.-R.: A high-resolution hindcast of wind and waves for the North Sea, the Norwegian Sea, and the Barents Sea, J. Geophys. Res., 116, C05019, https://doi.org/10.1029/2010JC006402, 2011.

Smith, J. A.: Wave current interactions in finite depth, J. Phys. Oceanogr., 36, 1403–1419, https://doi.org/10.1175/JPO2911.1, 2006.

Wang, X. L., Zwiers, F. W., and Swail, V. R.: North Atlantic Ocean wave climate change scenarios for the twenty-first century, J. Clim., 17, 2368–2383, https://doi.org/10.1175/1520-0442(2004)017<2368:NAOWCC>2.0.CO;2, 2004.

Whitham, G.: Mass, momentum and energy flux in water waves, J. Fluid Mech., 12, 135–147, https://doi.org/10.1017/S0022112062000099, 1962.

3

An atmosphere–wave regional coupled model: improving predictions of wave heights and surface winds in the southern North Sea

Kathrin Wahle[1], Joanna Staneva[1], Wolfgang Koch[1], Luciana Fenoglio-Marc[2], Ha T. M. Ho-Hagemann[1], and Emil V. Stanev[1]

[1]Institute of Coastal Research, Helmholtz-Zentrum Geesthacht, Geesthacht, Germany
[2]Institute of Geodesy and Geoinformation, University of Bonn, Bonn, Germany

Correspondence to: Kathrin Wahle (kathrin.wahle@hzg.de) and Joanna Staneva (joanna.staneva@hzg.de)

Abstract. The coupling of models is a commonly used approach when addressing the complex interactions between different components of earth systems. We demonstrate that this approach can result in a reduction of errors in wave forecasting, especially in dynamically complicated coastal ocean areas, such as the southern part of the North Sea – the German Bight. Here, we study the effects of coupling of an atmospheric model (COSMO) and a wind wave model (WAM), which is enabled by implementing wave-induced drag in the atmospheric model. The numerical simulations use a regional North Sea coupled wave–atmosphere model as well as a nested-grid high-resolution German Bight wave model. Using one atmospheric and two wind wave models simultaneously allows for study of the individual and combined effects of two-way coupling and grid resolution. This approach proved to be particularly important under severe storm conditions as the German Bight is a very shallow and dynamically complex coastal area exposed to storm floods. The two-way coupling leads to a reduction of both surface wind speeds and simulated wave heights. In this study, the sensitivity of atmospheric parameters, such as wind speed and atmospheric pressure, to the wave-induced drag, in particular under storm conditions, and the impact of two-way coupling on the wave model performance, is quantified. Comparisons between data from in situ and satellite altimeter observations indicate that two-way coupling improves the simulation of wind and wave parameters of the model and justify its implementation for both operational and climate simulations.

1 Introduction

Wind forcing is considered one of the largest error sources in wave modelling. In numerical atmospheric models, wind stress is parameterised by the drag coefficient usually considered spatially uniform over water. In reality however the wind waves extract energy and momentum from the atmosphere as they grow under the influence of wind. This effect is greater for young sea states and high wind speed in comparison to decaying sea and calm atmospheric conditions. Under such conditions, the drag coefficient cannot be considered independent of the sea state and uniform in time and space. This dependence needs to be accounted for in coupled atmosphere–wave models. Jenkins et al. (2012) demonstrated that the wave field alters the ocean's aerodynamic roughness and the air–sea momentum flux depending on the relationship between the surface wind speed and the propagation speed of wave crests (the wave age). Based on high-resolution coupled simulations, Doyle (1995) demonstrated that young ocean waves increase the effective surface roughness, decrease the 10 m wind speed, and modulate heat and moisture transports between the atmosphere and ocean. As a result of this boundary layer modification Doyle (1995) concluded that the mesoscale structures associated with cyclones are perturbed. The impact of sea surface roughness was investigated in studies by Bao et al. (2002) and Desjardins et al. (2000). As shown by Lionello et al. (1998), the two-way wave–atmosphere coupling attenuates the depth of the pressure minimum. In particular, non-linearities increase under extreme conditions, which can modify the intensity of storms due to feedbacks between waves and the atmosphere. This feedback needs to be accounted for in coupled models as strong winds cause the drag coefficient of the sea surface to increase, leading to a reduction of wind speeds and modification of wind directions (Warner et al., 2010). These effects feed back into the airflow, wind speed, and turbulence profile in the boundary layer. Zweers et al. (2010) showed that the used atmospheric model overestimates the surface drag for high wind speeds and underestimates the intensity of hurricane winds. Zweers et al. (2010) proposed an approach of calibrating the boundary layer parameterisation using a one-way coupled model. They tested a new parameterisation that decreased the surface drag for two hurricanes in the Caribbean. This new drag parameterisation leads to much stronger forecasted hurricanes, which were in good agreement with observations.

The coupling between atmospheric and wind wave models was first introduced operationally in 1998 by the European Centre for Medium-Range Weather Forecasts (ECMWF). The method based on the theoretical work of Janssen (1991) contributed to an improvement of both atmospheric and surface wave forecasts on the global scale. Waves were recently considered in operational coupled model systems, such as that of Meteo-France (Voldoire et al., 2013). Breivik et al. (2015) incorporated the effects of surface waves into ocean dynamics via ocean side stress, turbulent kinetic energy due to wave breaking, and the Stokes–Coriolis force in the ECMWF system.

The effect of coupling on model predictions becomes more important (Janssen et al., 2004) with increasing grid resolution, which therefore emphasises the need for coupling on the regional scales. Spatial and temporal changes in the wave and wave energy propagation are not yet sufficiently addressed in high-resolution regional atmospheric models. The shallow water terms in the wave equations (depth and current refraction, bottom friction and wave breaking) play a dominant role near coastal areas, especially during storm events, where the wave breaking term prevents unrealistically high waves near the coast. The spray caused by breaking waves modulates the atmosphere boundary layer. Air–sea interaction is also of great importance in regional climate modelling. Rutgersson et al. (2010, 2012) introduced two different parameterisations in a European climate model. One parameterisation uses roughness length and includes only the effect of a growing sea, as proposed by Janssen (1991). The other uses wave age and introduced the reduction of roughness due to swell. In both cases, these parametrisations affected the long-term averages of atmospheric parameters notably and demonstrated that the swell has an important impact on mixing in the boundary layer. Järvenoja and Tuomi (2002) emphasised the necessity to use wind data with fine temporal discretisation in the wave model in the Baltic Sea and found that the impact of the coupled model on the meteorological part of the model can mainly be seen in predicted surface winds. For the Mediterranean

Sea, Cavaleri et al. (2012) found that reduced wind speeds were compensated by a limited deepening of the pressure fields of atmospheric cyclones. Lionello et al. (2003) demonstrated the importance of the atmosphere–wave interaction by studying the sea surface roughness feedback on momentum flux. A coupled ocean–atmosphere–wave–sediment transport (COAWST) modelling system has been developed for the coastal ocean (Warner et al., 2010; Kumar et al., 2012). For the Balearic Sea, Renault et al. (2012) compared atmospheric and oceanic observations and showed that the use of COAWST improved their simulations, especially for storm events. Recently, high-resolution, regional, and fully coupled models have been further developed, as shown by Katsafados et al. (2016), who used the Mediterranean Sea as an example. They focused on air–sea momentum fluxes under conditions of extremely strong and time-variable winds and demonstrated that by including the sea state dependent drag coefficient, effects on a wave spectrum and their feedback on momentum flux lead to improved model predictions. For the southern North Sea (the German Bight area), Staneva et al. (2016) showed the effect of wave-induced forcing on sea level variability and hydrodynamics, although wave–atmosphere interaction processes were not considered.

Model outputs can be validated against in situ and space-based observational data from satellite altimetry. Particularly challenging for the significant wave height estimations are coastal data, due to land and calm water interference in the altimeter footprint and in low sea states (Fenoglio-Marc et al., 2015). Analyses of the differences between altimeter and in situ measurements over time intervals of several months provide an estimate of the accuracy of altimeter data relative to in situ data assumed as ground truth. Significant wave heights derived from satellite altimetry over an interval of 10 years (2002–2012) have been compared to wave height measurements from several waveriders in Passaro et al. (2015). Fenoglio-Marc et al. (2015) considered 2 years (2012–2013) of the CryoSat-2 satellite mission to estimate the accuracy of both significant wave height and wind speed.

In this study, we aim at a quantification of the effects of coupling of wave and atmospheric models, also during extreme storm events. We compare simulations between coupled and stand-alone models that we validate with newly available space-based observational data. In the one-way coupled set-up, the wind wave model only receives wind data from the atmospheric model. In the two-way coupled set-up, the wind wave model sends the computed sea surface roughness back to the atmospheric model. Then, we statistically assess the impact of the two-way coupling and validate the two set-ups against available in situ and remote sensing data. Our novel contribution here is that we simultaneously run (via a coupler) a regional North Sea coupled wave–atmosphere model together with a nested-grid high resolution in the German Bight wave model (one atmospheric model and two wind wave models). Using this configuration allows us to study the individual and combined effects of (1) model coupling and (2) grid resolution, especially under severe storm conditions, which is a challenging aspect for wave modelling at the German Bight because it is a very shallow and dynamically complex coastal area.

The paper is structured as follows. In Sect. 2 we describe the models used, the coupling and specification of different model set-ups, the period of model integration, and available data for validation. Afterwards, we validate the models against satellite and in situ measurements in Sect. 3. Section 4 discusses the impact of two-way coupling. The final section summarises our findings and also provides an outlook for future research.

2 Model description and set-up

2.1 The COSMO atmospheric model

The atmospheric model used in this study is the COSMO-CLM (CCLM) version 4.8 non-hydrostatic regional climate model (Rockel et al., 2008; Baldauf et al., 2011). The model is developed and applied by a number of national weather services affiliated with the Consortium for Small-Scale Modeling (COSMO; see also http://www2.cosmo-model.org/). Its climate model, COSMO-CLM (CCLM), is used by the Climate Limited-area Modelling Community (http://www.clm-community.eu/). CCLM is based on the primitive thermo-hydrodynamical equations that describe compressible flow in a moist atmosphere. The model equations are formulated in rotated geographical coordinates with generalised terrain following vertical coordinates. The model uses the primitive momentum equations. The continuity equation is replaced by a prognostic equation for perturbation pressure (i.e. pressure deviation from a reference state representing a time-independent dry atmosphere at rest, which is prescribed as horizontally homogeneous, vertically stratified and in hydrostatic balance).

In our set-up, we use a spatial resolution of $\sim 10\,\mathrm{km}$ and 40 vertical levels to discretise the area around the North Sea and Baltic Sea (Fig. 1a). Forcing and boundary condition data are taken from the coastDat-2 hindcast database for the North Sea (Geyer, 2014) covering the period 1948–2013 with a spatial resolution of $\sim 24\,\mathrm{km}$ (0.22°) and a temporal resolution of 6 h.

2.2 The WAM wave model

WAM Cycle 4.5.4 is an update of the third-generation WAM Cycle4 wave model (Komen et al., 1994). The basic physics and numerics are maintained in the new release. The source function integration scheme of Hersbach and Janssen (1999) and the reformulated wave model dissipation source function of Bidlot et al. (2007) and Janssen (2008) are incorporated. Depth-induced wave breaking (Battjes and Janssen, 1978) has been included as an additional source function. Depth and/or current fields can be non-stationary.

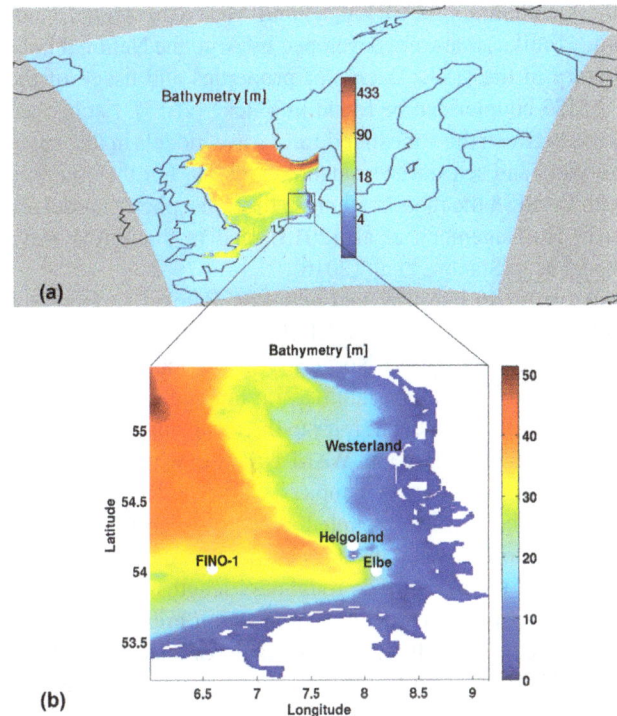

Figure 1. (a) Bathymetry (m) of the North Sea embedded in the COSMO model area (using a logarithmic scale) and (b) bathymetry (m) of the German Bight as used in the WAM model. The positions of four waverider buoys used for the validation are indicated, too.

The nested-grid set-up includes a regional wave model for the North Sea with a spatial resolution of $\sim 5\,\mathrm{km}$ (Fig. 1a) and a finer wave model for the German Bight with a resolution of $\sim 900\,\mathrm{m}$ (Fig. 1b). These models (described in Staneva et al., 2015) use a directional resolution of 15° and 30 frequencies with an equidistant relative resolution ranging from 0.04 to 0.66. The boundary values for the North Sea model are taken from the EWAM (European WAM) regional model of the German Weather Service (DWD). The forcing wind data are provided by CCLM (see Sect. 2.1). The German Bight wave model uses boundary values of the outer North Sea model and accounts additionally for depth-induced wave breaking and depth refraction. The sea state dependent roughness length, according to Janssen (1991), has already been implemented in WAM-4.5.4. Thus for the present study, the model only needed to be adapted for usage with the OASIS3-MCT coupler (see Sect. 2.3).

2.3 Coupling of models

WAM and CCLM are coupled via the OASIS3-MCT version 2.0 coupler (Valcke et al., 2013). The name OASIS3-MCT is a combination of OASIS3 (the Ocean, Atmosphere, Sea, Ice, and Soil model coupler version 3) from the European Centre for Research and Advanced Training in Scientific Computation (CERFACS) and MCT (the Model Coupling Toolkit) that were developed by Argonne National Laboratory in the USA. Details of properties and usage of the OASIS3 coupler can be found in Valcke (2013). Exchanged fields between the atmospheric and wave models in this study are wind and sea surface roughness length. For the coupling with OASIS3 the modifications in the atmospheric model are as in Ho-Hagemann et al. (2013), and in the WAM wave model as in Staneva et al. (2016).

We perform one-way and two-way coupled simulations. In the one-way coupled model, the atmospheric model provides wind data for the North Sea wave model via OASIS. This is equivalent to the familiar forcing of a wave model by 10 m wind fields. We will refer to the results of these simulations as COSMO-1wc and WAM-NS-1wc, where "1wc" and 'NS' stand for "one-way coupled" and "North Sea", respectively. In the two-way coupled model, the North Sea wave model is forced with winds provided by the atmospheric model and the sea surface roughness lengths are sent back to the atmospheric model, which in return might change the wind speeds. We will refer to the results of these simulations as COSMO-2wc and WAM-NS-2wc, respectively. The coupling time step is 3 min for all the simulations. This short time step is a great advantage when modelling fast moving storms in comparison to using stand-alone wave models forced by winds, which are usually available in hourly time steps.

The high-resolution German Bight wave model, which also runs simultaneously with CCLM and the North Sea WAM, is forced in the two simulations by the CCLM wind and the boundary data provided by the North Sea WAM set-up. We will refer to the two differently forced set-ups as WAM-GB-1wc and WAM-GB-2wc. In the second experiment roughness information is sent to the atmospheric model by WAM-NS-2wc, while it is not in the first experiment. Compared to previous atmosphere–wave coupling research, our study is novel as we are able to simultaneously run a high-resolution coastal model (the German Bight one) that uses winds and lateral forcing provided by the coupled regional atmosphere (COSMO-2wc) and wave (WAM-NS-2wc) models.

2.4 Study period and data availability

The coupled wave–atmosphere model system described in the previous section was used to simulate a 3-month period from October to December 2013. This period was chosen because it includes the time when storm Xaver passed over the study area on 6 December 2013. This was one of the most severe storms of the last decade, which originated south of Greenland and rapidly deepened as it moved eastwards from Iceland over the Norwegian Sea to southern Sweden and further to the Baltic Sea and Russia. At the German Bight, the arrival of Xaver coincided in time with a high tide. Because of the high tide and wind gusts of greater than $130\,\mathrm{km\,h^{-1}}$,

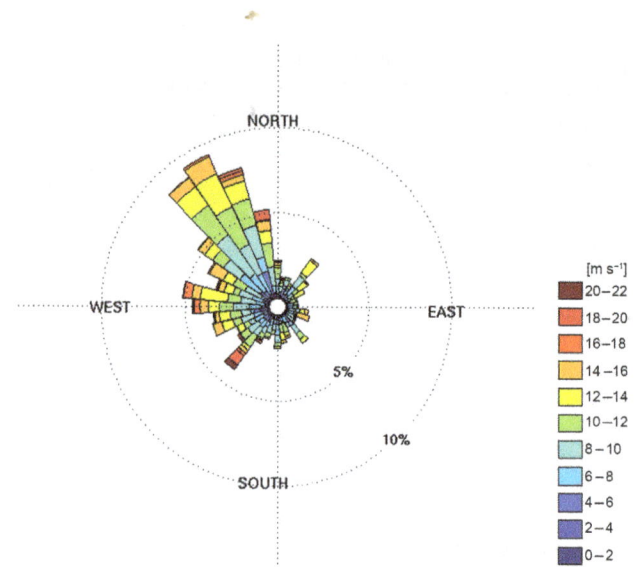

Figure 2. Distribution of frequency and wind speeds in $\mathrm{m\,s^{-1}}$ (see colour bar) and wind direction at the FINO-1 waverider buoy for the period of 1 October until 31 December 2013.

an extreme weather warning was given to the coastal areas of north-western Germany (Deutschländer et al., 2013). This storm event was also exceptional because of its long duration of nearly 2 days. The surge height reached $\sim 2.5\,\mathrm{m}$, with its maximum at low tide. During Xaver, two surge maxima were observed (Staneva et al., 2016). Fenoglio-Marc et al. (2015) described the first surge maximum as a locally generated surge. They found that the surge derived from the tide gauge records at Aberdeen and Lowestoft stations had only one maximum, reaching the eastern North Sea coastal areas (anticlockwise propagation) approximately 10 h later than at Lowestoft (easternmost UK coast). This caused the second storm surge maximum, which was detected by the measurements in the German Bight (surge generated further away and propagated to the study area). As demonstrated by Staneva et al. (2016), the wave-induced mechanisms contributed to a persistent increase in the surge after the first maximum (with a slight overestimation after the second peak).

In the present study, we perform statistical analyses for the whole integration period and investigate the period of the Xaver extreme storm event in more detail. The distribution of wind speeds and directions over the selected period as seen in the waverider data from the FINO-1 in situ platform (see Fig. 1b for its location) is shown in Fig. 2. North-westerly winds are generally dominant, but strong winds (higher than $20\,\mathrm{m\,s^{-1}}$) came from the west and southwest as the Xaver storm moved eastwards. South-easterly and north-easterly winds are rarely observed at the FINO-1 station.

To validate our experiments, we use wind speed and significant wave height data measured by satellite altimeters SARAL/AltiKa, Jason-2 and CryoSat-2 over the North Sea (see Fig. 3 with the tracks of the different satellites

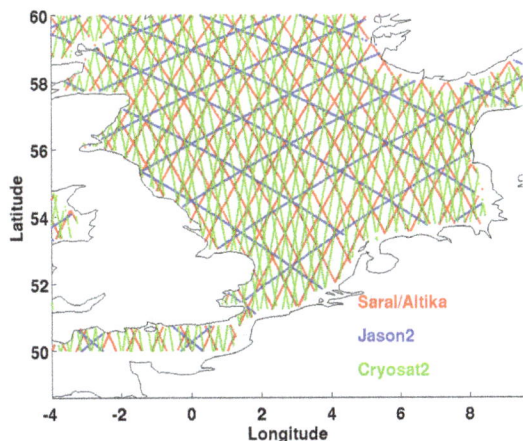

Figure 3. Tracks of all satellites during the study period (1 October until 31 December 2013).

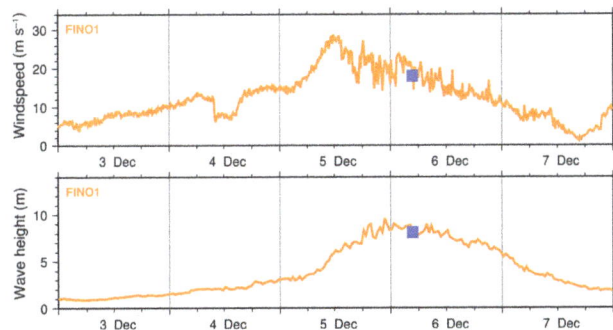

Figure 4. Time series during 5 days, which include storm Xaver of the observations in station FINO-1 (orange line) together with the Saral/Altika observation (blue full square). Top panel: wind speed $(m\,s^{-1})$; and bottom panel: significant wave height (m). SARAL/AltiKa passed over the German Bight during storm Xaver when the surge was at its maximum (the data during the overflight are plotted with a full blue mark).

over the 3-month study period). The first two carry classical pulse-limited altimeters that operate in low-resolution mode (LRM), while the CryoSat-2 altimeter operates either in LRM or in synthetic aperture radar (SAR) mode, also called delay Doppler altimetry (DDA), depending on the operational mask. In our region of analysis the mask was always in SAR mode and the CryoSat-2 data used are the pseudo-LRM (PLRM) data extracted from the RADS database (Scharroo et al., 2013). The accuracy and precision of PLRM data are slightly lower than LRM and SAR data (Smith and Scharroo, 2015). The altimeter satellites observe along their ground-track offshore up to a few kilometres from the coast (Fig. 3). Their ground track pattern and repeat period are different for each of the three missions, as the same location is revisited by each mission every 27, 10, and 350 days (Chelton et al., 2001). The SARAL/AltiKa data are of special interest in our study because this satellite passed over the German Bight during storm Xaver when the surge was at its maximum (Fenoglio-Marc et al., 2015). The in situ wave data of four directional waveriders in the German Bight are provided by the Federal Maritime and Hydrographic Agency (BSH) (see Fig. 1b for the buoy locations). The wind speed measurements close to the shore of the island of Sylt, near the Westerland buoy location, and on the island of Helgoland, are provided by the DWD. At station FINO-1 (see Fig. 1b for its location), there were also wind speed measurements available at 50 and 100 m a.s.l. (above sea level) for the selected period.

3   Validation of the results

3.1   Altimeter data

The long revisiting time of the same location and the global coverage could be considered intrinsic characteristics of the satellite altimetry. Therefore, a longer interval of analysis

is needed when analysing the agreement between altimeter and in situ measurements, collected from waveriders and anemometers. The tracks during the study period for the three different satellites are illustrated in Fig. 3. Wind speed and significant wave height data measured at the FINO-1 station during the 5-day period (2–7 December 2013). The nearest point observations of the SARAL/Altika satellite altimeter as it passed over the region at 05:45 UTC on 6 December (see also Fenoglio-Marc et al., 2015) are specified with the blue mark also in Fig. 4. The wave height and wind speed measured by the SARAL/Altika altimeter (blue symbol) during the Xaver storm are in good agreement with in situ observations.

The differences between the altimeter and in situ measurements over longer time intervals provide an estimate of the accuracy of the altimeter data relative to the in situ data assumed as ground truth. Fenoglio-Marc et al. (2015) considered wave height and wind speed derived from CryoSat-2 SAR altimetry data located in the open sea at a distances between 10 and 20 km from in situ stations of the network in the German Bight and found accuracies of about 15 cm for the wave height and $1.8\,m\,s^{-1}$ for the wind speed. They also found a good consistency between pseudo-conventional (PLRM) and SAR data in the open ocean, with standard deviations (SDs) between PLRM and SAR of 21 cm and $0.26\,m\,s^{-1}$ for wave height and wind speed, respectively. In situ analysis showed a higher accuracy in significant wave height for SAR compared to PLRM. As a demonstration, Fig. 5 shows the scatterplots for FINO-1 and CryoSat-2 SAR and PLRM measurements. For the significant wave height, SAR has higher accuracy than the standard PLRM (SDs with in situ data are 18 and 30 cm, respectively). For the wind speed the accuracies of SAR and PLRM are similar and equal to $1.9\,m\,s^{-1}$. The accuracy in the significant wave height from PLRM increases (SD is 19 cm) when a dedicated re-

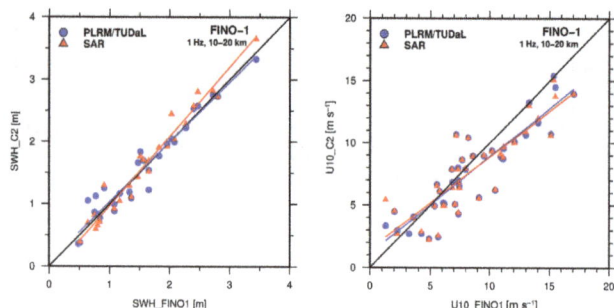

Figure 5. Comparison at station FINO-1 of in situ and altimeter-derived **(a)** significant wave height and **(b)** wind speed of in situ and co-located altimeter measurements at the FINO-1 station. Altimeter data are SAR (triangle) and PLRM (circle).

tracking procedure is applied (Fenoglio-Marc et al., 2015). Figure 5b shows an underestimation of wind speed of altimetry relative to the in situ data (the slope is below 0.8 in all cases).

3.2 Altimeter–model comparisons

In this section, we quantify the performance of one-way vs. two-way coupling by comparing the output of the atmospheric and wave models against remotely sensed data. Table 1 gives the statistics of the differences (bias and standard deviations) between the model and altimeter-derived values of wave height and wind speed over the selected 3-month period. The numbers of matched pairs (approximately 7000) of observations and simulations are also given in Table 1 for the different satellites.

For all three satellites, the standard deviation in the two-way coupled model is smaller than in the one-way coupled model. For Jason-2 and SARAL/Altika, the bias in the two-way coupled model is nearly halved compared to the bias in the one-way model. Measured values are lower than the modelled values in the one-way and two-way experiments.

For Cryosat-2, by contrast, the measured values are higher than the modelled values on average for both the wave height and wind speed. The biases between the CryoSat-2 data and the two-way model simulations (see the red shaded values in Table 1) are larger than the biases between the CryoSat-2 data and the one-way model runs. Fenoglio-Marc et al. (2015) also found that the CryoSat-2-derived wave height data overestimate the wave model data from the DWD. However, they found the opposite for the wind speed, i.e. the CryoSat-2-derived wind speed underestimates the COSMO winds from the DWD data. The difference between our results and Fenoglio-Marc et al. (2015) is due to the different data that have been used to force the atmospheric models by the DWD and this study.

To perform a spatial comparison between model simulations and the satellite data, we analysed individual tracks over the North Sea, and two of these are shown in Figs. 6

Table 1. Bias and standard deviation of validation of wind speed (m s^{-1}) and significant wave height (m) of the one- and two-way coupled models against the available satellite data over the whole period (measured minus modelled). Bold means an improvement of the two-way coupled model skills; italic means that the one-way coupled model skill is better than the ones of the two-way coupled model.

	Significant wave height (m)		Wind speed (m s^{-1})	
	One-way	Two-way	One-way	Two-way
Saral/AltiKa no. 6886				
Mean meas.	2.35		9.76	
Bias	−0.27	**−0.12**	−0.64	**−0.33**
SD	0.93	**0.86**	3.33	**3.16**
Jason-2 no. 6710				
Mean meas.	2.38		9.62	
Bias	−0.29	**−0.15**	−0.73	**−0.40**
SD	1.07	**1.01**	3.85	**3.75**
Cryosat-2 no. 7477				
Mean meas.	2.71		10.62	
Bias	0.18	*0.31*	0.39	*0.65*
SD	0.90	**0.87**	3.33	**3.18**

and 7. The satellite altimetry observations along the ground track at the time of the overflight at the German Bight last ∼ 38 s. The selected SARAL/AltiKa passes are very diverse, as one was taken under calm conditions (Fig. 6) and the other during storm Xaver (Fig. 7), which therefore provided an opportunity to compare measured and modelled wave heights and wind speeds along the satellite tracks under different atmospheric and wave conditions illustrated in Figs. 6 and 7. Under calm conditions, differences between the results of the one- and two-way coupling are very small (Fig. 6a). Both models (WAM-NS-1wc and WAM-NS-2wc) overestimate the measured wave height (red line) over a large part of the track. However, the increase in modelled wave height with increasing latitude appears to be consistent with the northward wind speed increase observed by the satellite data and simulated in the two simulations (Fig. 6b). During storm Xaver, the difference between the wave height in the WAM-NS-1wc and WAM-NS-2wc simulations (Fig. 7a) increases up to 1 m in the southern North Sea. The altimeter-derived quantities fluctuate greatly. However, the two-way coupled-model results are closer to the satellite data, in comparison to the ones in WAM-NS-1wc, except for the latitude of ∼ 56° N, where the significant wave height from the satellite measurements has a local peak. The modelled significant wave height (black lines) is much smoother than the satellite observations (red line), which can be explained by the fact that the model is not capable of resolving the small scales seen in the satellite observations. The corresponding wind speed does not grow at this latitude, neither for the measured nor for the modelled wind speeds. It is noteworthy that both model

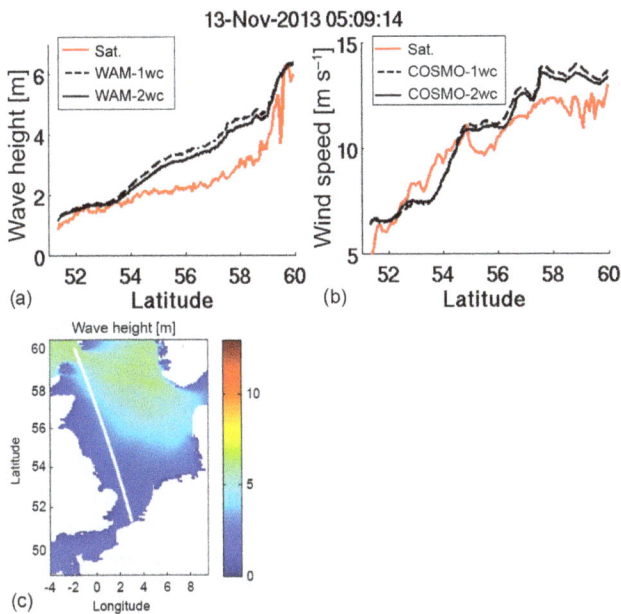

Figure 6. Time series of wave height (m) and wind speed (m s^{-1}) from the Saral/AltiKa data and as modelled by WAM-NS under calm weather conditions on 13 November 2013. The track of the satellite (the white line) is shown together with the model significant wave height at the time of the passage (bottom panel).

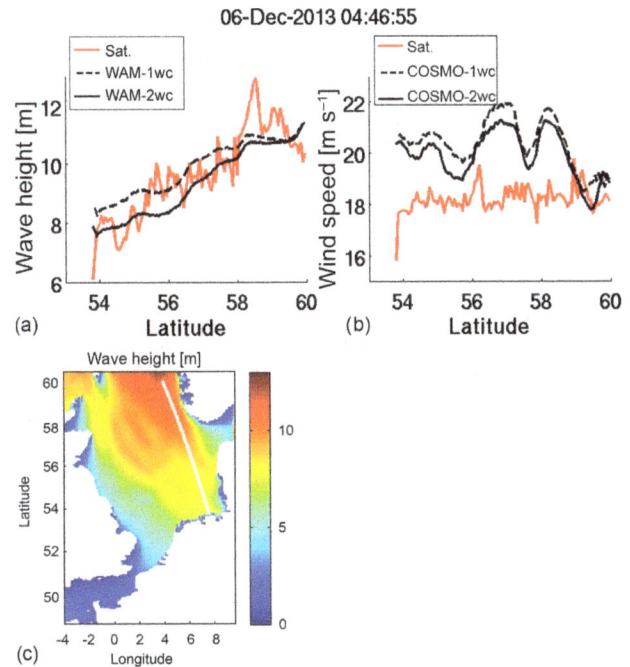

Figure 7. As Fig. 6 but for storm Xaver on 6 December 2013.

experiments missed the peak in measured significant wave height above 58° N (Fig. 7a).

The modelled wind speed fits well the altimeter-derived data during calm conditions in both experiments (COSMO-1wc/2wc, Fig. 6b). Northwards of 55° N, the wind speed is higher than 10 m s^{-1}, while the wind speed in the two-way coupled experiment (COSMO-2wc, full line) is slightly lower than in COSMO-1wc. During storm Xaver, the measured wind data fluctuate $\sim$ 18 m s^{-1}, whereas the modelled data show much higher values of $\sim$ 20 m s^{-1}, reaching $\sim$ 22 m s^{-1} at $\sim$ 57 and 59° N (Fig. 7b). This confirms the findings of Fenoglio-Marc et al. (2015), who had compared the same altimeter data with ERA-Interim, NOAA/GFC and COSMO/EU winds. They suggested that the low wind speeds derived from the altimeter are caused by an overestimation of the atmospheric attenuation of the radar power in the Ka-band. In fact a larger attenuation correction would result in too large a backscatter coefficient and hence reduced wind speed (Fenoglio-Marc et al., 2015). The correction in the SARAL/AltiKa products is larger than the correction based on surface pressure, near-surface temperature, and water vapour content (Lillibridge et al., 2014). Similar analyses along all tracks over the study period agree with the two examples demonstrated in Figs. 6 and 7. In general, the measured wind speeds were in slightly better agreement with the two-way coupled model results, which was also demonstrated by statistics presented in Table 1. The track during

the time of storm Xaver was the only track taken under such extreme conditions.

3.3 Validation against in situ measurements

Analyses of the temporal variability of the significant wave heights in the German Bight under stormy conditions allow us to investigate not only the impact of two-way coupling, but also the role of the horizontal resolution. Figure 8 illustrates the time variability of the significant wave height (top panels) and the wind speed (bottom panels) at the Helgoland and Westerland stations (see Fig. 1b for locations) from observations (black line) and the different model runs during storm Xaver.

The wind fields in both locations are very similar in the COSMO-1wc/2wc model runs; the peak of the storm is reduced from 26 to 22 m s^{-1}. By comparing the modelled and measured wind speeds, it is noticeable that the modelled wind speeds grow too early and too high at all locations at the beginning of the storm (see the bottom patterns in Fig. 8a and b for the Helgoland and Westerland examples). The storm characteristics are matched well at Helgoland but are slightly underestimated at Westerland. Still, the overall model performance at Westerland is satisfactory, considering the strongly fluctuating wind measurements. Similar behaviour is observed for the Elbe and FINO-1 (not shown here) wave buoy stations.

Throughout this period, the highest values of significant wave heights are simulated by the WAM-NS-1wc experiment. The lowest values, and closest to the observations, are

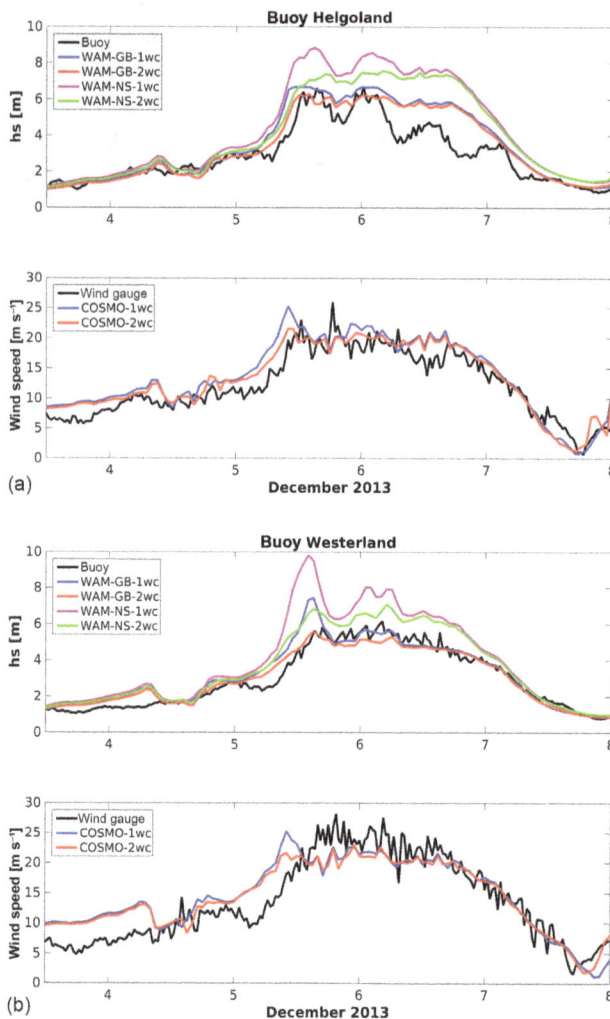

(a)

(b)

Figure 8. (a, b) Significant wave height (m, top panels) and wind speed (m s^{-1}, bottom panels) during storm Xaver at the Helgoland **(a)** and Westerland/Sylt **(b)** buoys.

Table 2. Wind speed (m s^{-1}) bias and standard deviation of the one- and the two-way coupled COSMO model data against the FINO-1 data over the whole period (measured minus modelled). Bold means an improvement of the two-way coupled model skills.

	Wind speed (m s^{-1}) at 50 m		Wind speed (m s^{-1}) at 100 m	
	One-way	Two-way	One-way	Two-way
Mean meas.	11.03		11.85	
Averaged difference	−0.67	**−0.41**	−0.23	**0.01**
rms difference	3.26	**3.17**	3.33	**3.22**

from the WAM-GB-2wc simulations (Fig. 8). At the beginning of December, during the calm atmospheric conditions, all model results are similar and fit relatively well with the in situ measurements. The differences in the wave growth between the different model simulations become notable after the storm onset. During the peak of the storm, the WAM-NS-1wc simulation overestimates the measured wave heights by ~3 m at the Helgoland station (water depth 30 m, Fig. 8a) and by ~4 m at the shallow water of the Westerland station (water depth 13 m, Fig. 8b). Compared to the in situ measurements, this peak occurs earlier in all simulations due to the time discrepancy between wind data and model time steps. The wave heights predicted by WAM-GB-2wc are in best agreement with the observations, especially for the Westerland station (Fig. 8b, the red line).

The influence of spatial resolution on the simulated characteristics can be clearly seen in the time series at the deep water buoy at Helgoland, for which the differences between simulated wave heights during storm Xaver reach ~1 to 1.5 m in the corresponding North Sea and German Bight simulations (Fig. 8a). This buoy is located in an area of large gradients in water depth (Fig. 1b), where the high-resolution model uses a finer bathymetry at coastal areas with a rather complex shore (such as at Helgoland) leading to a better simulation of wave heights.

At the shallow Westerland buoy station (Fig. 8b) the differences are additionally enhanced by the depth-induced wave breaking in the German Bight model. This can also be seen in the snapshots of wave height in the North Sea and German Bight models at the peak of the storm (Fig. 9a and b). Shoreward of the 15 m isobaths, the wave heights drop from 6 to 4 m in the German Bight model. In contrast, for the North Sea model, the 6 m high waves reach the south-eastern coast. The WAM-NS-1wc model run underperforms in comparison to the WAM-NS-2wc simulation at Westerland. This underperformance further proves the importance of two-way coupling for the coastal German Bight areas, where the model wind speed is even higher (by ~2 m s^{-1}) than at Helgoland. We admit that it is difficult to differentiate between the effects due to wave breaking and two-way coupling because both contribute to reducing the wave height under extreme weather conditions. Wave breaking plays a dominant role in very shallow water, especially during storm events, by preventing unrealistically high waves near the coast. For deep waters, the sea surface roughness feedback due to the two-way coupling plays a very important role (Fig. 8a). The importance of the two-way coupling is clearly demonstrated by comparing WAM-GB-2wc (the blue line) and WAM-GB-1wc (the red line) in Fig. 8. For all stations, the simulated significant wave height WAM-GB-2wc is reduced, especially during the Xaver peak, and is closer to the measurements.

The wind speed is validated against measured data from FINO-1 at 50 and 100 m height over the whole modelling period (Table 2). We find better agreement in the two-way coupled run. The bias in wind speed is negative for the one-way coupled set-up, i.e. the modelled wind speed is overestimated. The bias is significantly reduced due to the lower wind speed in the two-way coupled model. The root

Table 3. Significant wave height (m) bias and standard deviation of the one- and two-way coupled WAM German Bight model data against the available buoy data over the whole period (measured minus modelled). Bold means an improvement of the two-way coupled model skills; italic means that the one-way coupled model skill is better than the ones of the two-way coupled model.

Bouy name (depth)	FINO-1 (30 m)		Elbe (25 m)		Helgoland (30 m)		Sylt (13 m)	
Measured significant wave height (m)	1.95		1.42		1.63		1.45	
	1-way	2-way	1-way	2-way	1-way	2-way	1-way	2-way
Bias hs (m)	−0.14	**−0.03**	−0.07	**−0.01**	−0.13	**−0.03**	−0.15	−0.05
SD hs (m)	0.45	*0.50*	**0.49**	**0.49**	0.54	*0.55*	**0.59**	**0.59**

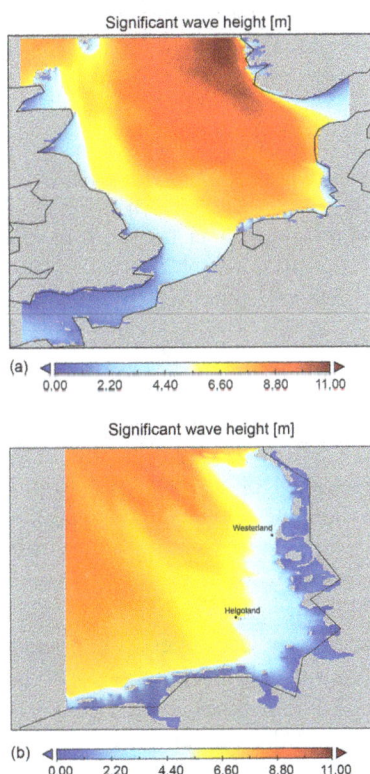

(a)

0.00   2.20   4.40   6.60   8.80   11.00

Significant wave height [m]

(b)

0.00   2.20   4.40   6.60   8.80   11.00

Figure 9. (a, b) Significant wave height (m) in the North Sea **(a)** and the German Bight **(b)** at the peak of storm Xaver (6 December 2013, 09:00 UTC) calculated by WAM-NS/GB-2wc.

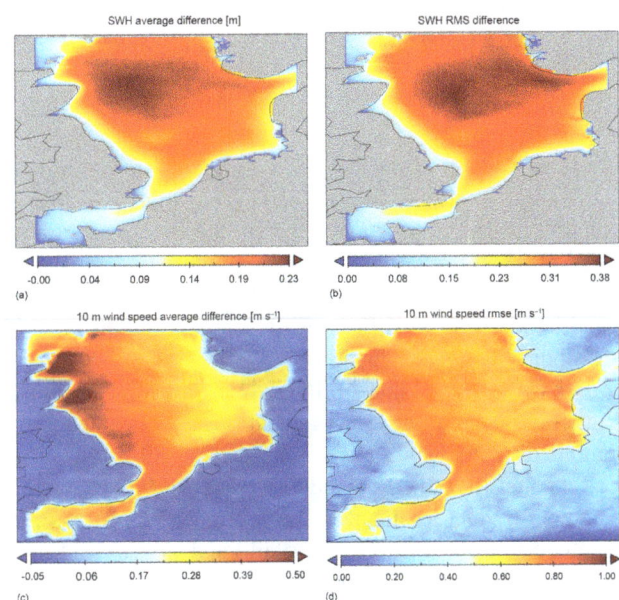

Figure 10. (a, c) Average difference and **(b, d)** rms difference (rms difference) of WAM modelled significant wave height (m, top panels) and COSMO modelled wind speed (m s^{-1}, bottom panels) when comparing one-way minus two-way coupled modelling results. The differences are calculated as averages over the whole 3-month period.

mean square (rms) difference is ~ 3 m s^{-1} in either case, but slightly reduced for the fully coupled set-up.

For a more quantitative validation of the WAM-GB-1wc/2wc results, we use four buoys (see Fig. 1b for their locations) at water depths of 13 to 30 m. Table 3 gives the statistics for significant wave height over the whole period (there are ~ 4000 matched pairs). For the four buoys and regardless of the type of coupling, the bias is slightly negative, i.e. the modelled data overpredict the measured values. The simulated significant wave heights are lower and the bias between the measurements and model results is significantly

reduced in the WAM-GB-2wc experiment. The standard deviation of the significant wave height of the two-way coupled simulation is similar to that of the one-way coupled simulations. Only for the FINO-1 station is the standard deviation increased by ~ 2.5 % in the two-way coupled model run.

4   Impact of the two-way coupling

In the following discussion, the impact of coupling is analysed for the North Sea, focusing on the spatial patterns under different physical conditions. The 3-month average of the significant wave height and wind speed are reduced significantly (Fig. 10) for the two-way coupling compared to the one-way coupling. This reduction results from an extraction

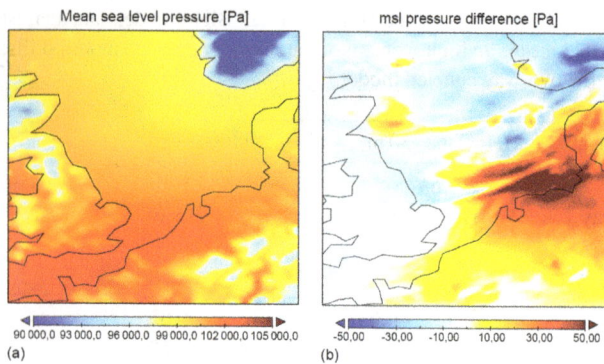

Figure 11. (a) COSMO pressure (Pa) at mean sea level height in the North Sea during storm Xaver and **(b)** mean sea level pressure differences when comparing one-way minus two-way coupled modelling).

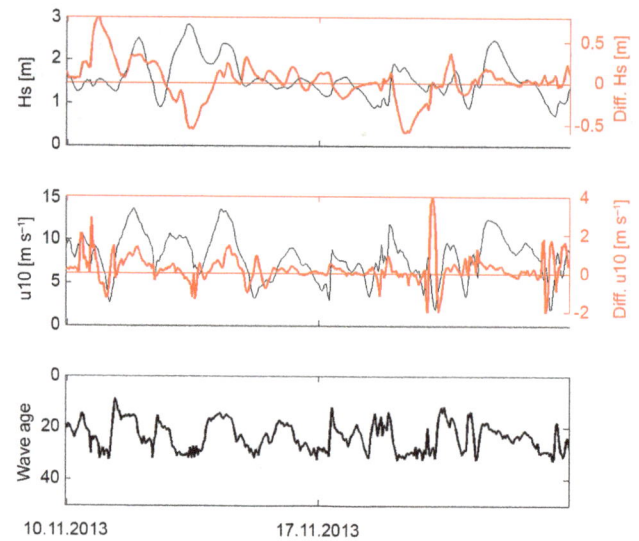

Figure 12. Time series of significant wave height (m, top panel), wind speed (m s^{-1}, middle panel) and wave age (bottom panel) from the two-way coupled German Bight set-up at FINO-1 for **(a)** a rather calm period with young wind sea and **(b)** during storm Xaver. Red lines in the top and middle panels show the differences between the one-way and two-way coupled models.

of energy and momentum from the atmosphere by waves. The average difference in wave height (Fig. 10a) is ~ 20 cm, which is a reduction of $\sim 8\%$ of the 3-month mean value (~ 2.3 m). The rms difference between the two simulations (Fig. 10b) is ~ 40 cm in the central North Sea. For the wind speeds, the averaged difference (Fig. 10c) is ~ 30 cm s^{-1}, corresponding to a reduction in wind speed of $\sim 3\%$ of the 3-month mean value (~ 10 m s^{-1}). The rms difference (Fig. 10d) between the two-way and one-way coupled simulations over the whole North Sea area is ~ 80 cm s^{-1}. The spatial patterns in the averaged differences in Fig. 10 can be explained by the dominant westerly winds (Fig. 2). As the wind comes from land (Great Britain) and strikes the North Sea, the differences in the wind speed between the two models are larger closer to the coast because of differences in sea surface roughness. Moving further east, the atmospheric boundary layer adapts in both cases to the winds over the sea, and there is less difference between the one- and two-way coupled models. For the wave height, the averaged differences close to the western coasts and in the English Channel are small because the fetch is too short for the waves to evolve.

The differences in the mean sea level pressure between COSMO-1wc and -2wc for the Xaver storm period are analysed in Fig. 10. At the peak of the storm (Fig. 11a) the mean sea level pressure is ~ 900 hPa over Norway and ~ 1000 hPa over the North Sea. Compared to the one-way coupled set-up, the pressure increased by ~ 50–100 Pa in the southeast (Fig. 11b). The slightly decreased pressure in the remaining part of the model area indicates a shift in the pressure low minimum, confirming the results of Cavaleri et al. (2012), who found similar patterns in the Mediterranean Sea under developing cyclones. As noted by Janssen and Viterbo (1996), the timescale of the wave impact on the atmospheric circulation is of the order of 5 days. However, our model area is too small to observe this impact. It is more plausible that our results are caused by wave–mean flow in-

teractions in the atmosphere. This effect of wave coupling on the atmospheric circulation will be analysed thoroughly in future experiments.

Another illustration of the influence of the coupling is given by the two time series at the FINO-1 station, each about 2 weeks long and taken under very different conditions. One period is in November, which was rather calm and contained young and developing wind seas (Fig. 12). The other period was in December, with several storms coming from the North Sea (including Xaver) with higher wave ages (Fig. 13). The differences in significant wave height and wind speed between the one- and two-way coupled models are mostly positive, i.e. both parameters are reduced in the two-way coupled model. The largest differences can be observed when the wave age (the ratio of phase velocity at the peak of the wave spectrum with friction velocity) is well below 20 and occurs before the maximum wave height has been reached (this can be well seen for Xaver, Fig. 13). Thus, the waves grow slower in the two-way coupled model. Negative differences seldom occur, only when the wave age increases rapidly (when the wind speeds approaches zero, the wave age diverges infinitely).

5   Summary and outlook

We developed a two-way coupled wave–atmosphere model for the North Sea including the possibility of nesting a coastal, high-resolution wave model, where the two models run simultaneously. The OASIS3-MCT coupling software

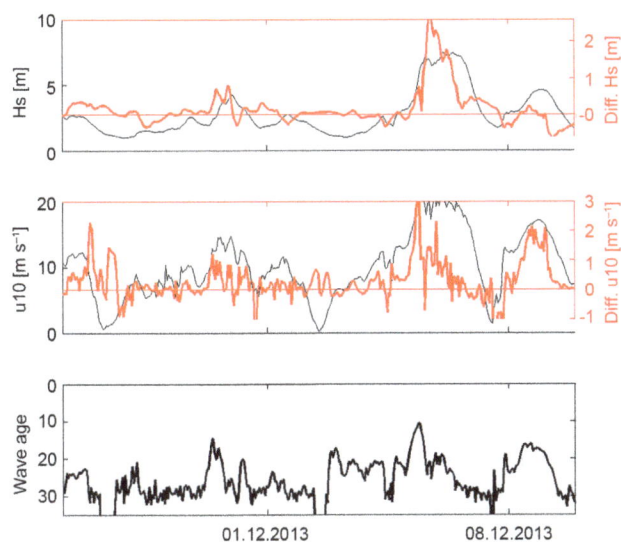

Figure 13. As Fig. 12 but during storm Xaver.

that we used allows for a parallel run of several models on different model grids. Simultaneous simulations of a regional North Sea coupled wave–atmosphere model together with a nested-grid high resolution in the German Bight wave model (one atmospheric model and two wind wave models) were performed. This enabled us to study the individual and combined effects of two-way coupling and grid resolution, especially under severe storm conditions. The sensitivity of atmospheric parameters such as wind speed and atmospheric pressure to wave-induced drag was quantified. Model intercomparisons gave encouraging results. Overall, the two-way coupled model results were in better agreement with the in situ and remotely sensed data of significant wave height and wind speed, in comparison to the one-way coupled model (COSMO drives WAM). New in this paper is the use of satellite altimetry, which provides complementary information to in situ data for the validation of models. We show that comparisons between the model results and satellite-derived parameters are satisfactory, except for a known degradation of wind speed under storm conditions, which is under investigation. The two-way coupling improved the modelled significant wave heights in the German Bight, which was demonstrated by the validation against in situ observations from four different buoys.

For storm event Xaver, the impact of the two-way coupling was of the highest significance. Wave heights decreased from ~ 8 to ~ 5 m due to the coupling, which matched buoy measurements very well. The corresponding wind speeds were lowered from ~ 22 to ~ 20 m s^{-1}. In addition to this extreme event, such large differences between one- and two-way coupled model results were only observed for young seas (wave age well below 20). We also found a slight spatial shift in the minimum of the cyclone mean sea level pressure together with a slight increase in the pressure field from the two-way

coupled model runs. These results may also have been caused by the wave–mean flow interactions in the atmosphere. This will be the subject of subsequent work, where we will study in more depth the consequences of coupling with other atmospheric parameters at sea level and the vertical structure of the planetary boundary layer.

Staneva et al. (2016) addressed the impact of coupling between wave and circulation models of the German Bight during extreme storm events. They demonstrated that the coupled model results revealed a closer match with observations than the stand-alone circulation model, especially during the Xaver extreme storm in December 2013. Staneva et al. (2016) also showed that the predicted surge of the coupled model is significantly enhanced during extreme storm events when accounting for wave–current interaction. We demonstrated that the potential uncertainties of shallow water in the wave model are due to both: inaccurate description in the bathymetry as well as the wave model source terms related to shallow water physics. Shallow water regions with the strongest wave–current interactions contribute to the coupled wave–atmosphere dynamics during extreme storm surge events. Depth and current refraction, bottom friction and wave breaking in the wave model play dominant roles in very shallow water. The model resolution is critical where the depth gradients are large. The improved model skills resulting from the new model developments justify further extension of the coupled model system by integrating atmosphere–wave–current interactions to further investigate the effects of coupling, especially on extreme storm events. Two-way coupling of wave and atmospheric models is an important component of a fully coupled ocean–atmosphere modelling system, as it resolves more adequately the interactions and exchanges in the atmospheric boundary layer. Accurate modelling of the boundary layer is of the utmost importance for long-range predictions.

Competing interests. The authors declare that they have no conflict of interest.

Acknowledgements. This publication has received funding from the European Union's H2020 Programme for Research, Technological Development and Demonstration under grant agreement no. H2020-EO-2016-730030-CEASELESS. The authors would like to thank Arno Behrens for providing the boundary values for the wave model from his Coastal Observing System for Northern and Arctic Seas (COSYNA) results. Beate Geyer extracted boundary values from the coastDat2 database for us. Markus Schultze supported us by setting up the atmospheric model and getting it started. Ha Ho-Hagemann is supported through the REKLIM German project and the Baltic Earth Programme. Luciana Fenoglio is supported by the European Space Agency (ESA) within the Climate Change Initiative (CCI). The authors are grateful to O. Krüger for his useful comments, I. Nöhren for assistance with the graphics and the BSH for providing the observational data.

Edited by: A. Sterl

References

Baldauf, M., Seifert, A., Förstner, J., Majewski, D., Raschendorfer, M., and Reinhardt, T.: Operational Convective-Scale Numerical Weather Prediction with the COSMO Model: Description and Sensitivities, Mon. Weather Rev., 139, 3887–3905, 2011.

Bao, J.-W., Michelson, S. A., and Wilczak, J. M.: Sensitivity of numerical simulations to parameterizations of roughness for surface heat fluxes at high winds over the sea, Mon. Weather Rev., 130, 1926–1932, 2002.

Battjes, J. and Janssen, J.: Energy loss and set-up due to breaking of random waves, Coast. Eng. Proc., 1, 569–587, 1978.

Bidlot, J., Janssen, P., and Abdalla, S.: A revised formulation of ocean wave dissipation and its model impact, Tech. Rep. Memorandum 509, ECMWF, Reading, UK, 2007.

Breivik, Ø., Mogensen, K., Bidlot, J. R., Balmaseda, M. A., and Janssen, P. A.: Surface wave effects in the NEMO ocean model: Forced and coupled experiments, J. Geophys. Res.-Oceans, 120, 2973–2992, 2015.

Cavaleri, L., Roland, A., Dutour Sikiric, M., Bertotti, L., and Torrisi, L.: On the coupling of COSMO to WAM, in: Proceedings of the ECMWF Workshop on Ocean-Waves, 25–27 June 2012, ECMWF, Reading, 2012.

Chelton, D., Ries, J., Haines, B., Fu, L., and Callahan, P.: Satellite Altimetry in Satellite Altimetry and Earth Sciences – A Handbook of Techniques and Applications, Academic Press, San Diego, 2001.

Desjardins, S., Mailhot, J., and Lalbeharry, R.: Examination of the impact of a coupled atmospheric and ocean wave system, part I, Atmospheric aspects, J. Phys. Oceanogr., 30, 385–401, 2000.

Deutschländer, T., Friedrich, K., Haeseler, S. and Lefebvre, C.: Severe storm XAVER across northern Europe from 5 to 7 December 2013, 2013 DWD report, Deutsche Wetter Dienst, Offenbach, Germany, 2013.

Doyle, J. D.: Coupled ocean wave/atmosphere mesoscale model simulations of cyclogenesis, Tellus A, 47, 766–788, 1995.

Fenoglio-Marc, L., Dinardo, S., Scharroo, R., Roland, A., Sikiric, M. D., Lucas, B., Becker, M. Benveniste, J., and Weiss, R.: The German Bight: A validation of CryoSat-2 altimeter data in SAR mode, Adv. Space Res., 55, 2641–2656, 2015.

Geyer, B.: High-resolution atmospheric reconstruction for Europe 1948–2012: coastDat2, Earth Syst. Sci. Data, 6, 147–164, doi:10.5194/essd-6-147-2014, 2014.

Hersbach, H. and Janssen, P. A. E. M.: Improvements of the short fetch behaviour in the WAM model, J. Atmos. Ocean. Tech., 16, 884–892, 1999.

Ho-Hagemann, H. T. M., Rockel, B., Kapitza, H., Geyer, B., and Meyer, E.: COSTRICE – an atmosphere–ocean–sea ice model coupled system using OASIS3, HZG Report 2013-5, Helmholtz-Zentrum Geesthacht, Geesthacht, 26 pp., 2013.

Janssen, P. A. E. M.: Quasi-linear theory of wind-wave generation applied to wave forecasting, J. Phys. Oceanogr., 21, 1631–1642, 1991.

Janssen, P. A. E. M.: Progress in ocean wave forecasting, J. Comput. Phys., 227, 3572–3594, 2008.

Janssen, P. A. E. M. and Viterbo, P.: Ocean waves and the atmospheric climate, J. Climate, 9, 1269–1287, 1996.

Janssen, P. A. E. M., Saetra, O., Wettre, C., Hersbach, H., and Bidlot, J.: Impact of the sea state on the atmosphere and ocean, Ann. Hydrogr., 772, 143–157, 2004.

Järvenoja, T. and Tuomi, L.: Coupled atmosphere–wave model for FMI and FIMR, Hirlam Newsletter, 40, 9–22, 2002.

Jenkins, A., Bakhoday Paskyabi, M., Fer, I., Gupta, A., and Adakudlu, M.: Modelling the effect of ocean waves on the atmospheric and ocean boundary layers, Energy Procedia, 24, 166–175, 2012.

Katsafados, P., Papadopoulos, A., Korres, G., and Varlas, G.: A fully coupled atmosphere–ocean wave modeling system for the Mediterranean Sea: interactions and sensitivity to the resolved scales and mechanisms, Geosci. Model Dev., 9, 161–173, doi:10.5194/gmd-9-161-2016, 2016.

Komen, G. J., Cavaleri, L., Donelan, M., Hasselmann, K., Hasselmann, S., and Janssen, P. A. E. M.: Dynamics and modelling of ocean waves, Cambridge University Press, Cambridge, UK, 560 pp., 1994.

Kumar, N., Voulgaris, G., Warner, J. C., and Olabarrieta, M.: Implementation of the vortex force formalism in the coupled ocean–atmosphere–wave–sediment transport (COAWST) modeling system for inner shelf and surf zone applications, Ocean Model., 47, 65–95, 2012.

Lillibridge, J. L., Scharroo, R., Abdalla, S., and Vandemark, D. C.: One- and two-dimensional wind speed models for Ka-band altimetry, J. Atmos. Ocean. Tech., 31, 630–638, doi:10.1175/JTECH-D-13-00167.1, 2014.

Lionello, P., Malguzzi, P., and Buzzi, A.: Coupling between the atmospheric circulation and the ocean wave field: An idealized case, J. Phys. Oceanogr., 28, 161–177, 1998.

Lionello, P., Elvini, E., and Nizzero, A.: Ocean waves and storm surges in the Adriatic Sea: intercomparison between the present and doubled CO_2 climate scenarios, Clim. Res., 23, 217–231, 2003.

Passaro, M., Fenoglio-Marc, L., and Cipollini, P.: Validation of Significant Wave Height From Improved Satellite Altimetry in the German Bight, IEEE T. Geosci. Remote, 53, 2146–2156, 2015.

Renault, L., Chiggiato, J., Warner, J. C., Gomez, M., Vizoso, G., and Tintoré, J.: Coupled atmosphere-ocean-wave simulations of a storm event over the Gulf of Lion and Balearic Sea, J. Geophys. Res., 117, C09019, doi:10.1029/2012JC007924, 2012.

Rockel, B., Will, A., and Hense, A.: The Regional Climate Model COSMO-CLM (CCLM), Meteorol. Z., 17, 347–348, 2008.

Rutgersson, A., Sætra, Ø., Semedo, A., Carlsson, B., and Kumar, R.: Impact of surface waves in a regional climate model, Meteorol. Z., 19, 247–257, 2010.

Rutgersson, A., Nilsson, E. O., and Kumar, R.: Introducing surface waves in a coupled wave–atmosphere regional climate model: Impact on atmospheric mixing length, J. Geophys. Res.-Oceans, 117, C00J15, doi:10.1029/2012JC007940, 2012.

Scharroo, R., Leuliette, E. W., Lillibridge, J. L., Byrne, D., Naeije, M. C., and Mitchum, G. T.: RADS: consistent multi-mission products, in: Proc. of the Symposium on 20 Years of Progress in Radar Altimetry, 24–29 September 2012, Venice-Lido, Italy, 2013.

Smith, W. H. F. and Scharroo, R.: Waveform aliasing in satellite radar altimetry, IEEE T. Geosci. Remote., 53, 1671–1681, doi:10.1109/TGRS.2014.2331193, 2015.

Staneva, J., Behrens, A., and Wahle, K.: Wave modelling for the German Bight coastal-ocean predicting system, J. Phys., 633, 233–254, doi:10.1088/1742-6596/633/1/012117, 2015.

Staneva, J., Wahle, K., Günther, H., and Stanev, E.: Coupling of wave and circulation models in coastal–ocean predicting systems: a case study for the German Bight, Ocean Sci., 12, 797–806, doi:10.5194/os-12-797-2016, 2016.

Valcke, S.: The OASIS3 coupler: a European climate modelling community software, Geosci. Model Dev., 6, 373–388, doi:10.5194/gmd-6-373-2013, 2013.

Valcke, S., Craig, T., and Coquart, L.: OASIS3-MCT User Guide, OASIS3-MCT 2.0, Technical Report, TR/CMGC/15/38, No. 1875, CERFACS/CNRS SUC URA, Toulouse, France, 2013.

Voldoire, A., Sanchez-Gomez, E., Mélia, D. S., Decharme, B., Cassou, C., Sénési, S., and Déqué, M.: The CNRM-CM5.1 global climate model: description and basic evaluation, Clim. Dynam., 40, 2091–2121, 2013.

Warner, J. C., Armstrong, B., He, R., and Zambon, J. B.: Development of a coupled ocean–atmosphere–wave–sediment transport (COAWST) modeling system, Ocean Model., 35, 230–244, 2010.

Zweers, N. C., Makin, V. K., de Vries, J. W., and Burgers, G.: A sea drag relation for hurricane wind speeds, Geophys. Res. Lett., 37, L21811, doi:10.1029/2010GL045002, 2010.

4

North Atlantic deep water formation and AMOC in CMIP5 models

Céline Heuzé

Department of Marine Sciences, University of Gothenburg, Box 115, 405 30 Göteborg, Sweden

Correspondence to: Céline Heuzé (celine.heuze@marine.gu.se)

Abstract. Deep water formation in climate models is indicative of their ability to simulate future ocean circulation, carbon and heat uptake, and sea level rise. Present-day temperature, salinity, sea ice concentration and ocean transport in the North Atlantic subpolar gyre and Nordic Seas from 23 CMIP5 (Climate Model Intercomparison Project, phase 5) models are compared with observations to assess the biases, causes and consequences of North Atlantic deep convection in models. The majority of models convect too deep, over too large an area, too often and too far south. Deep convection occurs at the sea ice edge and is most realistic in models with accurate sea ice extent, mostly those using the CICE model. Half of the models convect in response to local cooling or salinification of the surface waters; only a third have a dynamic relationship between freshwater coming from the Arctic and deep convection. The models with the most intense deep convection have the warmest deep waters, due to a redistribution of heat through the water column. For the majority of models, the variability of the Atlantic Meridional Overturning Circulation (AMOC) is explained by the volumes of deep water produced in the subpolar gyre and Nordic Seas up to 2 years before. In turn, models with the strongest AMOC have the largest heat export to the Arctic. Understanding the dynamical drivers of deep convection and AMOC in models is hence key to realistically forecasting Arctic oceanic warming and its consequences for the global ocean circulation, cryosphere and marine life.

1 Introduction

Global fully coupled climate models are a key tool to study current and future climate change, but although they clearly improve from one generation to the next, they still suffer from many biases (Flato et al., 2013). In particular the horizontal resolution of the ocean, around 1° (Table 1), is too coarse for explicitly representing eddies, freshwater plumes and overflows. Yet all these processes are necessary to correctly generate deep water formation (Marshall and Schott, 1999).

Deep water formation occurs around Antarctica and in the North Atlantic (Killworth, 1983). It is vital for ventilation of the ocean and for the global ocean circulation, but also for heat and carbon storage (e.g. Sabine et al., 2004; Lozier et al., 2008; Schmittner and Lund, 2015). Moreover, in the North Atlantic, deep water formation is tied to the strength of the Atlantic Meridional Overturning Circulation (AMOC, Böning et al., 2006), which transports heat to the Arctic (Spielhagen et al., 2011). This oceanic heat in turn melts the sea ice and Greenland floating glaciers from below (e.g. Polyakov et al., 2010; Straneo and Heimbach, 2013). Hence, the North Atlantic is a crucial area to assess the ability of current-generation climate models to represent deep water formation.

In this paper, we compare present-day deep water formation in 23 state-of-the-art global climate models that participated in the Climate Model Intercomparison Project phase 5 (CMIP5, Taylor et al., 2012). We assess their biases in the representation of deep convection in Sect. 3, explore the possible causes of these biases in Sect. 4, notably buoyancy forcings and sea ice, and estimate the consequences of their biases on the AMOC and heat export to the Arctic in Sect. 5. To the best of our knowledge, similar tests have been done on the previous generation of climate models (CMIP3, de Jong et al., 2009) and in ocean-only simulations (CORE-II, Danabasoglu et al., 2014), but not yet on CMIP5 models. Yet the magnitude of biases in CMIP5 models has to be known in order to properly simulate changes to the Arctic using the current generation of models, and also to evaluate improvements when CMIP6 model simulations become available (Eyring et al., 2016).

Table 1. List of CMIP5 models (Taylor et al., 2012): modelling groups, model names, ocean resolution in the North Atlantic (longitude/latitude/number of depth levels), type of vertical grid in the ocean (z is geopotential, z^* is geopotential with free sea surface, σ is terrain following, σ_2 is isopycnic, and H denotes an hybrid grid), and sea ice component. Stars * indicate the models whose pre-industrial control run is used in Sect. 5.2.

Modelling group	Model name	Resolution ($x/y/L$)	Grid	Sea ice model
CSIRO and Bureau of Meteorology, Australia	ACCESS1-0	1°/1°/50	z	CICE v4
Beijing Climate Center, China Meteorological Administration	bcc-csm1-1*	1°/1°/40	z	SIS
Canadian Centre for Climate Modelling and Analysis	CanESM2*	1.5°/1.5°/40	z	CanSIM1
National Center for Atmospheric Research	CCSM4 CESM1-CAM5	1°/0.5°/60 1°/0.5°/60	z z	CICE v4 CICE v4
Centro Euro-Mediterraneo sui Cambiamenti Climatici	CMCC-CM CMCC-CMS	2°/2°/31 2°/2°/31	z z	LIM2 LIM2
Centre National de Recherches Météorologiques/Centre Européen de Recherche et Formation Avancée en Calcul Scientifique	CNRM-CM5A*	0.7°/0.7°/42	z	GELATO v5
CSIRO and Queensland Climate Change Centre of Excellence	CSIRO-Mk3-6-0*	1.8°/0.9°/31	z	Component of Mk3
LASG, Institute of Atmospheric Physics, Chinese Academy of Sciences and CESS, Tsinghua University	FGOALS-g2	1°/1°/30	z^*	CICE v4
NOAA Geophysical Fluid Dynamics Laboratory	GFDL-CM3 GFDL-ESM2G* GFDL-ESM2M*	1°/1°/50 1°/1°/63 1°/1°/50	z^* σ_2 z^*	SISp2 SISp2 SISp2
NASA Goddard Institute for Space Studies	GISS-E2-R*	1.25°/1°/32	z^*	Russell sea ice
Met Office Hadley Centre	HadGEM2-CC* HadGEM2-ES*	1°/1°/40 1°/1°/40	z z	based on CICE based on CICE
Institut Pierre-Simon Laplace	IPSL-CM5A-LR* IPSL-CM5A-MR*	2°/2°/31 2°/2°/31	z z	LIM2 LIM2
JAMSTEC Atmosphere and Ocean Research Institute (The University of Tokyo), and National Institute for Environmental Studies	MIROC5* MIROC-ESM-CHEM*	0.5°/0.5°/50 1.4°/1.4°/44	H σ-z H σ-z	Component of COCO3.4 Component of COCO3.4
Max-Planck-Institut für Meteorologie	MPI-ESM-LR* MPI-ESM-MR	1.5°/1.5°/40 0.4°/1.5°/40	z z	Component of MPI-OM Component of MPI-OM
Norwegian Climate Centre	NorESM1-M	1.125°/1.125°/53	H σ_2-z	CICE v4

2   Data and methods

2.1   CMIP5 models

The output of 23 CMIP5 models (Taylor et al., 2012), listed in Table 1, were used in this study. In the North Atlantic, all models have approximately the same horizontal grid spacing, varying around 1° in both latitude and longitude. The coarsest resolution is 2° (for the CMCC models) and the highest is 0.4° (for MPI-ESM-MR). Most models have a z-level vertical grid with an average of 40 levels (Table 1). Although four models were run on a different type of grid (isopycnic, terrain following or hybrid), their output were submitted on a regular z-level grid.

In this study, 15 models use only three different sea ice components (Table 1): the Los Alamos sea ice model (CICE; Hunke and Lipscomb, 2008), the GFDL sea ice simulator (SIS; Delworth et al., 2006) and the Louvain-la-Neuve sea ice model (LIM; Fichefet and Maqueda, 1997). The other climate models mostly use the sea ice component of their respective ocean models. Although each climate model has a unique configuration, comparing models which share components – as we do in Sect. 4.2 – can indicate what causes a misrepresentation.

We are interested in the mean, present state of the ocean and hence use 20 years of monthly historical run from January 1986 to the end of the historical run in December 2005. The monthly pre-industrial control run was used to remove possible model drift. We also use the control run from 1986 to 2100 to study lagged correlations in a subset of 14 models for which such long runs were available (indicated with a star in Table 1); see Sect. 5.2. Only one ensemble member per model was used, r1i1p1, for it was the only one common to all the models at the date of download (July 2016).

2.2   Observational-based products

Three observational-based analysis products are used for assessing the models' representation of the present-day ocean. They are not the most recent climatologies, but have been chosen as representative of the 1986–2005 period studied here with the climate models. The observed monthly climatology of mixed layer depth (MLD) is that of de Boyer Montégut et al. (2004), available at http://www.ifremer.fr/cerweb/deboyer/mld/home.php. It was created using a density criterion of $0.03 \, \text{kg m}^{-3}$ over more than 4 million hydrographic profiles, taken from 1941 to 2002, interpolated onto a regular $2° \times 2°$ horizontal grid.

The temperature and salinity of the observed water column are given by the World Ocean Atlas 2009 (WOA09, Locarnini et al., 2013; Zweng et al., 2013, http://www.nodc.noaa.gov/OC5/WOA09/pr_woa09.html). It includes over 9 million quality-controlled hydrographic profiles. The monthly climatology is limited to the top 1500 m of the ocean, and hence the seasonal climatology is used here. It is provided as a regular $1° \times 1° \times 33$ level grid.

Finally, we use the HadISST monthly sea ice concentration measurements (Rayner et al., 2003, http://www.metoffice.gov.uk/hadobs/hadisst/), from January 1986 to December 2005, also provided as a regular $1° \times 1°$ grid. The observed sea ice extent is computed as the sum of the areas of the grid cells with a sea ice concentration larger than 15 %. To facilitate comparisons, the model output have been interpolated onto the common HadISST-WOA09 grid.

2.3   Methods

Some climate models provide a mixed layer depth output, but not the majority of them. For consistency amongst models and with the observations, we instead compute the monthly MLD for each model using the de Boyer Montégut et al. (2004) method. That is, using the monthly temperature and salinity model output to compute the density σ_θ, we define the MLD as the depth where the density exceeds that of the reference level (10 m) by $0.03 \, \text{kg m}^{-3}$.

Following observations, we consider that there is deep water formation or deep convection if the MLD exceeds 1000 m (e.g. Marshall and Schott, 1999; Våge et al., 2009). We divide the North Atlantic into two study areas where in the real ocean different deep waters form (Killworth, 1983): the Greenland–Iceland–Norwegian (GIN) seas (latitude 66–80° N, longitude 20° W to 20° E) and the subpolar gyre (SG, latitude 50–65° N, longitude 65–20° W; see orange boxes in Fig. 1a). The volume of deep water formed by each model is defined as the product of the grid cell area by the MLD, summed over all the grid cells with a MLD deeper than 1000 m in each of these two regions.

One of the buoyancy forcings whose impact on deep convection we study is the freshwater flux from the Arctic through the two sections closest to SG and GIN: the Davis and Fram straits, respectively. Following, for instance, Aagaard and Carmack (1989) these are computed as follow:

$$\text{FW} = \int_A (1 - S/S_{\text{ref}}) v \, dA, \tag{1}$$

where $S_{\text{ref}} = 34.8$ is a reference salinity, S is the monthly salinity field, v is the meridional velocity field, and A is the corresponding depth-longitude section. The coordinates considered for the Davis Strait are 66° N, 70–50° W; for the Fram Strait, 80° N and 20–15° E (cyan lines in Fig. 1a). Similarly, the heat flux through the Fram Strait was computed as follows:

$$Q = \int_A \rho_0 c_p \theta v \, dA, \tag{2}$$

where $\rho_0 = 1027 \, \text{kg m}^{-3}$ is a reference density of water, $c_p = 3.98 \, \text{kJ kg K}^{-1}$ is the specific heat capacity of water,

Figure 1. North Atlantic **(a)** climatological mixed layer depth of de Boyer Montégut et al. (2004) and **(b–x)** mean 1986–2005 winter MLD in the CMIP5 historical run. Orange boxes on **(a)** show the subpolar gyre (SG) and Greenland–Iceland–Norwegian (GIN) seas regions as defined in this study; cyan dashed lines are the Davis and Fram straits. Yellow dotted line on each panel indicates the 1000 m isobath; blueish green and magenta lines denote the mean March and September sea ice extent, respectively. Left number is the number of years, out of 20, with deep convection in SG; right number is deep convection in GIN.

and θ is the monthly temperature field. The other buoyancy forcings that are studied here are the local heat and salt changes by interaction with the atmosphere. These are defined as the month-to-month difference in heat and salt content, respectively, from the ocean surface to the MLD.

To assess the consequences of deep water formation, we study the hydrographic properties averaged over the same two depth ranges as de Jong et al. (2009):

- the Labrador Sea Water (LSW) layer, 750–1250 m depth;

– and the Northeast Atlantic Deep Water (NEADW) layer, 2000–2500 m depth.

We shall refer to water found at these two levels in models as North Atlantic Deep Water (NADW), with no further distinction between LSW and NEADW. We do not attempt to define NADW using temperature, salinity or density criteria as is done in observations (e.g. Weaver et al., 1999), since such criteria are not adapted to models that we expect to feature temperature, salinity or density biases. The monthly AMOC is obtained by integrating the meridional velocity at 30° N through the Atlantic basin from coast to coast, and then over depth using the bottom of the ocean as the reference level. The AMOC is defined as the maximum southward transport (Cheng et al., 2013).

3 The representation of North Atlantic deep water formation in CMIP5 models

3.1 Comparison with observations

Deep convection occurs in the North Atlantic in two main areas: in the subpolar gyre, and in the Greenland–Iceland–Norwegian seas (Fig. 1a). It has been measured and found to extend deeper than 2000 m (Marshall and Schott, 1999), but it does not occur every year in the real ocean. In fact, over the 1986–2005 period of this study, deep convection occurred in the subpolar gyre only from 1987 to 1994 and in winter 1999–2000 (Yashayaev, 2007; Våge et al., 2009); in the GIN seas, only in winter 1988 (Marshall and Schott, 1999). Hence, the climatology made of observations shows relatively shallow mean mixed layers that do not exceed 1000 m (Fig. 1a). Still, some models are clearly convecting too deep, with MLD reaching from the surface to the sea floor: GFDL-CM3, GISS-E2-R, IPSL-CM5A-MR and MPI-ESM-MR in the SG area (Fig. 1l, o, s, w) and GISS-E2-R and IPSL-CM5A-MR again as well as both MIROC models in the GIN area (Fig. 1o, s, t, u).

Most models exhibit very deep 20-year mean mixed layers, over large areas, and convect in both regions nearly every year. In the SG region, the models can be split into three different groups based on the location of the deep convection centre:

– the models that convect mostly in the Labrador Sea, or northern part of SG: CCSM4, CESM1-CAM5, CNRM-CM5, FGOALS-g2, HadGEM2-CC, HadGEM2-ES, MIROC5, MPI-ESM-LR and MPI-ESM-MR (Fig. 1e, f, i, k, p, q, t, v, w)

– the models that convect too far in the south: ACCESS1-0, bcc-csm1-1, CanESM2, CMCC-CM, CMCC-CMS, GFDL-ESM2G, GFDL-ESM2M, IPSL-CM5A-LR, IPSL-CM5A-MR, MIROC-ESM-CHEM, and NorESM1-M (Fig. 1b, c, d, g, h, m, n, r, s, u, x)

– the models that convect everywhere: CSIRO-Mk3-6-0, GFDL-CM3 and GISS-E2-R (Fig. 1j, l, o)

In fact, in the SG area, 9 out of 23 models convect at the correct location. The majority of models convect at the wrong location, and in particular too far in the south, which is a common feature in climate models (e.g. Treguier et al., 2005; Jungclaus et al., 2005). In both SG and GIN, the location of deep MLD seems constrained by the winter sea ice extent (blue line in Fig. 1); this will be further discussed in Sect. 4.2.

Unlike the real North Atlantic Ocean and its "deep convection seesaw" (Oka et al., 2006), i.e. the alternation between deep convection in SG and in the GIN seas, most models convect in both regions at the same time, every year of the study period. The exceptions are as follows:

– in SG, CanESM2, both CMCCs and CSIRO-Mk3-6-0 convect only 75 % of the years (left numbers, Fig. 1d, g, h, j), and CNRM-CM5 less than 50 % (Fig. 1i);

– in GIN, CNRM-CM5 again, both HadGEM2s, and IPSL-CM5A-LR convect 75 % of the years (right numbers, Fig. 1i, p, q, r), and CMCC-CM and FGOALS-g2 less than 50 % (Fig. 1g, k).

No CMIP5 model from this study has a variability similar to that observed in the real North Atlantic during 1986–2005, but two models, CMCC-CM and CNRM-CM5, exhibit more variability, in both seas, than the other models.

A full assessment of the impact of resolution and model code changes is not possible with the limited data used here. In fact, this is the motivation for the CORE-II (Danabasoglu et al., 2014) and upcoming OMIP (Griffies et al., 2016) exercises. It is nonetheless interesting to note how differently models which share components behave. For example, the following was observed:

– CMCC-CM and CMCC-CMS differ only in the configuration of the atmospheric code, yet CMCC-CMS has a far more intense deep convection region in the GIN seas (Fig. 1g, h). HadGEM2-CC and HadGEM2-ES also differ only slightly in their atmospheric code (HadGEM2-ES includes tropospheric chemistry), yet their deep convection behaviours are not obviously different (Fig. 1p, q).

– IPSL-CM5-LR and IPSL-CM5-MR differ in the resolution of their common atmospheric component (IPSL-CM5-MR is the highest), and the mean MLD is deeper, over a larger area in IPSL-CM5-MR (Fig. 1r, s). In the meantime, although CCSM4 and CESM1-CAM5 also have different atmosphere models but the same ocean code, their deep convection behaviours in both seas are equivalent (Fig. 1e, f).

– GFDL-ESM2M is more similar to GFDL-ESM2G in deep convection characteristics despite their different ocean components than to GFDL-CM3 whose ocean model code is the same as that of GFDL-ESM2M.

In summary, choices of ocean or atmosphere model codes and resolutions cannot be directly linked to specific deep convection behaviours. All models from this study convect too often, too deep and over too large an area when compared to observations. Nine models are relatively realistic, though, regarding the location of deep convection; among these nine models, four of them (CNRM-CM5, FGOALS-g2, HadGEM2-CC and HadGEM2-ES) also exhibit some temporal variability instead of wrongly convecting each year, and can hence be deemed "the most accurate models".

3.2 Has deep convection representation improved since CMIP3?

In a study of eight CMIP3 models, de Jong et al. (2009) found that deep convection was too shallow in the Labrador Sea, while Drijfhout et al. (2008) found deep convection to be too deep, over too large an area in a region corresponding to the southern part of our SG. Half of the models presented some variability in the mean maximum MLD, an indication that they did not convect every year in SG. To the best of our knowledge, no study has assessed the performance of CMIP3 models with respects to MLD in the GIN seas, although the occasional map of this region by Carman and McClean (2011) does show a large spread in maximum depth and area of deep convection among the 10 models of their study.

CMIP5 models have improved compared to their CMIP3 counterparts, since deep convection in the GIN seas is more localised for the majority of them. Out of our sample of 24, 9 models have realistic MLD in the Labrador Sea, at the correct location, and 4 of them even have a realistic variability. Most CMIP5 models also convect less deeply than the CMIP3 models did; most of the models in the present study convect only to 2000 m on average, whereas most mean CMIP3 MLDs in the subpolar gyre extended to the sea floor.

However, some problems remain. The majority of models in our study convect at the wrong location in the subpolar gyre, too far south and/or over too large an area extending south of Iceland. CMIP5 and CMIP3 models alike convect too often, or rather more often than the real ocean did over the same period. And a minority of CMIP5 models have MLDs that are far too deep.

Why are some of these biases still present in CMIP5 models? Can they be caused by other biases that have not been improved and/or specific model components? We investigate these questions now, in Sect. 4.

4 Across-model possible causes of deep water formation misrepresentations

4.1 Heat and salt

The aim of the present paper is not to determine the dynamics of deep convection in all the individual CMIP5 models, since that would require access to 100-to-1000 year simulations (as was done in the Southern Ocean by Martin et al., 2013, for example). Instead, we verify whether specific biases in the models are consistently associated with misrepresentations of deep water formation. We concentrate on features that have been highlighted in observations or in other modelling studies as potential triggers for deep convection: stratification, freshwater import from the Arctic, local buoyancy forcings and sea ice (Marshall and Schott, 1999).

In this section, we concentrate on the buoyancy biases: both freshwater import from the Arctic and local processes. No across-model relationship was found, but rather different behaviours for different models were observed, which are summarised in Table 2. The sign conventions (see Sect. 2.3) are as follows:

– negative for the vertical density gradient (bold font, Table 2) means that the smaller the density gradient, i.e. the less stratified the water column, the deeper the mixed layer;

– negative for "freshwater from Davis or Fram straits" (italic bold font) means that the more freshwater is flowing southward from the Arctic, the less deep convection;

– positive for "Heat loss" or "Salt gain" (italic font) means that the stronger the local surface cooling or salinification, the deeper the convection.

Fourteen models out of 23 show a significant relationship, in the sense one would expect, between the vertical density gradient and deep convection in the subpolar gyre, and thirteen in the GIN seas, but only nine in both regions (Table 2, figures in bold font). This relationship MLD – stratification does not correlate with the model MLD biases. For example, CCSM4 and CESM1-CAM have similar deep convection depth and area in the subpolar gyre (Fig. 1e and f), but only CCSM4 has a significant correlation between MLD and vertical density gradient. Moreover, there is no apparent relationship between stratification, MLD, and the vertical mixing parameterisation. In particular, the parameterisation designed by Fox-Kemper et al. (2011) to improve the mixed layer representation (present in the models marked with a black bullet point, Table 2) does not consistently result in better performances than those models without the parameterisation. This lack of relationship between MLD biases and vertical mixing parameterisation was already found by Huang et al. (2014) for the summer MLD.

Nine models out of 23 show a negative correlation between the freshwater coming from the Arctic via the Davis Strait

Table 2. For each model, for the subpolar gyre SG (left) and the GIN seas (right), significant correlation (if any) between the time series of the winter mixed layer depth and the local vertical density gradient $\frac{\partial \rho}{\partial z}$, the Arctic freshwater export via Davis (left) or Fram (right) straits, the local heat exchange with the atmosphere and the local surface salinity change. Different font styles highlight correlation that could explain deep water formation in each model. Sign conventions: negative with $\frac{\partial \rho}{\partial z}$ or freshwater means that low stratification or freshwater transport correspond to deep mixed layers; positive correspond to heat loss or salt gain if an ocean surface cooling or salinification corresponds to deep mixed layers. Models featuring the Fox-Kemper et al. (2011) mixed layer parameterisation are indicated by a black bullet after their name.

Model	Subpolar gyre (SG)				Nordic seas (GIN)			
	$\frac{\partial \rho}{\partial z}$	Davis	Heat loss	Salt gain	$\frac{\partial \rho}{\partial z}$	Fram	Heat loss	Salt gain
ACCESS1-0·	**−0.88**	*− 0.43*	*0.50*	–	–	*− 0.44*	*0.41*	−0.40
bcc-csm1-1	**−0.51**	0.54	*0.51*	*0.43*	**−0.89**	0.55	*0.42*	*0.48*
CanESM2	**−0.63**	*− 0.53*	–	−0.43	**−0.73**	–	*0.57*	–
CCSM4·	**−0.78**	–	*0.84*	–	–	–	*0.62*	−0.51
CESM1-CAM5·	–	0.51	*0.49*	–	**−0.60**	–	*0.41*	−0.40
CMCC-CM	**−0.81**	–	–	–	–	–	–	–
CMCC-CMS	**−0.59**	–	*0.48*	*0.41*	**−0.55**	0.86	–	−0.45
CNRM-CM5	–	*− 0.85*	–	*0.91*	–	0.64	–	–
CSIRO-Mk3-6-0	**−0.72**	–	*0.66*	–	**−0.70**	–	–	*0.67*
FGOALS-g2	**−0.83**	*− 0.39*	*0.52*	–	–	–	–	–
GFDL-CM3	–	0.73	*0.47*	*0.52*	**−0.83**	*− 0.54*	*0.50*	−0.45
GFDL-ESM2G·	**−0.81**	–	*0.50*	*0.43*	**−0.64**	0.68	*0.49*	−0.40
GFDL-ESM2M·	**−0.59**	–	−0.71	*0.61*	**−0.85**	*− 0.55*	*0.75*	−0.74
GISS-E2-R	–	0.41	*0.51*	*0.42*	–	0.55	−0.39	−0.45
HadGEM2-CC	**−0.59**	*− 0.55*	*0.48*	–	**−0.47**	–	–	–
HadGEM2-ES	**−0.59**	–	–	*0.41*	**−0.73**	–	*0.72*	−0.39
IPSL-CM5A-LR	**−0.64**	–	–	–	**−0.67**	–	*0.55*	−0.55
IPSL-CM5A-MR	–	0.46	−0.48	*0.70*	**−0.74**	–	−0.49	−0.48
MIROC5	–	*− 0.64*	−0.61	*0.53*	–	0.51	–	–
MIROC-ESM-CHEM	**−0.80**	*− 0.49*	–	*0.60*	–	–	–	–
MPI-ESM-LR	–	*− 0.40*	−0.43	*0.55*	–	–	–	*0.62*
MPI-ESM-MR	–	*− 0.48*	–	−0.42	**−0.56**	0.48	–	–
NorESM1-M	–	0.45	−0.69	–	–	–	−0.49	−0.51

and the MLD in the subpolar gyre (Table 2, first and fourth columns). For three of these models, CanESM2, MIROC-ESM-CHEM and MPI-ESM-MR, it is even the only meaningful relationship. In the GIN seas, the negative relationship freshwater from the Arctic – MLD is present in only three models. Note that only one model, ACCESS1-0, has a correlation in both regions. A mysterious positive correlation, i.e. the stronger the freshwater import the deeper the MLD, is found for six models in SG and six models in GIN, with two models common to both regions (bcc-csm1-1 and GISS-E2-R). Correlation does not mean causation, so it is possible that in these models MLD and freshwater import are linked to a third process, for example the sea ice extent.

The relationship between local heat loss to the atmosphere and MLD is more consistent: 11 models exhibit a positive correlation in SG, compared to 10 in the GIN seas (Table 2, italicised figures in third and seventh columns). For these models, as is the case in the real ocean (Killworth, 1983), the stronger the surface cooling, the deeper the MLD. For five models (GFDL-ESM2M, IPSL-CM5A-MR, MIROC5, MPI-ESM-LR and NorESM1-M), the opposite relationship

is found: deep convection corresponds to a surface heat gain. It could be that in these models, deep convection is triggered so fast that we see its result, the mixing up of subsurface warm water (Marshall and Schott, 1999), when other models are still in the preconditioning phase. Output at a higher temporal resolution than the current monthly data would be required to study this question.

Finally, 12 models have deep MLD in association with a salinification of the surface waters in SG, compared to only 3 in the GIN seas (italicised figures in fourth and last columns of Table 2). In fact in the GIN seas, the opposite relationship is encountered the most often, for 12 of the models. CMCC-CMS also has a positive relationship with the freshwater from the Arctic, suggesting that a more-complex-than-thought freshwater cycle in the GIN seas could be linked to deep convection. For the 11 other models, this negative relationship remains unexplained as monthly outputs are not of a high enough resolution for such a study.

In summary, in the real North Atlantic, deep convection is mainly controlled by local surface buoyancy forcings: heat loss to the atmosphere in SG (Marshall and Schott, 1999), and haline convection in the GIN seas (Rudels and Quad-fasel, 1991). In CMIP5 models, no consistent behaviour was found. CSIRO-Mk3-6-0 is the only model which seems to have the same drivers of deep convection as the observations, and a clear distinction between the two regions. Half of the models exhibit unexpected relationships, showing that higher-resolution outputs are required to study their dynamics. And five models had no significant correlation; their deep convection hence is probably controlled by something else. We now check if this could be the sea ice.

4.2 Sea ice

The link between deep convection and sea ice is evident in the GIN seas, where in the real ocean ice crystals have to form in the surface layer and then rise while saline droplets sink, triggering convection (Rudels and Quadfasel, 1991). The relationship has also been identified in the models from the CORE-II experiments. Danabasoglu et al. (2014) found that models with less sea ice had a salty bias at the surface and hence deeper MLDs.

In the current study, no across-model relationship was found between sea ice extent and deep convection in CMIP5 models. The maximum extent, the seasonal cycle and the variability yielded no significant result. However, we do observe that deep convection follows the winter sea ice edge (blue lines in Fig. 1), in agreement with Danabasoglu et al. (2014).

In fact, the majority of models that do not convect in the Labrador Sea are ice-covered in this region in winter. Bcc-csm1-1, CMCC-CM, CMCC-CMS, both GFDL-ESM2G and GFDL-ESM2M, IPSL-CM5A-LR, IPSL-CM5A-MR, and MIROC-ESM-CHEM have a sea ice cover between Greenland and North America that extends significantly further south and east than in observations (Fig. 1c, g, h, m, n, r, s, u). Similarly, in the GIN seas, the location of deep convection is immediately east of the sea ice. The models which are most ice-covered in the GIN seas, notably bcc-csm1-1 and the three GFDL models, have deep convection more in the east and south than the observations (Fig. 1c, l, m, n).

The models with the most accurate representation of deep convection, at least the most accurate location, seem to be the ones with the most accurate winter sea ice extent, as was found by Loder et al. (2015) in a subset of six CMIP5 models. In this study, 15 out of the 23 models share only three different sea ice components (Table 1). Seven of them in particular use CICE: ACCESS1-0, CCSM4, CESM1-CAM5, FGOALS-g2, both HadGEM2 models, and NorESM1-M. Although ACCESS1-0 and NorESM1-M convect too far south in the subpolar gyre (Fig. 1b, x), these seven models are amongst the most accurate in this study. Future model intercomparison effort should consider studying the effect of the sea ice model on the ocean. In fact, such is the plan of the upcoming SIMIP (Notz et al., 2016).

The present study does not mean to identify the driving mechanism for deep convection in the North Atlantic in CMIP5 models. In fact, it has proven that such an exercise is not possible with this type of output, and that dedicated modelling exercises should be performed instead. In a final result section, we shall see why they should indeed be performed, i.e. which impacts a misrepresentation of deep convection has on the water column and ocean circulation.

5 Why inaccurate North Atlantic deep water formation is a problem

5.1 Consequences on the water column

Following de Jong et al. (2009), Fig. 2 shows the across-model relationship between mean MLD and water property biases at two depth ranges representative of the North Atlantic deep waters in the subpolar gyre. We find no consistent significant relationship between the density bias and the MLD. For example, the models with the deepest MLD are not the densest. In SG in both layers (Fig. 2a, b), models with the smallest biases tend to be those with a mean MLD deeper than 2000 m, although bcc-csm1-1, FGOALS-g2 and GISS-E2-R are notable exceptions, with biases larger than $0.2 \, \text{kg} \, \text{m}^{-3}$. As was already the case in CMIP3 models (de Jong et al., 2009), there is no clear relationship between the water column density and deep convection, but there is a relationship with temperature in the subpolar gyre (Fig. 2c, d). At both depth levels, the deeper the mean mixed layer, the warmer the model. The relationship is the strongest below 2000 m ($R = 0.59$, Fig. 2d), where the temporal spread in the temperature values is also lower. As was to be expected in a region where salinity dominates the density signal, the salinity biases resemble the density biases (Fig. 2e, f). As such, no relationship is found between the salinity and the MLD in the subpolar gyre.

Similarly, in the GIN seas, the models that are the most accurate in density seem to correspond to deep mixed layers (Fig. 3a, b), with bcc-csm1-1 being again an exception. In the GIN seas, no significant relationship could be found between the temperature and the MLD, at either depth (Fig. 3c, d). It can be noted in particular that the models with the warmest temperature biases do not have much in common in the GIN seas (Fig. 1): FGOALS-g2 has shallow convection over an extended area, GFDL-CM3 convects to a moderate depth in a region too far south and east, and MIROC-ESM-CHEM convects far too deep but at the correct location. A similar result is found for salinity biases in the GIN seas (Fig. 3e, f). There is no significant across-model relationship between MLD and salinity biases, and the most extreme biases are encountered for similar MLD. FGOALS-g2, the saltiest, and CNRM-CM5, the freshest, both have a mean MLD of approximately 1000 m (Fig. 3e).

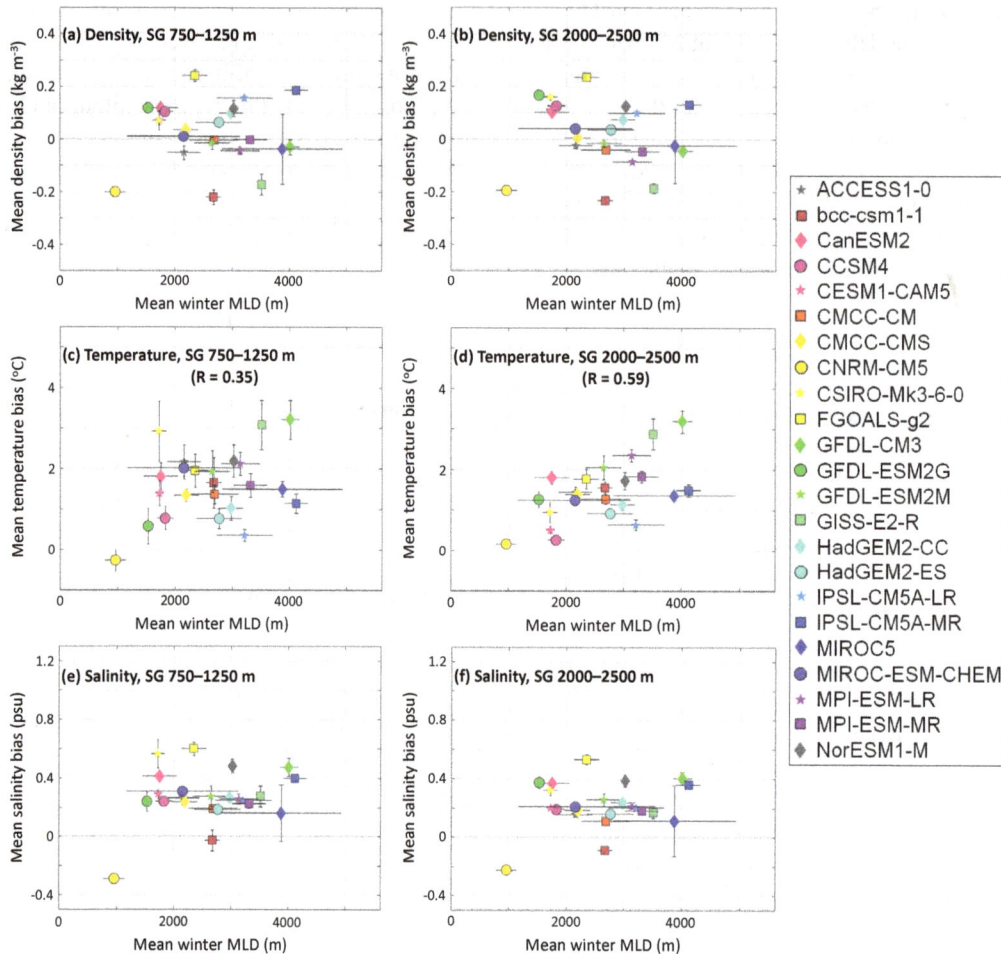

Figure 2. Across-model relationship between the 20-year mean density **(a, b)**, temperature **(c, d)** and salinity **(e, f)** bias at the two depth levels representative of NADW (columns) and the 20-year mean winter MLD, in the subpolar gyre SG.

It can be noted that the majority of models have relatively accurate NADW densities at both depths: 13 models are within $0.1\,\mathrm{kg\,m^{-3}}$ of the observations in the SG area (Fig. 2a, b), compared to 17 in the GIN seas (Fig. 3a, b). In fact, most models have a warm and salty bias in both seas (Figs. 2 and 3), but those compensate in density.

In summary, dense water formation is not associated with specific density biases, and the only significant correlation is linked to deep warm biases. To explain this seemingly counter-intuitive finding, we assess using Fig. 4 how the temperature, shown in panel (a), and density, shown in panel (b), are reorganised from month to month through the water column, and show only one model. Each year, deep convection occurs at two times (Fig. 4a):

– first, a warming from the surface, where the warming is the strongest, to approximately 500 m depth

– then, when the MLD is maximum, a cooling from the surface to a certain depth, and a warming below that depth.

For the events with very deep MLD such as those of 1987 to 1990 and 1993 in Fig. 4, the cooling happens through most of the ML, whereas during shallower events the cooling is limited to the top 500 m of the water column. In fact, during deep convection, heat is merely reorganised through the water column.

Density does increase during deep convection events (Fig. 4b), but also decreases as deep convection is triggered and in the months before. In agreement with the temperature results, it also decreases during deep convection at the depth levels where temperature increases. In fact, in the subpolar gyre in CMIP5 models, deep convection allows the mixing through the water column of the comparatively warm and salty pool that sits around 500 m. Hence, deep convection is associated with a warming of the deep waters, but like in CMIP3 models this warming is compensated for by salinity so that there is no consensus regarding density (de Jong et al., 2009).

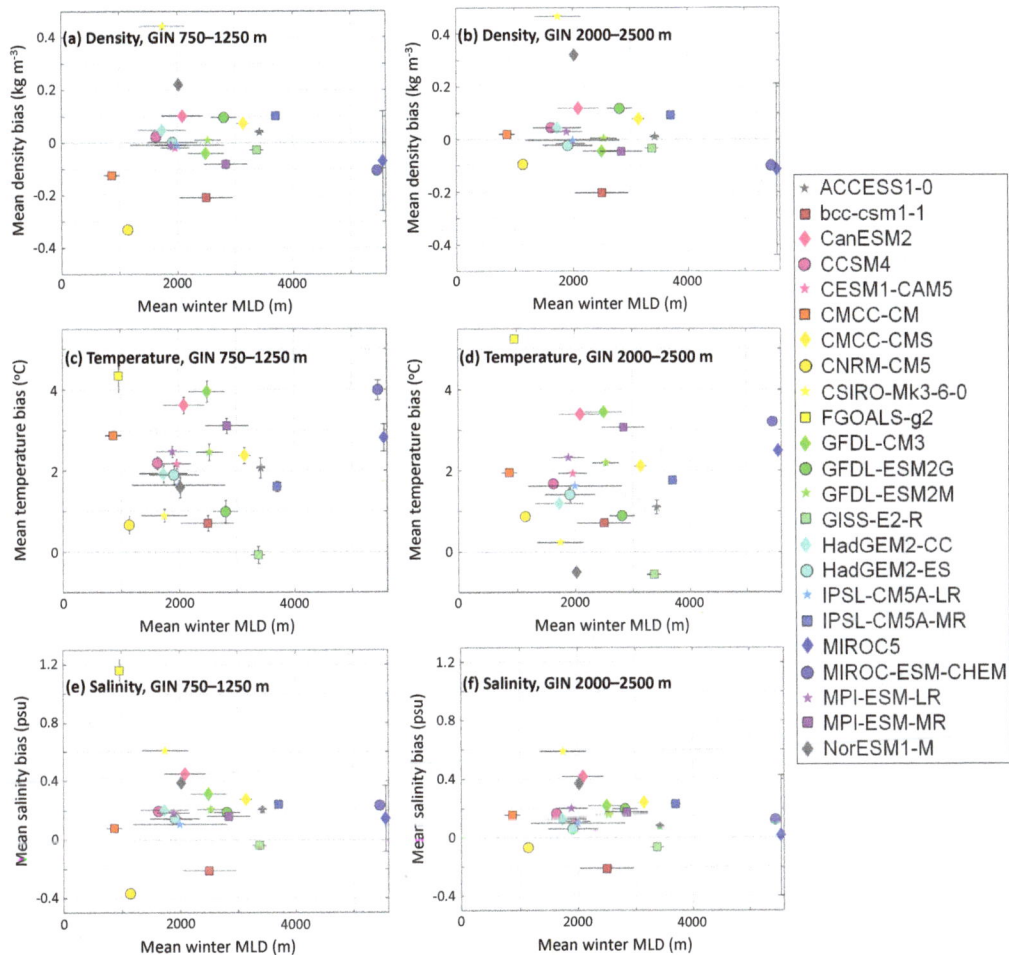

Figure 3. Same as Fig. 2 but for the GIN seas.

5.2 Consequences on the AMOC and heat export to the Arctic

In CCSM4, Jahn and Holland (2013) found that less deep convection in the North Atlantic leads to a reduced AMOC. Similarly, in CORE-II experiments, deep mixed layers were associated with large AMOC (Danabasoglu et al., 2014). In the current paper, no across-CMIP5 model relationship was found between the mean winter MLD or volume of deep convection in either region or the AMOC.

This lack of a result was actually not that surprising. In the real North Atlantic, the AMOC is the result of the combined effects of both deep convection regions (Yashayaev, 2007). Moreover, there is a lag between deep convection and the subsequent AMOC strength (Jahn and Holland, 2013). We evaluated such a lag on a subset of 14 CMIP5 models for which we could obtain 100-year pre-industrial control runs (Fig. 5). The majority of these models exhibit a significant correlation with the AMOC when deep convection in the gyre lags by 0 to 2 years before (Fig. 5, y axis). There

are two maxima, associated with a lag in the GIN seas of 0 to 1 year, but also 14 years before the AMOC (Fig. 5, x axis).

The mechanism linking deep water formation in both regions, the AMOC and poleward heat transport has been tested on several previous occasions using coupled models (e.g. Delworth et al., 1993; Menary et al., 2012; Lohmann et al., 2014). To the best of our knowledge, the most extensive such study was conducted by Ba et al. (2014) and included multi-centennial runs of 10 coupled models. They found that for most models, deep convection was associated with subsequent AMOC with a lag of a year in the SG and 10 years in the GIN seas; our results are consistent with their findings. The large range of values associated with the GIN seas' deep convection could be due to the wrong representation of overflows, caused by the too-coarse resolutions of the models (Jungclaus et al., 2013). This issue is known, and hence possible solutions such as pipe parameterisations that artificially transport the water undisturbed down an overflow area have been designed (Danabasoglu et al., 2010) and are expected to lead to better representations of the AMOC in future CMIPs.

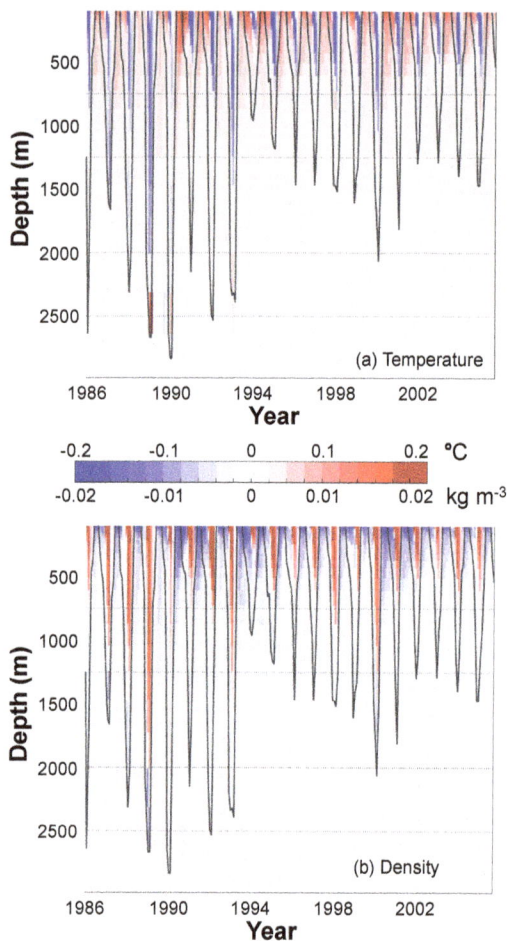

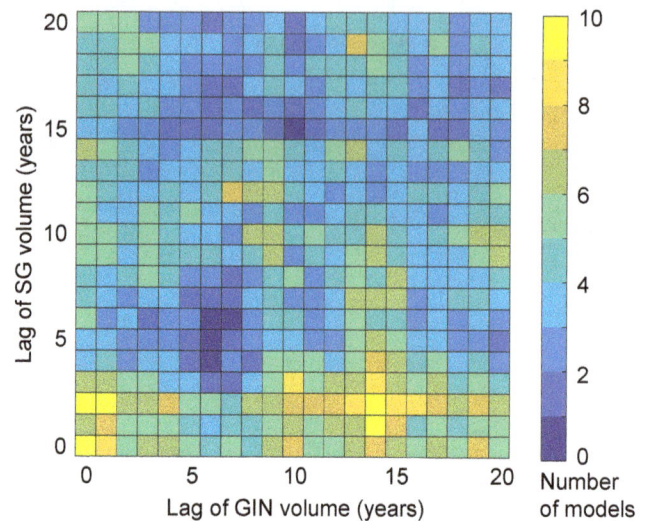

Figure 5. Number of models out of the 14-model subset where a significant correlation was found between the AMOC and the sum of the volumes of deep convection SG + GIN, for different lags of SG (vertical) and GIN (horizontal); deep convection before AMOC.

Figure 4. For only one model, CanESM2, Hovmöller diagram showing the difference from 1 month to the next of the temperature (a) and density (b) profiles with depth in the subpolar gyre. Dark grey line represents the SG mixed layer depth. Black dotted vertical lines highlight the depth levels representative of NADW: 750, 1250, 2000 and 2500 m.

correctly model the amount of oceanic heat that enters the Arctic through the Fram Strait.

6 Conclusions

So there is a relationship between deep convection and the strength of the AMOC in CMIP5 models. In other models, the stronger the AMOC, the more heat is sent northwards to the Arctic (e.g. Jahn and Holland, 2013; Danabasoglu et al., 2014). We find the same result in CMIP5 models (Fig. 6). The mean heat flux through the Fram Strait is not clearly related to the mean volume of deep convection in the subpolar gyre (Fig. 6a), the GIN seas (Fig. 6b), or the sum of the two (Fig. 6c). But there is a strong robust across-model relationship between the AMOC and the heat flux: the stronger the AMOC, the more heat is exported to the Arctic through the Fram Strait (Fig. 6d).

In most CMIP5 models, there is a dynamical relationship between deep water formation and the AMOC. There is also a relationship between the AMOC and the heat export to the Arctic. So not only the volumes but also the temporal variability of deep convection need to be better represented to

CMIP5 models have improved their representation of deep convection in the North Atlantic compared to CMIP3 models (de Jong et al., 2009). Nearly half of them convect at the correct location, and a third of them with some variability - as do the observations. The rest convects too often, too deep, and too far south in the subpolar gyre (Fig. 1). The cause for deep convection bias is model-dependent. The depth is linked to stratification and buoyancy forcings for more than half of the models (Table 2), as the area and location are to the sea ice extent (Fig. 1). In particular, models with the same sea ice component, CICE, seem to have the most accurate sea ice extent in the subpolar gyre and Greenland–Iceland–Norwegian seas, and the most accurate deep convection there. Surprisingly, some models exhibited counter-intuitive relationships, contrary to observations, between freshwater fluxes, local buoyancy forcings and mixed layer depth (Table 2). Dedicated studies should be performed by the modelling community to assess the causes of such spurious relationships, investigating the role of the vertical mixing parameterisation (Fox-Kemper et al., 2011) or the representation of the Canadian Archipelago (Komuro and Hasumi, 2005) for example. We found that deep convection leads to a redistribution of heat through the water column, so that the models with the most intense convection are in fact the warmest (Fig. 2); nothing consistent was found regarding density because of salinity mixing. Finally, the stronger the deep convection in CMIP5

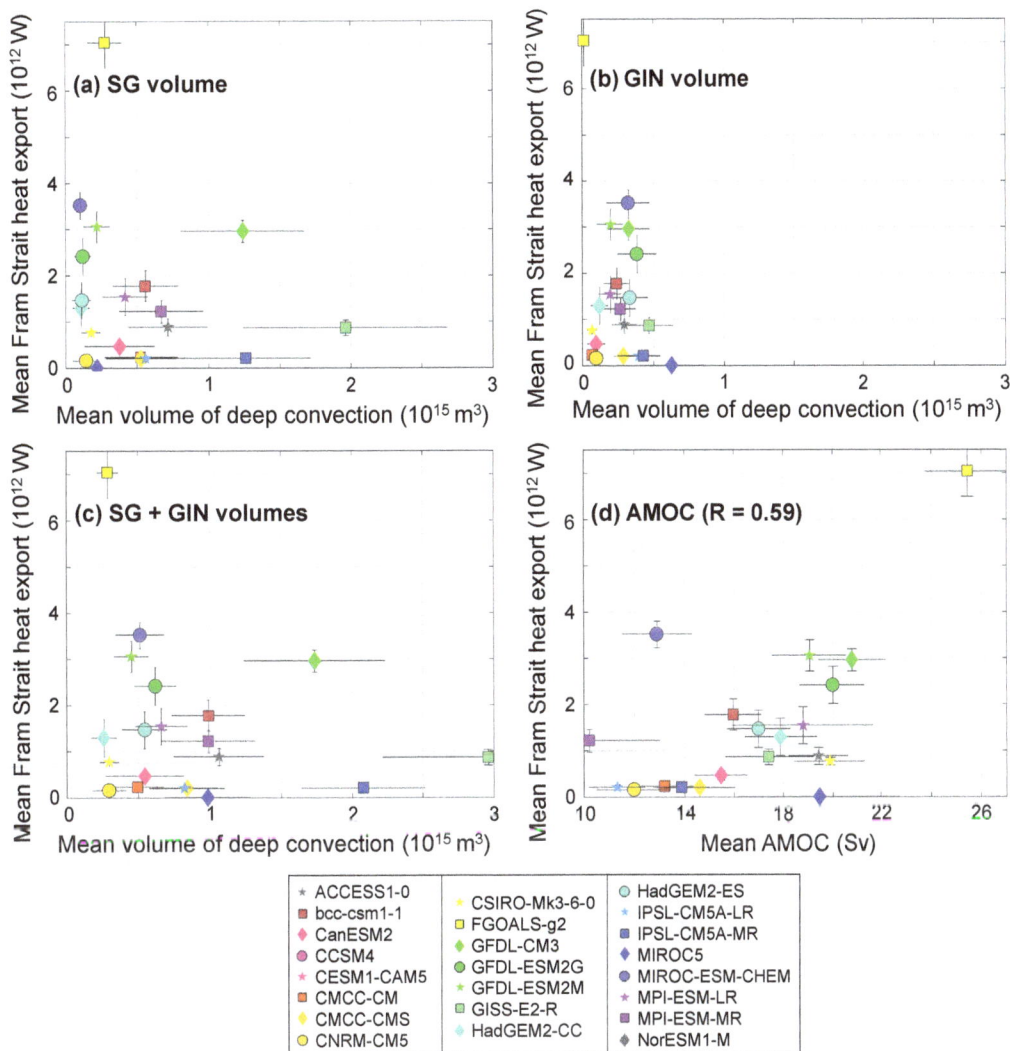

Figure 6. Across-model relationship between the 20-year mean heat export to the Arctic through the Fram Strait and the 20-year mean (**a**) volume of deep convection in SG; (**b**) volume of deep convection in GIN; (**c**) total volume of deep convection SG + GIN; (**d**) AMOC.

models, the stronger their Atlantic Meridional Overturning Circulation 2 years later (Fig. 5), and in turn the stronger the heat export to the Arctic (Fig. 6).

These results should be taken as they are: correlations, not full dynamical studies. Dedicated experiments, performed on an ocean at rest, over centuries, would be needed to assess what triggers deep convection in each model, and would probably require output at a higher time resolution than monthly means. Similarly, the relationship between the choice of a sea ice model and the accuracy of deep water formation would need a proper sea ice MIP to be checked. Fortunately, a SIMIP exercise is indeed planned for CMIP6 (Notz et al., 2016). Only then can we accurately assess the heat transport to the Arctic and its future change, and hence predict the demise of Arctic sea ice and Greenland floating glaciers.

Competing interests. The author declares that they have no conflict of interest.

Acknowledgements. We acknowledge the World Climate Research Programme's Working Group on Coupled Modelling, which is responsible for CMIP, and we thank the climate modelling groups (whose models are listed in Table 1 of this paper) for producing and making available their model output. The author would like to thank Matthew Palmer of UK Met Office for his willingness to share model outputs (that were eventually not used for this paper) and Anna Wåhlin for advice and helpful comments, as well as Peter R. Gent and an anonymous reviewer, whose suggestions notably improved the quality of this paper.

Edited by: Matthew Hecht

References

Aagaard, K. and Carmack, E. C.: The role of sea ice and other fresh water in the Arctic circulation, J. Geophys. Res.-Oceans, 94, 14485–14498, https://doi.org/10.1029/JC094iC10p14485, 1989.

Ba, J., Keenlyside, N. S., Latif, M., Park, W., Ding, H., Lohmann, K., Mignot, J., Menary, M., Otterå, O. H., Wouters, B., Salas y Melia, D., Oka, A., Bellucci, A., and Volodin, E.: A multi-model comparison of Atlantic multidecadal variability, Clim. Dynam., 43, 2333–2348, 2014.

Böning, C. W., Scheinert, M., Dengg, J., Biastoch, A., and Funk, A.: Decadal variability of subpolar gyre transport and its reverberation in the North Atlantic overturning, Geophys. Res. Lett., 33, L21S01, https://doi.org/10.1029/2006GL026906, 2006.

Carman, J. C. and McClean, J. L.: Investigation of IPCC AR4 coupled climate model North Atlantic mode water formation, Ocean Model., 40, 14–34, https://doi.org/10.1016/j.ocemod.2011.07.001, 2011.

Cheng, W., Chiang, J. C. H., and Zhang, D.: Atlantic Meridional Overturning Circulation (AMOC) in CMIP5 Models: RCP and historical simulations, J. Climate, 26, 7187–7197, https://doi.org/10.1175/JCLI-D-12-00496.1, 2013.

Danabasoglu, G., Large, W. G., and Briegleb, B. P.: Climate impacts of parameterized Nordic Sea overflows, J. Geophys. Res.-Oceans, 115, C11005, https://doi.org/10.1029/2010JC006243, 2010.

Danabasoglu, G., Yeager, S. G., Bailey, D., Behrens, E., Bentsen, M., Bi, D., Biastoch, A., Böning, C., Bozec, A., Canuto, V. M., and Cassou, C.: North Atlantic simulations in coordinated ocean-ice reference experiments phase II (CORE-II) – Part I: mean states, Ocean Model., 73, 76–107, https://doi.org/10.1016/j.ocemod.2013.10.005, 2014.

de Boyer Montégut, C., Madec, G., Fisher, A. S., Lazar, A., and Iudicone, D.: Mixed layer depth over the global ocean: An examination of profile data and a profile-based climatology, J. Geophys. Res., 109, C12003, https://doi.org/10.1029/2004JC002378, 2004.

de Jong, M. F., Drijfhout, S. S., Hazeleger, W., van Aken, H. M., and Severijns, C. A.: Simulations of hydrographic properties in the northwestern North Atlantic Ocean in coupled climate models, J. Climate, 22, 1767–1786, 2009.

Delworth, T., Manabe, S., and Stouffer, R. J.: Interdecadal variations of the thermohaline circulation in a coupled ocean-atmosphere model, J. Climate, 6, 1993–2011, https://doi.org/10.1175/1520-0442(1993)006<1993:IVOTTC>2.0.CO;2, 1993.

Delworth, T. L., Broccoli, A. J., Rosati, A., Stouffer, R. J., Balaji, V., Beesley, J. A., Cooke, W. F., Dixon, K. W., Dunne, J., Dunne, K. A., and Durachta, J. W.: GFDL's CM2 Global Coupled Climate Models – Part I: Formulation and simulation characteristics, J. Climate, 19, 643–674, 2006.

Drijfhout, S., Hazeleger, W., Selten, F., and Haarsma, R.: Future changes in internal variability of the Atlantic Meridional Overturning Circulation, Clim. Dynam., 30, 407–419, 2008.

Eyring, V., Bony, S., Meehl, G. A., Senior, C. A., Stevens, B., Stouffer, R. J., and Taylor, K. E.: Overview of the Coupled Model Intercomparison Project Phase 6 (CMIP6) experimental design and organization, Geosci. Model Dev., 9, 1937–1958, https://doi.org/10.5194/gmd-9-1937-2016, 2016.

Fichefet, T. and Maqueda, M. A. M.: Sensitivity of a global sea ice model to the treatment of ice thermodynamics and dynamics, J. Geophys. Res., 102, 12609–12646, 1997.

Flato, G., Marotzke, J., Abiodun, B., Braconnot, P., Chou, S. C., Collins, W. J., Cox, P., Driouech, F., Emori, S., Eyring, V., and Forest, C.: Climate Change 2013: The Physical Science Basis, Contribution of Working Group I to the Fifth Assessment Report of the Intergovernmental Panel on Climate Change, Evaluation of Climate Models, Cambridge University Press, Cambridge, UK, New York, NY, USA, 2013.

Fox-Kemper, B., Danabasoglu, G., Ferrari, R., Griffies, S. M., Hallberg, R. W., Holland, M. M., Maltrud, M. E., Peacock, S., and Samuels, B. L.: Parameterization of mixed layer eddies, III: Implementation and impact in global ocean climate simulations, Ocean Model., 39, 61–78, 2011.

Griffies, S. M., Danabasoglu, G., Durack, P. J., Adcroft, A. J., Balaji, V., Böning, C. W., Chassignet, E. P., Curchitser, E., Deshayes, J., Drange, H., Fox-Kemper, B., Gleckler, P. J., Gregory, J. M., Haak, H., Hallberg, R. W., Heimbach, P., Hewitt, H. T., Holland, D. M., Ilyina, T., Jungclaus, J. H., Komuro, Y., Krasting, J. P., Large, W. G., Marsland, S. J., Masina, S., McDougall, T. J., Nurser, A. J. G., Orr, J. C., Pirani, A., Qiao, F., Stouffer, R. J., Taylor, K. E., Treguier, A. M., Tsujino, H., Uotila, P., Valdivieso, M., Wang, Q., Winton, M., and Yeager, S. G.: OMIP contribution to CMIP6: experimental and diagnostic protocol for the physical component of the Ocean Model Intercomparison Project, Geosci. Model Dev., 9, 3231–3296, https://doi.org/10.5194/gmd-9-3231-2016, 2016.

Huang, C., Qiao, F., and Dai, D.: Evaluating CMIP5 simulations of mixed layer depth during summer, J. Geophys. Res.-Oceans, 119, 2568–2582, 2014.

Hunke, E. C. and Lipscomb, W. H.: CICE: the Los Alamos sea ice model user's manual, version 4, Tech. rep., Los Alamos National Laboratory, 2008.

Jahn, A. and Holland, M. M.: Implications of Arctic sea ice changes for North Atlantic deep convection and the meridional overturning circulation in CCSM4 CMIP5 simulations, Geophys. Res. Lett., 40, 1206–1211, https://doi.org/10.1002/grl.50183, 2013.

Jungclaus, J. H., Haak, H., Latif, M., and Mikolajewicz, U.: Arctic-North Atlantic interactions and multidecadal variability of the meridional overturning circulation, J. Climate, 18, 4013–4031, https://doi.org/10.1175/JCLI3462.1, 2005.

Jungclaus, J. H., Fischer, N., Haak, H., Lohmann, K., Marotzke, J., Matei, D., Mikolajewicz, U., Notz, D., and Storch, J. S.: Characteristics of the ocean simulations in the Max Planck Institute Ocean Model (MPIOM) the ocean component of the MPI Earth system model, J. Adv. Model. Earth Sy., 5, 422–446, https://doi.org/10.1002/jame.20023, 2013.

Killworth, P. D.: Deep convection in the World Ocean, Rev. Geophys. Space Phys., 21, 1–26, https://doi.org/10.1029/RG021i001p00001, 1983.

Komuro, Y. and Hasumi, H.: Intensification of the Atlantic deep circulation by the Canadian Archipelago throughflow, J. Phys. Ocean., 35, 775–789, https://doi.org/10.1175/JPO2709.1, 2005.

Locarnini, R. A., Mishonov, A. V., Antonov, J. I., Boyer, T. P., Garcia, H. E., Baranova, O. K., Zweng, M. M., Paver, C. R., Reagan, J. R., Johnson, D. R., Hamilton, M., and Seidov, D.: World Ocean Atlas 2013, Volume 1: Temperature, Tech. rep., NOAA Atlas Nesdis, 2013.

Loder, J., van der Baaren, A., and Yashayaev, I.: Climate comparisons and change projections for the Northwest Atlantic from six CMIP5 models, Atmos. Ocean, 53, 529–555, 2015.

Lohmann, K., Jungclaus, J. H., Matei, D., Mignot, J., Menary, M., Langehaug, H. R., Ba, J., Gao, Y., Otterå, O. H., Park, W., and Lorenz, S.: The role of subpolar deep water formation and Nordic Seas overflows in simulated multidecadal variability of the Atlantic meridional overturning circulation, Ocean Sci., 10, 227–241, https://doi.org/10.5194/os-10-227-2014, 2014.

Lozier, M. S., Leadbetter, S., Williams, R. G., Roussenov, V., Reed, M. S. C., and Moore, N. J.: The spatial pattern and mechanisms of heat-content change in the North Atlantic, Science, 319, 800–803, https://doi.org/10.1126/science.1146436, 2008.

Marshall, J. and Schott, F.: Open-ocean convection: Observations, theory and models, Rev. Geophys., 37, 1–64, https://doi.org/10.1029/98RG02739, 1999.

Martin, T., Park, W., and Latif, M.: Multi-centennial variability controlled by Southern Ocean convection in the Kiel Climate Model, Clim. Dynam., 40, 2005–2022, https://doi.org/10.1007/s00382-012-1586-7, 2013.

Menary, M., Park, W., Lohmann, K., Vellinga, M., Palmer, M., Latif, M., and Jungclaus, J.: A multimodel comparison of centennial Atlantic meridional overturning circulation variability, Clim. Dynam., 38, 2377–2388, 2012.

Notz, D., Jahn, A., Holland, M., Hunke, E., Massonnet, F., Stroeve, J., Tremblay, B., and Vancoppenolle, M.: The CMIP6 Sea-Ice Model Intercomparison Project (SIMIP): understanding sea ice through climate-model simulations, Geosci. Model Dev., 9, 3427–3446, https://doi.org/10.5194/gmd-9-3427-2016, 2016.

Oka, A., Hasumi, H., Okada, N., and Suzuki, T. T., and Suzuki, T.: Deep convection seesaw controlled by freshwater transport through the Denmark Strait, Ocean Model., 15, 157–176, https://doi.org/10.1016/j.ocemod.2006.08.004, 2006.

Polyakov, I. V., Timokhov, L. A., Alexeev, V. A., Bacon, S., Dmitrenko, I. A., Fortier, L., Frolov, I. E., Gascard, J. C., Hansen, E., Ivanov, V. V., and Laxon, S.: Arctic Ocean warming contributes to reduced polar ice cap, J. Phys. Oceanogr., 40, 2743–2756, https://doi.org/10.1175/2010JPO4339.1, 2010.

Rayner, N. A., Parker, D. E., Horton, E. B., Folland, C. K., Alexander, L. V., Rowell, D. P., Kent, E. C., and Kaplan, A.: Global analyses of sea surface temperature, sea ice, and night marine temperature since the late nineteenth century, J. Geophys. Res., 108, 4407, https://doi.org/10.1029/2002JD002670, 2003.

Rudels, B. and Quadfasel, D.: Convection and deep water formation in the Arctic Ocean-Greenland Sea system, J. Marine Syst., 2, 435–450, https://doi.org/10.1016/0924-7963(91)90045-V, 1991.

Sabine, C. L., Feely, R. A., Gruber, N., Key, R. M., Lee, K., Bullister, J. L., Wanninkhof, R., Wong, C., Wallace, D. W., Tilbrook, B., and Millero, F. J.: The oceanic sink for anthropogenic CO_2, Science, 305, 367–371, https://doi.org/10.1126/science.1097403, 2004.

Schmittner, A. and Lund, D. C.: Early deglacial Atlantic overturning decline and its role in atmospheric CO_2 rise inferred from carbon isotopes ($\delta^{13}C$), Clim. Past, 11, 135–152, https://doi.org/10.5194/cp-11-135-2015, 2015.

Spielhagen, R. F., Werner, K., Sørensen, S. A., Zamelczyk, K., Kandiano, E., Budeus, G., Husum, K., Marchitto, T. M., and Hald, M.: Enhanced Modern Heat Transfer to the Arctic by Warm Atlantic Water, Science, 331, 450–453, https://doi.org/10.1126/science.1197397, 2011.

Straneo, F. and Heimbach, P.: North Atlantic warming and the retreat of Greenland's outlet glaciers, Nature, 504, 36–43, https://doi.org/10.1038/nature12854, 2013.

Taylor, K. E., Stouffer, R. J., and Meehl, G. A.: An overview of CMIP5 and the experiment design, B. Am. Meteorol. Soc., 93, 485–498, https://doi.org/10.1175/BAMS-D-11-00094.1, 2012.

Treguier, A. M., Theetten, S., Chassignet, E. P., Penduff, T., Smith, R., Talley, L., Beismann, J. O., and Böning, C.: The North Atlantic Subpolar Gyre in Four High-Resolution Models, J. Phys. Oceanogr., 35, 757–774, https://doi.org/10.1175/JPO2720.1, 2005.

Våge, K., Pickart, R. S., Thierry, V., Reverdin, G., Lee, C. M., Petrie, B., Agnew, T. A., Wong, A., and Ribergaard, M. H.: Surprising return of deep convection to the subpolar North Atlantic Ocean in winter 2007–2008, Nat. Geosci., 2, 67–72, https://doi.org/10.1038/ngeo382, 2009.

Weaver, A. J., Bitz, C. M., Fanning, A. F., and Holland, M. M.: Thermohaline circulation: High-latitude phenomena and the difference between the Pacific and Atlantic, Annu. Rev. Earth Pl. Sc., 27, 231–285, https://doi.org/10.1146/annurev.earth.27.1.231, 1999.

Yashayaev, I.: Hydrographic changes in the Labrador Sea, 1960–2005, Prog. Oceanogr., 73, 242–276, https://doi.org/10.1016/j.pocean.2007.04.015, 2007.

Zweng, M. M., Reagan, J. R., Antonov, J. I., Locarnini, R. A., Mishonov, A. V., Boyer, T. P., Garcia, H. E., Baranova, O. K., Johnson, D. R., Seidov, D., and Biddle, M. M.: World Ocean Atlas 2013, Volume 2: Salinity, Tech. rep., NOAA Atlas Nesdis, 2013.

5

Spatial distribution of turbulent mixing in the upper ocean of the South China Sea

Xiao-Dong Shang[1], **Chang-Rong Liang**[1,2], **and Gui-Ying Chen**[1]

[1]State Key Laboratory of Tropical Oceanography, South China Sea Institute of Oceanology, Chinese Academy of Sciences, Guangzhou 510301, China
[2]University of Chinese Academy of Sciences, Beijing 100049, China

Correspondence to: Xiao-Dong Shang (xdshang@scsio.ac.cn)

Abstract. The spatial distribution of the dissipation rate (ε) and diapycnal diffusivity (κ) in the upper ocean of the South China Sea (SCS) is presented from a measurement program conducted from 26 April to 23 May 2010. In the vertical distribution, the dissipation rates below the surface mixed layer were predominantly high in the thermocline where shear and stratification were strong. In the regional distribution, high dissipation rates and diapycnal diffusivities were observed in the region to the west of the Luzon Strait, with an average dissipation rate and diapycnal diffusivity of 8.3×10^{-9} W kg^{-1} and 2.7×10^{-5} m^2 s^{-1}, respectively, almost 1 order of magnitude higher than those in the central and southern SCS. In the region to the west of the Luzon Strait, the water column was characterized by strong shear and weak stratification. Elevated dissipation rates ($\varepsilon > 10^{-7}$ W kg^{-1}) and diapycnal diffusivities ($\kappa > 10^{-4}$ m^2 s^{-1}), induced by shear instability, occurred in the water column. In the central and southern SCS, the water column was characterized by strong stratification and weak shear and the turbulent mixing was weak. Internal waves and internal tides generated near the Luzon Strait are expected to make a dominant contribution to the strong turbulent mixing and shear in the region to the west of the Luzon Strait. The observed dissipation rates were found to scale positively with the shear and stratification, which were consistent with the MacKinnon–Gregg model used for the continental shelf but different from the Gregg–Henyey scaling used for the open ocean.

1 Introduction

Turbulent mixing is a crucial mechanism that controls the distribution of nutrients, sediments, freshwater, and pollutants throughout the water column (Sandstrom and Elliott, 1984). The magnitude and distribution of diapycnal diffusivity are important for large-scale ocean circulation (Saenko and Merryfield, 2005). Assuming a balance between vertical advection and vertical diffusion for tracers, Munk (1966) reported that a global average diapycnal diffusivity of 10^{-4} m^2 s^{-1} is required to maintain gross oceanic stratification and overturning circulation (Tsujino et al., 2000). However, diapycnal diffusivity from turbulent mixing in the open ocean thermocline only ranges from 5×10^{-6} to 3×10^{-5} m^2 s^{-1} (Gregg, 1998; Polzin et al., 1995). Therefore, it has been argued that elevated turbulent mixing concentrated over rough topography (Ledwell et al., 2000; Wu et al., 2011) would aid in explaining this discrepancy. In the past decade, elevated diapycnal diffusivities, i.e., O (10^{-4} m^2 s^{-1}) or higher, have been found in mixing hotspots such as seamounts (Carter et al., 2006; Lueck and Mudge, 1997), ridges (Klymak et al., 2006a; Lee et al., 2006), and canyons (Carter and Gregg, 2002). However, these elevated mixing events are highly localized. Whether such topographically enhanced mixing is sufficiently intense or widespread to significantly increase the basin-wide average remains unclear. Using a simple averaging scheme, Kunze and Toole (1997) suggested that topographically induced mixing was insufficient to support a basin-averaged diffusivity of O (10^{-4} m^2 s^{-1}) above a 3000 m depth in the North Pacific.

Compared with the open ocean, less attention has been given to marginal seas. In recent years, observations (Tian et al., 2009) indicated that turbulent mixing in marginal seas could make an important contribution to ocean mixing. The South China Sea (SCS), one of the largest marginal seas of the Pacific, connects to the Pacific through the Luzon Strait. Measurements and numerical simulations (Alford et al., 2015; Chang et al., 2006; Lien et al., 2005) indicated that energetic internal tides and internal waves generated near the Luzon Strait propagate into the SCS and facilitate turbulent mixing. Considerable effort has been put forth to explore the characteristics of turbulent mixing in the SCS. Using fine-scale parameterization, Tian et al. (2009) reported a turbulent mixing distribution along a section from the northern SCS to the Pacific. They found that the diapycnal diffusivity in the upper 500 m of the northern SCS reached O (10^{-5} m^2 s^{-1}), almost 1 order of magnitude larger than that in the Pacific. Yang et al. (2016) explored the turbulent mixing in the SCS with a fine-scale parameterization and found diapycnal diffusivity in the northern SCS as large as O (10^{-3} m^2 s^{-1}). In addition to these parameterizations, some direct measurements from microstructure profilers are also available. A direct observation of turbulent dissipation was reported by Laurent (2008), who found a dissipation rate as high as 10^{-6} W kg^{-1} over the shelf break of the northern SCS. Lozovatsky et al. (2013) reported a regional mapping of the averaged dissipation rate in the upper pycnocline of the northern SCS and found values in the Luzon Strait as high as 10^{-7} W kg^{-1}. Yang et al. (2014) conducted direct measurements of turbulence along a section across the continental shelf and slope in the northern SCS. Their results show that the averaged dissipation rate over the shelf reached 10^{-7} W kg^{-1}, which is an order of magnitude larger than that over the slope. There is no doubt that these studies have greatly aided our knowledge of turbulent mixing in the SCS. However, the direct microstructure measurements are localized and scattered, with most of them focusing on the northern SCS. Few microstructure measurements have been conducted in the central and southern SCS. Where the strong turbulent mixing takes place in the SCS and what drives the turbulent mixing are not fully understood. In this work, we present direct microstructure measurements that cover the upper ocean of the SCS and explore the features and regimes of the turbulent mixing. Energy sources for the turbulent mixing are also discussed.

In addition, there is a lack of studies assessing parameterizations in the SCS. Fine-scale parameterizations are aimed at reproducing the dissipation rate in terms of more easily observed or modeled quantities, such as stratification and shear. Generally, microstructure measurements are fewer and more difficult than the fine-structure measurements (e.g., CTD and ADCP measurements), especially microstructure measurements in the deep sea. Therefore, to study the spatial and temporal distribution of turbulent mixing, researchers often resort to fine-scale parameterizations (Jing and Wu, 2010; Tian

et al., 2009; Wu et al., 2011). In addition, fine-scale parameterizations would provide a reference for modelers. Shelf sea models have success in reproducing the water column structure in seasonally stratified shelf seas (Holt and Umlauf, 2008; Simpson and Bowers, 1981). However, models need to calibrate a background mixing level to correctly predict the water column structure (Rippeth, 2005). The requirement of calibration reduces the success of models on shelf-wide scales since differing forcing mechanisms and mixing processes require specific methods and levels of tuning. This presents a clear challenge to oceanographic models. Before the water column structure in shelf seas can be modeled realistically, the distribution of mixing must be established and the major mixing processes parameterized. Confidence in future predictions is therefore dependent on an ocean turbulence model that can be validated against observed mixing or parameterized mixing, but not on the calibration of a background mixing level. In order to estimate the turbulent mixing without microstructure measurements, we assess two fine-scale parameterizations with microstructure data and investigate which one works better and why it works better. We begin in Sect. 2 with a description of our measurements and methods. In Sect. 3 we explore the features and regimes of the turbulent mixing, and assess two fine-scale parameterizations. We discuss the turbulent mixing and fine-scale parameterizations in Sect. 4. A summary of our results is presented in Sect. 5.

2 Measurements and methods

The field experiment was performed from 26 April to 23 May 2010 (local time) prior to the South China Sea summer monsoon (SCSSM) onset. A total of 82 stations were conducted in the experiment (Fig. 1a). Direct turbulence measurements were collected with the Turbulence Ocean Microstructure Acquisition Profiler (TurboMAP). TurboMAP is a quasi-free-falling instrument that measures turbulent parameters ($\partial u/\partial z$ and $\partial T/\partial z$), bio-optical parameters (in vivo fluorescence and backscatter), and hydrographic parameters (conductivity, temperature, and depth; Wolk et al., 2002). TurboMAP carries seven environmental sensors and a three-axis accelerometer that measures tilt and vibrations. The turbulent velocity fluctuations are measured with two standard shear probes. Conductivity (C) and temperature (T) are measured with a combined $C-T$ sensor consisting of a platinum wire thermometer and an inductive conductivity cell. Depth is measured with a semiconductor strain gauge pressure transducer, and the instrument's sinking velocity is computed from the rate of change of the pressure signal. All sensors are sampled at a rate of 256 Hz. TurboMAP was deployed at a speed of 0.5–0.7 m s^{-1} and the maximum deployment depth was approximately 800 m. It took 10–30 min to complete each profile at shallow stations and approximately 1 h at deep stations. Continuous time series of velocity at

5 min intervals and 16 m vertical spacing between 38 and 982 m were obtained from a shipboard acoustic Doppler current profiler (ADCP). At stations where the water depth was more than 982 m, the current velocity cannot be referenced to the sea floor. The movement of the ship was determined from GPS data and absolute value of current velocity was estimated. CTD casts were conducted to provide measurements of temperature and salinity for comparison. At stations where the water depth was less than 800 m, CTD was deployed to 5 m above the seafloor. At stations where the water depth was larger than 800 m, the maximum deployment depth of CTD ranged from 800 to 1500 m. Data obtained from six moorings (Fig. 1, yellow squares) were used to perform a brief analysis of the internal wave field in the SCS. Moorings 1–3 were deployed over the continental shelf/slope and moorings 4–6 were deployed in the deep basin. More information regarding the moorings is given in Table 1.

Figure 2a shows the depth profile of shear $\partial u / \partial z$. At depths of 190–200 m, the shear signal shows variations with peak levels around $0.6\,\mathrm{s}^{-1}$, corresponding to dissipation rates of $10^{-7}\,\mathrm{W\,kg}^{-1}$. The velocity shear decreases below 200 m to peak values of $0.02\,\mathrm{s}^{-1}$, corresponding to dissipation rates of $10^{-10}\,\mathrm{W\,kg}^{-1}$. Dissipation spectra $\psi(k)$ computed from the shear signal in Fig. 2a are shown in Fig. 2b–g along with the corresponding scaled Nasmyth universal spectra (Nasmyth, 1970). The shape of the measured spectra agrees well with the universal spectrum except in the wavenumber regions affected by vibration noise caused by the strumming of the suspension wires in the flow (Wolk et al., 2002). The spectra are computed using Welch's averaged periodogram method with a fast Fourier transformation (FFT) length of 2 m, corresponding to a consecutive segment of approximately 1700 data points. The dissipation rates ε based on the measured spectra range from 10^{-10} to $10^{-7}\,\mathrm{W\,kg}^{-1}$, and they are computed by integrating the measured shear spectrum

$$\varepsilon = 7.5\nu \langle \left(\frac{\partial u}{\partial z} \right)^2 \rangle = 7.5\nu \int_{k_1}^{k_2} \psi(k)\,\mathrm{d}k,$$

where ν is the kinematic viscosity and $\langle \rangle$ denotes the spatial average. k_1 and k_2 are the integration limits. The lower integration limit k_1 is set to 1 cpm and the upper limit k_2 is the highest wavenumber that is not contaminated by vibration noise. Energy density in the low wavenumber area around 1 cpm is not well estimated because of the limited length of the data segments and the length of the profiler itself as the profiler tends to follow the larger-scale flow. The noise level of the TurboMAP profiler is $\varepsilon \sim 10^{-10}\,\mathrm{W\,kg}^{-1}$ (Matsuno and Wolk, 2005; Wolk et al., 2002). Diapycnal diffusivity (Osborn, 1980) was calculated based on the dissipation rate (ε) and stratification ($N^2 = -(g/\rho)\partial\rho/\partial z$) using

$$\kappa = \Gamma\varepsilon/N^2,$$

where the mixing efficiency Γ is set to 0.2 (Oakey, 1982). The shear variance, $S^2 = (\Delta\overline{U}/\Delta z)^2 + (\Delta\overline{V}/\Delta z)^2$, was calculated with $\Delta z = 16\,\mathrm{m}$, where $\overline{U}$ and $\overline{V}$ are the respective zonal and meridional components of the mean horizontal velocity obtained from the shipboard ADCP. The mean velocity is averaged over the time intervals of the TurboMAP measurements.

3   Results

3.1   Water mass properties

Intrusion of water from the Pacific can influence the evolving water properties in the SCS. It has been confirmed by in situ measurements and models (Shaw, 1991; Wu and Hsin, 2012) that there is a strong intrusion of water from the Pacific into the SCS through the Luzon Strait. Two well-defined water masses are active in this process (Qu et al., 2000): high-salinity North Pacific Tropical Water (NPTW) and low-salinity North Pacific Intermediate Water (NPIW). For simplicity, we divide the observations into four regions (Fig. 1): region 1 is located to the west of the Luzon Strait, region 2 is located to the northeast of Hainan Island, region 3 is located in the central SCS, and region 4 is located in the southern SCS. Figure 3 shows the $T - S$ curves of the SCS and western Pacific. Temperature and salinity data in the western Pacific (18.5–22.5° N, 124.5–128.5° E) were obtained from the World Ocean Database 2013 (http://www.nodc.noaa.gov/OC5/woa13/woa13data.html). The $T - S$ curve in the western Pacific shows a reversed "S" shape with NPTW and NPIW clearly identified (Fig. 3, black dashed curve). NPTW and NPIW correspond to the maximum salinity layer at $\sigma_\theta \in (22.5–25.5)\,\mathrm{kg\,m}^{-3}$ and minimum salinity layer at $\sigma_\theta \in (25.5–27.5)\,\mathrm{kg\,m}^{-3}$, respectively. In the maximum salinity layer (22.5–25.5 kg m^{-3}), the water column in region 1 had a salinity maximum of 34.8 psu that approaches the maximum value of the NPTW. Salinity decreased gradually from the Luzon Strait to Hainan Island (region 2) and to the central and southern SCS (region 3 and region 4). This trend is reversed in the minimum salinity layer (25.5–27.5 kg m^{-3}), where the salinity slightly increased from the Luzon Strait to Hainan Island and to the central and southern SCS. The salinity minimum in the Pacific was found to be lower than that in the SCS. The reverse S shape becomes remarkably weak from the northern SCS to the southern SCS, a change to which turbulent mixing occurring in the SCS might have made a significant contribution.

3.2   Microstructure measurements

Figure 4a shows the distribution of the dissipation rate with the thermocline boundaries overlain. Different criteria have been used to define the top of the thermocline (z_t) in terms of either temperature or density. Here, we defined the top of the thermocline as the depth at which the potential temperature

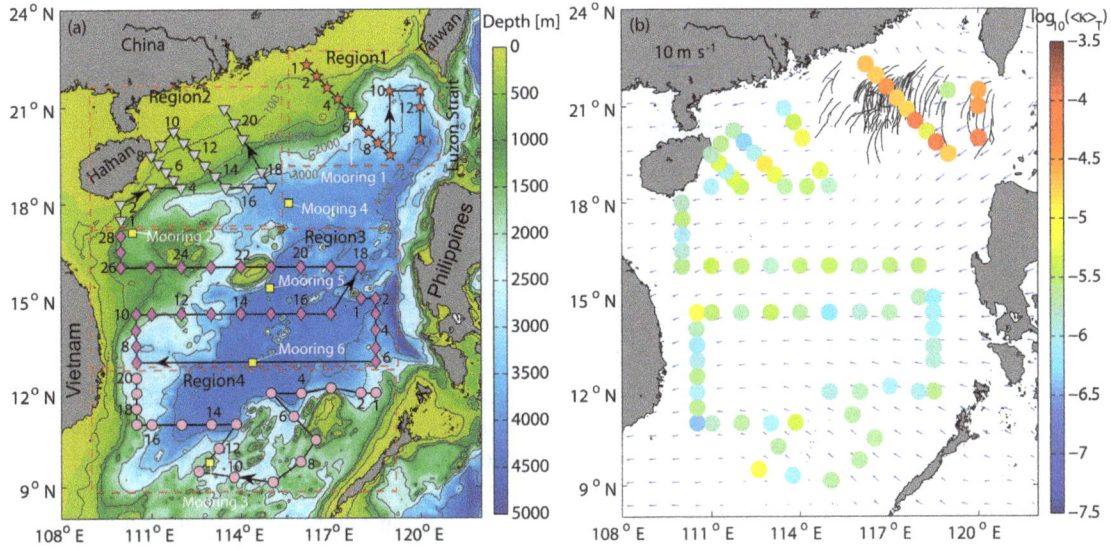

Figure 1. (a) Bottom topography of the SCS and observation stations (symbols). Red stars indicate the stations located in region 1. Gray triangles indicate the stations located in region 2. Magenta diamonds indicate the stations located in region 3. Pink dots indicate the stations located in region 4. Station numbers (i.e., 1, 2, 4 ...) are indicated in each region. The arrows indicate the order of the measurement. The yellow squares indicate the locations of the moorings. **(b)** Spatial distribution of $\langle \kappa \rangle_T$ (color dots). The blue vector gives the averaged 10 m wind speed during the cruises. Black curves in the northern SCS are internal wave packets derived from satellite images by Zhao et al. (2004).

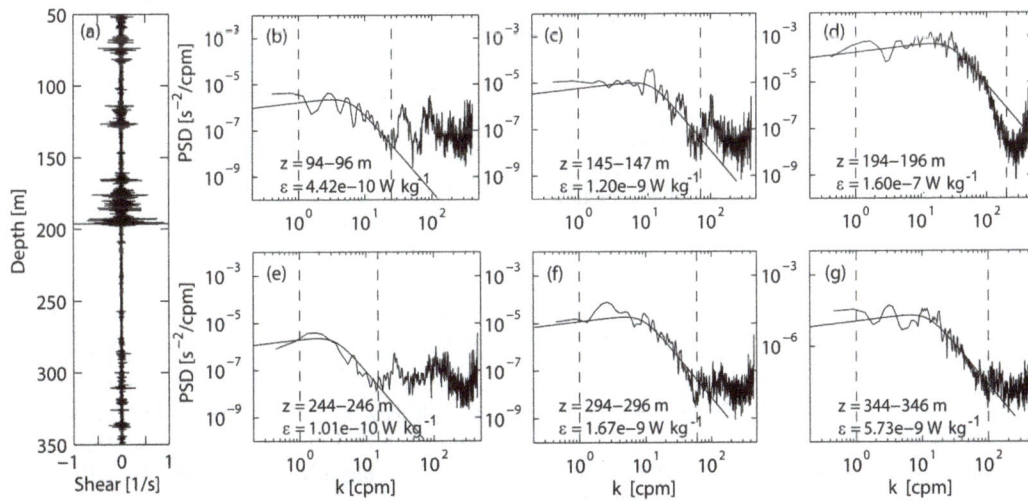

Figure 2. Examples of **(a)** micro-shear and **(b–g)** shear spectra at different depths. The integration bounds (vertical dashed lines) and Nasmyth spectra (smooth curves) are shown.

change from the surface temperature is 0.5 °C. The bottom of the thermocline (z_b) is defined as the depth at which the temperature gradient is equal to 0.05 °C m^{-1}. The surface mixed layers are slightly deep in region 1 compared with the other regions. The average depths of the surface mixed layer in regions 1–4 are 35.2, 14.7, 19.4, and 26.8 m, respectively.

In the surface mixed layer, strong turbulence was accompanied by high dissipation rates (Fig. 4a), which may be attributed to various factors, such as wind stirring, buoyancy flux, and surface waves. Below the surface mixed layer, high dissipation rates (Fig. 4a) were observed in the ther-

mocline, with the average ε in the thermocline reaching 4.6×10^{-9} W kg^{-1}, which was 5 times larger than the value of 8.2×10^{-10} W kg^{-1} below the thermocline. Strong shear (Fig. 4d) also occurred in the thermocline, with an averaged S^2 in the thermocline of 3.3×10^{-5} s^{-2}, which was 5 times larger than that below the thermocline (6.5×10^{-6} s^{-2}). The strong spatial correlation between dissipation and shear implies that shear played an important role in driving the dissipation. Contrary to the dissipation rates, the diapycnal diffusivities (Fig. 4b) in the thermocline were slightly weaker than that below the thermocline. The high diapycnal diffusiv-

Table 1. Information about the moorings.

Mooring	Latitude (° N)	Longitude (° E)	Water depth (m)	Measurement depth range (m)	Measurement duration (d/m/yr)	Time interval (min)	Bin size (m)
Mooring 1	20.74	117.75	1260	13–454	01/08/14–27/09/14	2	16
Mooring 2	17.10	110.39	1410	6–478	04/05/09–04/09/10	60	8
Mooring 3	9.79	112.74	1680	40–416	25/05/09–10/11/10	60	8
Mooring 4	18.01	115.60	3790	60–370	09/04/98–05/10//98	60	10
Mooring 5	15.34	114.96	4265	30–270	07/10/98–11/04//99	60	10
Mooring 6	12.98	114.38	4370	30–270	09/10/98–12/04//99	60	10

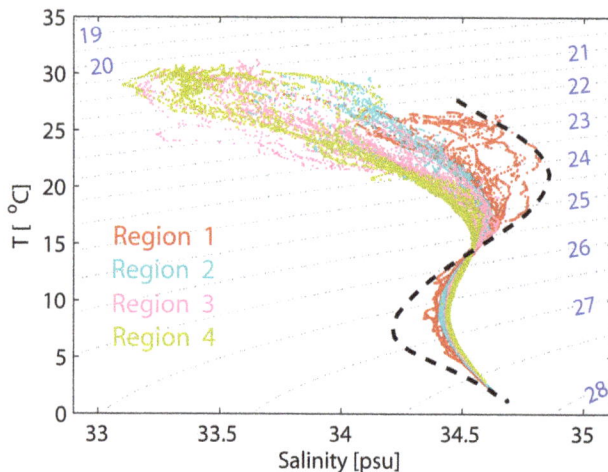

Figure 3. Relation of potential temperature versus salinity (with the potential density σ_θ in kg m^{-3} contours overlaid) of all stations. The black dashed curve shows the relation for potential temperature versus salinity of the western Pacific for reference.

ities below the thermocline were mainly due to the relatively weak stratification (Fig. 4c). The average N^2 below the thermocline was 8.4×10^{-5} s^{-2}, 4 times smaller than the value of 3.4×10^{-4} s^{-2} in the thermocline.

Turbulent mixing in region 1 displayed a different feature from that of the other regions. In region 1, turbulence was more active than that in other regions, with the maximum dissipation rate reaching 10^{-6} W kg^{-1} (Fig. 4a) and the maximum diapycnal diffusivity exceeding 10^{-3} m^2 s^{-1} (Fig. 4b). In addition, region 1 had weak stratification but strong shear compared with other regions (Fig. 4c and d). Most of the water column in region 1 was occupied by a Richardson number of order 1, almost 2 orders of magnitude smaller than that in the other regions (Fig. 4e). Richardson number $Ri = N^2/S^2$ was estimated following MacKinnon and Gregg (2005). A 2 m buoyancy frequency and 16 m shear were used in the calculation. The resolutions of shear used in previous literatures range from 2 m to 16 m (MacKinnon and Gregg, 2003b, 2005; van der Lee and Umlauf, 2011; Xie et al., 2013; Yang et al., 2014). High resolution of shear (2–4 m) was used on the shelf area to resolve small-

scale internal waves and low resolution of shear (8–16 m) was often used in deep water to cover more water depth. Although the Richardson number calculated on 16 m shear might be overestimated, it does not affect the comparison of the Richardson number in different regions too much. One prominent feature in region 1 is that some turbulent patches with elevated dissipation rates ($\varepsilon > 10^{-7}$ W kg^{-1}) and diapycnal diffusivities ($\kappa > 10^{-4}$ m^2 s^{-1}) were observed below the surface mixed layer. These turbulent patches often occurred at depths where the Richardson number was below 0.25, for example, station 6 (between 175 and 195 m), station 8 (between 80 and 100 m), and station 11 (between 175 and 195 m) (indicated by the arrows in Fig. 4), which suggests that elevated dissipation rates and diapycnal diffusivities in the turbulent patches are likely to result from shear instability. More detail regarding the shear instability will be discussed in the following text. Compared with region 1, turbulent mixing in regions 2–4 was relatively weak, with an average ε and κ in the upper 500 m (not including the surface mixed layer) of 1.1×10^{-9} W kg^{-1} and 3.7×10^{-6} m^2 s^{-1}, respectively. These two values are almost 1 order of magnitude smaller than those ($\varepsilon \sim 8.3 \times 10^{-9}$ W kg^{-1} and $\kappa \sim 2.7 \times 10^{-5}$ m^2 s^{-1}) in region 1. Weak turbulent mixing in regions 2–4 is likely to be associated with the strong stratification and weak shear. N^2 (Fig. 4c) was greater than S^2 (Fig. 4d) in regions 2–4, with most of the water column occupied by a large Richardson number ($Ri > 10$; Fig. 4e).

To further understand the changing pattern of turbulence in the SCS, we now look in detail at the profiles of various quantities at three stations in different regions (Fig. 5): station 6 was from region 1, station 22 was from region 3, and station 6 was from region 4. At station 6 in region 1 (Fig. 5a–e), the shear variance was slightly smaller than the buoyancy frequency squared over most of the water column (Fig. 5b). However, the shear variance exceeded the buoyancy frequency squared at some depths; for example, the shear variance was greater than the buoyancy frequency squared at a depth of 185 m, pushing the Richardson number below 0.25, which implies shear instability (Fig. 5c). Small overturns were also found in the density profile at depths of 175 to 195 m (Fig. 5a, the inset). The dissipation rates (Fig. 5d) and diapycnal diffusivities (Fig. 5e) at the corre-

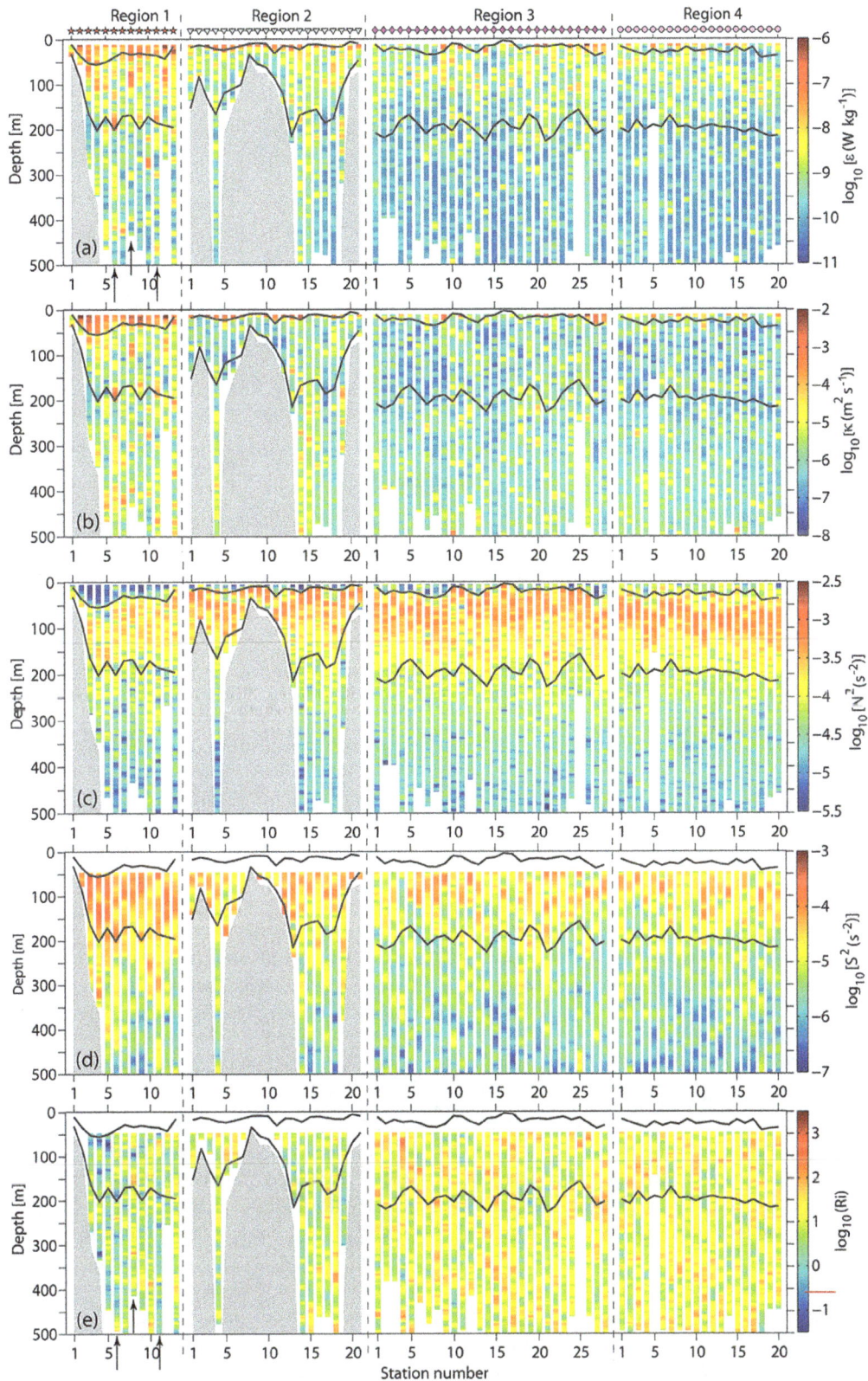

Figure 4. (a) Dissipation rate (ε), (b) diapycnal diffusivity (κ), (c) buoyancy frequency squared (N^2), (d) shear variance (S^2), and (e) Richardson number (Ri) from all of the stations. The gray shading indicates the bathymetry. In (a)–(e) the boundaries of the thermocline are indicated (gray curves). The red line on the color bar of (e) represents $Ri = 0.25$. The vertical dashed lines divide the stations into four regions with the symbols (red stars, gray triangles, magenta diamonds, and pink dots) shown at the top of (a). These symbols correspond to the station symbols in Fig. 1a.

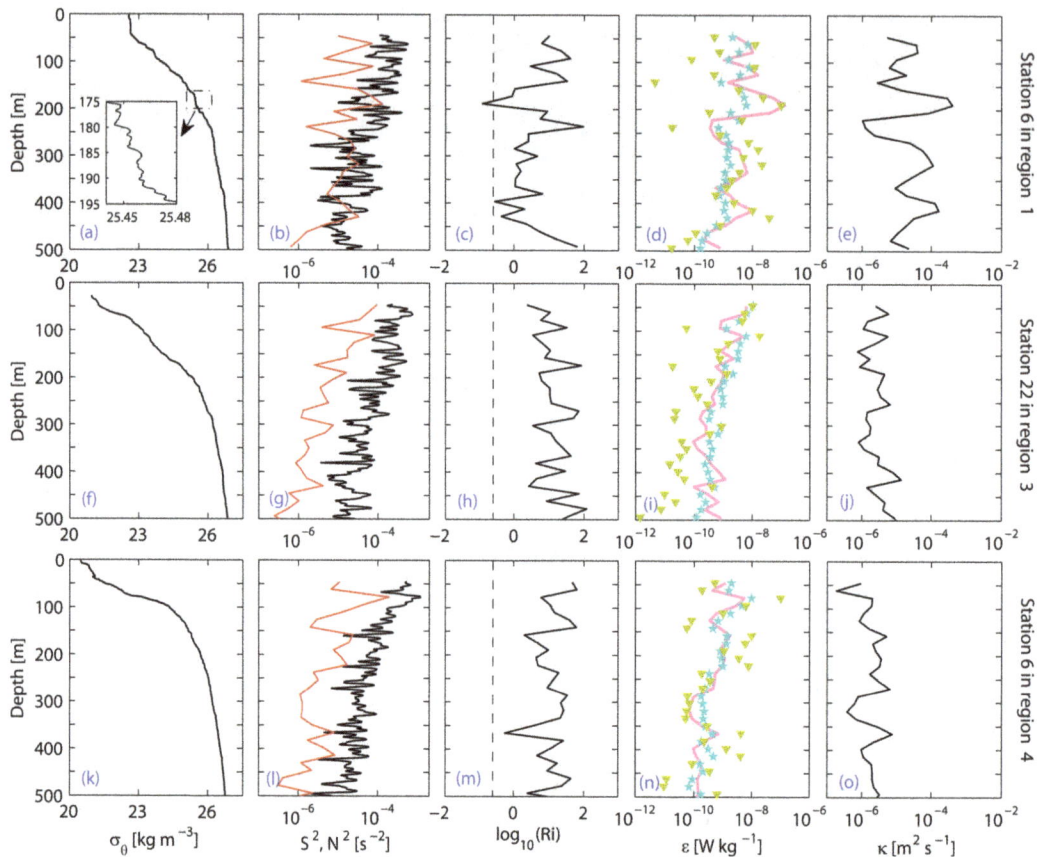

Figure 5. From top to bottom, three sets of profiles are from station 6 in region 1 (**a, b, c, d, e**), station 22 in region 3 (**f, g, h, i, j**), and station 6 in region 4 (**k, l, m, n, o**). For each station, quantities plotted are (from left to right) potential density, shear variance (red) and buoyancy frequency squared (black), Richardson number (the vertical line indicates $Ri = 0.25$), observed (pink curves) and MG model (stars) and GH model (triangles) dissipation rates, and observed diapycnal diffusivity. The observed dissipation rate and diapycnal diffusivity have been vertically averaged over the 16 m ADCP bins. The inset in (**a**) enlarges the density profile to show the overturns.

sponding depths (175–195 m) were elevated by more than 1 order of magnitude, with the diapycnal diffusivities reaching 5.0×10^{-4} $\text{m}^2\,\text{s}^{-1}$, 1 to 2 orders of magnitude higher than the levels in an open ocean thermocline. The dissipation rates induced by shear instability contributed significantly to the turbulent mixing in the water column. Nearly 45 % of the total dissipation rates in the upper 500 m (not including the surface mixed layer) was contributed by the elevated dissipation rates from the turbulent patch. The second and third sets of profiles were from region 3 (Fig. 5f–j) and region 4 (Fig. 5k–o), respectively. The buoyancy frequency squared was higher than the shear variance (Fig. 5g and l), and no Richardson numbers below 0.25 were observed (Fig. 5h and m). The water column was occupied by dissipation rates ranging from 10^{-10} to 10^{-9} $\text{W}\,\text{kg}^{-1}$ (Fig. 5i and n) and diapycnal diffusivities of 10^{-6} to 10^{-5} $\text{m}^2\,\text{s}^{-1}$ (Fig. 5j and o), comparable to the levels in an open ocean thermocline.

The above analysis indicates that high dissipation rates mainly occurred in the thermocline and the distribution of thermocline dissipation was spatially non-uniform. Turbulent

mixing in the thermocline can be driven by various factors, such as surface wind, internal waves, and internal tides. In order to find out whether the turbulent mixing in the thermocline is driven by a single forcing or multiple forcing, we explore the probability density function (PDF) of dissipation rates estimated from a non-parametric PDF estimator (histogram). The PDFs of dissipation rates (Fig. 6) in the four regions do not show sharp shapes with a single significant peak. Instead, they show flat shapes with multiple peaks, especially the PDFs in regions 1 and 4, which suggests that the turbulent mixing in the thermocline is driven by multiple forcing. To further explore the energy sources to the thermocline dissipation, we calculate the averaged dissipation rate $\langle \varepsilon \rangle_{\text{T}}$ and the averaged diapycnal diffusivity $\langle \kappa \rangle_{\text{T}}$ in the thermocline. $\langle \varepsilon \rangle_{\text{T}}$ and $\langle \kappa \rangle_{\text{T}}$ are given by

$$\langle \varepsilon \rangle_{\text{T}} = 1/(z_{\text{b}} - z_t) \int_{z_{\text{b}}}^{z_{\text{t}}} \varepsilon \, dz \quad \text{and} \quad \langle \kappa \rangle_{\text{T}} = 1/(z_{\text{b}} - z_t) \int_{z_{\text{b}}}^{z_{\text{t}}} \kappa \, dz,$$

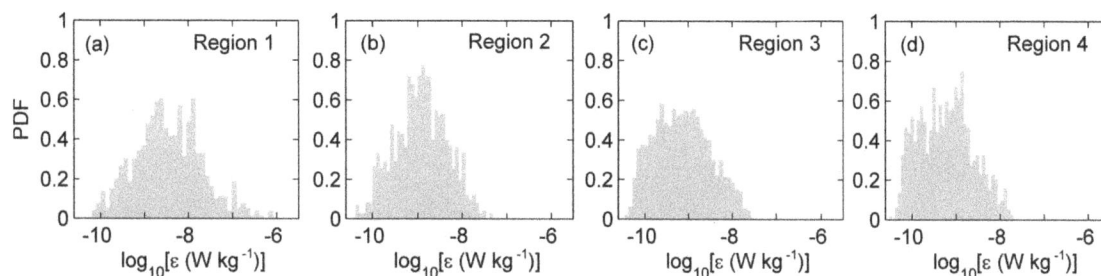

Figure 6. Probability density functions of dissipation rates in **(a)** region 1, **(b)** region 2, **(c)** region 3, and **(d)** region 4.

where z_b and z_t are the bottom and top of the thermocline, respectively. The dissipation rates and diapycnal diffusivities affected by the surface mixed layer were excluded before calculating $\langle \varepsilon \rangle_T$ and $\langle \kappa \rangle_T$. Figure 7b shows the averaged dissipation rate in the thermocline. $\langle \varepsilon \rangle_T$ decreased toward the south from O $(10^{-8}\,\mathrm{W\,kg^{-1}})$ in region 1 to O $(10^{-9}\,\mathrm{W\,kg^{-1}})$ in region 4. In region 1, $\langle \varepsilon \rangle_T$ ranged from 1.8×10^{-9} to $5.0 \times 10^{-8}\,\mathrm{W\,kg^{-1}}$, with a mean value of $1.8 \times 10^{-8}\,\mathrm{W\,kg^{-1}}$, which was 7 times, 9 times, and 12 times higher than the mean values of region 2 $(2.5 \times 10^{-9}\,\mathrm{W\,kg^{-1}})$, region 3 $(2.1 \times 10^{-9}\,\mathrm{W\,kg^{-1}})$, and region 4 $(1.5 \times 10^{-9}\,\mathrm{W\,kg^{-1}})$, respectively. Elevated $\langle \kappa \rangle_T$ was also observed in region 1 (Fig. 7c). The average $\langle \kappa \rangle_T$ in region 1 was $3.5 \times 10^{-5}\,\mathrm{m^2\,s^{-1}}$, which was an order of magnitude greater than the values of region 2 $(3.3 \times 10^{-6}\,\mathrm{m^2\,s^{-1}})$, region 3 $(2.2 \times 10^{-6}\,\mathrm{m^2\,s^{-1}})$, and region 4 $(2.1 \times 10^{-6}\,\mathrm{m^2\,s^{-1}})$. One prominent feature in the northern SCS is that the mean of $\langle \kappa \rangle_T$ in region 1 was 11 times higher than the value in region 2, while the mean of $\langle \varepsilon \rangle_T$ in region 1 was only 7 times higher than that the value in region 2. This difference mainly resulted from the weak stratification in region 1 (Fig. 4c).

Microstructure measurements at different stations were taken at different times and the measurement time might be one of the factors that affect the variability of $\langle \varepsilon \rangle_T$ and $\langle \kappa \rangle_T$. Strong turbulent mixing generally occurs during spring tides (Peters and Bokhorst, 2000). Thus it is possible that microstructure measurements in region 1 were taken during spring tides and those in regions 2–4 were taken during neap tides, and the elevated turbulent mixing in region 1 may result from a different measurement time. To rule out this possibility, we obtained the barotropic tides from the global inverse tide model (TPXO; Egbert and Erofeeva, 2002), which give us the time information of spring–neap tides during the period of observation. Only the barotropic tides at $18°$ N, $114°$ E were extracted because the bias in the arrival of spring–neap tides in different locations of the SCS is small (no longer than 3 h, not shown). The 14-day spring–neap cycles were well represented in the extracted barotropic tides (Fig. 7d). A comparison of $\langle \varepsilon \rangle_T$ and $\langle \kappa \rangle_T$ to the extracted tides suggests that elevated turbulent mixing in region 1 was not attributed to the measurement time; for example,

stations in regions 1 and 3 spanned neap and spring tides (see Fig. 7d, stars and diamonds), but the averaged $\langle \varepsilon \rangle_T$ and $\langle \kappa \rangle_T$ in region 1 were still an order of magnitude greater than the values in region 3 (Fig. 7b and c).

Surface wind is an important energy source for the turbulence in the ocean (Brainerd and Gregg, 1993; Burchard and Rippeth, 2009; Matsuno et al., 2005; Shay and Gregg, 1986), and indirectly enhances the turbulence in the thermocline through inertial-gravity wave motion generated by surface wind stress. To find out whether surface wind affects the turbulence in the thermocline significantly, we estimate the wind energy flux. The wind energy flux (Yang et al., 2014) at a height of 10 m, E_{10}, is given by $E_{10} = \rho_a C_D U_{10}^3$, where $\rho_a = 1.2\,\mathrm{kg\,m^{-3}}$ is the air density, C_D is the drag coefficient with a value of 1.14×10^{-3} (Large and Pond, 1981), and U_{10} is the wind speed at 10 m. The wind speed data during the observation come from the European Centre for Medium-Range Weather Forecasts (http://apps.ecmwf.int/datasets/data/interim-full-daily/levtype=sfc/). The variability of E_{10} is shown in Fig. 7a. Winds were light $(U_{10} < 9\,\mathrm{m\,s^{-1}})$ at all the stations with $E_{10} < 1.0\,\mathrm{W\,m^{-2}}$ except for stations 11–13 in region 1. The influence of wind stress on the variability of turbulence in the thermocline was small, as one can see from Fig. 7a–c that the variability of $\langle \varepsilon \rangle_T$ and $\langle \kappa \rangle_T$ did not follow the variability of E_{10}. The values of E_{10} $(10^{-1}\,\mathrm{W\,m^{-2}})$ at stations 1–9 in region 4 were an order of magnitude larger than that at stations 1–7 in region 1, while the values of $\langle \varepsilon \rangle_T$ and $\langle \kappa \rangle_T$ at stations 1–9 in region 4 were almost an order of magnitude smaller than that at stations 1–7 in region 1. Evidence can also be found from the comparison between $\langle \kappa \rangle_T$ and averaged wind speed during the cruises (Fig. 1b). The average winds were evenly distributed over the SCS, which is significantly different from the spatial distribution of $\langle \kappa \rangle_T$. These observations suggest that the contribution of surface winds to the observed strong turbulence in region 1 was small. Measurements from Matsuno and Wolk (2005) also indicate that the contribution of surface winds to the turbulence below the surface mixing layer was small during light winds and that only when the wind speed reached $10\,\mathrm{m\,s^{-1}}$ would wind stirring have made a notable contribution to the turbulence below the surface mixing layer.

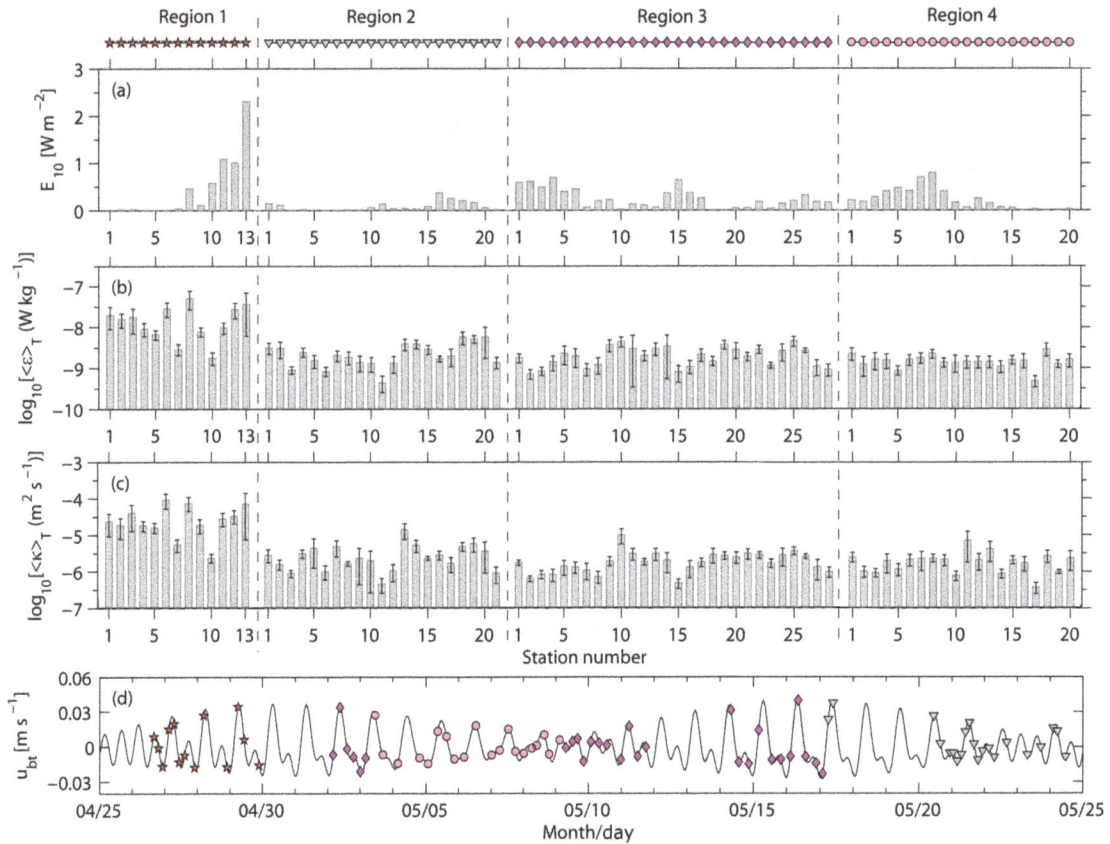

Figure 7. (a) Wind energy flux E_{10} for each station during the TurboMAP measurement. **(b)** The average dissipation rate $\langle \varepsilon \rangle_T$ and **(c)** average diapycnal diffusivity $\langle \kappa \rangle_T$ in the thermocline. The vertical bars in **(b)** and **(c)** indicate the 95 % bootstrapped confidence interval. The vertical dashed lines divide the stations into four regions with symbols (red stars, gray triangles, magenta diamonds, and pink dots) shown at the top of **(a)**. These symbols correspond to the station symbols in Fig. 1a. **(d)** Time series of the barotropic tidal velocity (u_{bt}) predicted from TPXO 7.1 with the station symbols overlain.

Internal waves and internal tides are candidates that contribute to the elevated turbulent mixing in region 1. It is known that internal waves can provide large amounts of energy for turbulence in the ocean (Alford et al., 2015). Internal waves are unevenly distributed throughout the SCS. Most of the internal waves and internal tides originate in the Luzon Strait and propagate northwestwards through the deep water zone near the Luzon Strait to the continental shelf (Guo and Chen, 2014; Klymak et al., 2006b; Lien et al., 2005; Ramp et al., 2004; Zhao, 2014; Zhao et al., 2004). Internal wave packets derived from satellite images by Zhao et al. (2004) are shown in Fig. 1b for reference. Most of the internal wave packets occurred on the continental shelf in region 1 where $\langle \kappa \rangle_T$ can be 10^{-5}–10^{-4} m^2 s^{-1}, almost an order of magnitude greater than that on the adjacent continental shelf in region 2. A report based on mooring data (Lien et al., 2014) indicates that internal waves would induce strong shear during propagation. Strong shear was also found in region 1 in our measurement (Fig. 4d). These observations suggested that internal waves and internal tides generated near the Luzon

Strait are expected to make a dominant contribution to the elevated turbulence in region 1.

3.3 Parameterizations of turbulence

In this section we evaluate two models for parameterizing the dissipation rate in terms of more easily observed or modeled quantities, such as stratification and shear. One wave–wave interaction parameterization (Gregg, 1989; MacKinnon and Gregg, 2003a) in the open ocean is the Gregg–Henyey scaling (known as the GH model) given by

$$\varepsilon_{GH} = \alpha_0 \left[f \cosh^{-1} \left(\frac{N_0}{f} \right) \right] \left(\frac{S^4}{S_{GM}^4} \right) \left(\frac{N^2}{N_0^2} \right) \text{ and}$$

$$S_{GM}^4 = \beta_0 \left(\frac{N^2}{N_0^2} \right)^2,$$

where $\alpha_0 = 1.8 \times 10^{-6}$ J kg^{-1}, f is the Coriolis frequency, S is the low-frequency/low-mode resolved shear, N_0 is a reference buoyancy frequency, $\cosh^{-1}$ denotes the inverse hyperbolic cosine function, and $\beta_0 = 1.66 \times 10^{-10}\,\mathrm{s}^{-4}$. Another analytical model (MacKinnon and Gregg, 2003a) is the MacKinnon–Gregg model (known as the MG model) given by

$$\varepsilon_{\mathrm{MG}} = \varepsilon_0 \left(\frac{N}{N_0} \right) \left(\frac{S}{S_0} \right),$$

where $S_0 = N_0 = 5.1 \times 10^{-3}\,\mathrm{rad\,s}^{-1}$ and ε_0 is an adjustable constant that gives the model dissipation rate the same cruise average as the observational data. The adjustable constant ε_0 shows great variability in different regions and seasons, spanning from 10^{-10} to more than $10^{-8}\,\mathrm{W\,kg}^{-1}$ (MacKinnon and Gregg, 2005; Palmer et al., 2008; van der Lee and Umlauf, 2011; Xie et al., 2013). This regional and temporal variability of ε_0 strongly suggests the importance of different physical processes for setup and maintenance of the background levels of turbulent dissipation. Here, we assess the two models for parameterization of the turbulence in the northern SCS (dissipation data are from the stations in region 1 and region 2), central SCS (dissipation data are from the stations in region 3), and southern SCS (dissipation data are from the stations in region 4). Different values of parameter ε_0 are selected for the parameterizations due to their different mixing backgrounds: $\varepsilon_0 = 1.65 \times 10^{-9}\,\mathrm{W\,kg}^{-1}$ for the northern SCS, $\varepsilon_0 = 0.96 \times 10^{-9}\,\mathrm{W\,kg}^{-1}$ for the central SCS, and $\varepsilon_0 = 0.50 \times 10^{-9}\,\mathrm{W\,kg}^{-1}$ for the southern SCS. All of the data affected by the surface mixed layers or bottom mixed layers were excluded for the parameterizations. To reduce the bias introduced by the different vertical resolutions of the shear and stratification data, 16 m buoyancy frequency and 16 m shear were used in the parameterization; i.e., density was first interpolated onto the ADCP grid and N^2 was computed from finite differencing. Accordingly, the dissipation rates were vertically averaged over the 16 m ADCP bins.

Figure 8 shows the distribution of dissipation rates (observed and modeled) in N^2 and S^2 space. The observed dissipation rates in the SCS (Fig. 8, left column) increase with increasing buoyancy frequency and shear. The GH model fails to reproduce these kinematic relationships (Fig. 8, right column). The dependence of $\varepsilon_{\mathrm{GH}}$ on shear is too strong, with the dissipation rates underestimated in weak shear. $\varepsilon_{\mathrm{GH}}$ also varies inversely with the buoyancy frequency for a given level of shear, contrary to the observation (Fig. 8, left column). Instead, the MG model dissipation rates (Fig. 8, middle column) display a pattern qualitatively consistent with the observed data (Fig. 8, left column). Both the observed and MG model dissipation rates scale positively with shear and the buoyancy frequency. In the northern SCS, the turbulence was more complicated than the predictions of the MG model. The MG model (Fig. 8b) underestimates the elevated dissipation rates that scattered in Fig. 8a; for example, the MG model underestimates the elevated dissipation rates at ($N^2 = 6.5 \times 10^{-5}$, $S^2 = 5.0 \times 10^{-6}\,\mathrm{s}^{-2}$), ($N^2 = 1.0 \times 10^{-4}$, $S^2 = 1.0 \times 10^{-5}\,\mathrm{s}^{-2}$), and ($N^2 = 7.9 \times 10^{-5}$, $S^2 = 2.0 \times 10^{-5}\,\mathrm{s}^{-2}$).

Figure 9 shows the dissipation rate binned in terms of stratification or shear alone. They are equivalent to integrating the two-dimensional plots in Fig. 8 horizontally and vertically. Both models reproduce the slope of the dissipation rate versus the buoyancy frequency ($\varepsilon \propto N^2$; Fig. 9a, c, and e), though the GH model dissipation rates are too large on average. However, the two models show large differences in the trend of the dissipation rate versus shear (Fig. 9b, d, and f). The MG model successfully captures the essential kinematic relationship between the dissipation rate and shear, whereas the GH model dissipation rates have a much steeper relationship with shear. Comparing the three regions, we find that the confidence intervals of the observed dissipation rates in the northern SCS (Fig. 9a and b) were wider than those in the central and southern SCS (Fig. 9c–f). In addition, the observed dissipation rates in the northern SCS were slightly larger and showed greater fluctuations than the MG model dissipation rates (Fig. 9a and b). The wide confidence intervals and high observed dissipation rates in Fig. 9a and b mainly resulted from the elevated dissipation rates scattered in Fig. 8a. The MG model underestimated these elevated dissipation rates (comparing Fig. 8a with Fig. 8b). To explore these underestimations, we directly compared the model dissipation rates with the observed dissipation rates at three selected stations (Fig. 5, fourth column). For the stations from regions 3 and 4 (Fig. 5i and n), the relationships between the observed dissipation rates (pink curves) and the GH model dissipation rates (triangles) were poor, with the GH model dissipation rates deviating from the observed data by 1 order of magnitude. Instead, the MG model dissipation rates (stars) fared better than the GH model dissipation rates against the observed data. For station 6 from region 1 (Fig. 5d), the GH model dissipation rates also failed to overlap the observed data. Instead, the MG model dissipation rates agreed quite well with the observed data, except for the elevated dissipation rates induced by shear instability; for example, the MG model underestimated the elevated dissipation rates at depths of 175 to 195 m by more than 1 order of magnitude. The elevated dissipation rates scattered in Fig. 8a mainly resulted from the dissipation rates induced by shear instability. However, the GH model dissipation rates seemed to agree with the elevated dissipation rates induced by shear instability (175–195 m). This agreement might be due to the fact that dissipation rates resulting from shear instability depend on the Richardson number, and the GH model also demonstrates Richardson number dependency.

To assess the efficacy of the models in estimating the dissipation rates, we show a direct comparison of observed dissipation rates versus modeled dissipation rates in Fig. 10. It can be seen from Fig. 10 that the MG model predicts the magnitude of the dissipation rates better than the GH model. The linear fittings of the data, $\log_{10}(\varepsilon_{\mathrm{OB}}) = \alpha \log_{10}(\varepsilon_{\mathrm{M}}) + \beta$, are also shown in Fig. 10, where ε_{M} represent the modeled dissipation rates. Parameters α and β are given in Table 2. We also test the linear regression with a t-test (Rice, 2006),

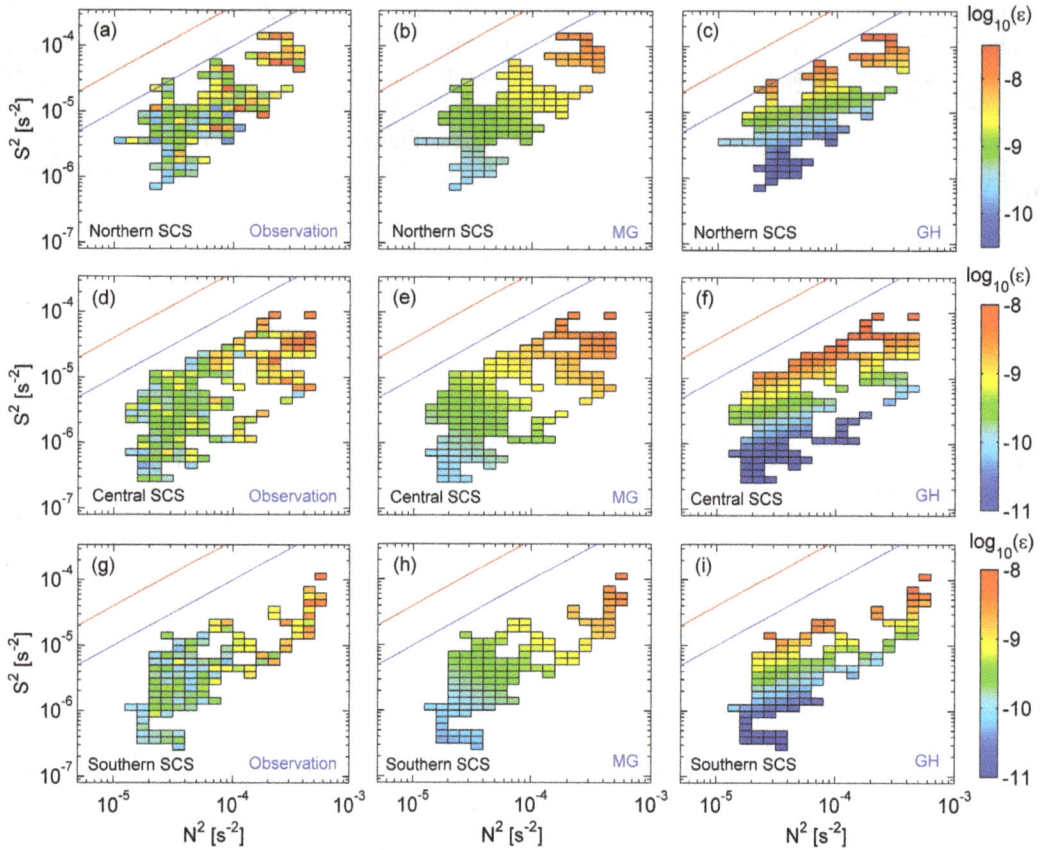

Figure 8. Dissipation rates of observation ε_{OB} (left column), MG model ε_{MG} (middle column), and GH model ε_{GH} (right column) averaged in bins of 16 m buoyancy frequency squared (N^2) and 16 m shear variance (S^2). All data affected by the surface mixed layers or bottom mixed layers were excluded. **(a)**–**(c)** show the results of the stations in the northern SCS, **(d)**–**(f)** show the results of the stations in the central SCS, and **(g)**–**(i)** show the results of the stations in the southern SCS. The boundaries of $Ri = 0.25$ (oblique red lines) and $Ri = 1$ (oblique blue lines) are shown for reference.

and the results are given in Table 2. For the MG model, the two-tailed test p-values for the northern, central, and southern SCS are 0.9966, 0.988, and 0.9651, respectively. These values are larger than 0.05, which suggests no significant difference between the observed and fitted values. However, for the GH model, the two-tailed test p-values for the northern (0.0087), central (0.0053), and southern (0.0476) SCS are smaller than 0.05, which suggests that there are significant differences between the observed and fitted values. The coefficients of determination (R^2) are also given in Table 2. R^2 is the ratio of the difference between the variance of the observed values and the variance of the residuals from the fit to the variance of the observed values (Rice, 2006). It can be interpreted as the proportion of the variability of the observed values that can be explained by the fitted values. The values of R^2 from the MG model are larger than those from the GH model, which suggests that the MG model predicts the observed data better than the GH model. For the MG model, large R^2 in the central and southern SCS suggest that the MG model works better in the central and southern SCS than in the northern SCS. Small R^2 in the northern SCS is

mainly due to the elevated dissipation rates induced by shear instability since the MG model largely underestimates these dissipation rates.

4   Discussion

Our observations indicate that turbulent mixing in the upper ocean of the SCS is spatially non-uniform, with strong turbulent mixing found in the northern SCS. This spatial pattern is consistent with the mixing distribution reported by Yang et al. (2016). Our estimates of diapycnal diffusivity ($\sim 10^{-5}\,\mathrm{m^2\,s^{-1}}$) in region 1 are similar to those ($\sim 10^{-5}\,\mathrm{m^2\,s^{-1}}$) of Tian et al. (2009) but almost 2 orders of magnitude smaller than those ($\sim 10^{-3}\,\mathrm{m^2\,s^{-1}}$) reported by Yang et al. (2016); these different values might be attributed to various factors such as estimation methods and observation seasons. Diapycnal diffusivities from Tian et al. (2009) and Yang et al. (2016) were estimated with Gregg–Henyey–Polzin parameterizations, which depends on reference dissipation. Different reference dissipations chosen in the param-

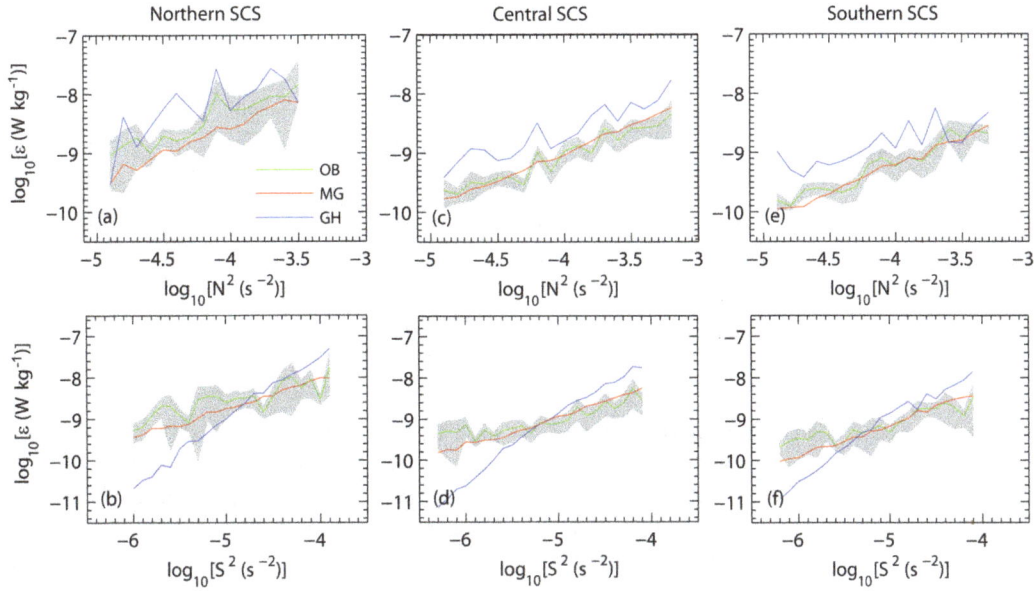

Figure 9. Average dissipation rate calculated in bins of buoyancy frequency squared (N^2) and shear variance (S^2) for the northern SCS (**a–b**), central SCS (**c–d**), and southern SCS (**e–f**). The green, red, and blue curves are the results of the observation, MG model, and GH model, respectively. The grey shading indicates the 95 % bootstrapped confidence interval for the observed dissipation rates.

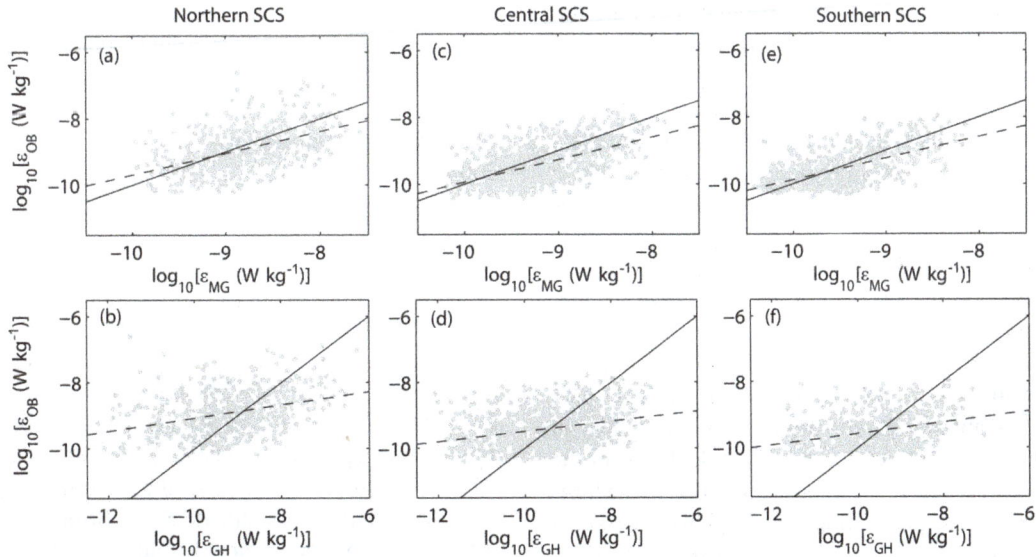

Figure 10. Observed dissipation (ε_{OB}) plotted against modeled (top) MG and (bottom) GH dissipation for the northern SCS (**a–b**), central SCS (**c–d**), and southern SCS (**e–f**). The solid lines indicate the one-to-one relation: $\log_{10}(\varepsilon_{OB}) = \log_{10}(\varepsilon_{GII})$ or $\log_{10}(\varepsilon_{OB}) = \log_{10}(\varepsilon_{MG})$. The dash lines indicate the linear fittings of the data.

eterization can make the estimated diapycnal diffusivity different. In addition, the data used in the parameterization of Yang et al. (2016) span from 2005 to 2012 and cover all the year round, while the microstructure data in our observation just cover 1 month. Seasonal and inter-annual variations of internal waves in the SCS (Huang et al., 2008; Yang et al., 2009) might affect the turbulent mixing.

The GH model and the MG model were derived from the eikonal model of Henyey et al. (1986) which is applicable to parameterize the dissipation controlled by wave–wave interactions that transfer energy from large-scale waves to small-scale waves (MacKinnon and Gregg, 2005). The GH model is based on the assumption that the waves are statistically stationary, with the energy of small-scale waves and the shear of the large-scale waves maintaining a particular relation-

Table 2. Results of the linear regression and t-test.

Location	Model	α	β	t value	Sig. (two-tailed)	R^2
Northern SCS	MG	0.6517	−3.178	0.0042	0.9966	0.2373
	GH	0.2	−7.0805	−2.628	0.0087	0.1098
Central SCS	MG	0.6721	−3.2202	−0.0151	0.988	0.3352
	GH	0.1546	−7.9507	−2.7892	0.0053	0.1267
Southern SCS	MG	0.646	−3.421	−0.0437	0.9651	0.3799
	GH	0.1724	−7.8608	−1.9827	0.0476	0.1687

ship through the Garrett–Munk (GM) spectrum (Garrett and Munk, 1975). It is typically evaluated for the internal wave field with the GM spectral shape (Gregg, 1989). The MG model was first proposed by MacKinnon and Gregg (2003) to parameterize the turbulence over the continental shelf. It is found to be suitable for the wave field of the continental shelf in which the energy and shear are dominated by the near-inertial motions, internal tides, or low-frequency internal waves (MacKinnon and Gregg, 2003a; Palmer et al., 2008; van der Lee and Umlauf, 2011). Recently it was found that the MG model also successfully parameterizes the turbulent mixing in the upper layer of the deep sea (Xie et al., 2013).

Statistical analysis shows the dissipation rates in the SCS to be proportional to both the shear and buoyancy frequencies, in marked contrast to the predictions of the GH model, but consistent with the predictions of the MG model. The disagreement of the GH model might be associated with the wave field in the SCS. Previous studies (Polzin et al., 1995; Wijesekera et al., 1993) have indicated that the predictions of the GH model would exhibit departure from the observed dissipation by more than 1 order of magnitude in regions where the wave field deviates from the GM spectrum. Thus, it is appropriate to examine the wave field in the SCS. Data obtained from six moorings deployed in the SCS (Fig. 1a, yellow squares) were used to estimate the horizontal kinetic energy spectra. Though the data were obtained from different periods, they reflected the main characteristics of the wave field in the SCS. The spectra (Fig. 11) show significant peaks in the local inertial (f) and tidal frequencies (diurnal O_1 and K_1; semidiurnal M_2); these peaks imply that energy was primarily dominated by the near-inertial motions and internal tides. Within the internal wave band, significant peaks were also observed at higher tidal harmonic frequencies such as D_3, D_4, and D_5 (respectively, about three, four, and five cycles per day). These higher tidal harmonic frequencies mainly result from nonlinear interaction between internal waves (van Haren, 2003; van Haren et al., 2002; Xie et al., 2010). These energetic internal tides and harmonic internal waves cannot be well described by the GM spectrum. Furthermore, the spectra deviated from the GM spectrum at

high frequencies ($\sigma > 3$ cpd), which is especially evident in the spectra of the moorings from the northern SCS (mooring 1) and southern SCS (mooring 3). These observations are not supportive of the assumption that the GH model is based on. In contrast, some of our observations support the MG model, such as the wave field being dominated by near-inertial waves and internal tides, and the dissipation rates scale positively with shear and stratification. Overall, the MG model succeeds in parameterizing the turbulence in the SCS, except for some elevated dissipation rates induced by shear instability. The MG model tends to underestimate these elevated dissipation rates. This is not surprising because the MG model, which is based on wave–wave interactions, represents bulk averages of turbulent properties and does not reproduce individual shear instability events (MacKinnon and Gregg, 2005).

5   Summary

We analyzed observations of turbulent dissipation and mixing in the SCS with microstructure data obtained from 26 April to 23 May 2010. The observations are divided into four regions: region 1 is located to the west of the Luzon Strait, region 2 is located to the northeast of Hainan Island, region 3 is located in the central SCS, and region 4 is located in the southern SCS. Strong turbulent mixing was observed in region 1, with the mean $\langle \varepsilon \rangle_T$ reaching 1.8×10^{-8} W kg^{-1}, which is 9 times and 12 times larger than the values in the central (2.1×10^{-9} W kg^{-1}) and southern (1.5×10^{-9} W kg^{-1}) SCS, respectively. Elevated $\langle \kappa \rangle_T$ were also found in region 1, i.e., O (10^{-5} m^2 s^{-1}), which is almost an order of magnitude higher than the values of the central and southern SCS. The turbulent mixing in different regions displays different mixing features, to which shear variance and stratification have made a significant contribution. In region 1, the shear was stronger and the stratification was weaker than those in other regions. Shear instability events occasionally occurred in these conditions and produced elevated dissipation and diapycnal diffusivity. Although the turbulent patches induced by shear instability were occasional and sparse, they significantly contributed to the tur-

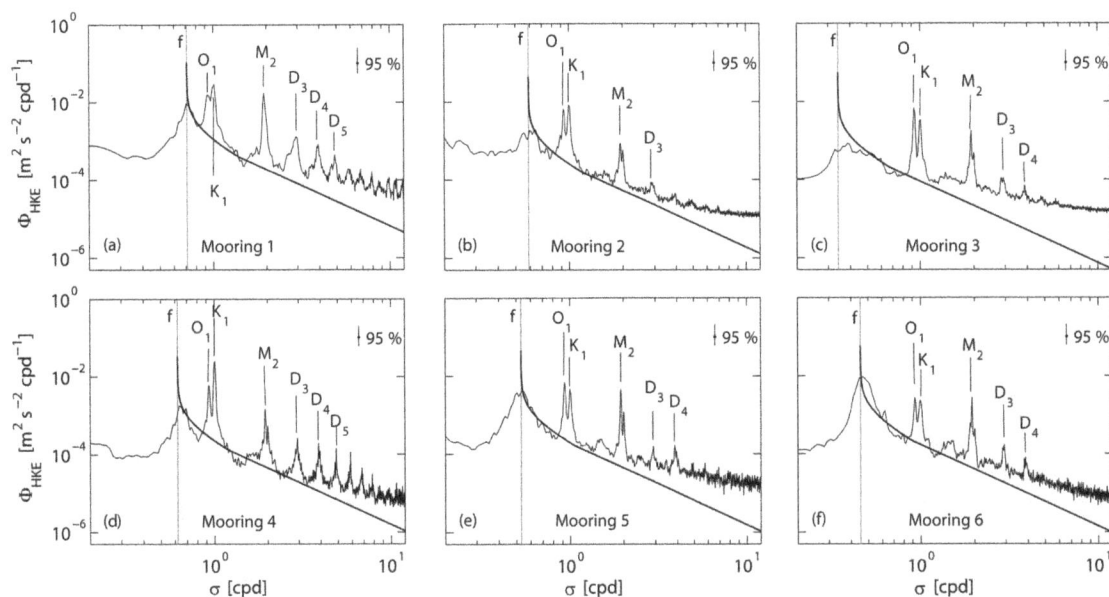

Figure 11. (a–f) Rotary spectra (clockwise plus counterclockwise) of horizontal kinetic energy for the six moorings deployed in the SCS (1 cpd = 1 day^{-1}). Spectra are averaged over $z \in [60:270]$ m. The canonical Garrett and Munk spectrum is shown for reference (smooth curve). The vertical lines represent various frequencies (f, O_1, K_1...). The 95 % statistical significance level is indicated by the vertical bar in the upper-right corner.

bulent mixing in the water column. In regions 2–4, the water column was characterized by weak shear and strong stratification. Shear was no longer sufficient to produce subcritical Richardson numbers and the turbulence was weak. The strong spatial correlation between high dissipation rates and strong shear presented in the thermocline in region 1 suggests that shear was one of the important drivers of the elevated turbulent mixing. The analysis of surface winds, internal waves, and barotropic tides indicates that the spatial distribution of turbulent mixing with elevated dissipation rates and diapycnal diffusivity concentrated in region 1 does not result from the measurement time or surface winds. The energetic internal waves and internal tides generated near the Luzon Strait are expected to make a dominant contribution to create this mixing pattern. Unfortunately, we have only one profile of microstructure measurement and short time series of current velocity obtained by the shipboard ADCP for each station; thus, it is impossible to separate the internal waves of various frequencies and explore their respective contributions to the dissipation. In order to resolve the internal waves in various frequencies, a long time series of fine-scale current velocities is required. We suggest further observations be done with frequent microstructure measurements and long time series of current velocity measurements to identify the dominant mixing mechanism in the northern SCS.

To predict realistic climate and circulation, mixing must be accurately represented in ocean models. Mapping of the dissipation rates throughout the ocean is a daunting task. However, this task can be made considerably easier if mixing can be estimated from more easily observed or modeled quan-

tities, such as shear, stratification, and latitude. Two models (the GH model and MG model) were evaluated for parameterizing the dissipation rate in the SCS. Statistical analysis shows the dissipation in the SCS to be proportional to both the shear and buoyancy frequencies, in marked contrast to the predictions of the GH model, but consistent with the predictions of the MG model. The replication of the turbulence behavior greatly depends on the correct choice of model and appropriate tuning of the free parameters. The resolution of the shear and stratification is another factor in determining the success of models in parameterizing the turbulence (Palmer et al., 2013). Although the MG model can reproduce the dissipation in the SCS for our chosen vertical resolution (16 m), whether the distribution of the observed dissipation would change with finer resolution of shear and stratification is still an open problem. However, at least on the scale of internal waves (16 m), the MG model is clearly a better model than the GH model for the parameterization of turbulence in the upper ocean of the SCS, which provides a useful reference for modelers. Additional data with higher resolution are required to robustly fix this model in the near future.

Author contributions. Xiao-Dong Shang and Gui-Ying Chen designed and carried out the experiments. Chang-Rong Liang prepared the manuscript with contributions from all co-authors.

Competing interests. The authors declare that they have no conflict of interest.

Acknowledgements. This work is supported by the National Natural Science Foundation of China: 41630970, 41376022, 41676022, and 41521005. The data we used are from the South China Sea Institute of Oceanology.

Edited by: John M. Huthnance

References

Alford, M. H., Peacock, T., MacKinnon, J. A., Nash, J. D., Buijsman, M. C., Centuroni, L. R., Chao, S. Y., Chang, M. H., Farmer, D. M., Fringer, O. B., Fu, K. H., Gallacher, P. C., Graber, H. C., Helfrich, K. R., Jachec, S. M., Jackson, C. R., Klymak, J. M., Ko, D. S., Jan, S., Johnston, T. M. S., Legg, S., Lee, I. H., Lien, R. C., Mercier, M. J., Moum, J. N., Musgrave, R., Park, J. H., Pickering, A. I., Pinkel, R., Rainville, L., Ramp, S. R., Rudnick, D. L., Sarkar, S., Scotti, A., Simmons, H. L., St Laurent, L. C., Venayagamoorthy, S. K., Wang, Y. H., Wang, J., Yang, Y. J., Paluszkiewicz, T., and Tang, T. Y.: The formation and fate of internal waves in the South China Sea, Nature, 521, 65–69, https://doi.org/10.1038/nature14399, 2015.

Brainerd, K. E. and Gregg, M. C.: Diurnal Restratification and Turbulence in the Oceanic Surface Mixed-Layer: 1. Observations, J. Geophys. Res.-Oceans, 98, 22645–22656, https://doi.org/10.1029/93jc02297, 1993.

Burchard, H. and Rippeth, T. P.: Generation of Bulk Shear Spikes in Shallow Stratified Tidal Seas, J. Phys. Oceanogr., 39, 969–985, 2009.

Carter, G. S. and Gregg, M. C.: Intense, variable mixing near the head of Monterey Submarine Canyon, J. Phys. Oceanogr., 32, 3145–3165, 2002.

Carter, G. S., Gregg, M. C., and Merrifield, M. A.: Flow and mixing around a small seamount on Kaena Ridge, Hawaii, J. Phys. Oceanogr., 36, 1036–1052, https://doi.org/10.1175/Jpo2924.1, 2006.

Chang, M. H., Lien, R. C., Tang, T. Y., D'Asaro, E. A., and Yang, Y. J.: Energy flux of nonlinear internal waves in northern South China Sea, Geophys. Res. Lett., 33, 155–170, https://doi.org/10.1029/2005GL025196, 2006.

Egbert, G. D. and Erofeeva, S. Y.: Efficient inverse modeling of barotropic ocean tides, J. Atmos. Ocean. Tech., 19, 183–204, 2002.

Garrett, C. and Munk, W.: Space-Time Scales of Internal Waves - Progress Report, J. Geophys. Res., 80, 291–297, https://doi.org/10.1029/Jc080i003p00291, 1975.

Gregg, M. C.: Scaling Turbulent Dissipation in the Thermocline, J. Geophys. Res.-Oceans, 94, 9686–9698, https://doi.org/10.1029/Jc094ic07p09686, 1989.

Gregg, M. C.: Estimation and geography of diapycnal mixing in the stratified ocean, Coast. Estuar. Stud., 54, 305–338, 1998.

Guo, C. and Chen, X.: A review of internal solitary wave dynamics in the northern South China Sea, Prog. Oceanogr., 121, 7–23, https://doi.org/10.1016/j.pocean.2013.04.002, 2014.

Henyey, F. S., Wright, J., and Flatte, S. M.: Energy and Action Flow through the Internal Wave Field – an Eikonal Approach, J. Geophys. Res.-Oceans, 91, 8487–8495, https://doi.org/10.1029/Jc091ic07p08487, 1986.

Holt, J. and Umlauf, L.: Modelling the tidal mixing fronts and seasonal stratification of the Northwest European Continental shelf, Cont. Shelf Res., 28, 887–903, 2008.

Huang, W., Johannessen, J., Alpers, W., Yang, J., and Gan, X.: Spatial and temporal variations of internal wave sea surface signatures in the northern South China Sea studied by spaceborne SAR imagery, in Proc. SeaSAR, Frascati, Italy, 21–25 January, 1–6, 2008.

Jing, Z. and Wu, L.: Seasonal variation of turbulent diapycnal mixing in the northwestern Pacific stirred by wind stress, Geophys. Res. Lett., 37, 137–139, 2010.

Klymak, J. M., Moum, J. N., Nash, J. D., Kunze, E., Girton, J. B., Carter, G. S., Lee, C. M., Sanford, T. B., and Gregg, M. C.: An estimate of tidal energy lost to turbulence at the Hawaiian Ridge, J. Phys. Oceanogr., 36, 1148–1164, https://doi.org/10.1175/Jpo2885.1, 2006a.

Klymak, J. M., Pinkel, R., Liu, C. T., Liu, A. K., and David, L.: Prototypical solitons in the South China Sea, Geophys. Res. Lett., 33, L11607, https://doi.org/10.1029/2006gl025932, 2006b.

Kunze, E. and Toole, J. M.: Tidally driven vorticity, diurnal shear, and turbulence atop Fieberling Seamount, J. Phys. Oceanogr., 27, 2663–2693, https://doi.org/10.1175/1520-0485(1997)027<2663:TDVDSA>2.0.CO;2, 1997.

Large, W. G. and Pond, S.: Open Ocean Momentum Flux Measurements in Moderate to Strong Winds, J. Phys. Oceanogr., 11, 324–336, https://doi.org/10.1175/1520-0485(1981)011<0324:Oomfmi>2.0.Co;2, 1981.

Laurent, L. S.: Turbulent dissipation on the margins of the South China Sea, Geophys. Res. Lett., 35, L23615, https://doi.org/10.1029/2008gl035520, 2008.

Ledwell, J. R., Montgomery, E. T., Polzin, K. L., St Laurent, L. C., Schmitt, R. W., and Toole, J. M.: Evidence for enhanced mixing over rough topography in the abyssal ocean, Nature, 403, 179–182, https://doi.org/10.1038/35003164, 2000.

Lee, C. M., Kunze, E., Sanford, T. B., Nash, J. D., Merrifield, M. A., and Holloway, P. E.: Internal tides and turbulence along the 3000 m isobath of the Hawaiian Ridge, J. Phys. Oceanogr., 36, 1165–1183, https://doi.org/10.1175/Jpo2886.1, 2006.

Lien, R. C., Tang, T. Y., Chang, M. H., and D'Asaro, E. A.: Energy of nonlinear internal waves in the South China Sea, Geophys. Res. Lett., 32, L05615, https://doi.org/10.1029/2004GL022012, 2005.

Lien, R. C., Henyey, F., Ma, B., and Yang, Y. J.: Large-Amplitude Internal Solitary Waves Observed in the Northern South China Sea: Properties and Energetics, J. Phys. Oceanogr., 44, 1095–1115, https://doi.org/10.1175/Jpo-D-13-088.1, 2014.

Lozovatsky, I., Liu, Z., Fernando, H. J. S., Hu, J., and Wei, H.: The TKE dissipation rate in the northern South China Sea, Ocean Dynam., 63, 1189–1201, 2013.

Lueck, R. G. and Mudge, T. D.: Topographically induced mixing around a shallow seamount, Science, 276, 1831–1833, https://doi.org/10.1126/science.276.5320.1831, 1997.

MacKinnon, J. A. and Gregg, M. C.: Mixing on the late-summer New England shelf – Solibores, shear, and stratification, J. Phys. Oceanogr., 33, 1476–1492, https://doi.org/10.1175/1520-0485(2003)033<1476:MOTLNE>2.0.CO;2, 2003a.

MacKinnon, J. A. and Gregg, M. C.: Shear and baroclinic energy flux on the summer New England shelf, J. Phys. Oceanogr., 33, 1462–1475, https://doi.org/10.1175/1520-0485(2003)033<1462:Sabefo>2.0.Co;2, 2003b.

MacKinnon, J. A. and Gregg, M. C.: Spring mixing: Turbulence and internal waves during restratification on the New England shelf, J. Phys. Oceanogr., 35, 2425–2443, https://doi.org/10.1175/Jpo2821.1, 2005.

Matsuno, T. and Wolk, F.: Observations of turbulent energy dissipation rate epsilon in the Japan Sea, Deep-Sea Res. Pt. II, 52, 1564–1579, https://doi.org/10.1016/j.dsr2.2004.06.037, 2005.

Matsuno, T., Shimizu, M., Morii, Y., Nishida, H., and Takaki, Y.: Measurements of the turbulent energy dissipation rate around the shelf break in the East China Sea, J. Oceanogr., 61, 1029–1037, https://doi.org/10.1007/s10872-006-0019-9, 2005.

Munk, W. H.: Abyssal recipes, paper presented at Deep Sea Research and Oceanographic Abstracts, Elsevier, 13, 707–730, 1966.

Nasmyth, P. W.: Oceanic turbulence, Retrospective Theses and Dissertations, 1919–2007, 1970.

Oakey, N. S.: Determination of the Rate of Dissipation of Turbulent Energy from Simultaneous Temperature and Velocity Shear Microstructure Measurements, J. Phys. Oceanogr., 12, 256–271, https://doi.org/10.1175/1520-0485(1982)012, 1982.

Osborn, T. R.: Estimates of the Local-Rate of Vertical Diffusion from Dissipation Measurements, J. Phys. Oceanogr., 10, 83–89, https://doi.org/10.1175/1520-0485(1980)010<0083:EOTLRO>2.0.CO;2, 1980.

Palmer, M. R., Rippeth, T. P., and Simpson, J. H.: An investigation of internal mixing in a seasonally stratified shelf sea, J. Geophys. Res.-Oceans, 113, C12005, https://doi.org/10.1029/2007jc004531, 2008.

Palmer, M. R., Polton, J. A., Inall, M. E., Rippeth, T. P., Green, J. A. M., Sharples, J., and Simpson, J. H.: Variable behavior in pycnocline mixing over shelf seas, Geophys. Res. Lett., 40, 161–166, https://doi.org/10.1029/2012gl054638, 2013.

Peters, H. and Bokhorst, R.: Microstructure observations of turbulent mixing in a partially mixed estuary. Part I: Dissipation rate, J. Phys. Oceanogr., 30, 1232–1244, 2000.

Polzin, K. L., Toole, J. M., and Schmitt, R. W.: Finescale Parameterizations of Turbulent Dissipation, J. Phys. Oceanogr., 25, 306–328, https://doi.org/10.1175/1520-0485(1995)025<0306:FPOTD>2.0.CO;2, 1995.

Qu, T., Mitsudera, H., and Yamagata, T.: Intrusion of the North Pacific waters into the South China Sea, J. Geophys. Res.-Oceans, 105, 6415–6424, https://doi.org/10.1029/1999jc900323, 2000.

Ramp, S. R., Tang, T. Y., Duda, T. F., Lynch, J. F., Liu, A. K., Chiu, C. S., Bahr, F. L., Kim, H. R., and Yang, Y. J.: Internal solitons in the northeastern South China Sea. Part I: Sources and deep water propagation, IEEE J. Oceanic Eng., 29, 1157–1181, https://doi.org/10.1109/JOE.2004.840839, 2004.

Rice, J. A.: Mathematical Statistics and Data Analysis, Nelson Education, 421–425, 2006.

Rippeth, T. P.: Mixing in seasonally stratified shelf seas: a shifting paradigm, Philosophical Transactions, 363, 2837, 2005.

Saenko, O. A. and Merryfield, W. J.: On the effect of topographically enhanced mixing on the global ocean circulation, J. Phys. Oceanogr., 35, 826–834, https://doi.org/10.1175/Jpo2722.1, 2005.

Sandstrom, H. and Elliott, J. A.: Internal Tide and Solitons on the Scotian Shelf – a Nutrient Pump at Work, J. Geophys. Res.-Oceans, 89, 6415–6426, https://doi.org/10.1029/Jc089ic04p06415, 1984.

Shaw, P. T.: The Seasonal-Variation of the Intrusion of the Philippine Sea-Water into the South China Sea, J. Geophys. Res.-Oceans, 96, 821–827, https://doi.org/10.1029/90jc02367, 1991.

Shay, T. J. and Gregg, M. C.: Convectively Driven Turbulent Mixing in the Upper Ocean, J. Phys. Oceanogr., 16, 1777–1798, https://doi.org/10.1175/1520-0485(1986)016<1777:Cdtmit>2.0.Co;2, 1986.

Simpson, J. H. and Bowers, D.: Models of stratification and frontal movement in shelf seas, Deep-Sea Res. Pt. I, 28, 727–738, 1981.

Tian, J. W., Yang, Q. X., and Zhao, W.: Enhanced Diapycnal Mixing in the South China Sea, J. Phys. Oceanogr., 39, 3191–3203, https://doi.org/10.1175/2009jpo3899.1, 2009.

Tsujino, H., Hasumi, H., and Suginohara, N.: Deep Pacific Circulation Controlled by Vertical Diffusivity at the Lower Thermocline Depths, J. Phys. Oceanogr., 30, 2853–2865, 2000.

van der Lee, E. M. and Umlauf, L.: Internal wave mixing in the Baltic Sea: Near-inertial waves in the absence of tides, J. Geophys. Res.-Oceans, 116, C10016, https://doi.org/10.1029/2011jc007072, 2011.

van Haren, H.: On the polarization of oscillatory currents in the Bay of Biscay, J. Geophys. Res.-Oceans, 108, 3290, https://doi.org/10.1029/2002jc001736, 2003.

van Haren, H., Maas, L., and van Aken, H.: On the nature of internal wave spectra near a continental slope, Geophys. Res. Lett., 29, 1615, https://doi.org/10.1029/2001gl014341, 2002.

Wijesekera, H., Padman, L., Dillon, T., Levine, M., Paulson, C., and Pinkel, R.: The Application of Internal-Wave Dissipation Models to a Region of Strong Mixing, J. Phys. Oceanogr., 23, 269–286, 1993.

Wolk, F., Yamazaki, H., Seuront, L., and Lueck, R. G.: A new free-fall profiler for measuring biophysical microstructure, J. Atmos. Ocean. Tech., 19, 780–793, https://doi.org/10.1175/1520-0426(2002)019<0780:ANFFPF>2.0.CO;2, 2002.

Wu, C. R. and Hsin, Y. C.: The forcing mechanism leading to the Kuroshio intrusion into the South China Sea, J. Geophys. Res.-Oceans, 117, 9, https://doi.org/10.1029/2012jc007968, 2012.

Wu, L. X., Jing, Z., Riser, S., and Visbeck, M.: Seasonal and spatial variations of Southern Ocean diapycnal mixing from Argo profiling floats, Nat. Geosci., 4, 363–366, https://doi.org/10.1038/Ngeo1156, 2011.

Xie, X. H., Shang, X. D., and Chen, G. Y.: Nonlinear interactions among internal tidal waves in the northeastern South China Sea, Chin. J. Oceanol. Limn., 28, 996–1001, https://doi.org/10.1007/s00343-010-9064-8, 2010.

Xie, X. H., Cuypers, Y., Bouruet-Aubertot, P., Ferron, B., Pichon, A., Lourenco, A., and Cortes, N.: Large-amplitude internal tides, solitary waves, and turbulence in the central Bay of Biscay, Geophys. Res. Lett., 40, 2748–2754, https://doi.org/10.1002/Grl.50533, 2013.

Yang, Q. X., Tian, J. W., Zhao, W., Liang, X. F., and Zhou, L.: Observations of turbulence on the shelf and slope of northern South China Sea, Deep-Sea Res. Pt. I, 87, 43–52, https://doi.org/10.1016/j.dsr.2014.02.006, 2014.

Yang, Q. X., Zhao, W., Liang, X., and Tian, J.: Three-Dimensional Distribution of Turbulent Mixing in the South China Sea, J. Phys. Oceanogr., 46, 769–788, https://doi.org/10.1175/JPO-D-14-0220.1, 2016.

6

Synoptic fluctuation of the Taiwan Warm Current in winter on the East China Sea shelf

Jiliang Xuan[1], **Daji Huang**[1,2], **Thomas Pohlmann**[3], **Jian Su**[3], **Bernhard Mayer**[3], **Ruibin Ding**[2,1], **and Feng Zhou**[1,2]

[1]State Key Laboratory of Satellite Ocean Environment Dynamics, Second Institute of Oceanography, State Oceanic Administration, Hangzhou, China
[2]Ocean College, Zhejiang University, Zhoushan, China
[3]Institute of Oceanography, University of Hamburg, Hamburg, Germany

Correspondence to: Jiliang Xuan (xuanjl@sio.org.cn) and Daji Huang (djhuang@sio.org.cn)

Abstract. The seasonal mean and synoptic fluctuation of the wintertime Taiwan Warm Current (TWC) were investigated using a well-validated finite volume community ocean model. The spatial distribution and dynamics of the synoptic fluctuation were highlighted. The seasonal mean of the wintertime TWC has two branches: an inshore branch between the 30 and 100 m isobaths and an offshore branch between the 100 and 200 m isobaths. The Coriolis term is much larger than the inertia term and is almost balanced by the pressure gradient term in both branches, indicating geostrophic balance of the mean current. Two areas with significant fluctuations of the TWC were identified during wintertime. One of the areas is located to the north of Taiwan with velocities varying in the cross-shore direction. These significant cross-shore fluctuations are driven by barotropic pressure gradients associated with the intrusion of the Taiwan Strait Current (TSC). When a strong TSC intrudes to the north of Taiwan, the isobaric slope tilts downward from south to north, leading to a cross-shore current from the coastal area to the offshore area. When the TSC intrusion is weak, the cross-shore current to the north of Taiwan is directed from offshore to inshore. The other area of significant fluctuation is located in the inshore area between the 30 and 100 m isobaths. The fluctuations are generally strong both in the alongshore and cross-shore directions, in particular at the latitudes 26.5 and 28° N. Wind affects the synoptic fluctuation through episodic events. When the northeasterly monsoon prevails, the southwestward Zhe-Min coastal current dominates the inshore area associated with a deepening of the mixed layer. When the winter monsoon is weakened or the southwesterly wind prevails, the northeastward TWC dominates in the inshore area.

1 Introduction

On the East China Sea (ECS) shelf, the mean path of the Taiwan Warm Current (TWC) has two branches: the inshore branch along the 50 m isobath and the offshore branch along the 100 m isobath (Su and Pan, 1987). The summer TWC has been well studied because the current is stationary and strong, with an average speed of $0.3 \, \mathrm{m \, s^{-1}}$ (Guan, 1978; Fang et al., 1991; Isobe, 2008; Yang et al., 2011, 2012). The spatial structure and temporal variation of the wintertime (December to March) TWC are less known due to its weak mean surface velocity, according to a climatological structure of the surface current in the ECS mapped by Qiu and Imasato (1990).

The wintertime TWC on the ECS shelf shows synoptic fluctuations (Cui et al., 2004; Zhu et al., 2004; Zeng et al., 2012; Huang et al., 2016). These synoptic fluctuations show some features common to those over other continental shelves; i.e., they have periods between 3 and 15 days and are associated with coastal sea level changes, which can be explained by local winds or by coastal trapped waves (Huyer, 1990; Brink, 1991; Huthnance et al., 1986). Huang et al. (2016) showed that the wind was a main physical factor, which caused the temporal variation of the wintertime currents at the synoptic scale in the coastal area of the ECS. However, the dominant physical factors of the TWC fluctu-

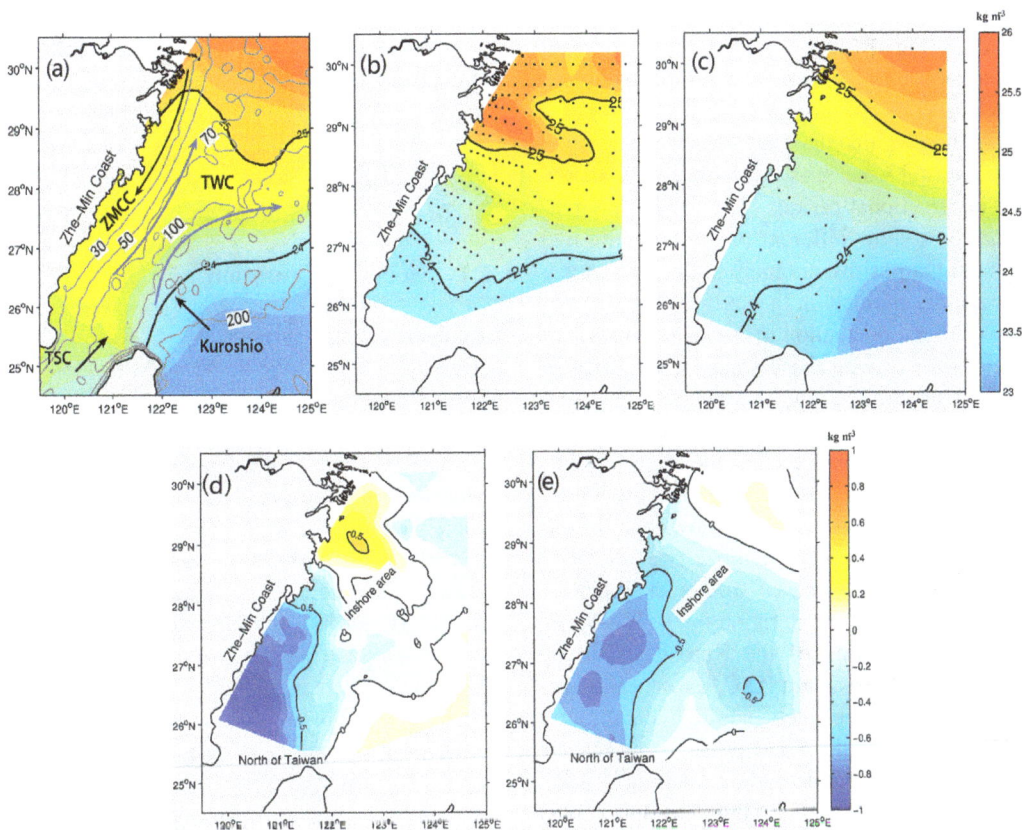

Figure 1. Density (σ_t, kg m^3) distributions at 50 m depth derived from the GDEM climatological data in February (**a**), an ocean survey from 1 to 27 February 2007 (**b**), an ocean survey from 3 to 16 February 2007 (**c**) and the density anomalies between the GDEM data and the two surveys (**d, e**). The two blue arrows indicate the two TWC branches in winter. The 30, 50, 70, 100 and 200 m isobaths are indicated with gray lines in (**a**).

ations still lack study; the fluctuations on the whole shelf of the ECS may be complicated due to the complex bottom topography, alternating wind forcing and conjunction of several current systems such as the Kuroshio Current, the Taiwan Strait Current (TSC) and the Zhe-Min Coastal Current (ZMCC). These synoptic fluctuations are also known to influence the regional material transport, especially when the amplitude of the fluctuations is comparable to, or even larger than, the mean current. On the ECS shelf, some recent observations have shown that the TWC has an episodic wintertime feature (Zhu et al., 2004) and the variations of the TWC in winter have an amplitude as large as 0.2 m s^{-1} (Zeng et al., 2012). Moreover, it has been observed that the variations of the TWC in winter cause a cross-shore current, which is closely linked to the alongshore component (Huang et al., 2016). Therefore, we focus on studying the spatial patterns of synoptic fluctuations to better understand the role of the wintertime TWC on the cross-shore water exchange.

A comparison between the wintertime climatological density (Fig. 1a) and synoptic density distributions observed during two surveys (Fig. 1b and c) suggests that two distinct areas with significant synoptic fluctuations exist. The clima-

tological density is taken from the Generalized Digital Environment Model (GDEM; Carnes, 2009) data, and the two surveys were carried out in February 2007 by two research vessels. Because the isopycnal lines are closely related to geostrophic currents, we can infer the strength of the TWC from the horizontal gradient of the isopycnals between $24\sigma_t$ and $25\sigma_t$ contours (Fig. 1a). This accounts for the fact that in winter the water mass of TWC is located in this density range (according to the hydrography analysis of Su et al., 1994). The two-branch structure of the TWC can be inferred from the wintertime climatological density. In this paper, we defined that the near-coast area is the area between the coast and 30 m isobath where the ZMCC occurs, the inshore area is the area between the 30 and 100 m isobaths where the TWC inshore branch dominates and the offshore area is the region between the 100 and 200 m isobaths where the TWC offshore branch prevails. According to the hydrographic data analysis and numerical interpretation by Su and Pan (1987), the TWC inshore and offshore branches mainly occur close to those specific isobaths. However, these two branches were missing during the two synoptic surveys (Fig. 1b and c), indicating strong synoptic fluctuations of the TWC on the ECS shelf.

Furthermore, the density anomalies between the two surveys and the GDEM data (Fig. 1d and e) indicate that the most significant fluctuations are located north of Taiwan and in the inshore area. Both surveys show negative density anomalies north of Taiwan, indicating that the TWC was weak and that more low-density coastal water was transported to the ECS shelf during the observational periods. The density anomalies in the inshore area show different patterns for the two synoptic surveys, with a positive anomaly in the first survey (Fig. 1d) and a negative anomaly in the second (Fig. 1e), indicating a strong synoptic fluctuation in the inshore area.

Candidate factors for driving these synoptic fluctuations are local wind, surface cooling, and the upstream currents of the Kuroshio Current and the TSC. As discussed by Huyer (1990), wind is often considered as the major driving mechanism of synoptic fluctuations of the wintertime TWC. The northeasterly monsoon wind in winter blows against the northeastward TWC and produces a southwestward ZMCC (Chuang and Liang, 1994; Oey et al., 2010). Zhu et al. (2004) suggested that the occurrence and duration of the TWC are associated with the meandering of the Kuroshio Current north of Taiwan. The northeastward TSC, as an upstream flow of the TWC, also influences the synoptic fluctuation of the wintertime TWC. Hong et al. (2011) and Hu et al. (2010) summarized that the temporal and spatial variation of TSC is modulated by strong wind forcing, complex topography and circulation in the northern South China Sea as well as coastal water input and the Kuroshio intrusion. Guan and Fang (2006) showed evidence that the TSC and the TWC merge in the area between the Taiwan Strait and the Zhe-Min coastal region. Takahashi and Morimoto (2013) pointed out that the temporal variation of the TWC is characterized by the propagation of vorticity anomalies originating from northeast of the Taiwan Strait, which further demonstrated that the fluctuations of TWC were associated with its upstream currents such as the TSC.

To explore the spatial distribution of synoptic fluctuations of the wintertime TWC on the ECS shelf, current data with high resolution in both space and time are required. Previous studies on the wintertime TWC were based on cruise surveys (Su and Pan, 1987; Chen et al., 1994; Chen and Wang, 1999), anchored mooring observations (Zhu et al., 2004; Zeng et al., 2012; Huang et al., 2016) and numerical simulations (Guo et al., 2003, 2006; Yang et al., 2011, 2012; Xuan et al., 2012a, 2016). The observation data are limited in terms of temporal and spatial coverage; hence, they cannot fully reveal the synoptic fluctuations of the TWC and their regional differences. Numerical simulations provide a promising approach for studying the overall structure and driving mechanisms of synoptic fluctuations of the TWC in more detail.

In this study, the Finite Volume Coastal Ocean Model (FVCOM; Chen et al., 2003) is used to investigate wintertime TWC synoptic fluctuations and their mechanisms. The rest of this paper is organized as follows. In Sect. 2, we provide a description of methods and validation. The mean distribu-

tion, synoptic fluctuations and dynamic diagnostics of the wintertime TWC are given in Sect. 3. The impact of synoptic fluctuation on water exchange is further discussed in Sect. 4, followed by conclusions in Sect. 5.

2 Methods and validation

2.1 Model configuration

To investigate the currents (TWC, Kuroshio Current, ZMCC, etc.) and their synoptic fluctuations on the ECS shelf, a three-dimensional (3-D) unstructured-grid (Fig. 2, left panel) FVCOM is developed for the entire Bohai, Yellow and East China Seas (part of the Japan/East Sea, and part of the Pacific Ocean). A regional refinement of the resolution (approximately 3 km) is specified around the ECS shelf break at the 200 m isobaths, where a strong excursion of the Kuroshio Current also occurs. The General Bathymetric Chart of the Oceans (GEBCO) provides high-resolution (approximately 1 km) bathymetric data (Smith and Sandwell, 1997). In all, 20 vertical layers with 76 954 triangle cells were specified in the water column in a sigma-stretched coordinate system.

The driving forces of the numerical simulation include tides, river discharge, surface heat fluxes, wind and open-boundary conditions. Harmonic constants of 11 major tidal constituents (M_2, S_2, N_2, K_2, K_1, O_1, P_1, Q_1, M_4, MS_4 and MN_4) were used; these are based on the Oregon State University global inverse tidal model TPXO.7.0 (Egbert et al., 1994; Egbert and Erofeeva, 2002). The daily mean river discharge of the Changjiang and Huanghe was taken from publicly available observation data at the Datong hydrometric station (http://yu-zhu.vicp.net/). Other rivers were not included because of their small discharges; e.g., the Qiantang River, with the largest runoff from the Zhejiang coast, has a climatological mean discharge in winter of about $230\,\text{m}^3\,\text{s}^{-1}$, which is nearly negligible compared to the Changjiang winter discharge of about $11\,500\,\text{m}^3\,\text{s}^{-1}$. The daily mean heat fluxes were from the objectively analyzed air–sea fluxes (Yu and Weller, 2007), and the 3-hourly wind stress and 10 m wind speed data were from the ERA-Interim re-analysis (Dee et al., 2011). The open-boundary conditions, including daily temperature, salinity and fluxes at the Taiwan Strait, the western Pacific Ocean and the Japan/East Sea, were obtained from the Hybrid Coordinate Ocean Model (Bleck, 2002) and interpolated onto the FVCOM model grid points. The temporal resolution of all the driving force fields is better than or equal to 1 day, which is essential to resolve synoptic fluctuations.

The hindcast outputs of sea surface height, temperature, salinity and velocities for the 5 years of simulation from 2009 to 2013 are used, following three spin-up years (2006–2008) initiated with the temperature and salinity taken from the Hybrid Coordinate Ocean Model and velocity set to zero. The initial conditions are ramped-up over a period of 30 days and

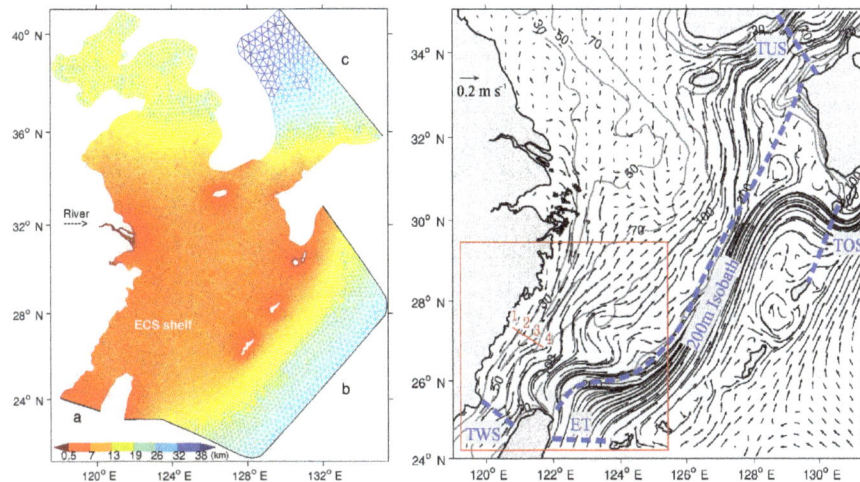

Figure 2. The FVCOM model grid (left panel) and the surface mean flow in the ECS in winter (right panel). The colors in the left panel show the grid length (km). The letters a–c indicate the three open boundaries at the Taiwan Strait, the northwest Pacific Ocean and the Japan/East Sea, respectively. The blue dashed lines (right panel) show some important straits around shelf boundary, including the Taiwan Strait (TWS), the East Taiwan Channel (ET), the Tsushima Strait (TUS), the Tokara Strait (TOS) and shelf break at the 200 m isobath. The red rectangle shows the study area of the wintertime TWC. The four red numbers off the Zhe-Min coast show the four mooring sites observed from 5 January to 28 February 2009.

at the lateral boundaries a sponge layer was used with the same method as Chen et al. (2008). The model time step was 15 seconds for the 2-D barotropic mode and 90 seconds for the 3-D baroclinic mode. All of the output fields were processed with a tidal filter (Godin, 1972) to remove tidal oscillations (considering that the major timescale of synoptic fluctuations in this study area is 3–15 days).

Since the currents in 2009 could partly be validated by means of available observational data (see Sect. 2.2), the currents from 1 January to 28 February 2009 were selected for analysis of the wintertime TWC.

2.2 Validation of the mean currents and synoptic fluctuations

The mean currents, e.g., the Kuroshio Current, the TWC, and the ZMCC, were calculated by averaging the outputs of January and February 2009. We validated the mean currents in terms of circulation structure, boundary fluxes, and coastal currents.

The FVCOM has reproduced almost all of the known circulation structure in the ECS in winter. The surface mean currents (Fig. 2) show three major currents: the Kuroshio Current, the TWC and the ZMCC. The Kuroshio Current, with a speed of about $1 \, \mathrm{m \, s^{-1}}$, enters the ECS just northeast of Taiwan and flows along the shelf break up to the northern area and ultimately leaves the ECS through the Tokara Strait. Both the route and strength of the Kuroshio are comparable with those reported in the literature (Guan, 1978; Qiu and Imasato, 1990). The TWC has two northeastward branches, one inshore (between the 30 and 100 m isobaths) and another

offshore (between the 100 and 200 m isobaths), which is consistent with Su and Pan (1987). The southwestward-directed ZMCC in the nearshore area from the Changjiang Estuary to the Taiwan Strait agrees well with that reported in previous studies (Guan and Mao, 1982; Zeng et al., 2012).

The simulated volume transports across the Taiwan Strait, the East Taiwan Channel, the Tsushima Strait, the Tokara Strait and the shelf break of the 200 m isobath were validated using results from the literature (Table 1). The simulated transports were accurate enough to reproduce volume transport (1.22 Sv) through the Taiwan Strait, which is closer to the observation value (1.20 Sv) from Isobe (2008) than former model results. The volume transports across the Taiwan Strait and the Tokara Strait, as well as the cross-shore exchange, affected the path and magnitude of the TWC. The annual-mean transport across the 200 m isobath toward the shelf is 1.66 Sv, which is balanced by the inflow from the Taiwan Strait (1.22 Sv) and the outflow through the Tsushima Strait (2.85 Sv).

Figure 3 shows a comparison between simulation and observation results for the alongshore currents and the cross-shore currents on the ECS shelf. The observational data were obtained from four mooring surveys (Fig. 2, red stations) off the Zhe-Min coast (Zeng et al., 2012). The observed and simulated currents were both averaged for the observational period, which was from 1 January to 28 February 2009. Using the same method as in Huang et al. (2016), we defined the positive alongshore current direction as from the southwest (218°) to the northeast (38°), which is the mean tangential direction of the isobaths on the southwestern shelf of the ECS. The positive cross-shore direction is from the north-

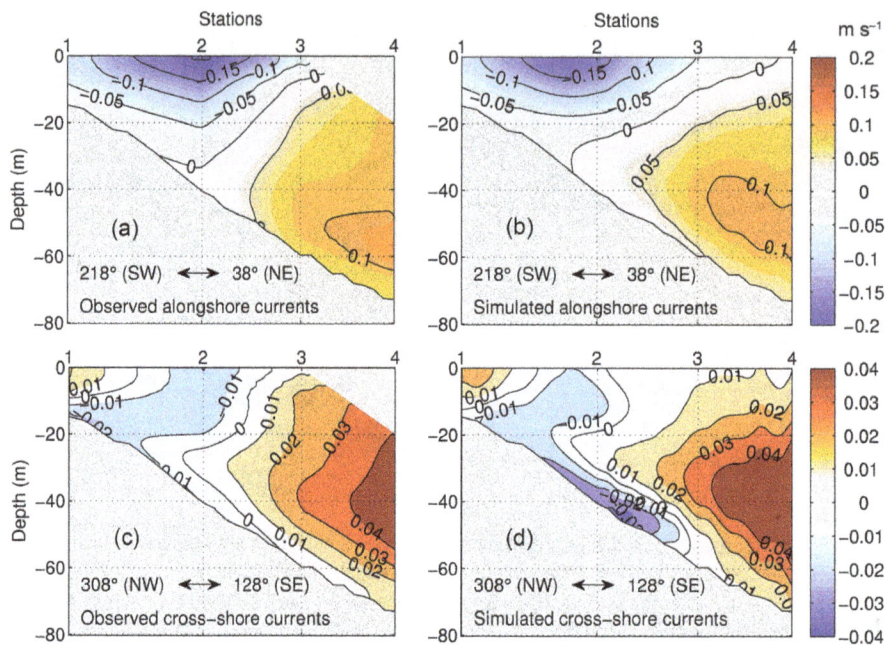

Figure 3. Validations of the wintertime TWC (warm color) along the section off the Zhe-Min coast (the short line with four red numbers in Fig. 2): **(a)** observed alongshore currents, **(b)** simulated alongshore currents, **(c)** observed cross-shore currents and **(d)** simulated cross-shore currents. Note, an enlarged color scale is used for the cross-shore component to have a clear view of its weak structure.

west (308°) to the southeast (128°), normal to the isobaths. The alongshore components (Fig. 3a and b) show that the ZMCC flows southwestward parallel to the coast in winter, with a maximum speed of $0.15\,\mathrm{m\,s^{-1}}$ along the 30 m isobath. The TWC flows northeastward with a speed of $0.05\,\mathrm{m\,s^{-1}}$, and the core is located in the lower layer at about 50 m at station 4. The cross-shore component (Fig. 3c and d) is much weaker than the alongshore components, and it shows a complex spatial pattern. It flows offshore in the upper layer and onshore in the lower layer at station 1. Moreover, it mainly flows onshore at station 2, and it flows offshore in the entire water column at stations 3 and 4. Altogether, the simulated pattern and magnitude both of the alongshore and cross-shore components are in good agreement with the observations. However, there are some differences between the observed and simulated results; for example, the simulated ZMCC occupies a broader space than that in the observations. This may have been caused by the relatively low number of observational stations.

Synoptic fluctuations of the TWC inshore branch during January and February 2009 were also validated against the mooring results (Fig. 4). Since the TWC shows a strong signature at station 4, the time series of the alongshore currents and cross-shore currents in the whole water column of station 4 were used for the validation. To eliminate the influence of local effects, the simulated currents were averaged in a $10 \times 10\,\mathrm{km^2}$ area around station 4. Both the observed and simulated results show that the TWC fluctuates with a period of 3–15 days. The simulated TWC (Fig. 4a, warm color) ap-

peared stronger ($> 0.1\,\mathrm{m\,s^{-1}}$) on 7, 12, 18, 21, 26 and 29 January, and 10, 14, 19, 22 and 25 February, which agrees well with data from the observations (Fig. 4b). The time series of the simulated cross-shore component (Fig. 4c) are virtually in phase with the observations (Fig. 4d). The magnitude of the cross-shore fluctuations is comparable to the alongshore fluctuations. This is different to the anisotropic characteristic of the mean currents (Fig. 3), for which the alongshore component is nearly 1 order of magnitude larger than the cross-shore component.

2.3 EOF analysis of synoptic fluctuations

The empirical orthogonal function (EOF) method (Emery and Thomson, 2001), as a statistical method, has been used to understand synoptic fluctuations of the wintertime TWC. The simulated currents from 1 January to 28 February 2009 were selected and their anomalies were calculated. Then, using the Matlab EOF function, the current vectors were separated into several orthogonal modes to show the spatial and temporal variations. Because the first two leading modes explain 91 % of the total variance, only these two modes were used for the analysis.

The spatial distributions of the two leading EOF modes were used to analyze the regional difference of the synoptic fluctuations. To investigate the driving force of the two EOF modes, the temporal variation was compared to the potential influence factors, such as wind, upstream currents and net surface heat flux.

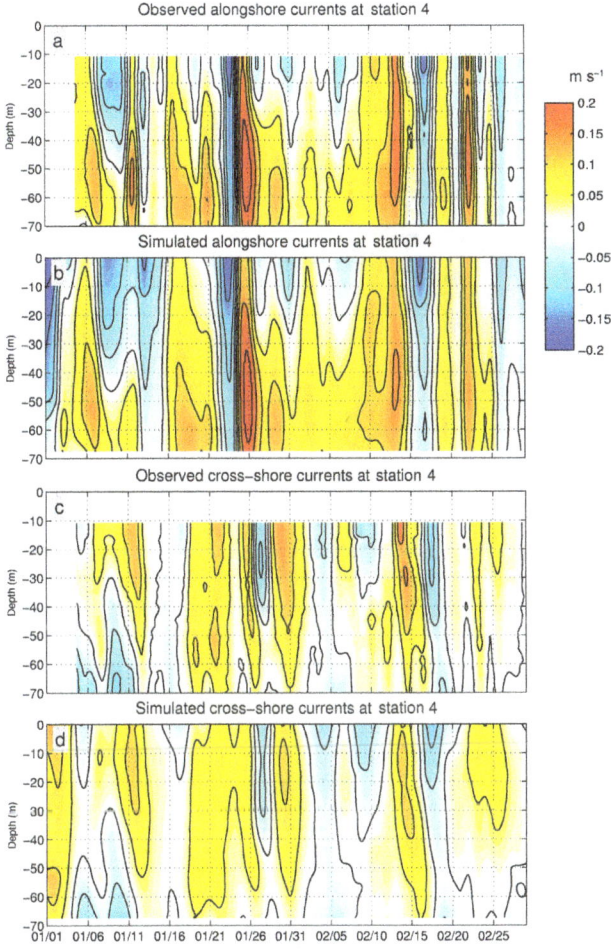

Figure 4. Validations of the wintertime TWC fluctuations: (**a**) observed alongshore currents, (**b**) simulated alongshore currents, (**c**) observed cross-shore currents and (**d**) simulated cross-shore currents. The observation data come from station 4 in Fig. 1 and the simulated data have the same position and period as the observation data.

2.4 Momentum analysis

The driving mechanisms of the synoptic fluctuations were further analyzed using the momentum equation. First, the momentum balance as implemented in FVCOM (Chen et al., 2003) is shown in Eq. (1). The three terms on the left-hand side represent local acceleration, Coriolis acceleration and advection, and the three terms on the right-hand side represent pressure gradient, friction and diffusion.

$$\frac{\partial V}{\partial t} - 2\boldsymbol{\Omega} \times V + (V \cdot \nabla V) = -\frac{1}{\rho_0}\nabla P + \frac{\partial}{\partial z}\left(K_m \frac{\partial V}{\partial z}\right) + F, \quad (1)$$

where V is velocity, $\boldsymbol{\Omega}$ is the Earth's rotation angular velocity, ρ_0 is the average density, P is pressure, K_m is the vertical eddy viscosity coefficient and F is horizontal diffusion.

Second, according to the hydrostatic approximation used in FVCOM (as shown in Eq. 2), the pressure gradient is given

Table 1. Annual-mean volume transports ($\mathrm{Sv} = 10^6\,\mathrm{m}^3\,\mathrm{s}^{-1}$) through various sections. The sections are shown in Fig. 2 using blue dashed lines.

Section	Present model	Previous estimates
Taiwan Strait	1.22	1.2 (Isobe, 2008)
		1.8 (Wang et al., 2003)
		1.09 (Wu and Hsin, 2005)
		1.03 (Yang et al., 2011)
		1.72 (Guo et al., 2006)
		0.5 (Hung et al., 2003)
		1.10 (X. Liu et al., 2014b)
Tsushima Strait	2.85	2.65 (Isobe, 2008)
		3.03 (Guo et al., 2006)
		2.70 (Yang et al., 2011)
		2.52 (X. Liu et al., 2014b)
200 m isobath	1.66	1.46 (Guo et al., 2006)
		0.87 (C. Liu et al., 2014a)
		3.0 (Teague et al., 2003)
		2.74 (Lee and Matsuno, 2007)
East Taiwan Channel	22.71	21.50 (Johns et al., 2001)
		23.00 (Teague et al., 2003)
		23.83 (Guo et al., 2006)
		28.4 (Hsin et al., 2013)
		21.37 (Yang et al., 2011)
		20.74 (X. Liu et al., 2014b)
Tokara Strait	23.20	23.4 (Feng et al., 2000)
		20.00 (Teague et al., 2003)
		20.66 (Yang et al., 2011)
		24.42 (X. Liu et al., 2014b)

as the product of density times the gravitational acceleration. This results in Eq. (3), which indicates that pressure gradient can be decomposed into the effects of the barotropic and baroclinic components, as shown in Eq. (4).

$$\frac{\partial P}{\partial z} = \rho g, \quad (2)$$

$$P_z = \int_z^\eta \rho g\,\mathrm{d}z = \int_z^\eta \left(\rho_0 + \rho'\right)g\,\mathrm{d}z = \rho_0 g(z+\eta) + \int_z^\eta \rho' g\,\mathrm{d}z, \quad (3)$$

$$\nabla P = \rho_0 g \nabla \eta + \nabla\left(\int_z^\eta \rho' g\,\mathrm{d}z\right), \quad (4)$$

where ρ is density, ρ' is density anomaly, g is the gravitational acceleration and η is sea surface height.

Finally, the momentum equation is vertically integrated to estimate momentum balance for the water column. Since the horizontal diffusion is a comparatively small term, it is neglected for simplicity.

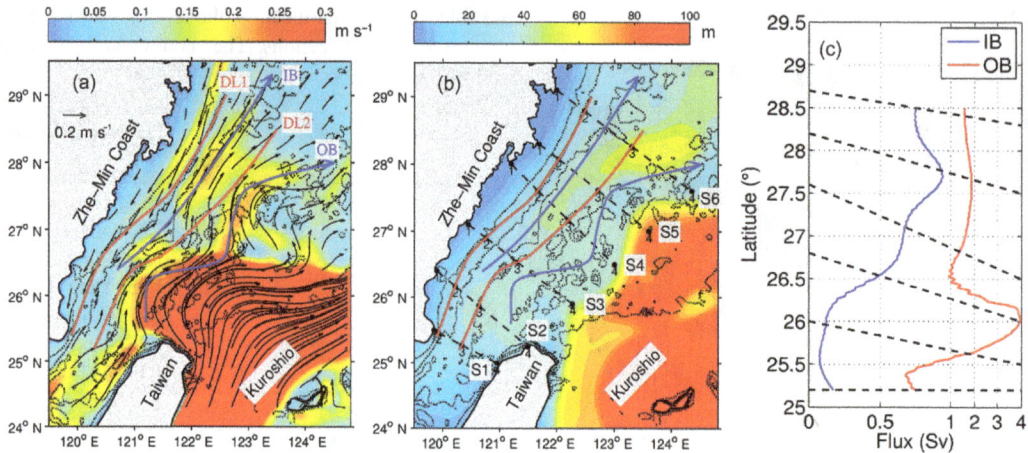

Figure 5. (a) Distribution of flow axes in the ECS in winter. The black arrows show the maximum velocity $(\mathrm{m\,s^{-1}})$ in the vertical profile (VMV) and the color shows the speed of the VMV. The two blue arrows labeled IB and OB represent the flow axes of the inshore branch and offshore branch. The red line DL1 represents the dividing line between the coastal current and inshore branch, and the red line DL2 separates the two TWC branches. **(b)** Depth (m) of flow axes in the ECS are shown by color. Sections S1–S6 were selected to study the wintertime TWC. **(c)** Flux of inshore branch (blue) and offshore branch (red) at different latitudes. Dashed lines show the positions of Sections S1–S6. Note, the scale is not linear.

$$\underbrace{\int_{-H}^{0} \frac{\partial V}{\partial t}}_{\text{Acceleration}} + \underbrace{\int_{-H}^{0} -2\mathbf{\Omega} \times V}_{\text{Coriolis}} + \underbrace{\int_{-H}^{0} (V \cdot \nabla V)}_{\text{Advection}}$$

$$= \underbrace{\underbrace{-gH\nabla\eta}_{\text{Barotropic}} - \underbrace{\int_{-H}^{0} \nabla \left(\int_{z}^{\eta} \rho' g \,\mathrm{d}z \right)}_{\text{Baroclinic}}}_{\text{Total Pressure}}$$

$$+ \underbrace{\rho_a C_D |U|U}_{\tau_a} - \underbrace{k_b |V_b|V_b}_{\tau_b}, \qquad (5)$$

where τ_a is wind stress and τ_b is bottom stress, ρ_a is the density of air, U is the wind speed at 10 m above sea surface, C_D is a drag coefficient at the sea surface (which varies with wind speed U), k_b is a bottom friction coefficient ($k_b = 0.005$) and V_b is the simulated velocity at the bottom.

3   Results

3.1   Mean distribution of TWC in winter

Since the observational results (Su and Pan, 1987; Zeng et al., 2012) show that both branches of the wintertime TWC are flowing in the subsurface, we use the vertical maximum velocity (VMV) and its corresponding depth as two indices to quantify the strength of the subsurface currents (Fig. 5).

As stated above, the distribution of the VMV shows two branches of the TWC (Fig. 5a). The inshore branch (Fig. 5a,

blue arrow of IB), which was located between the 30 and 100 m isobaths, followed a straight route from the northwest of Taiwan to the northern ECS shelf. The offshore branch (Fig. 5a, blue arrow of OB) existed near the 100 m isobath and had two meanders. The two meanders turn to the cross-shore direction along latitudes 26.5 and 28° N. These two branches are further illustrated in the distributions of current speed along the six TWC cross sections (S1–S6), which were located at critical points in the two meanders (Fig. 6). From the VMV structure, it can be inferred that the intrusions of the TSC and the Kuroshio Current both affected the origin of the offshore branch (Fig. 6, S1–S3).

We further examined the subsurface current core using the depth of the VMV (Fig. 5b). We found that the VMV of the TWC was located 40–60 m below the surface at the inshore branch and 20–40 m below the surface at the offshore branch. Figure 6 shows the VMV positions in the subsurface layer; furthermore, it illustrates that the depth of the subsurface VMV in the inshore branch was deeper than that in the offshore branch. The difference can be explained by the combined effects of baroclinicity and wind friction. Assuming a relatively spatially homogeneous heat loss, different cooling occurs, due to the smaller heat capacity of the shallow coastal water compared to the deeper offshore waters, hence generating a northwestward horizontal density gradient leading to a northeastward thermal current (vertical current shear) according to the thermal wind relationship, resulting in an upward-increasing northeastward flow. The northeasterly wind in winter weakens the northeastward TWC, particularly in the upper layer, which leads to the formation of the subsurface VMV. Therefore, the fact that the depth of the subsurface current core in the inshore branch

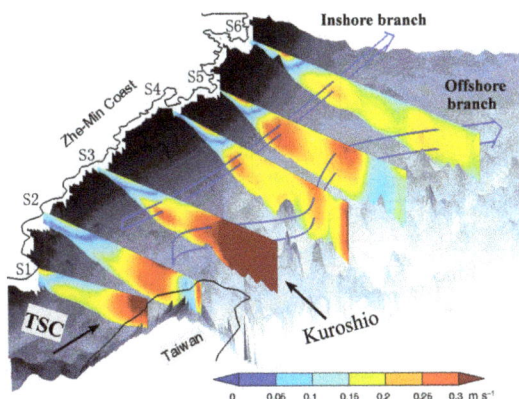

Figure 6. Distributions of current speed along the six sections S1–S6 in winter. The blue arrow on the left indicates the inshore branch according to the velocity cores from section S3 to S6. The blue arrow on the right indicates the offshore branch according to the velocity cores from section S2 to S6. TSC is the Taiwan Strait Warm Current.

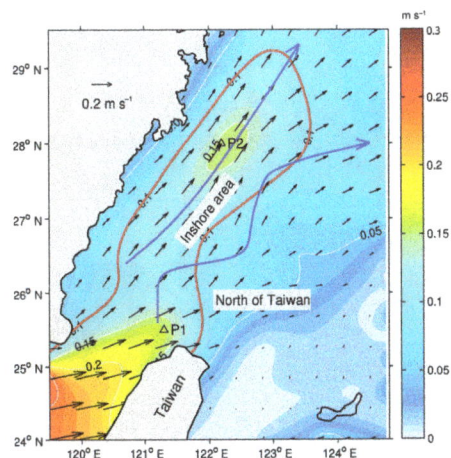

Figure 7. Current standard deviation in the layer of the VMV. The black arrows indicate the major axis of the ellipse, which represent the standard deviation of the current. The color shading shows the respective magnitude. The two blue arrows indicate the two TWC branches. The red curve indicates the area where the current standard deviation is larger than $0.1 \, \mathrm{m \, s^{-1}}$ and the branches' representative points (P1 and P2) are selected for later analysis.

is greater than that in the offshore branch indicates weaker baroclinicity or stronger wind friction on the inshore branch than on the offshore branch.

The magnitude of the wintertime TWC was obtained by flux analysis. Two dividing lines (Fig. 5a, red lines) were defined as the boundaries for the ZMCC: the TWC inshore branch, and the TWC offshore branch, which had the weakest flows. The flux of each branch (Fig. 5c) was calculated using the horizontal integration between the boundaries and the vertical integration in the water column. The inshore branch intensifies along its way and becomes significant north of $26.5° \, \mathrm{N}$, showing particularly strong flow velocities between 27.5 and $28.0° \, \mathrm{N}$. In this area, the subsurface current was much stronger from S4 to S5 than in the other areas (Fig. 6). The flux in the entire offshore branch was large, particularly north of Taiwan.

3.2 Synoptic fluctuations

The observations (Fig. 4) have demonstrated that the synoptic fluctuation in the TWC inshore branch (near $121.5° \, \mathrm{E}$, $27.0° \, \mathrm{N}$) is significant. We further investigated the regional difference of fluctuations in the two TWC branches in winter 2009 using the following three steps: (i) two regions with significant fluctuations are identified by the current standard deviations of the VMV (Fig. 7) and the corresponding temporal variation of vertical structures at their extremes (Fig. 8); (ii) each of the two significant fluctuations is decomposed into EOF components (Fig. 9); and (iii) the influence factors, such as wind, upstream currents, and net surface heat flux, are investigated by examining their correlations with the first two leading EOF components (Figs. 10 and 11).

The current standard deviations (Fig. 7) show that prominent fluctuations occurred in two regions: north of Taiwan

and the inshore area. The standard deviations of VMV at the two regions were larger than $0.1 \, \mathrm{m \, s^{-1}}$ (comparable to the mean currents). In the area north of Taiwan, the fluctuation was located in the origin area of the TWC offshore branch. The fluctuation in this region was in phase with the fluctuation in the Taiwan Strait, indicating that the TSC played an important role in generating the fluctuation north of Taiwan (to a greater extent than did the Kuroshio intrusion). The TWC fluctuation had a strong cross-shore component, which means the fluctuation transported the water north of Taiwan to both the inshore and offshore branches. In the inshore area, the fluctuations were influencing a wide region between the 30 and 100 m isobaths, with a magnitude that was sometimes larger than the mean flow (Fig. 5a). These strong fluctuations led to an episodic occurrence of the TWC inshore branch, as observed at the site off the Zhe-Min coast (Fig. 4, red color). When the TWC inshore branch was weakened due to these fluctuations, the ZMCC might even dominate a wide region out to the 100 m isobath, especially at the surface (Fig. 4, blue color).

The vertical structures of the fluctuations north of Taiwan and in the inshore area at two representative points and their relation with the upper mixed layer depth are further analyzed (Fig. 8). The major component (the alongshore current) of the TWC in each of the two regions (P1 and P2, Fig. 7) is used to show the vertical structure of the fluctuation. The depths of the upper mixed layer were determined by a Richardson number criterion (Mellor and Durbin, 1975; Grachev et al., 2013; Richardson et al., 2013), i.e., where the critical Richardson number equals 0.25 in this paper (as in Xuan et al., 2012b). The mean depth of the upper mixed

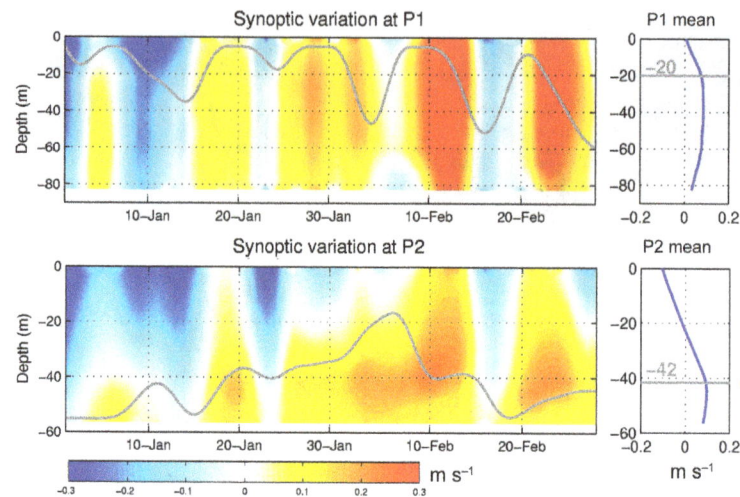

Figure 8. Variation of alongshore currents (m s^{-1}, shown by color scale) for the entire water column north of Taiwan (P1) and in the inshore area (P2) and their relation with upper mixed layer depth. The positive velocity (warm color) indicates the occurrence of the TWC. The gray solid lines show the depth of the upper mixed layer.

layer north of Taiwan (20 m) was much shallower than the mean depth of the inshore area (42 m). However, the TWC (Fig. 8, warm color) fluctuated with significant variations of the upper mixed layer depth (Fig. 8, gray lines) in both areas. When the upper mixed layer deepened, the northeastward TWC (Fig. 8, warm color) was weakened or even replaced by the southwestward ZMCC, and vice versa. Wind and surface cooling, which both drive the mixed layer depth, can affect the TWC fluctuation.

The TWC fluctuations were further decomposed into EOF modes. The first two leading EOF modes account for 54 and 37 % of the total variances (Fig. 9), associated with the two prominent fluctuations north of Taiwan and in the inshore area (Fig. 7). Both EOF modes had a maximum fluctuation larger than 0.2 m s^{-1} (comparable to the mean currents). The spatial pattern of the first EOF mode (EOF1; Fig. 9a) shows that the fluctuation continued from the Taiwan Strait to the area north of Taiwan, indicating that the fluctuation north of Taiwan was related to the TSC and not to the Kuroshio Current. The alongshore component also showed a strong fluctuation in the Taiwan Strait, which means that the TSC episodically intruded the shelf. The cross-shore component revealed a fluctuation north of Taiwan that was larger than 0.1 m s^{-1}. This cross-shore fluctuation impacted on the trajectory of the Taiwan Strait (TWS) water, synoptically flowing into the TWC inshore branch, offshore branch or Kuroshio Current.

The spatial pattern of the second EOF mode (EOF2; Fig. 9b) shows a synoptic fluctuation in the inshore area. The area with alongshore fluctuation (Fig. 9d) larger than 0.1 m s^{-1} was located between the 30 and 100 m isobaths, which demonstrates that the TWC could episodically affect this area. In addition, there were cross-shore fluctuations in the inshore area (Fig. 9f), mostly along the latitudes 26.5 and 28° N. The latitudes of larger cross-shore fluctuations agreed

well with the latitudes where the TWC offshore branch of the mean currents (Fig. 5a) turned to the cross-shore direction. This indicated that the cross-shore transports were most significant at the latitudes 26.5 and 28° N, according to both the mean currents and the synoptic fluctuations.

Figure 10 shows the temporal variation of EOF1 and its relation to the north–south component of wind speed, net surface heat flux, the TSC and the Kuroshio Current. We found a close correlation between EOF1 and TSC ($R = 0.86$), demonstrating that the TSC played the most important role in generating the TWC fluctuation north of Taiwan. The EOF1 and TSC were positively correlated, meaning that a larger TSC intrusion north of Taiwan leads to a cross-shore current from the coastal area to the offshore area and that a weak TSC intrusion causes a cross-shore current from offshore to inshore north of Taiwan.

Figure 11 shows the temporal variation of EOF2 and its relation with the north–south component of wind speed, net surface heat flux, the TSC and the Kuroshio Current. It can be seen that EOF2 and wind are well correlated ($R = 0.89$), indicating the important role of wind in generating the TWC fluctuation in the inshore area. The northeasterly monsoon would greatly enhance the southwestward ZMCC, which would then replace the northeastward TWC in the inshore area.

3.3 Dynamic diagnostics

The wintertime (January and February 2009) mean of the water column momentum balance (Fig. 12) is used to show the overall distribution of the fundamental forces over the ECS shelf. The Coriolis force (Fig. 12a) is mainly balanced by the total pressure (Fig. 12b) in both branches, indicating the dominant role of geostrophic balance in the wintertime

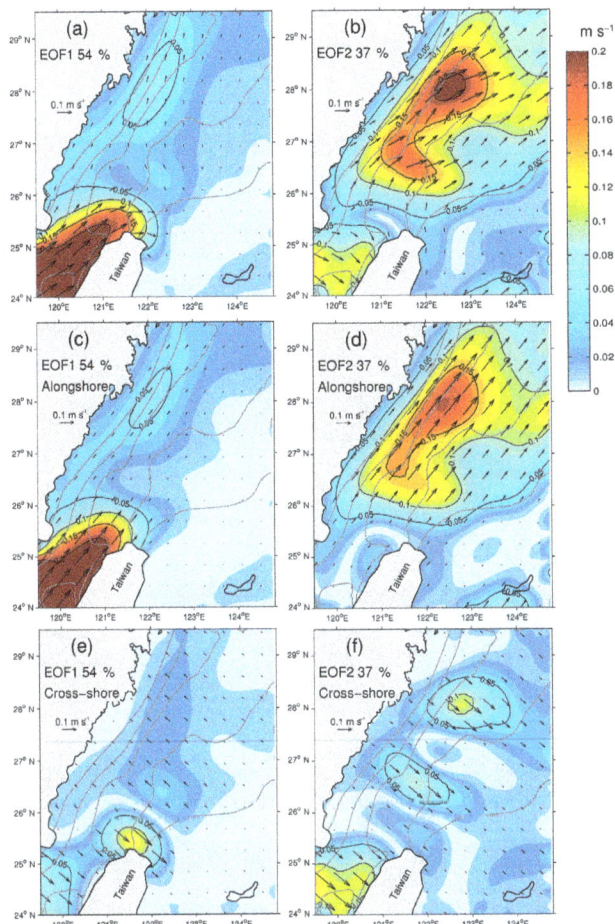

Figure 9. The spatial pattern of the first (EOF1; left panels) and second (EOF2; right panels) leading modes of the VMV in the ECS: **(a)** EOF1 currents, **(b)** EOF2 currents, **(c)** EOF1 alongshore component, **(d)** EOF2 alongshore component, **(e)** EOF1 cross-shore component and **(f)** EOF2 cross-shore component (all shown by black arrows with the color representing the magnitude). The 30, 50, 70, 100 and 200 m isobaths are indicated with gray lines.

TWC. However, the wind-induced surface friction plays an important role in the TWC, especially in the inshore area and the Taiwan Strait (Fig. 12c). The bottom friction has an impact north of Taiwan and in the shallow Taiwan Strait, in particular when significant Kuroshio intrusion enhances the bottom flow (Fig. 12d). The effects of advection and acceleration are predominantly local indicated by mostly incoherent small-scale distributions (Fig. 12e and f); therefore, they can be ignored when studying the large-scale current of the wintertime TWC.

The variation of the driving forces at two representative points P1 and P2 was used to analyze the dynamics of synoptic fluctuations north of Taiwan and in the inshore area. Regarding the results from the EOF analysis, the three force terms, namely, Coriolis, total pressure and wind (Fig. 13), were selected to investigate the effect of the TSC on the fluc-

tuation north of Taiwan (Fig. 9a) and the effect of wind on the fluctuation in the inshore area (Fig. 9b).

In the area north of Taiwan, the cross-shore fluctuations were induced by the TSC intrusion. The variation of alongshore Coriolis force (Fig. 13a, black line) was much greater than the cross-shore Coriolis force (Fig. 13b, black line), which means that the fluctuation north of Taiwan was mainly in the cross-shore direction. The Coriolis force (Fig. 13a, black line) was mainly balanced by the total pressure (Fig. 13a, blue line), which means the currents fluctuations north of Taiwan are dominated by geostrophic balance. As mentioned in Sect. 3.2, the TWC fluctuation north of Taiwan was associated with the TSC rather than with the Kuroshio Current. Therefore, in the shallow coastal area the TSC mainly caused variations in the depth-independent barotropic pressure gradients, which further generated the cross-shore fluctuation. The mechanism can be interpreted as follows. When a larger TSC intrusion occurred, the isobaric slope tilted downward from south to north, generating a cross-shore current from the coastal area to the offshore area. On the contrary, when the TSC intrusion was weak, the Kuroshio intrusion from offshore to inshore dominated north of Taiwan.

Wind friction (Fig. 13c and d) was a fundamental factor in generating the fluctuations in the inshore area. Although the geostrophic balance dominated in the inshore branch for most of the time, the episodically strong winter monsoon had an important role in generating the TWC fluctuations. The northwestward direction Coriolis force (Fig. 13c, black line) shows that the southwestward ZMCC occurred on 12, 22 January and 14 February 2009, and was associated with a northeasterly wind (Fig. 13c, red line). It indicates that strong northeasterly monsoon in winter can reduce or even stop the northeastward TWC in the inshore area, causing the intermittency of the TWC inshore branch.

4 Discussion

Simulated results in the winters (December–March) of the years 2010 to 2013 (Fig. 14) show that general structures of the TWC in the other winters were similar to that in winter 2009 (Figs. 5 and 9), which indicates that the results from the winter 2009 can be regarded as representative for the winter situation. The two TWC branches and the two areas of strong fluctuations were present in all winters from 2009 to 2013, although their strength showed a certain interannual variability in accordance with the changing surface forcing and boundary fluxes.

The wintertime TWC, which is manifested by two subsurface branches and significant synoptic fluctuations, has a very different structure when compared with the stationary and surface summertime TWC reported in previous studies (Guan, 1978; Fang et al., 1991; Isobe, 2008). The synoptic events, with timescales of 3–15 days, play a dominant role

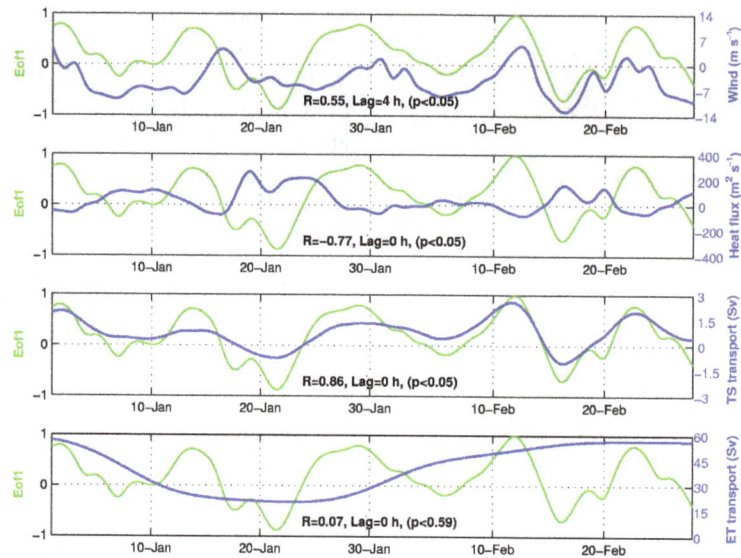

Figure 10. Temporal variation of EOF1, north–south component of wind speed, surface net heat flux, and TSC flux across the TWS section, and Kuroshio flux across the ET section. Their linear correlation coefficients R and time lags are also indicated in each panel. The p value is a declining indicator, which indicates the impact significance of the linear correlation coefficients R, whereby R has statistical significance and the confidence level is larger than 95 % when the p value is less than 0.05.

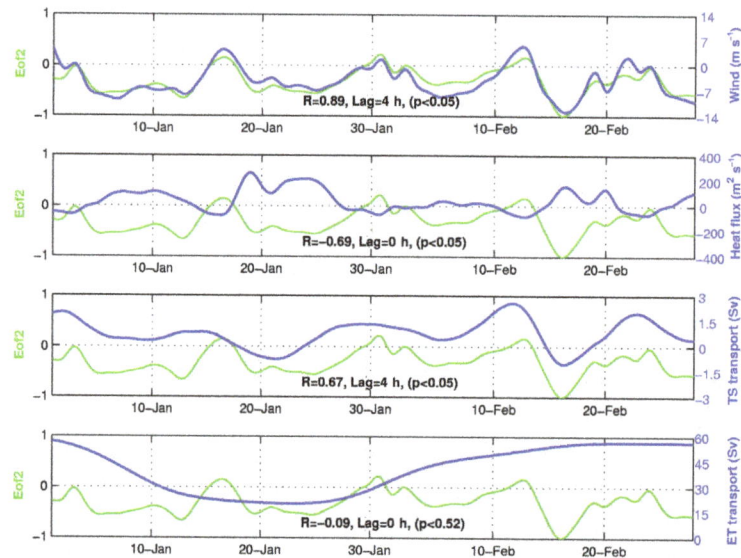

Figure 11. Temporal variation of EOF2, north–south component of wind speed, surface net heat flux, and TSC flux across the TWS section, and Kuroshio flux across the ET section. Their linear correlation coefficients and time lags are also indicated in each panel.

on the horizontal advective transports. According to Ledwell et al. (1998) synoptic variations are much more effective on the horizontal transport than variations on shorter timescales. The synoptic fluctuations modulate the spatial structure of the wintertime TWC, especially when their magnitudes are comparable with that of the mean currents, such as the two prominent fluctuations north of Taiwan and in the inshore area (Fig. 7). Therefore, the two prominent fluctuations will

be discussed next in terms of their contributions to the along-shore and cross-shore transports.

4.1 Cross-shore transport north of Taiwan induced by the TSC

In the area north of Taiwan, the TSC intrusion generated strong fluctuations of the TWC in the cross-shore direction (Fig. 9a). When a larger TSC intrusion occurred, the iso-baric slope tilted downward from south to north, generating

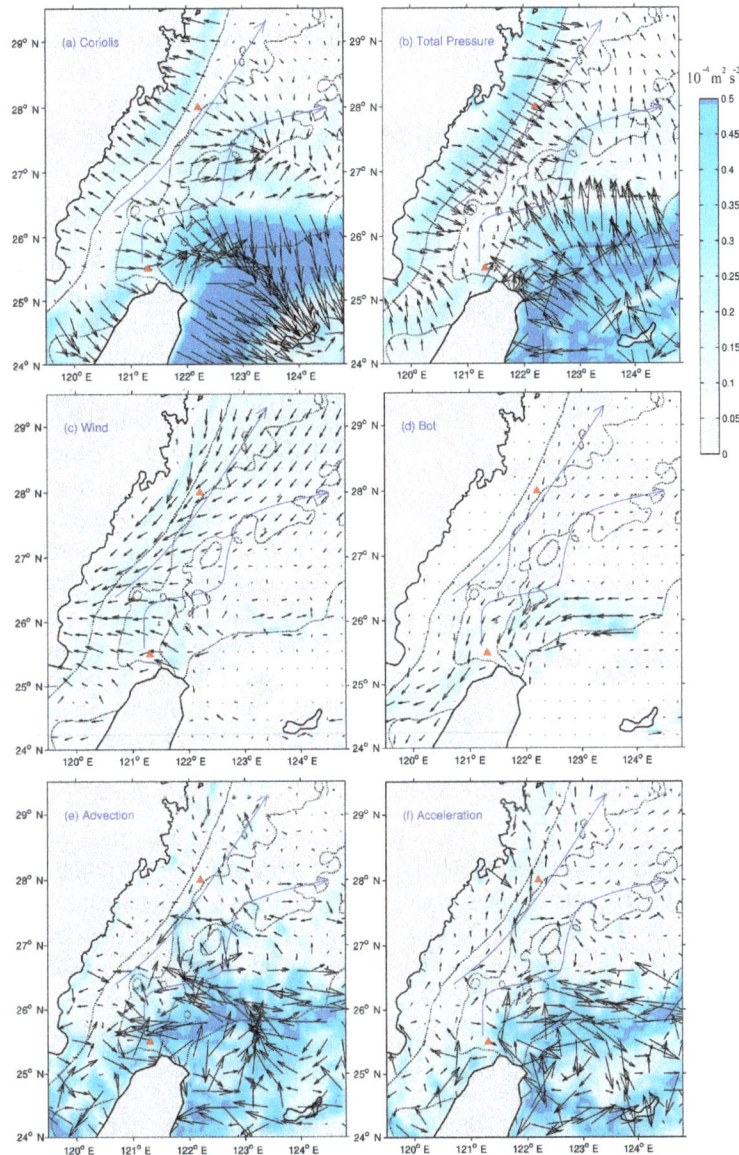

Figure 12. The effects of Coriolis force (**a**), total pressure (**b**), surface friction (**c**), bottom friction (**d**), advection (**e**) and local acceleration (**f**) for water column in winter according to Eq. (5) (shown by black arrows with the color representing the magnitude; units: 10^{-4} m^2 s^{-2}). The two blue arrows indicate the two TWC branches. The two triangles indicate the two regions with significant fluctuation north of Taiwan (P1) and in the inshore area (P2).

a cross-shore current from the coastal area to the offshore area. Compared to the reported summer route that transports Taiwan Strait water to the inshore area between the 30 and 100 m isobaths (Guan, 1978; Fang et al., 1991; Isobe, 2008; Yang et al., 2011, 2012), our results showed that most Taiwan Strait water was transported to the TWC offshore branch and to the Kuroshio area as a result of the cross-shore fluctuations induced by the synoptic TSC intrusion.

A numerical tracer simulation was used to analyze the role of the cross-shore fluctuation in the transport of the TSC water and the Kuroshio water north of Taiwan. In order to demonstrate the characteristics of the flow patterns more clearly, artificial tracers are released in the model domain and transported by the velocity field provided by the FV-COM simulation. The tracer running was part of the FVCOM simulation; therefore, all the abovementioned dynamics were involved, e.g., tide, wind and boundary forces. The release location and start date of the particles were configured as follows. Two sections, one in the Taiwan Strait (Fig. 15a, black dots) and another in the East Taiwan Channel (Fig. 15b, black dots), were selected as the source locations for the water masses of the TSC and the Kuroshio, respectively. The particles were released on 1 January 2009 and tracked until 31 March 2009 (a total of 90 days).

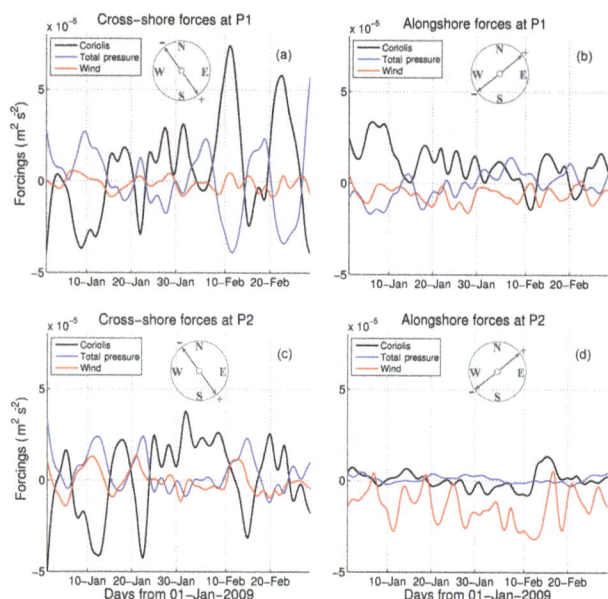

Figure 13. Variations in Coriolis force, total pressure, and wind in the cross-shore direction at P1 (**a**), the alongshore direction at P1 (**b**), the cross-shore direction at P2 (**c**) and the alongshore direction at P2 (**d**) according to Eq. (5). The gray pointers indicate the alongshore and cross-shore directions of dynamical effects in the Earth coordinate system.

Figure 15a shows the traces originating from the TSC area. Unlike the traditional route, where the TSC water flows from the Taiwan Strait to the inshore area between the 30 and 100 m isobaths, most particles (Fig. 15a, gray lines) were concentrated in the offshore branch under the effect of cross-shore fluctuation. Two particles were selected to show the inshore route (Fig. 15a, red line) and offshore route (Fig. 15a, blue line), with both passing the area north of Taiwan. When the two particles arrived at the area north of Taiwan, the behavior of the tracers, according to specific velocity conditions (Fig. 15c), was very different; a northwestward transport occurred on 25 January for the inshore particles (Fig. 15c) and a northeastward transport occurred on 12 February for the offshore particles (Fig. 15c). The velocity conditions in the area north of Taiwan corresponded to the variation of the Taiwan Strait flux (Fig. 10), which shows that the Taiwan Strait flux on 12 February was much greater than on 25 January. Therefore, it can be concluded that the TSC intrusion induced an offshore transport north of Taiwan.

Figure 15b shows the traces originating from the Kuroshio area. In the same way as the TSC water, the Kuroshio water was also transported to the northern shelf via both the inshore branch and the offshore branch. The separation of the two branches north of Taiwan was caused by cross-shore fluctuations of the currents. When the two particles arrived at the area north of Taiwan, a northwestward transport occurred on Feb. 2 for the inshore particles (Fig. 15c) and a northeastward transport occurred on Feb. 12 for the offshore particles

(Fig. 15c). This means that the offshore transport induced by the TSC also had an effect on the distribution of Kuroshio water north of Taiwan. Liu et al. (2016) showed that the winter TSC originated from a small branch of Kuroshio intrusion into the Luzon Strait. Our results complement this picture, since they show that most TSC particles flow into the TWC offshore branch under the influence of cross-shore fluctuation.

Our results may underestimate the impact of Kuroshio intrusion on the fluctuation of the TWC northeast of Taiwan, especially at the seasonal and interannual timescales. Wei et al. (2013) demonstrated that the annual and interannual variations of the Kuroshio volume transport are large. In addition, Zhou et al. (2015) pointed out that the annual and interannual variations of the Kuroshio intrusion northeast of Taiwan are prominent. X. Liu et al. (2014b) presented supportive evidence that the Kuroshio intrusion, from east of Taiwan to the onshore area north of Taiwan, is closely related to the Kuroshio volume transport. This relation between the Kuroshio intrusion and the Kuroshio volume transport had been interpreted by Su and Pan (1987) as the β effect because of the sudden change in topography northeast of Taiwan. Our results show that the intraseasonal variation of the Kuroshio intrusion and the Kuroshio volume transport was negligible compared with the TSC variation at the same timescale, indicating that the synoptic fluctuation of TWC north of Taiwan is mainly induced by the TSC. However, because FVCOM uses sigma coordinates in the vertical that are prone to errors in regions of steep topography, our results may underestimate the fluctuations at the shelf break, in particular to the northeast of Taiwan where Kuroshio intrusion occurs.

4.2 Water exchange in the inshore area induced by wind

In the inshore area, the synoptic fluctuations of the TWC (Fig. 9b) caused by wind were generally strong in the alongshore direction and regionally important (along the latitudes 26.5 and 28° N) in the cross-shore direction. The alongshore fluctuations showed that the TWC inshore branch occurred episodically. This episodic occurrence of the TWC agrees with the results from a previous study based on four mooring surveys off the Zhe-Min coast (Zeng et al., 2012). The mechanism of the episodic occurrence of the TWC was mainly associated with the winter monsoon, which agrees with the analysis of observational data by Huang et al. (2016). However, the overall magnitude of the TWC fluctuation, and its role in the cross-shore flux, is still not fully understood due to the short-term nature of the observational data.

We investigated the magnitude of TWC fluctuation, and its role in the water exchange, in the inshore area. Previous studies (Su and Pan, 1987; Zeng et al., 2012) showed that the TWC flows between the 50 and 100 m isobaths, whereas the ZMCC water dominates the coastal area west of the 50 m isobath in the surface layer. As mentioned when

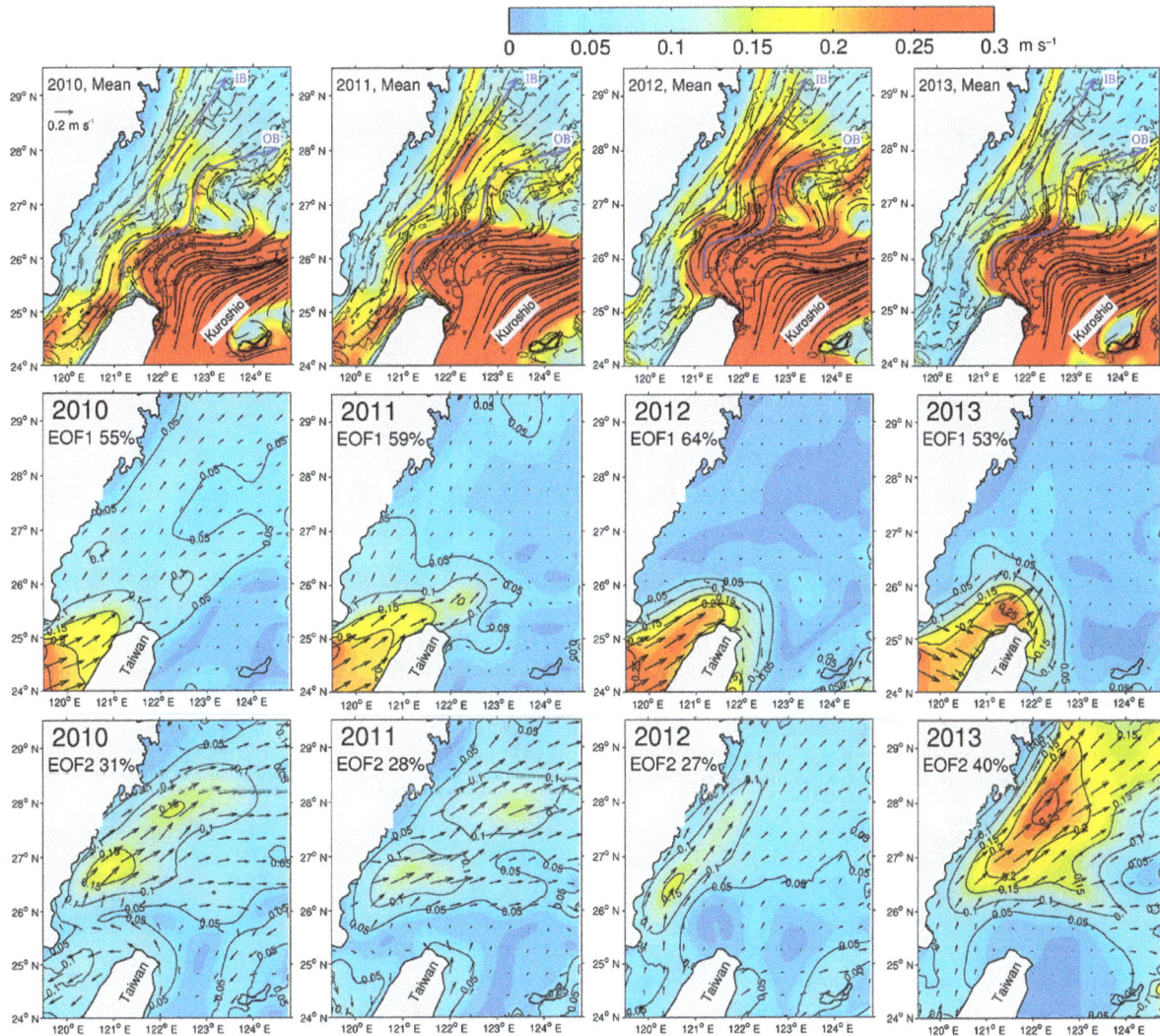

Figure 14. Mean currents (upper panels) and synoptic fluctuations (EOF1 in middle panels and EOF2 in bottom panels) in the winters of 2010–2013. The black arrows in the upper panels show the velocity (m s^{-1}) in the layer of VMV with the color representing the current speed. The two blue arrows with label IB and OB represent the flow axes of the inshore branch and offshore branch, respectively. The black arrows in the middle panels and bottom panels represent the EOF components (m s^{-1}) with their magnitude represented by color scales.

discussing Fig. 9d, the strongest TWC could reach the coastal area as close as the 30 m isobath, being stronger than those reported in the literature. Moreover, the area with large fluctuations spanned the area between the 30 and 100 m isobaths (Fig. 9b), indicating that water between the 30 and 100 m isobaths may be either ZMCC or TWC water.

The episodic occurrence of the TWC inshore branch is directly related to the relative importance of the southwestward ZMCC (Fig. 16, blue arrows) and the northeastward TWC (Fig. 16, red arrows). In this paper, only wind-induced synoptic fluctuations are considered, not short-term extreme storm events. When the winter monsoon (the northeasterly wind) prevails, the ZMCC occupies most of the inshore area and the TWC inshore branch weakens (Fig. 16a). On the contrary, the TWC inshore branch can intrude into the near-coast area

under southwesterly wind conditions (Fig. 16b). The boundary between the coastal current and the TWC may shift from the 100 m isobaths to the 30 m isobath in the cross-shore direction, covering the entire area of the TWC inshore branch.

Our results further reveal that strong wind-induced cross-shore fluctuations occur in the inshore area (Fig. 9f). This cross-shore fluctuation has a significant ecological impact because of the connected nutrient transport (Zhao and Guo, 2011). Ren et al. (2015) observed a cross-shore flux in the inshore area, which was triggered by the transition of northeasterly to southwesterly winds. Their observed features can be further interpreted with our result that wind-induced fluctuations can affect the cross-shore water transport in the inshore area.

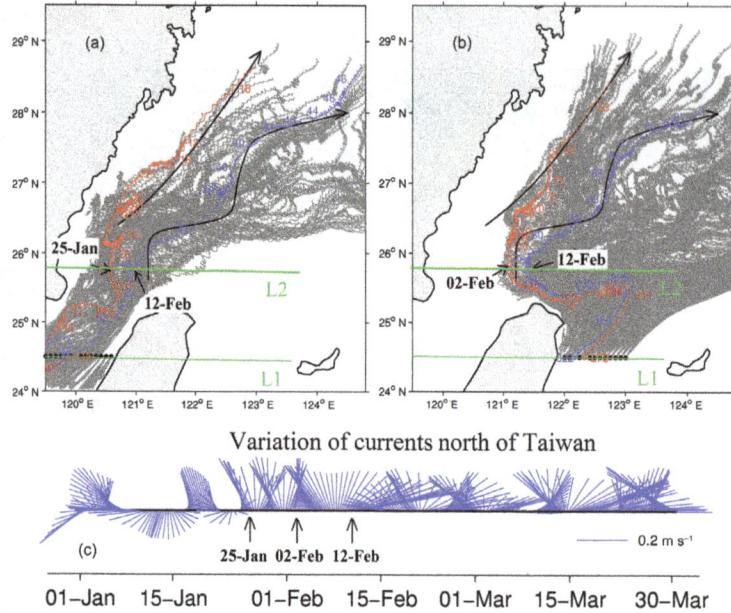

Figure 15. Traces of TSC water **(a)** and Kuroshio water **(b)** in winter, with the variation of surface currents north of Taiwan **(c)**. The green lines L1 and L2 indicate the starting latitude of the tracers (24.5° N) and the latitude which is representative for synoptic fluctuations north of Taiwan (25.8° N), respectively. The black dots represent the release locations of tracers originated from line L1. The gray lines show the entire trajectories of the tracers. The red lines and blue lines are selected trajectories, which are close to the inshore branch and offshore branch, respectively. The dates show the times when selected tracers cross the latitude indicated by line L2. The numbers are the depths of the tracers, which are labeled at an interval of 6 days. The two black arrows represent the two TWC branches.

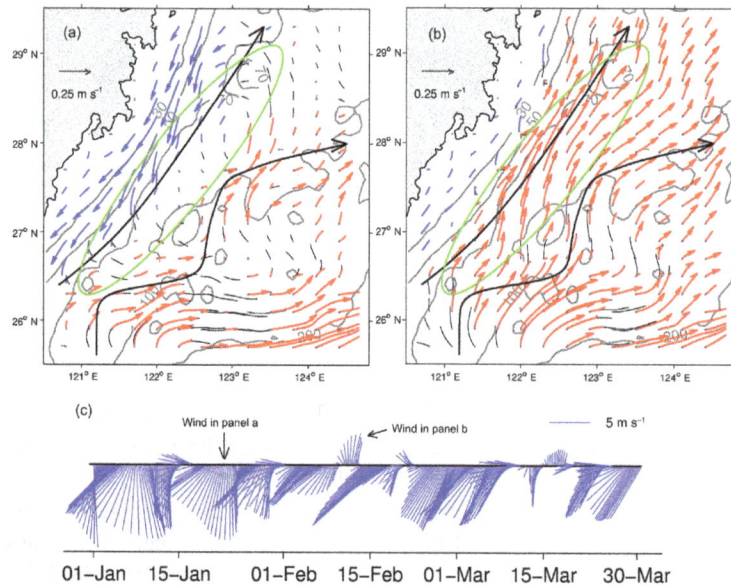

Figure 16. The VMV under the northerly wind **(a)** and southerly wind **(b)**. **(c)** shows the variation of wind in winter. Blue vectors and red vectors show the southwestward coastal current and the northeastward TWC, respectively. Gray contours indicate the 30, 50, 70, and 100 m isobaths. The two black arrows represent the two TWC branches. The green ellipse indicates the inshore area with significant fluctuation.

The largest cross-shore fluctuations were located at the latitudes 26.5 and 28° N (Fig. 9f), which agreed well with the latitudes where the TWC offshore meanders occurred in the mean currents (Fig. 5a). Thus, the offshore transports were most significant along the latitudes 26.5 and 28° N according to both the mean currents and the synoptic fluctuations. The offshore transport may be associated with the offshore-penetrating fronts of coastal water in the ECS. Many remote sensing images (He et al., 2010; Bai et al., 2013) have exhibited offshore-penetrating fronts that crossed the 70 m isobath and played an important role in cross-shore material exchange, but the mechanisms of the offshore-penetrating fronts are still under debate. Yuan et al. (2005) pointed out that both downwelling- and upwelling-favorable winds are associated with the occurrence of the offshore-penetrating front. Ren et al. (2015) suggested that the penetrating front is generated by the transition of northeasterly to southwesterly winds. Wu (2015) suggested that the offshore-penetrating front is the response of buoyant coastal water to an along-isobath undulation of the ambient pycnocline, which is controlled by a temperature stratification of the water column. Our study offers a new interpretation; i.e., the penetrating front is generated through the wind-induced fluctuations and the TWC offshore meanders.

5 Conclusions

The FVCOM model was able to reproduce the wintertime TWC in 2009 reasonably well, as shown by a validation in terms of the overall structure of the surface mean currents, the ECS boundary fluxes and data from four mooring stations. The validation showed that the simulated TWC was comparable to the observed results, not only in terms of the mean currents but also in terms of the synoptic fluctuations.

The wintertime TWC showed two branches: one inshore and another offshore. The inshore branch covered an area between the 30 and 100 m isobaths and flowed northeastward via a straight route. The offshore branch was located between the 100 and 200 m isobaths and showed two prominent meanders. It was shown that the Coriolis force was nearly balanced by the pressure gradient in both branches, indicating the dominant role of the geostrophic balance for the mean current in both branches.

Two regions with significant synoptic fluctuations, north of Taiwan and the inshore area, were investigated using the EOF method. The first two leading modes explained 91 % of the total variance. EOF1 showed that fluctuations occurred in the cross-shore direction south of 26° N. These fluctuations were mainly associated with variation of the TSC flux. EOF2 showed significant fluctuation between the 30 and 100 m isobaths. These fluctuations caused the episodic existence of the TWC inshore branch in the alongshore direction and cross-shore fluctuations mainly at latitudes 26.5 and 28° N, which were mainly associated with the variation of wind speed.

We also studied the different dynamic reasons for the fluctuations in the two regions. In the area north of Taiwan, the TSC and Kuroshio converged to initiate the TWC. A barotropic pressure anomaly was generated by TSC intrusion from the Taiwan Strait causing a barotropic pressure gradient in the alongshore direction; this explains why the synoptic fluctuations in this area occurred in the cross-shore direction. Additionally, the wind had a strong effect on the

synoptic fluctuations in the inshore area. The northeasterly monsoon enhanced the southwestward ZMCC and replaced the TWC in the inshore area. This situation is reversed during the southwesterly wind.

The synoptic fluctuations north of Taiwan and in the inshore area are important for both the alongshore and cross-shore transports. Due to the fluctuation north of Taiwan, the mixed water of the TSC and the Kuroshio was transported to both the inshore area and the offshore area, whereas most Taiwan Strait water was transported to the offshore area in winter. The inshore fluctuation not only caused an episodic occurrence of the TWC in the alongshore direction, which affected the alongshore transport of ZMCC water and TWC water between the 30 and 100 m isobaths, but also impacted the cross-shore transports along latitudes 26.5 and 28° N.

Competing interests. The authors declare that they have no conflict of interest.

Acknowledgements. The authors sincerely thank John M. Huthnance and the three anonymous reviewers for insightful suggestions that improved this manuscript. This study was jointly supported by the Sino-German cooperation in ocean and polar research under the grant BMBF-03F0701A (CLIFLUX), the National Natural Science Foundation of China (U1609201, 41621064, 41306025), the grant from the scientific research fund of the Second Institute of Oceanography, SOA (QNYC201603), and the project of State Key Laboratory of Satellite Ocean Environment Dynamics, the Second Institute of Oceanography (SOEDZZ1512).

Edited by: J. M. Huthnance

References

Bai, Y., Pan, D., Cai, W. J., He, X., Wang, D., Tao, B., and Zhu, Q.: Remote sensing of salinity from satellite-derived CDOM in the Changjiang River dominated East China Sea, J. Geophys. Res.-Oceans, 118, 227–243, 2013.

Bleck, R.: An oceanic general circulation model framed in hybrid isopycnic-Cartesian coordinates, Ocean Model., 37, 55–88, 2002.

Brink, K. H.: Costal trapped waves and wind-induced currents over the continental shelf, Ann. Rev. Fluid Mech., 23, 389–412, 1991.

Carnes, M. R.: Description and evaluation of GDEM-V3.0, NRL Rep. NRL/MR/7330-09-9165, Nav. Res. Lab., Washington, D.C., 2009.

Chen, C., Beardsley, R. C., Limeburner, R., and Kim, K.: Comparison of winter and summer hydrographic observations in the Yellow and East China seas and adjacent Kuroshio during 1986, Cont. Shelf Res., 14, 909–929, 1994.

Chen, C., Liu, H., and Beardsley, R. C.: An unstructured, finite-volume, three-dimensional, primitive equation ocean model: application to coastal ocean and estuaries, J. Atmos. Ocean. Tech., 20, 159–186, 2003.

Chen, C., Xue, P., Ding, P., Beardsley, R. C., Xu, Q., Mao, X., Gao, G., Qi, J., Li, C., Lin, H., Cowles, G., and Shi, M.: Physical mechanisms for the offshore detachment of Changjiang Diluted Water in the East China Sea, J. Geophys. Res., 113, C02002, doi:10.1029/2006JC003994, 2008.

Chen, C. T. A. and Wang, S. L.: Carbon, alkalinity and nutrient budget on the East China Sea continental shelf, J. Geophys. Res.-Oceans, 104, 20675–20686, 1999.

Chuang, W. S. and Liang, W. D.: Seasonal variability of intrusion of the Kuroshio water across the continental shelf northeast of Taiwan, J. Oceanogr., 50, 531–542, 1994.

Cui, M., Hu, D., and Wu, L.: Seasonal and intraseasonal variations of the surface Taiwan Warm Current, Chin. J. Oceanol. Limnol., 22, 271–277, 2004.

Dee, D. P., Uppala, S. M., Simmons, A. J., Berrisford, P., Poli, P., Kobayashi, S., Andrae, U., Balmaseda, M. A., Balsamo, G., Bauer, P., Bechtold, P., Beljaars, A. C. M., van de Berg, L., Bidlot, J., Bormann, N., Delsol, C., Dragani, R., Fuentes, M., Geer, A. J., Haimberger, L., Healy, S. B., Hersbach, H., Hólm, E. V., Isaksen, L., Kållberg, P., Köhler, M., Matricardi, M., McNally, A. P., Monge-Sanz, B. M., Morcrette, J.-J., Park, B.-K., Peubey, C., de Rosnay, P., Tavolato, C., Thépaut, J.-N., and Vitart, F.: The ERA-Interim reanalysis: configuration and performance of the data assimilation system, Q. J. Roy. Meteorol. Soc., 137, 553–597, doi:10.1002/qj.828, 2011.

Egbert, G. D. and Erofeeva, S. Y.: Efficient inverse modeling of barotropic ocean tides, J. Atmos. Ocean. Tech., 19, 183–204, 2002.

Egbert, G. D., Bennett, A., and Foreman, M.: TOPEX/Poseidon tides estimated using a global inverse model, J. Geophys. Res., 99, 24821–24852, doi:10.1029/94JC01894, 1994.

Emery, W. J. and Thomson, R. E.: Data analysis methods in physical oceanography, Second and revised version, Elsevier Science B.V., Amsterdam, the Netherlands, 658 pp., 2001.

Fang, G., Zhao, B., and Zhu, Y.: Water volume transport through the Taiwan Strait and the continental shelf of the East China Sea measured with current meters, in: Oceanography of Asian Marginal Seas, edited by: Takano, K., Elsevier, New York, 345–358, doi:10.1016/S0422-9894(08)70107-7, 1991.

Feng, M., Mitsudera, H., and Yoshikawa, Y.: Structure and Variability of the Kuroshio Current in Tokara Strait, J. Phys. Oceanogr., 30, 2257–2276, 2000.

Godin, G.: The Analysis of Tides, University of Toronto Press, Toronto, 264 pp., 1972.

Grachev, A. A., Andreas, E. L., Fairall, C. W., Guest, P. S., and Persson, P. O. G.: The critical Richardson number and limits of applicability of local similarity theory in the stable boundary layer, Bound.-Lay. Meteorol., 147, 51–82, 2013.

Guan, B. and Fang, G.: Winter counter-wind currents off the southeastern China coast: A review, J. Oceanogr., 62, 1–24, 2006.

Guan, B. and Mao, H.: A note on circulation of the East China Sea, Chin. J. Oceanol. Limnol., 1, 5–16, 1982.

Guan, B. X.: A sketch of the current system of the East China Sea, in: Collected Papers of the Continental Shelf of the East China Sea, Inst. of Oceanol., Chin. Acad. of Sci., Qingdao, China, 126–133, 1978.

Guo, X. Y., Hukuda, H., Miyazawa, Y., and Yamagata, T.: A triply nested ocean model for simulating the Kuroshio – Roles of horizontal resolution on JEBAR, J. Phys. Oceanogr., 33, 146–169, 2003.

Guo, X. Y., Miyazawa, Y., and Yamagata, T.: The Kuroshio onshore intrusion along the shelf break of the East China Sea: The origin of the Tsushima Warm Current, J. Phys. Oceanogr., 36, 2205–2231, doi:10.1175/JPO2976.1, 2006.

He, L., Li, Y., Zhou, H., and Yuan, D.: Variability of cross-shelf penetrating fronts in the East China Sea, Deep-Sea Res., 57, 1820–1826, 2010.

Hong, H., Chai, F., Zhang, C., Huang, B., Jiang, Y., and Hu, J.: An overview of physical and biogeochemical processes and ecosystem dynamics in the Taiwan Strait, Cont. Shelf Res., 31, 3–12, 2011.

Hsin, Y. C., Qiu, B., Chiang, T. L., and Wu, C. R.: Seasonal to interannual variations in the intensity and central position of the surface Kuroshio east of Taiwan, J. Geophys. Res.-Oceans, 118, 4305–4316, 2013.

Hu, J., Kawamura, H., Li, C., Hong, H., and Jiang, Y.: Review on current and seawater volume transport through the Taiwan Strait, J. Oceanogr., 66, 591–610, 2010.

Huang, D., Zeng, D., Ni, X., Zhang, T., Xuan, J., Zhou, F., Li, J., and He, S.: Alongshore and cross-shore circulations and their response to winter monsoon in the western East China Sea, Deep-Sea Res. Pt. II, 124, 6–18, doi:10.1016/j.dsr2.2015.01.001, 2016.

Hung, J. J., Chen, C. H., Gong, G. C., Sheu, D. D., and Shiah, F. K.: Distributions, stoichiometric patterns and cross-shelf exports of dissolved organic matter in the East China Sea, Deep-Sea Res. Pt. II, 50, 1127–1145, 2003.

Huthnance, J. M., Mysak, L. A., and Wang, D. P.: Coastal trapped waves, in: Baroclinic Processes on Continental Shelves, Coastal and Estuarine Sciences, edited by: Mooers, C. N. K., American Geophysical Union, Washington, D.C., 1–18, 1986.

Huyer, A.: Shelf circulation, in: The Sea, Vol. 9: Ocean Engineering Science, edited by: Mehaute, B. L. and Hames, D. M., Wiley, New York, 423–466, 1990.

Isobe, A.: Recent advances in ocean-circulation research on the Yellow Sea and East China Sea shelves, J. Oceanogr., 64, 569–584, doi:10.1007/s10872-008-0048-7, 2008.

Johns, W. E., Lee, T. N., Zhang, D., Zantopp, R., Liu, C. T., and Yang, Y.: The Kuroshio east of Taiwan: Moored transport observations from the WOCE PCM-1 array, J. Phys. Oceanogr., 31, 1031–1053, 2001.

Ledwell, J. R., Watson, A. J., and Law, C. S.: Mixing of a tracer in the pycnocline, J. Geophys. Res., 103, 21499–21529, doi:10.1029/98JC01738, 1998.

Lee, J. S. and Matsuno, T.: Intrusion of Kuroshio water onto the continental shelf of the East China Sea, J. Oceanogr., 63, 309–325, 2007.

Liu, C., Wang, F., Chen, X., and VonStorch, J. S.: Interannual variability of the Kuroshio onshore intrusion along the East China Sea shelf break: Effect of the Kuroshio volume transport, J. Geophys. Res.-Oceans, 119, 6190–6209, doi:10.1002/2013JC009653, 2014.

Liu, T., Xu, J., He, Y., Lü, H., Yao, Y., and Cai, S.: Numerical simulation of the Kuroshio intrusion into the South China Sea by a passive tracer, Acta Oceanol. Sin., 35, 1–12, doi:10.1007/s13131-016-0930-x, 2016.

Liu, X., Dong, C., Chen, D., and Su, J.: The pattern and variability of winter Kuroshio intrusion northeast of Taiwan, J. Geophys. Res.-Oceans, 119, 5380–5394, doi:10.1002/2014JC009879, 2014.

Mellor, G. L. and Durbin, P. A.: The structure and dynamics of the ocean surface mixed layer, J. Phys. Oceanogr., 5, 718–728, 1975.

Oey, L. Y., Hsin, Y. C., and Wu, C. R.: Why does the Kuroshio northeast of Taiwan shift shelfward in winter?, Ocean Dynam., 60, 413–426, 2010.

Qiu, B. and Imasato, N.: A numerical study on the formation of the Kuroshio countercurrent and the Kuroshio Branch Current in the East China Sea, Cont. Shelf Res., 10, 165–184, doi:10.1016/0278-4343(90)90028-K, 1990.

Ren, J. L., Xuan, J., Wang, Z. W., Huang, D., and Zhang, J.: Cross-shelf transport of terrestrial Al enhanced by the transition of northeasterly to southwesterly monsoon wind over the East China Sea, J. Geophys. Res.-Oceans, 120, 5054–5073, doi:10.1002/2014JC010655, 2015.

Richardson, H., Basu, S., and Holtslag, A. A. M.: Improving stable boundary-layer height estimation using a stability-dependent critical bulk Richardson number, Bound.-Lay. Meteorol., 148, 93–109, 2013.

Smith, W. H. F. and Sandwell, D. T.: Global sea floor topography from satellite altimetry and ship depth soundings, Science, 277, 1956–1962, 1997.

Su, J. L. and Pan, Y. Q.: On the shelf circulation north of Taiwan, Acta Oceanol. Sin., 6, 1–20, 1987.

Su, J. L., Pan, Y. Q., and Liang, X. S.: Kuroshio intrusion and Taiwan warm current, Oceanology of China Seas, Springer Netherlands, 59–70, 1994.

Takahashi, D. and Morimoto, A.: Mean field and annual variation of surface flow in the East China Sea as revealed by combining satellite altimeter and drifter data, Prog. Oceanogr., 111, 125–139, doi:10.1016/j.pocean.2013.01.007, 2013.

Teague, W., Jacobs, G., Ko, D., Tang, T., Chang, K. I., and Suk, M. S.: Connectivity of the Taiwan, Cheju, and Korea straits, Cont. Shelf Res., 23, 63–77, 2003.

Wang, Y., Jan, S., and Wang, D.: Transports and tidal current estimates in the Taiwan Strait from shipboard ADCP observations (1999–2001), Estuar. Coast. Shelf Sci., 57, 193–199, 2003.

Wei, Y., Huang, D., and Zhu, X. H.: Interannual to decadal variability of the Kuroshio Current in the east china sea from 1955 to 2010 as indicated by in-situ hydrographic data, J. Oceanogr., 69, 571–589, 2013.

Wu, C. R. and Hsin, Y. C.: Volume transport through the Taiwan Strait: a numerical study, Terr. Atmos. Ocean. Sci., 16, 377–391, 2005.

Wu, H.: Cross-shelf penetrating fronts: A response of buoyant coastal water to ambient pycnocline undulation, J. Geophys. Res., 120, 5101–5119, doi:10.1002/2014JC010686, 2015.

Xuan, J., Huang, D., Zhou, F., Zhu, X. H., and Fan, X.: The role of wind on the detachment of low salinity water in the Changjiang Bank in summer, J. Geophys. Res.-Oceans, 117, C10004, doi:10.1029/2012JC008121, 2012a.

Xuan, J., Zhou, F., Huang, D., Zhu, X. H., Xing, C., and Fan, X.: Modelling the timing of major spring bloom events in the central Yellow Sea, Estuar. Coast. Shelf Sci., 113, 283–292, 2012b.

Xuan, J., Yang, Z., Huang, D., Wang, T., and Zhou, F.: Tidal residual current and its role in the mean flow on the Changjiang Bank, J. Mar. Syst., 154, 66–81, doi:10.1016/j.jmarsys.2015.04.005, 2016.

Yang, D., Yin, B., Liu, Z., and Feng, X.: Numerical study of the ocean circulation on the East China Sea shelf and a Kuroshio bottom branch northeast of Taiwan in summer, J. Geophys. Res.-Oceans, 116, C05015, doi:10.1029/2010JC006777, 2011.

Yang, D., Yin, B., Liu, Z., Bai, T., Qi, J., and Chen, H.: Numerical study on the pattern and origins of Kuroshio branches in the bottom water of southern East China Sea in summer, J. Geophys. Res.-Oceans, 117, C02014, doi:10.1029/2011JC007528, 2012.

Yu, L. and Weller, R. A.: Objectively Analyzed air–sea heat Fluxes (OAFlux) for the global oceans, B. Am. Meteorol. Soc., 88, 527 539, 2007.

Yuan, D., Qiao, F., and Su, J.: Cross-shelf penetrating fronts off the southeast coast of China observed by MODIS, Geophys. Res. Lett., 32, L19603, doi:10.1029/2005GL023815, 2005.

Zeng, D. Y., Ni, X., and Huang, D.: Temporal and spatial variability of the Zhe-Min Coastal Current and the Taiwan Warm Current in winter in the southern Zhejiang coastal sea, Sci. Sin. Terrae., 42, 1123–1134, 2012.

Zhao, L. and Guo, X.: Influence of cross-shelf water transport on nutrients and phytoplankton in the East China Sea: A model study, Ocean Sci., 7, 27–43, doi:10.5194/os-7-27-2011, 2011.

Zhou, F., Xue, H., Huang, D., Xuan, J., Ni, X., Xiu, P., and Hao, Q.: Cross shelf exchange in the shelf of the East China Sea, J. Geophys. Res.-Oceans, 120, 1545–1572, doi:10.1002/2014JC010567, 2015.

Zhu, J., Chen, C., Ding, P., Li, C., and Lin, H.: Does the Taiwan Warm Current exist in winter?, Geophys. Res. Lett., 31, L12302, doi:10.1029/2004GL019997, 2004.

7

Observation of dominance of swells over wind seas in the coastal waters of the Gulf of Mannar, India

M. M. Amrutha and V. Sanil Kumar

Ocean Engineering Division, CSIR-National Institute of Oceanography (Council of Scientific and Industrial Research), Dona Paula, Goa 403 004, India

Correspondence to: V. Sanil Kumar (sanil@nio.org)

Abstract. Wind seas typically dominate over swell seas in coastal gulfs. Waves measured at a location having a water depth of 12 m in the near-shore waters of the Gulf of Mannar during a 1-year period (1 May 2015 to 30 April 2016) are used to examine the predominance of wind seas and swells through spectral characterization. The study shows that even though the location is in a gulf, the annual average value (~ 0.84 m) of the significant wave height in this area is comparable to that along the coastal waters of the Indian subcontinent, but the annual maximum value (~ 1.7 m) recorded is much less than that (3 to 5 m) observed in those regions. Also, large seasonal variations are not observed in the wave height. The waves of the study region are under the control of sea breeze, with the maximum in the late evening hours and the minimum in the early morning hours. A 5 % increase in the forcing wind field during the monsoon period improved the comparison statistics between the model wave height and the measured values. A total of 53 % of the surface height variance in the study area is a result of swells from the southeast and south, and the remainder are wind seas from the east and southeast.

1 Introduction

The Gulf of Mannar (GoM) connects the Arabian Sea in the south to the Palk Bay in the north. Palk Bay is a shallow basin with a maximum water depth of ~ 13 m and connects to the Bay of Bengal at its northeastern end (Fig. 1). The western region of the GoM is a marine biosphere reserve and slight changes in the waves and meteorological parameters will have large impacts in this area. The seasonal reversal of the monsoon winds in the north Indian Ocean (Wyrtki, 1971) induces changes in the directionality of the surface waves (Sanil Kumar et al., 2012). The winds are from the southwest in the Indian summer monsoon which lasts from June to September and from the northeast during October to January (winter monsoon). As over the rest of India, the winds in the GoM too reverse with the season (Fig. 2). Winds over the region are much stronger in the summer monsoon (~ 5–7 m s^{-1}) than those during the winter monsoon (~ 3–5 m s^{-1}). Over this region, the winds during February–May are weak with a seasonal average value less than 3 m s^{-1}.

Arena and Guedes Soares (2009) grouped the sea states with bimodal spectra in three sets, i.e., swell dominated seas, wind-sea-dominated seas and mixed seas based on the ratio of the significant wave height of the swells and the wind seas. Generally, the swells propagating from distant storms and the local wind seas comprise the waves in the open ocean (Hanson and Phillips, 1999). Wind seas dominate in coastal regions, bays and gulfs (Hwang et al., 2011). When the spectrum is bimodal, the wind seas and the swells can have different directions and can alter the direction of the littoral drift. The high-frequency wave components govern the momentum flux between the ocean and atmosphere (Cavaleri et al., 2012). The long-period waves cause problems for navigation and offshore operations, and induce large motion on moorings (McComb et al., 2009); hence, it is necessary to know the occurrence of long-period waves at a location. Recently, many studies identified long-period waves in the eastern Arabian Sea (Sanil Kumar et al., 2012; Glejin et al., 2016; Amrutha et al., 2017). Sanil Kumar et al. (2003) observed that, in the Indian waters, the wave energy spectra contain multiple peaks for about 60 % of the year and when the signifi-

Table 1. Studies on waves in Indian waters based on measured wave data covering 1 year and above.

Location	Position	Water depth (m)	Period of data used	Aspects studied	Reference
Ratnagiri, eastern Arabian Sea	16.980° N, 73.258° E	13	1 January 2011–31 December 2011	Short-term statistics of waves	Amrutha and Sanil Kumar (2014)
			1 May 2010–30 April 2012	Seasonal and annual variations in waves, role of sea breeze and land breeze on waves	Glejin et al. (2013)
Eastern Arabian Sea	16.980° N, 73.258° E	13	1 January 2011–31 December 2012	Characteristics of long-period swells	Glejin et al. (2017)
	14.822° N, 74.052° E	15			
	14.304° N, 74.391° E	9			
Karwar, eastern Arabian Sea	14.822° N, 74.052° E	15	1 January 2011–31 December 2015	Interannual variations wave spectra	Anjali Nair and Sanil Kumar (2017)
Honnavar, eastern Arabian Sea	14.304° N, 74.391° E	9	1 January 2009–31 December 2015	Seasonal variation in wave characteristics and spatial and temporal variations of wave energy	Amrutha and Sanil Kumar (2016)
	14.307° N, 74.291° E	30	18 April 2014–18 August 2014 1 June 2015–31 July 2015		
Off Dhanushkodi, Gulf of Mannar	9.113° N, 79.407° E	12	9 February 2010 31 March 2011	Seasonal variations in wave characteristics and wave spectra	Gowthaman et al. (2013)
Off Dhanushkodi, Palk Bay	9.319° N, 79.434° E	12	9 February 2010–31 March 2011	Seasonal variations in wave characteristics and wave spectra	Gowthaman et al. (2013)
Puducherry, western Bay of Bengal	11.924° N, 79.851° E	15	1 January 2009–29 December 2011	Seasonal and annual variations in monsoon- and cyclone-induced waves; influence of sea and land breezes on waves	Johnson et al. (2013)
Gangavaram, western Bay of Bengal	17.633° N, 83.267° E	18	1 January 2010–31 December 2010	Spectral wave characteristics and seasonal variations	Sanil Kumar et al. (2014)
Gopalpur, northern Bay of Bengal	19.258° N, 84.907° E	23	1 June 2008–31 May 2009	Variations in wind-sea and swell characteristics	Patra et al. (2016)

cant wave height (H_{m0}) is higher than 2 m, they are generally single-peaked. The multi-peaked wave spectra observed in the coastal region of India are largely dominated by swells (Sanil Kumar et al., 2003). Most of the studies in the Indian waters are in the eastern Arabian Sea (Sanil Kumar et al., 2003, 2012; Glejin et al., 2016; Amrutha et al., 2017; Anjali Nair and Sanil Kumar, 2017) and the western Bay of Bengal (Sundar, 1986; Nayak et al., 2013; Patra and Bhaskaran, 2016). The studies on waves in Indian waters based on measured wave data covering 1 year and above are presented in Table 1. Based on 1-year measured data, Gowthaman et al. (2013) observed that swells are dominant in the northern GoM from January to April and the wind sea dominates in the remaining period of the year. Due to the lack of mea-

surements, our knowledge of the wave characteristics in the western part of the GoM is poor. Hence, in this paper, we describe the wave characteristics based on 1-year measured wave data in the western GoM. Apart from describing the seasonal variations, the present study identifies the predominant wave systems in the western GoM. The interannual variation in wind-sea and swell percentage in the surface variance based on numerical model results is studied at 8° N, 78.25° E. The change in wind-sea and swell percentage during 1 year along the longitude 78.25° E when the waves propagate from 7 to 8.5° N in the GoM is also examined.

The paper is structured as follows. In Sect. 2, the data used in the study and the details of numerical model and validation are described. Section 3 includes the results. First, wave

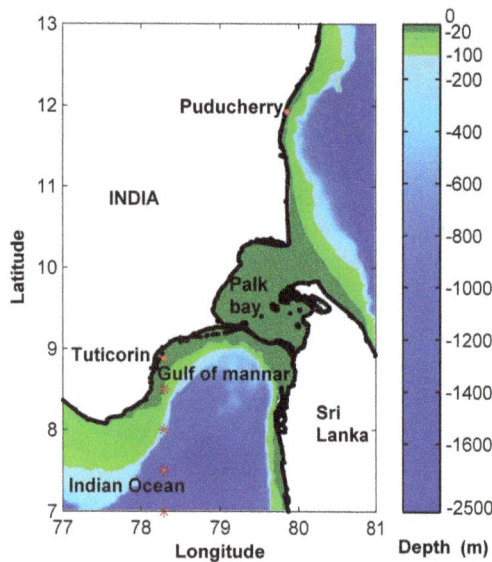

Figure 1. The location of the Waverider buoy mooring in the region of interest in the Gulf of Mannar. The bathymetry is from ETOPO1 (1 arcmin) global relief model (Amante and Eakins, 2009). The star symbols indicate the points considered for studying the percentage change in swells.

parameter statistics are introduced for subsequent use. Second, wave spectral characteristics are presented. Third, the interannual variations are presented based on model data and the results are discussed. The summary of the study is given in Sect. 4.

2   Data and methods

2.1   Data

This study uses waves measured from 1 May 2015 to 30 April 2016 in the western GoM at a location (latitude 8°52′52″ N; longitude 78°17′44″ E) having a water depth of 12 m. The wave data are recorded in a moored directional Waverider buoy continuously at 1.28 Hz for a 1-year period. Heave is measured with 1 cm resolution and with 3 % accuracy. A moored wave buoy may travel around a large crest in a short-crested sea or even be dragged through a large crest if it reaches the limit of its mooring line (Whittaker et al., 2016). Additionally, the Lagrangian buoy motion will still affect the wave measurements of an idealized buoy capable of perfectly following the free surface motions. Although the linear contributions to the free surface elevation measured by a surface-following and fixed sensor are equal, it is generally assumed that this Lagrangian motion will prevent the buoy from measuring the second harmonic component of steep deep-water waves obvious on a wave staff record (Longuet-Higgins, 1986). These effects are not considered in the present study. The wave spectrum is estimated from

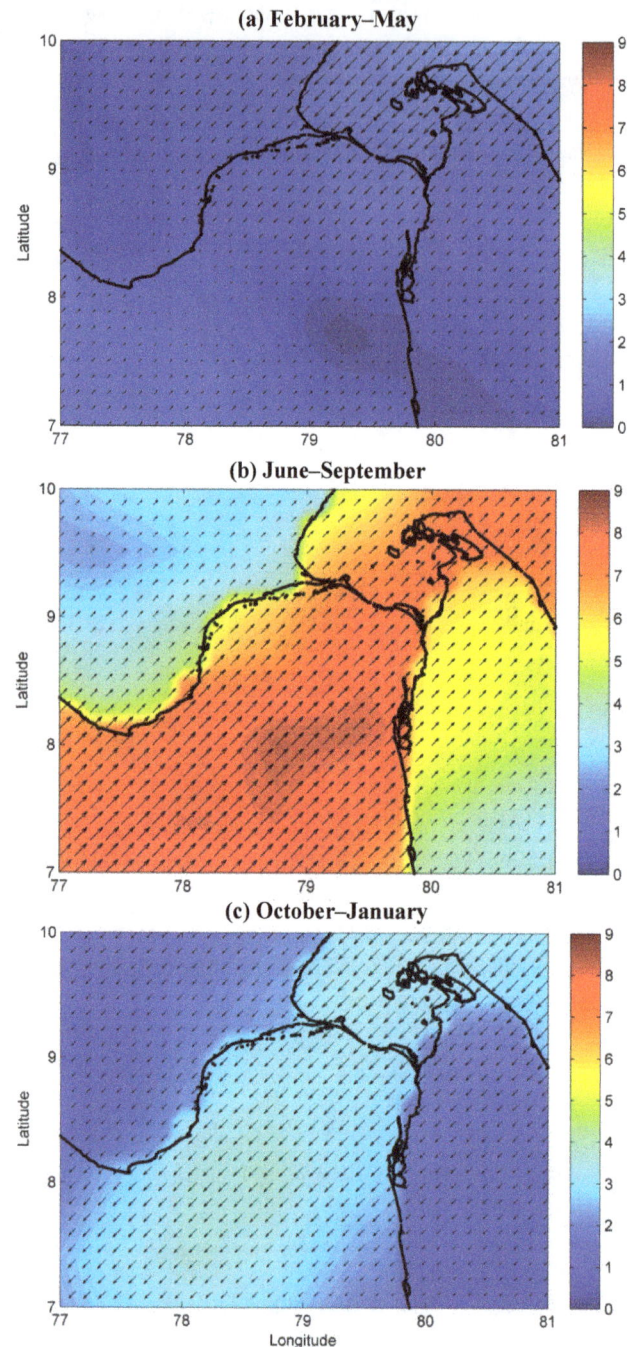

Figure 2. Wind field over the study area in different seasons; **(a)** pre-monsoon (February–May), **(b)** southwest monsoon (June–September) and **(c)** northeast monsoon (October–January). The wind field is from ERA-Interim reanalysis data and the wind speed is in m s^{-1}.

the measured buoy heave data through a fast Fourier transform of eight series, each consisting of 256 data points. The resolution of the wave spectrum is 0.005 Hz from 0.025 to 0.1 Hz and thereafter it is 0.01 up to 0.58 Hz. From the wave spectrum, significant wave height (H_{m0}), mean wave period

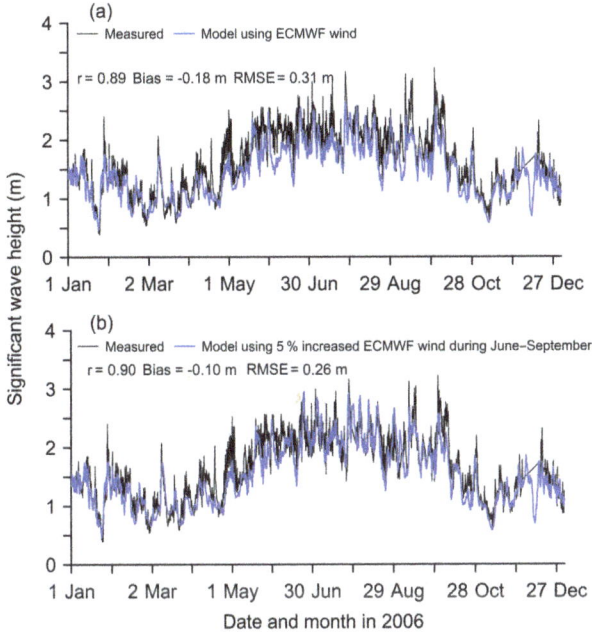

Figure 3. Time history of simulated significant wave height through numerical model forced by **(a)** ECMWF wind fields and **(b)** ECMWF wind speed increased by 5 % during June–September and its comparison with field measurements at 8.27° N, 78.56° E during January–December 2006.

(T_{m02}) and spectral peak period (T_p) are estimated for data covering 30 min, and the wave direction is obtained based on circular moments (Kuik et al., 1988). Other parameters estimated are the maximum spectral energy density, spectral peakedness parameter (Q_p) (Goda, 1970) and spectral narrowness parameter (v) (Longuet-Higgins, 1984). The directional spectra are based on the maximum entropy method (Lygre and Krogstad, 1986). Measurements reported in this article are in Coordinated Universal Time (UTC) and the local time is 5.5 h ahead of UTC. The method proposed by Portilla et al. (2009) is used to separate the wind seas and swells from the measured data. The 1-D separation algorithm is on the assumption that the energy at the peak frequency of a swell cannot be higher than the value of a Pierson–Moskowitz (PM) spectrum with the same frequency. If the ratio between the peak energy of a wave system and the energy of a PM spectrum at the same frequency is above a threshold value of 1, the system is considered to represent wind sea; otherwise, it is taken to be a swell. A separation frequency f_c is estimated following Portilla et al. (2009) and the swell and wind-sea parameters are obtained for frequencies ranging from 0.025 Hz to f_c and from f_c to 0.58 Hz, respectively. The deep-water (8.27° N, 78.56° E) wave data from 1 January to 31 December 2006 are used for comparison of the model results in the deep water.

2.2 Model

The third-generation spectral wave model WAVEWATCH III version 4.18 (Tolman, 1991, 2014) is used in the wave hindcast studies. The source term of the model consists of several parts: a wind–wave interaction, nonlinear wave–wave interactions, a dissipation (whitecapping) and wave–bottom interactions. The treatment of the nonlinear interactions defines a third-generation wave model which is modeled here using the discrete interaction approximation (DIA; Hasselmann et al., 1985). The source term package (ST4) of Ardhuin et al. (2010) is used as the input and dissipation source terms. The JONSWAP parameterization (BT1) is used as the empirical relation for bottom friction (Tolman, 2014). For the southern and large parts of the Indian Ocean domain, the model grid resolution is 0.5° × 0.5° (20–112° E and 70° S–35° N) and 0.1° × 0.1° for the north Indian Ocean (65–90° E and 5–25° N). The model is forced with ERA-Interim (Dee et al., 2011) surface wind fields at every 6 h interval with a spatial resolution of 0.5°. The resolution in wave direction is 10° and the wave frequencies are on a logarithmic scale from 0.04 to 0.5 Hz. The model output for the period of 1 May 2015 to 30 April 2016 at four locations along longitude 78.25° E is used to study the variations in the swell percentage when the waves propagate from 7 to 8.5° N. The model results (H_{m0}, mean wave period and mean wave direction) are also compared with the ECMWF wave data at longitude 78.25° E and latitudes 7, 7.5 and 8° N. Wind-sea and swell percentage from the model output extracted at 8° N is used for studying the trend in the H_{m0} and the wind-sea and swell percentage for 36 years (1980–2015). The trend is estimated based on the slope of the linear best-fit curve to the annual mean value for 36 years.

2.3 Error estimates

The error statistics for significant wave height, mean wave period, mean wave direction from the model against measured and ERA-Interim data were computed based on bias, root mean square error (RMSE), scatter index (SI) and Pearson's linear correlation coefficient (r) as defined below:

$$\text{Bias} = \frac{1}{N}\sum_{i=1}^{N}(A_i - B_i), \tag{1}$$

$$\text{RMSE} = \sqrt{\frac{1}{N}\sum_{i=1}^{N}(A_i - B_i)^2}, \tag{2}$$

$$\text{SI} = \frac{\text{RMSE}}{\overline{B}}, \tag{3}$$

$$r = \frac{\sum_{i=1}^{N}\left|(A_i - \overline{A})(B_i - \overline{B})\right|}{\sqrt{\sum_{i=1}^{N}(A_i - \overline{A})^2}\sqrt{\sum_{i=1}^{N}(B_i - \overline{B})^2}}, \tag{4}$$

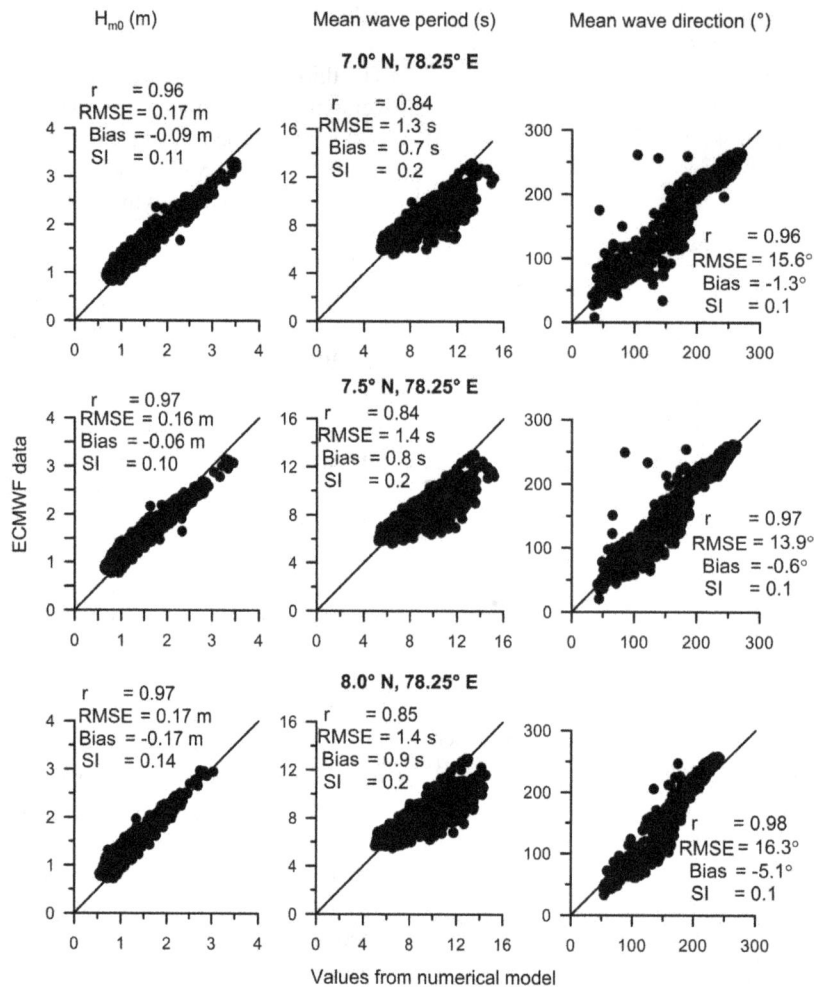

Figure 4. Correlation between model data and ECMWF ERA-Interim data of significant wave height, mean wave period and mean wave direction at different locations.

where A_i and B_i represent the parameter based on numerical model and measured data, N is the number of data points and the over bar represents the mean value.

2.4 Comparison of model output with measured data

The H_{m0} is extracted for the point 8.27° N, 78.56° E from the model and compared with the measured buoy data (Fig. 3). The comparison between model and measured H_{m0} for the year 2006 shows bias, RMSE, SI and r values of −0.18 m, 0.31 m, 0.19 and 0.89, respectively. The wave heights are underpredicted by the model and the underprediction is larger during the monsoon period. A 5 % increase in the forcing wind field during monsoon period improved the comparison statistics of bias, RMSE, SI and r values to −0.10 m, 0.26 m, 0.16 and 0.90, respectively

We compared the model H_{m0}, mean wave period and mean wave direction with ERA-Interim wave data for three deepwater points having the same longitude of 78.25° E and dif-

ferent latitudes of 7.0, 7.5 and 8.0° N (Fig. 4). It is found that the model H_{m0} is in comparatively good agreement with ERA-Interim H_{m0}: bias ranging from −0.17 to −0.06 m shows the underestimation by the model, RMSE has a range of 0.16 to 0.17 m, SI ranges from 0.10 to 0.14 and correlation coefficient is around 0.97. The statistical parameters (bias, RMSE, SI and r) for the wave period have ranges of 0.84 to 0.85 s, 1.3 to 1.4 s, 0.2, 0.84 to 0.85. Similarly, ranges for bias, RMSE and SI for mean wave direction are −5.1 to −0.6°, 13.9 to 16.3° and 0.1, respectively.

3 Result and discussions

3.1 Wave parameters statistics

Based on the maxima, mean value and standard deviation, the statistical analysis of the main wave characteristics is obtained. The directional wave parameter presented here is the

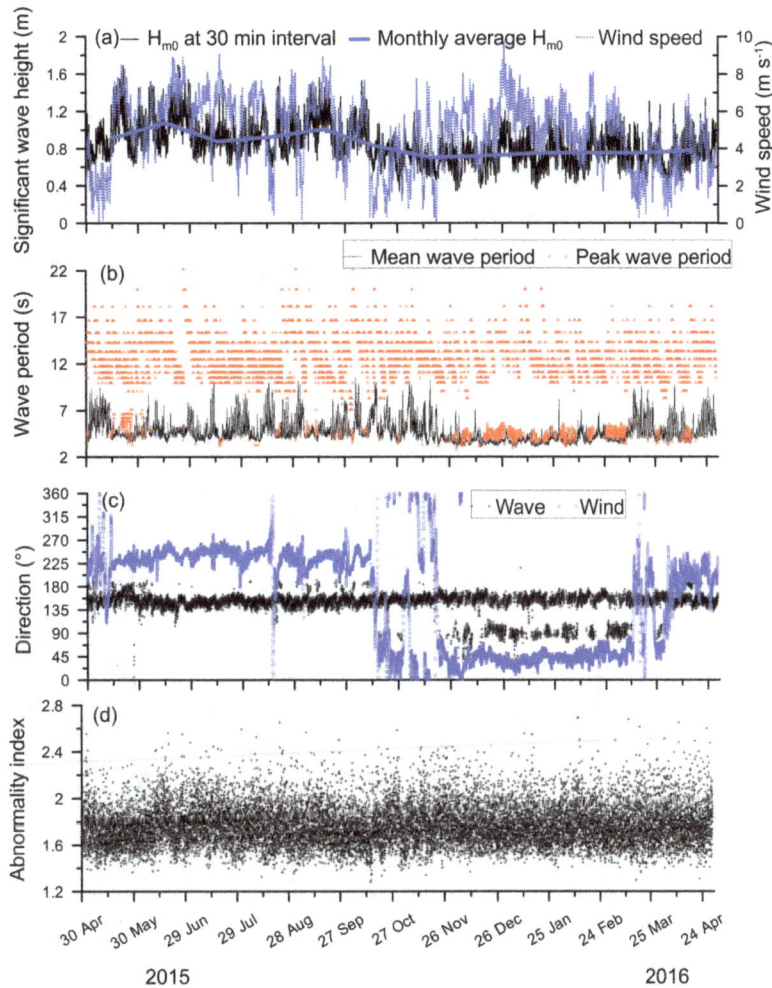

Figure 5. Time series plots of **(a)** significant wave height and wind speed, **(b)** mean wave period for data covering 30 min and peak wave period, **(c)** direction of wind and wave and **(d)** abnormality index. The monthly average significant wave height values are also shown.

Table 2. Statistics of each month: mean value, standard deviation, maximum and minimum of significant wave height along with swell and wind-sea percentage in the measured data.

Month	Mean (m)	Standard deviation (m)	Maximum (m)	Minimum (m)	No. of data	Swell (%)	Wind sea (%)
May 2015	0.91	0.22	1.69	0.46	1488	54.3	45.7
June 2015	1.08	0.19	1.70	0.70	1439	58.0	42.0
July 2015	0.88	0.12	1.54	0.61	1488	64.5	35.5
August 2015	0.93	0.17	1.57	0.60	1488	63.9	36.1
September 2015	1.01	0.19	1.56	0.55	1438	58.8	41.2
October 2015	0.82	0.20	1.53	0.34	1488	63.3	36.7
November 2015	0.70	0.13	1.19	0.34	1439	59.8	40.2
December 2015	0.72	0.17	1.21	0.35	1488	30.7	69.3
January 2016	0.75	0.15	1.27	0.36	1484	28.9	71.1
February 2016	0.75	0.16	1.15	0.42	1389	38.0	62.0
March 2016	0.75	0.16	1.31	0.43	1488	53.6	46.4
April 2016	0.78	0.17	1.34	0.37	1440	60.1	39.9
Annual average	0.84	0.21	1.70*	0.34*	1463	52.8	47.2

*Extremes.

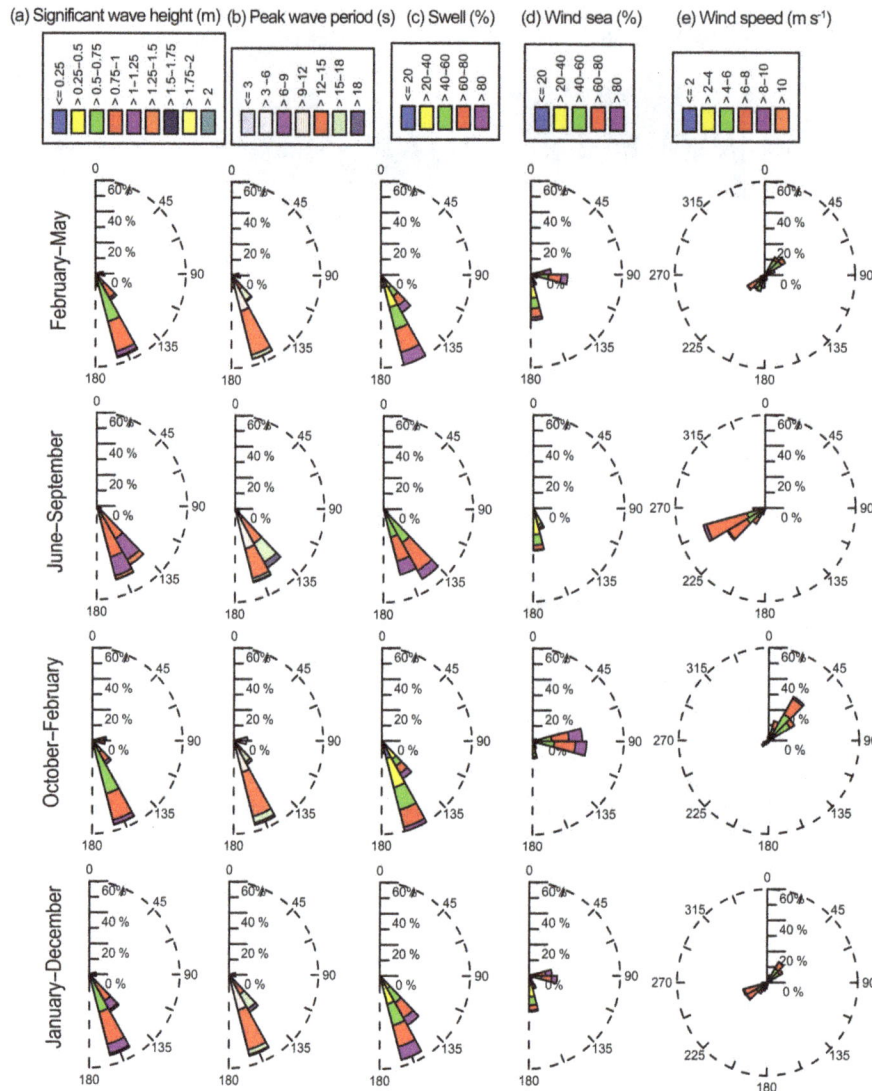

Figure 6. Wave roses from 1 May 2015 to 30 April 2016: **(a)** significant wave height and mean wave direction, **(b)** peak wave period and mean wave direction, **(c)** percentage of swell in the measured data and mean wave direction, **(d)** percentage of wind sea in the measured data and mean wave direction, **(e)** wind speed and direction. The plots represent the direction where the waves come from. The radius of the figure indicates the percentage of the time.

mean direction corresponding to the spectral peak. The annual maximum H_{m0} measured is 1.70 m and the mean value is 0.84 m (Table 2), whereas H_{m0} less than 0.34 m did not occur over the annual period. The monthly mean H_{m0} varied from 0.7 m in November to 1.08 m in June with a standard deviation of 0.12 m. Slightly higher monthly means of H_{m0} occurred during the months of May to September when means of 0.88 to 1.08 m are observed (Table 2). The lowest values occurred during months of November to April. Variation in monthly average H_{m0} in 1 year is small (< 0.4 m) in the study area, which is in contrast to the large variations (~ 2 m) in monthly average values observed in the eastern Arabian Sea (Sanil Kumar et al., 2014). The seasonal variations in the wave height along the Indian coast have been well char-

acterized and show a summer–winter pattern (Sanil Kumar et al., 2012). In the study area, seasonal mean H_{m0} is 0.97, 0.75 and 0.80 m during the summer, winter and fair-weather period (February–May) indicating that large seasonal variations are not observed in H_{m0}. The variation in wave height follows that of the local wind speed. Only during 20 % of the year, H_{m0} exceeded 1 m and no maximum wave height (H_{max}) more than 2.9 m was measured at this location. The Sri Lankan land mass is at a distance of 170 km in the northeast direction and 185 km in the southeast direction from the wave buoy location. In the north–northeast direction, the Rameswaram Island is present at a distance of 110 km. The study area is exposed to the Indian Ocean swells from south and southwest. Hence, the wave field at the study location

Table 3. Average wave parameters and amount of data in different spectral peak frequencies.

Peak frequency (f_p) range (Hz)	Number of data and %	H_{m0} (m)	T_{m02} (s)	Peak wave period (s)
$0.04 < f_p \leq 0.05$	139 (0.79)	0.91	4.98	20.14
$0.05 < f_p \leq 0.06$	1573 (8.96)	0.91	5.17	17.13
$0.06 < f_p \leq 0.07$	3838 (21.86)	0.85	4.97	14.71
$0.07 < f_p \leq 0.08$	4429 (25.23)	0.80	4.80	12.96
$0.08 < f_p \leq 0.10$	4921 (28.03)	0.82	4.71	11.06
$0.10 < f_p \leq 0.15$	368 (2.10)	0.80	4.50	8.69
$0.15 < f_p \leq 0.20$	477 (2.72)	1.05	4.03	5.30
$0.20 < f_p \leq 0.30$	1779 (10.13)	0.86	3.65	4.32
$0.30 < f_p \leq 0.50$	33 (0.19)	0.73	3.33	3.23

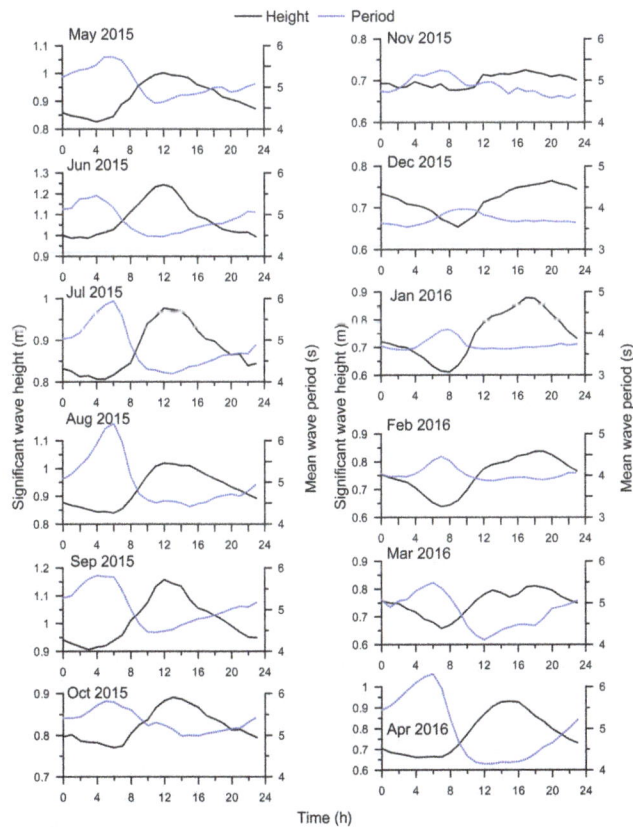

Figure 7. Variation of hourly averaged significant wave height and mean wave period in different months.

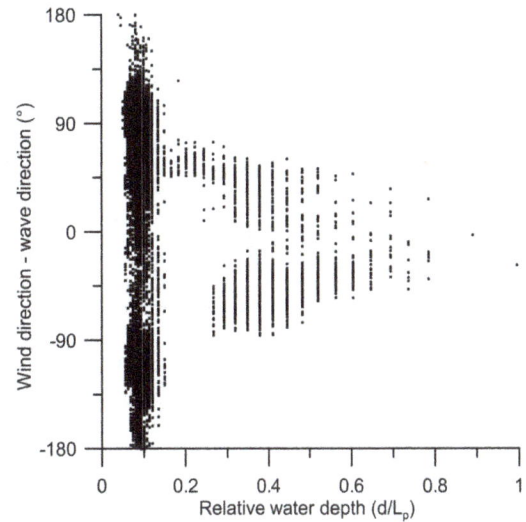

Figure 8. Variation of difference in wind and wave direction with relative water depth.

is partially restricted and high waves are not observed compared to those in the open-sea conditions along the western Bay of Bengal and the eastern Arabian Sea. Even though the study location is in a gulf and high waves are not observed, the annual mean H_{m0} observed at the study region (0.84 m) is comparable to the values (0.7 to 1.1 m) reported for other locations in the western Bay of Bengal and eastern Arabian Sea (Gowthaman et al., 2013; Sanil Kumar et al., 2013).

On 8 November 2015, a depression was formed in the Bay of Bengal and later it was upgraded to a deep depression and crossed the coast of Tamil Nadu near Puducherry with peak wind speeds of 15.3 m s^{-1} (55 km h^{-1}) and a minimum central pressure of 991 hPa (IMD, 2015). Even though the wave measurement location is only 370 km from the track of the depression, the influence of this deep depression is not observed in the measured wave data. This shows that the waves in GoM are not influenced by the storms north of Palk Bay. The gradual increase in wave height seen in May and June is associated with the summer monsoon (Fig. 5). The locally generated waves (wind sea) and the swells are separated to identify different wave components at the study location. Swell H_{m0} up to 1.23 m is recorded with a mean value of 0.58 m, whereas the mean wind-sea H_{m0} is 0.56 m with a maximum value of 1.62 m. The high wind-sea H_{m0} is observed in May with negligible swell (8–10 %) on that occasion; the swell was not always negligible in May.

A wide range (3 to 22 s) is observed in the peak wave period with a mean value of 12 s, indicating that the wave regime of the study area consists of short- to long-period waves. Even though wind seas and swells are present in the study area, the variation over an annual cycle in mean wave period is 3 to 11 s with the mean value of 4.7 s. For all months, the mean wave periods are still short relative to other areas in the western part of the Bay of Bengal and the eastern part of the Arabian Sea (5 to 6.5 s). Distribution of mean wave direction for the three seasons is similar for the swells throughout the year except in the southwest monsoon months. However, for the wind seas, a large variation in wave direction is observed from October to May. Short-period waves ($T_p < 6$ s) approach from east, northeast, southeast and south except in the southwest monsoon months. In the southwest monsoon period, the short-period waves ap-

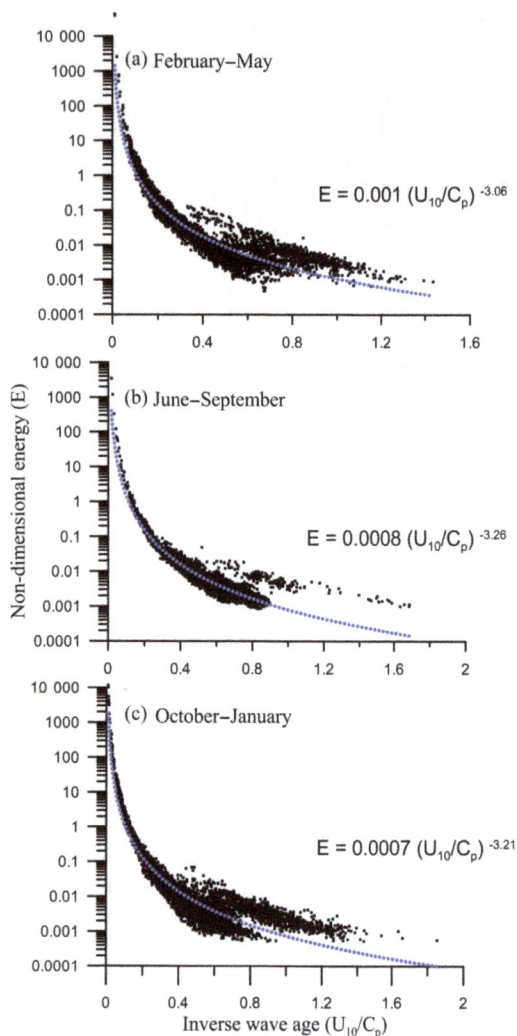

Figure 9. Variation of non-dimensional energy with inverse wave age in different periods: **(a)** February–May, **(b)** June–September and **(c)** October–January.

proach from southeast and south. Waves with a period of more than 8 s are mainly from south and southeast (Fig. 6). Over an annual cycle, 31.6 % of the time, long-period waves ($T_p > 14$ s) are also observed (Table 3) and these swell waves are produced from storms in the Southern Ocean and reach the Indian coast within 5 to 6 days (Amrutha et al., 2017).

A total of 53 % of the surface height variance at the study area over the annual cycle is a result of the south and southeast swells and the balance is the east and southeast wind seas (Table 2). The wave field at the study region shows the dominance of swells over the wind seas, which is in agreement with that reported for the areas around the Indian coast (Sanil Kumar et al., 2003, 2012, 2014; Glejin et al., 2013). However, in the southwest monsoon period, the seasonal average swell contribution is 61 %, whereas it varies from 70 to 79 % for locations around India (Sanil Kumar et al., 2014). Over the western GoM during most of the time, the wave

climate is characterized by sea–land breeze structure and is feeble during November and December (Fig. 7). The waves in the western GoM are under the control of wind seas generated by sea breeze in a diurnal pattern, with the maximum during the late evening and the minimum during the early morning, and are similar to those reported over the southwestern Bay of Bengal (Glejin et al., 2013).

Wave length linked with the mean wave period ranged from 14 to 107 m and the ratio of water depth and wave length (varied from 0.11 to 0.85) is more than 0.5 during 27 % of the time, indicating that for 27 % of the time the measured waves are in the deep-water condition (USACE, 1984). On separating the waves into wind seas and swells, it is observed that 97 % of the wind seas are in the deep-water condition, whereas 98 % of the time the swells satisfy the transitional water condition (ratio of water depth and wave length between 0.05 and 0.5). The water depth to wave length ratio shows that the wave height and the wave direction presented in this article will be influenced by the sea bed. Wind and wave directions corresponding to spectral peak differ by 20 to 120° during most of the time since the measurements are made close ($\sim$ 12 km) to the coast and the wave direction will be mostly aligned to the depth contour due to refraction, whereas such changes in wind direction are not expected (Fig. 5). When the difference between the wind and wave direction is more than 45°, the relative water depth based on spectral peak period (d/L_p) shows that the wave regime is in intermediate and shallow water (Fig. 8). During the deep-water regime, the difference between the wind and wave direction is less than 45°.

The wind is predominantly southwesterly from March to September and from northeast during October to February, and the average wind speed is $4.8\,\mathrm{m\,s^{-1}}$ (Fig. 5a and c). The nature of sea state can be recognized based on the wave age (C / U_{10}) and the steepness of wave (H_{m0} / L), where C is the phase speed corresponding to the mean wave period. Based on wave steepness, Thompson et al. (1984) grouped ocean waves as locally generated waves if the steepness values were greater than 0.025. The wave measurements in this study show that 61 % of the time wave steepness is greater than 0.025. An old sea is defined when wave age is greater than 25, and when the wave age is less than 10, it is a young sea. For the present data, wave age is less than 10 during 98 % of the time, indicating young seas with the presence of young swells. Donelan et al. (1993) identified that the value of the spectrum at full development corresponds to $U_{10} / C_p = 0.83$, where the spectral components above this value are classified as wind seas and those below as swells, and C_p is the phase speed of the waves corresponding to the spectral peak. For the study location, the inverse wave age is more than 0.83 for 7 % of the time (Fig. 9). Inverse wave age values are biased towards lower values with peaks in the range of 0.4–0.8, indicating a young-swell-driven wave regime along the study area.

Table 4. Percentage of single-peaked and multi-peaked wave spectra in different months along with classification as wind sea, swell or mixed, based on the spectral peak period.

Month	Single peak (%)				Multi-peak (%)			
	Total	Wind sea $(T_p < 6)$	Swell $(T_p > 8)$	Mixed $(6 < T_p < 8)$	Total	Wind-sea dominated $(T_p < 6)$	Swell dominated $(T_p > 8)$	Mixed $(6 < T_p < 8)$
May	40.2	0.8	39.4	0	59.8	9.7	45.6	4.6
June	55.8	0.1	55.7	0	44.2	1.7	42.1	0.4
July	57.8	0.0	57.8	0	42.2	1.1	41.1	0.0
August	48.3	0.0	48.3	0	51.7	3.3	48.4	0.0
September	51.3	0.1	51.2	0	48.7	5.6	43.0	0.2
October	48.0	0.1	47.9	0	52.0	4.2	46.9	0.9
November	47.0	0.0	47.0	0	53.0	3.3	49.8	0.0
December	15.0	0.5	14.4	0	85.0	34.2	50.8	0.0
January	9.5	0.5	9.0	0	90.5	31.4	59.1	0.0
February	25.1	0.3	24.8	0	74.9	27.5	47.4	0.0
March	43.3	0.0	43.3	0	56.7	21.8	34.8	0.0
April	56.5	0.1	56.5	0	43.5	8.5	35.0	0.0
Annual average	41.5	0.2	41.3	0.0	58.5	12.7	45.3	0.5

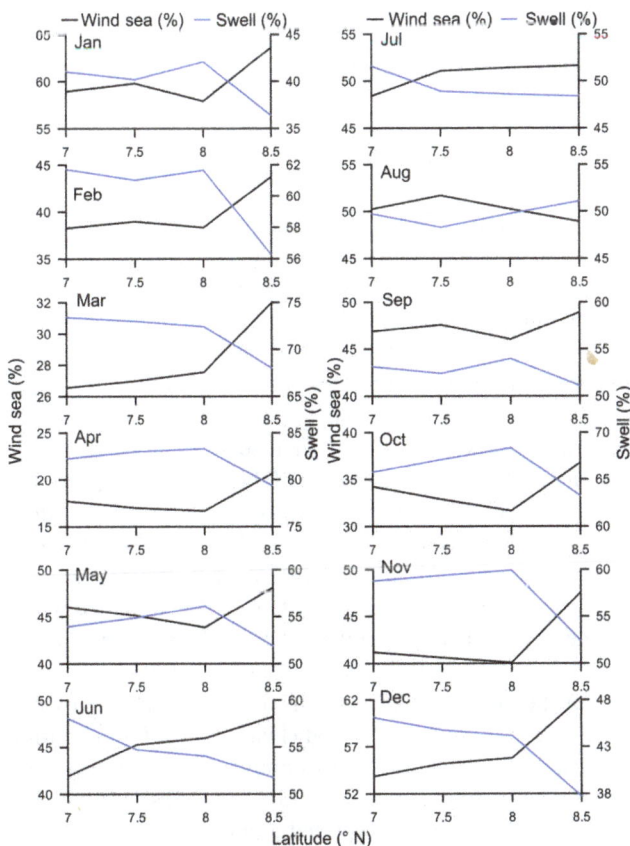

Figure 10. Swell and wind-sea percentage at 7, 7.5, 8 and 8.5° N latitudes in different months based on wave model results.

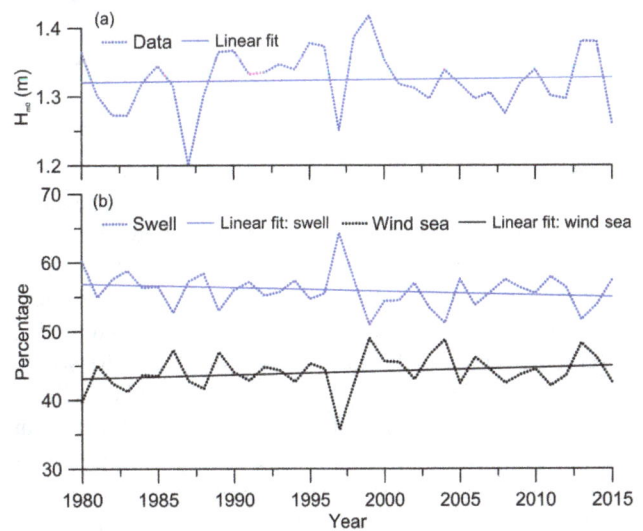

Figure 11. Variation in (a) significant wave height and (b) swell and wind-sea percentage at 8° N, 78.29° E during 1980–2015. Linear trend is also presented.

We have examined the variations in swells and wind seas of the surface height variance as the wave propagates from deep water to the shallow waters based on wave model results. The monthly average swell and wind-sea percentage along a longitude transect of 78.25° E at 7.0°, 7.5, 8.0 and 8.5° N latitudes is presented in Fig. 10. The study shows that, in all months, the percentage of swells decreased as the waves moved from open ocean to bay (7.0 to 8.5° N latitude) with an average decrease of ∼ 5 % in the swells, except

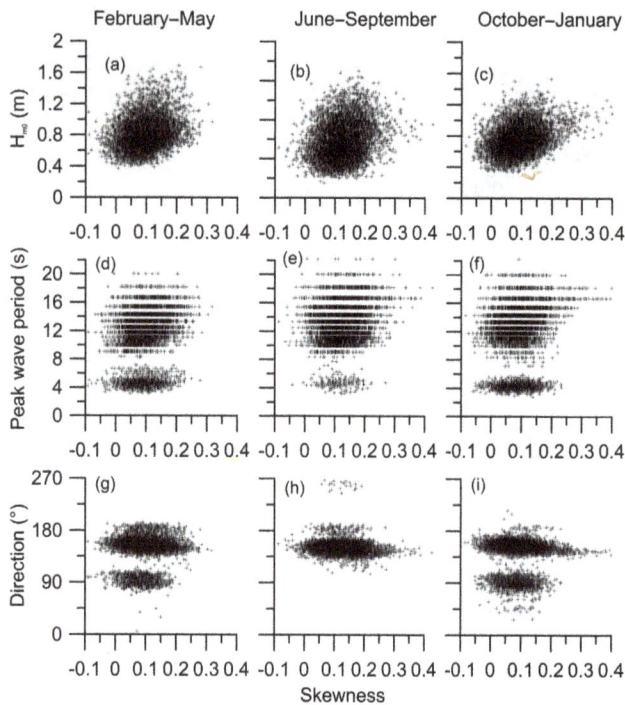

Figure 12. Variation of skewness with significant wave height, mean wave period and mean wave direction in different seasons.

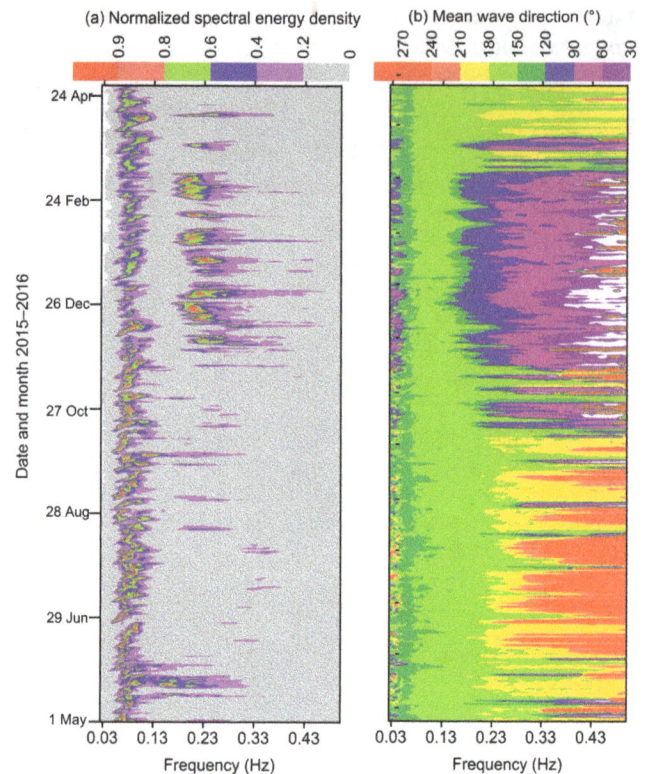

Figure 13. Contour plots of (**a**) normalized spectral energy density and (**b**) mean wave direction from 1 May 2015 to 30 April 2016.

in August. The linear trend of the wind-sea and swell percentage at 8.0° N, 78.25° E during 1980–2015 shows that the trend is slightly positive (0.05 %) for wind seas and negative (0.05 %) for the swells (Fig. 11b). The H_{m0} shows a negligible upward trend (2 cm yr^{-1}) during 1980–2015 (Fig. 11a). Even though the study area is in a gulf region, since its opening is exactly toward the southwest, the upward trend in H_{m0} observed is due to the increase in wave heights in the Southern Ocean (Hemer et al., 2010). Young et al. (2011) indicate a weak increase (0–0.25 % of annual mean value) of H_{m0} in the north Indian Ocean based on the 23-year period (1985–2008) of satellite altimeter measurements.

Nonlinearity in the surface elevations is reflected in sharpening of the wave crests and flattening of the wave troughs, and these effects are reflected in the skewness of the sea surface elevation (Toffoli, 2006). A positive skewness value indicates that the wave crests are bigger than the troughs, and zero skewness indicates linear sea states (Anjali Nair and Sanil Kumar, 2017). Figure 12 shows the variation of skewness with significant wave height, mean wave period and mean wave direction. The waves from the east are mainly the wind seas and gave low skewness values. The high skewness values are for long-period swells ($T_p > 16$ s) superimposed on the wind seas. The increase in nonlinearity with the increase in the H_{m0} is not predominant at this location (Fig. 12a–c). The abnormality index (H_{max}/H_{m0}) more than 2 is observed during 8.5 % of the time, but it is only 1.5 % for waves with H_{m0} more than 1 m (Fig. 5d).

3.2 Wave spectra

In order to have a better understanding of the wave systems in the study area, we show the characterization of waves through the analysis of each individual spectrum. The wave spectra are generally classified as exhibiting either one or two peaks (Henrique et al., 2015). The dominance of wind-sea or swell systems varied for both cases and are presented in this section. Measured data consist of single-peaked wind sea, single-peaked swell and wind-sea-dominated or swell-dominated multi-peaked spectra (Table 4). A majority of the data recorded are multi-peaked spectra (58.5 % of the time) and the multi-peaked spectra are swell dominated 45.3 % of time and wind sea dominated 12.7 % of the time. The peak frequency (f_p) is shown to be unstable when swell and wind-sea peak energies are similar. The single-peak spectra are mainly swell dominated. Gowthaman et al. (2013) observed that in the northern GoM, swells are predominant in GoM during the non-monsoon period (January–April), and during the rest of the year wind seas dominate. The multi-peaked spectra observed in the present study area are slightly higher than those (37 to 54 %) reported along the eastern Arabian Sea (Sanil Kumar et al., 2014).

The energy distribution of waves over a range of frequencies with time is studied by plotting the normalized wave spectral energy density in the time–frequency frame. Each

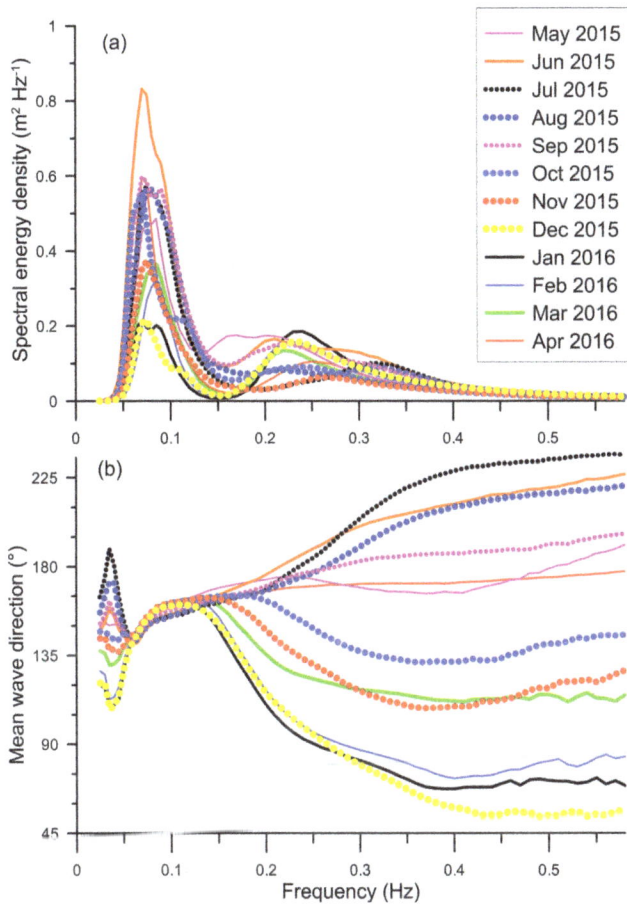

Figure 14. Monthly average wave spectrum **(a)** and mean wave direction **(b)** from May 2015 to April 2016.

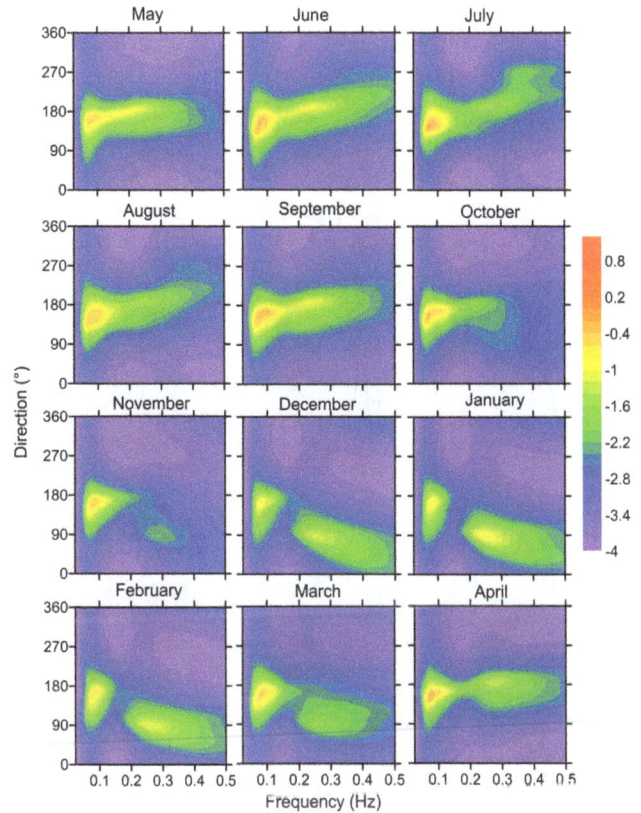

Figure 15. Monthly average directional wave spectrum during different months. The color bar is for spectral energy ($m^2 \, deg^{-1} \, Hz^{-1}$). The spectral energy is shown in logarithmic scale (base 10).

wave spectrum is normalized by the maximum wave spectral energy density of the respective spectrum. Normalized spectral energy density plots in the time–frequency field indicate the predominance of spectral energy in frequency bands 0.06–0.09 Hz (corresponding to swells) during most of the time except from November to March during which the energy is 0.18–0.24 Hz (corresponding to wind seas; Fig. 13). The monthly average wave spectrum shows that the wave spectrum is swell dominated in all months except December and January during which the wind seas dominate (Fig. 14). The peak of the swell part of the monthly averaged wave spectrum varied from 0.07 to 0.08 Hz. Gowthaman et al. (2013) observed that dominance of swells is at its maximum (98 %) during March and dominance of wind seas is at its maximum (94 %) during October in the northern GoM.

Waves of different frequency have different directions. Long-period swells ($T_p > 14$ s) and the intermediate-period waves ($14 > T_p > 6$ s) vary from 150 to 180°, whereas the wind-sea direction varies from southwest to northeast (Fig. 13). Waves with the period less than 4 s come from the northeast and east during November–March and from south to southwest during the remaining period (Fig. 14b).

The monthly average wave spectrum has a similar direction for the region from 0.06 to 0.13 Hz in all months, whereas the average monthly direction varies significantly for regions beyond this frequency range (Fig. 14b). For example, waves with the period less than 3 s come from the northeast during December–February, southwest during June–August and southeast during the remaining period. The monthly mean directional wave spectrum shows the spread of spectral energy in different frequencies and direction (Fig. 15). Two well-separated peaks in spectral energy are observed from November to April when the winter monsoon is active.

The wave spectra grouped into different frequency ranges show that the peak frequency of a large number of wave spectra (~ 75 %) falls between 0.06 and 0.1 Hz with an average H_{m0} of 0.82 m. The mean wave spectra for different peak frequency ranges show that for all groups, double-peaked wave spectra are observed (Fig. 16). The intensity of the secondary peak increased as the spectral peak frequency shifted from a low to a high frequency. The relative distance between the two peaks of the wave spectrum represented by the quotient between the mean wave period of the swell components and the mean period of the wind-sea part varied from 1.9 to 5.8 with a mean value of 3.6 and the larger values indi-

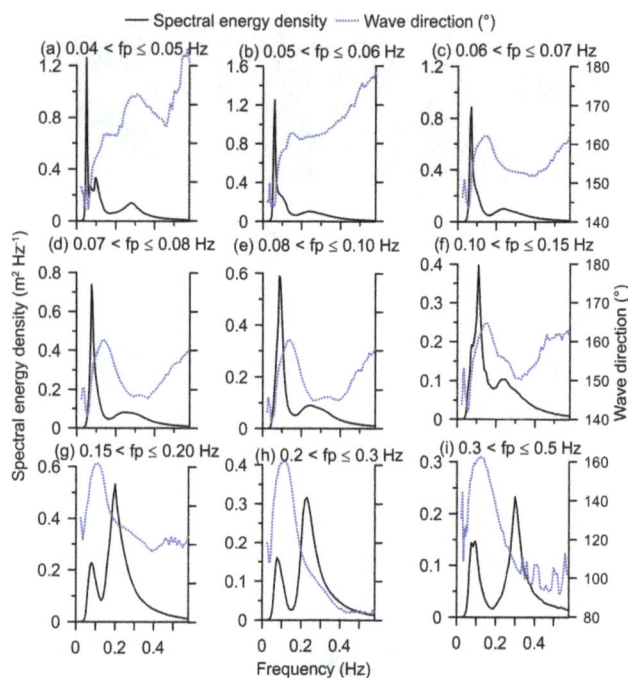

Figure 16. Plot of average spectral energy density and average mean wave direction of waves grouped under different peak frequency bins.

cate more distance between the two peaks of the spectrum. During the study period, the spectral narrowness parameter (ν) has an average value of ~ 0.64 and is marginally higher (~ 0.7 to 0.9) when a multi-modal wave spectrum consisting of high-frequency local waves and the swells from the south Indian Ocean are present. The values of the spectral peakedness parameter ranged between 2 and 3 for high waves, and most of the time the spectral peakedness parameter tends to be smaller since the spectral energy is distributed across the swell band.

4 Concluding remarks

The 1-year measured records of waves show that the waves are lower in the western Gulf of Mannar than in the eastern Arabian Sea, and the variation in the wave height in different seasons is also less in the study area. A total of 53 % of the surface height variance in the study area is a result of swells from the southeast and south and the remainder are wind seas from the east and southeast. The seasonal average swell contribution is less than that observed for other locations around India. For a majority of the time, multi-peaked spectra (58.5 % of the time) are observed. Even though the study area is in a gulf region, since its opening is exactly toward the southwest and the winds also blow from the southwest during the summer monsoon, the monthly mean wave spectrum is swell-dominated in all months, with the exception of December and January during which the wind seas

dominate. The percentage of swells in the surface variance decreased as the waves moved from open ocean to bay (7 to 8.5° N latitude) in all months except August. Over an annual cycle, long-period waves ($T_p > 14$ s) are observed 31.6 % of the time. The wave age of the recorded data is less than 10 during 98 % of the time, signifying that the measured waves are a young sea mixed with swells. An increase in nonlinearity with the increase in significant wave height is not prominent at this location. During 1980–2015, the significant wave height shows a negligible upward trend (2 cm yr^{-1}).

Competing interests. The authors declare that they have no conflict of interest.

Acknowledgements. The authors thankfully acknowledge the CSIR, New Delhi, for facilitating the research work and MoES, New Delhi, for the partial financial support given for this research. We thank T. M. Balakrishnan Nair, A. Nherakkol, Jeyakumar, INCOIS, for the help. J. Mohanraj, Kamaraj College, Tuticorin, provided the logistics during data collection and the deployment of the buoy. The deep-water wave data used for comparison of the model results are provided by National Institute of Ocean Technology, Chennai. We thank the editor John M. Huthnance and the two reviewers for the critical comments and suggestions which improved the contents of the paper. This work contributes part of the Ph.D. work of the first author. The authors acknowledge the high-performance computing resources made available at CSIR-NIO for conducting the research reported in this paper. This work has received NIO contribution no. 6087.

Edited by: John M. Huthnance

References

Amante, C. and Eakins, B. W.: ETOPO1 1 arc-minute global relief model: Procedures, data sources and analysis, NOAA Tech. Memo. NESDIS NGDC-24, Boulder, Colorado, USA, 19 pp., 2009.

Amrutha, M. M. and Sanil Kumar, V.: Spatial and temporal variations of wave energy in the nearshore waters of the central west coast of India, Ann. Geophys., 34, 1197–1208, https://doi.org/10.5194/angeo-34-1-2016, 2016.

Amrutha, M. M., Sanil Kumar, V., and Jesbin, G.: Observations of long-period waves in the nearshore waters of central west coast of India during the fall inter-monsoon period, Ocean Eng., 131, 244–262, https://doi.org/10.1016/j.oceaneng.2017.01.014, 2017.

Anjali Nair, M. and Sanil Kumar, V.: Wave spectral shapes in the coastal waters based on measured data off Karwar on the west coast of India, Ocean Sci., 13, 365–378, https://doi.org/10.5194/os-13-365-2017, 2017.

Ardhuin, F., Rogers, E., Babanin, A. V., Filipot, J. F., Magne, R., Roland, A., van der Westhuysen, A., Queffeulou, P., Lefevre, J. M., Aouf, L., and Collard, F.: Semiempirical dissipation source functions for ocean waves. Part I: Definition, calibration, and validation, J. Phys. Oceanogr., 40, 1917–1941, 2010.

Arena, F. and Guedes Soares, C.: Nonlinear high wave groups in bimodal sea states, J. Waterway, Port, Coast. Ocean Eng., 135, 69–79, 2009.

Cavaleri, L., Fox-Kemper, B., and Hemer, M.: Wind-waves in the coupled climate system, B. Am. Meteor. Soc., 93, 1651–1661, 2012.

Dee, D. P., Uppala, S. M., Simmons, A. J., Berrisford, P., Poli, P., Kobayashi, S., Andrae, U., Balmaseda, M. A., Balsamo, G., Bauer, P., Bechtold, P., Beljaars, A. C. M., van de Berg, L., Bidlot, J., Bormann, N., Delsol, C., Dragani, R., Fuentes, M., Geer, A. J., Haimberger, L., Healy, S. B., Hersbach, H., Hólm, E. V., Isaksen, L., Kållberg, P., Köhler, M., Matricardi, M., McNally, A. P., Monge-Sanz, B. M., Morcrette, J.-J., Park, B.-K., Peubey, C., de Rosnay, P., Tavolato, C., Thépaut, J.-N., and Vitart, F.: The ERA-Interim reanalysis: Configuration and performance of the data assimilation system, Q. J. Roy. Meteor. Soc., 137, 553–597, 2011.

Donelan, M. A., Dobsen, F. W., Smith, S. D., and Anderson, R. J.: On the dependence of sea surface roughness on wave development, J. Phys. Oceanogr., 23, 2143–2149, 1993.

Glejin, J., Sanil Kumar, V., Balakrishnan Nair, T. M., and Singh, J.: Influence of winds on temporally varying short and long period gravity waves in the near shore regions of the eastern Arabian Sea, Ocean Sci., 9, 343–353, https://doi.org/10.5194/os-9-343-2013, 2013.

Glejin, J., Sanil Kumar, V., Amrutha, M. M., and Singh, J.: Characteristics of long-period swells measured in the in the near shore regions of eastern Arabian Sea, Int. J. Nav. Arch. Ocean Eng., 8, 312–319, https://doi.org/10.1016/j.ijnaoe.2016.03.008, 2016.

Goda, Y.: Numerical experiments on wave statistics with spectral simulation, Report Port and Harbour Research Institute, Japan, 9, 3–57, 1970.

Gowthaman, R., Sanil Kumar, V., Dwarakish, G. S., Soumya, S. M., Singh, J., and Ashok Kumar, K.: Waves in Gulf of Mannar and Palk Bay around Dhanushkodi, Tamil Nadu, India, Curr. Sci. India, 104, 1431–1435, 2013.

Hanson, J. L. and Phillips, O. M.: Wind sea growth and dissipation in the open ocean, J. Phys. Oceanogr., 29, 1633–1648, 1999.

Hasselmann, S., Hasselmann, K., Allender, J. H., and Barnett, T. P.: Computations and parameterizations of the nonlinear energy transfer in a gravity-wave spectrum, Part II: parameterizations of the nonlinear energy transfer for application in wave models, J. Phys. Oceanogr., 15, 1378–1391, 1985.

Hemer, M. A., Church, J. A., and Hunter, J. R.: Variability and trends in the directional wave climate of the Southern Hemisphere, Int J. Climatol., 30, 475–491, 2010.

Henrique, R., Babanin, A. V., Schulz, E., Hemer, M. A., and Durrant, T. II.: Observation of wind-waves from a moored buoy in the Southern Ocean, Ocean Dynam., 65, 1275–1288, 2015.

Hwang, P. A., García-Nava, H., and Ocampo-Torres, F. J.: Dimensionally consistent similarity relation of ocean surface friction coefficient in mixed seas, J. Phys. Oceanogr., 41, 1227–1238, 2011.

IMD: Deep Depression over the Bay of Bengal (08-10 November 2015): A Report, Cyclone Warning Division, India Meteorological Department, New Delhi, http://www.rsmcnewdelhi.imd.gov.in/images/pdf/publications/ preliminary-report/DD_08112015.pdf (last access: 10 December 2016), 2015.

Johnson, G., Sanil Kumar, V., and Balakrishnan Nair, T. M.: Monsoon and cyclone induced wave climate over the near shore waters off Puduchery, south western Bay of Bengal, Ocean Eng., 72, 277–286, https://doi.org/10.1016/j.oceaneng.2013.07.013, 2013.

Kuik, A. J., Vledder, G. P., and Holthuijsen, L. H.: A method for the routine analysis of pitch and roll buoy wave data, J. Phys. Oceanogr., 18, 1020–1034, 1988.

Longuet-Higgins, M. S.: Statistical properties of wave groups in a random sea state, Phil. T. R. Soc. Lond. A, 312, 219–250, 1984.

Longuet-Higgins, M. S.: Eulerian and Lagrangian aspects of surface waves, J. Fluid Mech., 173, 683–707, 1986.

Lygre, A. and Krogstad, H. E.: Maximum entropy estimation of the directional distribution in ocean wave spectra, J. Phys. Oceanogr., 16, 2052–2060, 1986.

McComb, P., Johnson, D., and Beamsley, B.: Numerical model study to reduce swell and long wave penetration to Port Geraldton, Proceedings of the 2009 Pacific Coasts and Ports Conference, Wellington, New Zealand, 16–18 September 2009, 1–6 pp., 2009.

Nayak, S., Bhaskaran, P. K., Venkatesan, R., and Dasgupta, S.: Modulation of local wind waves at Kalpakkam from remote forcing effects of Southern Ocean swells, Ocean Eng., 64, 23–35, 2013.

Patra, A. and Bhaskaran, P. K.: Temporal variability in wind-wave climate and its validation with ESSO NIOT wave atlas for the head Bay of Bengal, Clim. Dynam., 1–18, https://doi.org/10.1007/s00382-016-3385-z, 2016.

Patra, S. K., Mishra, P., Mohanty, P. K., Pradhan, U. K., Panda, U. S., Ramana Murthy, M. V., Sanil Kumar, V., and Balakrishnan Nair, T. M.: Cyclone and monsoonal wave characteristics of northwestern Bay of Bengal: long-term observations and modeling, Nat. Hazards, 82, 1051–1073, https://doi.org/10.1007/s11069-016-2233-0, 2016.

Portilla, J., Ocampo-Torres, F. J., and Monbaliu, J.: Spectral Partitioning and Identification of Wind Sea and Swell, J. Atmos. Ocean. Technol., 26, 107–122, 2009.

Sanil Kumar, V., Anand, N. M., Kumar, K. A., and Mandal, S.: Multipeakedness and groupiness of shallow water waves along Indian coast, J. Coast. Res., 19, 1052–1065, 2003.

Sanil Kumar, V., Glejin, J., Dora, G. U., Sajiv, P. C., Singh, T., and Pednekar, P.: Variations in near shore waves along Karnataka, west coast of India, J. Earth Syst. Sci., 121, 393–403, https://doi.org/10.1007/s12040-012-0160-3, 2012.

Sanil Kumar, V., Dubhashi, K. K., Balakrishnan Nair, T. M., and Singh, J.: Wave power potential at few shallow water locations around Indian coast, Curr. Sci. India, 104, 1219–1224, 2013.

Sanil Kumar, V., Dubhashi, K. K., and Balakrishnan Nair, T. M.: Spectral wave characteristics off Gangavaram, Bay of Bengal, J. Oceanography, 70, 307–321, https://doi.org/10.1007/s10872-014-0223-y, 2014.

Sanil Kumar, V., Shanas, P. R., and Dubhashi, K. K.: Shallow water wave spectral characteristics along the eastern Arabian Sea, Nat. Hazards, 70, 377–394, https://doi.org/10.1007/s11069-013-0815-7, 2014.

Sundar, V.: Wave characteristics off the South East Coast of India, Ocean Eng., 13, 327–338, 1986.

Thompson, W. C., Nelson, A. R., and Sedivy, D. G.: Wave group anatomy of ocean wave spectra, Proceeding of 19th conference on Coastal engineering, Am. Soc. Civil Eng., 1, 661–677, 1984.

Toffoli, A., Onorato, M., and Monbaliu, J.: Wave statistics in unimodal and bimodal seas from a second-order model, Eur. J. Mech. B-Fluids, 25, 649–661, 2006.

Tolman, H. L.: A third-generation model for wind waves on slowly varying, unsteady, and inhomogeneous depths and currents, J. Phys. Oceanogr., 21, 782–797, 1991.

Tolman, H. L.: User manual and system documentation of WAVEWATCH-III version 4.18, NOAA/NCEP, Tech. Note, Maryland, USA, 311 pp., 2014.

USACE: Shore Protection Manual, Department of the Army, U.S. Corps of Engineers, Washington, DC, 3-81–3-84, 1984.

Whittaker, C. N., Raby, A. C., Fitzgerald, C. J., and Taylor, P. H.: The average shape of large waves in the coastal zone, Coast. Eng., 114, 253–264, 2016.

Wyrtki, K.: Oceanographic atlas of the international Indian Ocean expedition, Washington, DC, National Science Foundation, 1–531, 1971.

Young, I. R., Zieger, S., and Babanin, A. V.: Global trends in wind speed and wave height, Science, 332, 451–455, 2011.

8

Trapped planetary (Rossby) waves observed in the Indian Ocean by satellite borne altimeters

Yair De-Leon and Nathan Paldor

Fredy and Nadine Herrmann Institute of Earth Sciences, The Hebrew University of Jerusalem, Edmond J. Safra Campus, Givat Ram, Jerusalem, 9190401, Israel

Correspondence to: Nathan Paldor (nathan.paldor@huji.ac.il)

Abstract. Using 20 years of accurately calibrated, high-resolution observations of sea surface height anomalies (SSHAs) by satellite borne altimeters, we show that in the Indian Ocean south of the Australian coast the low-frequency variations of SSHAs are dominated by westward propagating, trapped, i.e., non-harmonic, Rossby (Planetary) waves. Our results demonstrate that the meridional-dependent amplitudes of the SSHAs are large only within a few degrees of latitude next to the southern Australian coast while farther in the ocean they are uniformly small. This meridional variation of the SSHA signal is typical of the amplitude structure in the trapped wave theory. The westward propagation speed of the SSHA signal is analyzed by employing three different methods of estimation. Each one of these methods yields speed estimates that can vary widely between adjacent latitudes but the combination of at least two of the three methods yields much smoother variation. The estimates obtained in this manner show that the observed phase speeds at different latitudes exceed the phase speeds of harmonic Rossby (planetary) waves by 140 to 200 % (which was also reported in previous studies). In contrast, the theory of trapped Rossby (planetary) waves in a domain bounded by a wall on its equatorward side yields phase speeds that approximate more closely the observed phase speeds in the study area.

1 Introduction

The analysis of observations of sea surface height anomalies (SSHAs), i.e., the deviation of the sea surface height from its mean value at any given point in the ocean, was carried out since the 1990s in various parts of the world ocean by various satellite borne altimeters. Chelton and Schlax (1996), for example, analyzed the first 3 years of altimetry data collected by the TOPEX/Poseidon satellite in the world ocean, Zang and Wunsch (1999) analyzed 5 years of TOPEX/Poseidon data in the North Pacific Ocean and Osychny and Cornillon (2004) analyzed 6 years of modified TOPEX/Poseidon data in the North Atlantic Ocean. Additional observational studies are summarized in Barron et al. (2009) and references therein.

In most parts of the ocean the satellite observations showed a ubiquitous and pronounced westward migration of SSHAs with amplitude of a few centimeters. This westward, rather than eastward, propagation led to the interpretation of these observations as a surface manifestation of the first baroclinic mode of planetary (also known as Rossby) waves that propagate westward (i.e., their phase speed is negative) in the ocean thermocline. Recent studies (e.g., Chelton et al., 2007, 2011), however, argue that the observed SSHA features belong to mesoscale eddies and are not surface manifestations of planetary waves in the thermocline but this change of view has no effect on the estimate of the westward propagation speed since these eddies propagate westward at the phase speed of long Rossby waves (Chelton et al., 2011; O'Brien et al., 2013; Polito and Sato, 2015; see also Nof, 1981, for theoretical estimate of eddy migration rate on the β-plane).

The quantification of the rate of westward propagation of the observed SSHA features is based on the construction of time–longitude (also known as Hovmöller) diagrams at a given latitude. The slopes of contours on these diagrams are proportional to the propagation speed of the SSHA features. These slopes can be calculated using methods that are commonly employed in image processing such as the

Radon transform (or its more recent alternative – the variance method) and the two-dimensional fast Fourier transform (2D FFT), which are described in details in Sect. 2.2 below.

Previous studies of the westward propagation of observed SSHAs in mid-latitudes have all yielded rates of westward propagation that are faster than the phase speeds predicted by the harmonic planetary wave theory (see below for details). Explanations for these underestimates by the harmonic theory were proposed, which are based on considerations that involve either the addition of mean zonal flows in the equations (Killworth el al., 1997 and see also Colin de Verdière and Tailleux, 2005, who emphasized the curvature effect of the mean flow rather the mean flow itself) or the influence of bottom topography (Tailleux and McWilliams, 2001) while Killworth and Blundell (2005) applied a combination of these two effects. Watanabe et al. (2016) showed that the standard linear wave theory can be tailored to fit the observations in the tropics by considering parameters such as effective β (that includes the meridional gradient of the background potential vorticity) and forcing by Ekman pumping. LaCasce and Pedlosky (2004) argued that due to baroclinic instability the wave structure is changed and becomes more barotropic so it propagates faster and no mean flow is required. Along similar lines, Hochet et al. (2015) suggested that the assumption that observations are of the first baroclinic mode cannot be made a priori, but the vertical structure is predicted from the altimetry data. Thus, they found that in some regions the vertical structure is more barotropic than baroclinic; therefore, the theoretical phase speed is larger and no discrepancy exists between theory and observations. By incorporating physical elements that are not included in the simple linear wave theory of the shallow water equations (e.g., velocity shear, nonlinear terms, topography, mean flows and juxtaposing barotropic and baroclinic modes) these (and other) past studies were successful in bridging some of the discrepancies found between the observed SSHA propagation speeds and the phase speeds of harmonic wave theory.

In contrast to the phase speed, other wave characteristics, such as the meridional variations of the SSHA amplitudes (which are predicted by the harmonic theory to be sinusoidal), have never been verified in these past studies. The reason is that in the framework of the harmonic theory (see more details below) the central latitude, ϕ_0, which determines the origin of the y (meridional) coordinate, is determined by the latitude of observation. Thus, observations of SSHAs at adjacent latitudes cannot be compared to one another since their y dependencies are determined by the same equations but with different origins so the same y coordinate denotes different points in the two sets of equations.

The traditional interpretation of these SSHA observations has employed the harmonic theory of westward propagating, low frequency waves that assumes the existence of a zonal channel that bounds the north–south extent on the β-plane. Under these assumptions zonally propagating wave solutions of the shallow water equations can be constructed and ex-

plicit expressions can be derived for both the zonal phase speed of the waves and the spatial structure of their amplitudes. The emerging spatial structure of the waves is oscillatory (harmonic) in both the zonal and meridional coordinates; i.e., the waves simply oscillate with wavenumber k in the zonal direction and wavenumber l in the meridional direction (Pedlosky, 1982; Cushman-Roisin, 1994; Vallis, 2006).

An alternative to the traditional harmonic theory is the trapped wave theory, which was developed on the mid-latitude β-plane by Paldor et al. (2007) and Paldor and Sigalov (2008). In this theory the meridional variation of the wave's amplitude is not harmonic but is given instead by the Airy function (see details in Sect. 4.1 below), and the requirement of two channel walls of the harmonic theory is replaced in this trapped wave theory by a single wall that marks the equatorward boundary of the domain. In sufficiently wide meridional ranges the phase speed of the trapped waves is higher than that of the corresponding harmonic waves by a factor of 2 to 4.

The current study employs the available series of SSHA observations sampled on a $1/4°$ spatial grid, which are compared to the theoretical phase speeds and meridional structures of the height field using the trapped and harmonic wave theories. The comparisons provide a measure of the relevance of the trapped and harmonic wave theories to the observed SSHA fields in the Indian Ocean.

This paper is organized as follows: Sect. 2 provides details of the observations and methods used for estimating the observed phase speed and in Sect. 3 we compare the theoretical and observational meridional variation of the height field in the Indian Ocean south of the Australian coast (which includes the Great Australian Bight). Section 4 describes theoretical expressions for the phase speed and the meridional structure of the height field of the harmonic and trapped wave theories that are compared with SSHA observations in the region of interest in Sect. 5. The paper ends in Sect. 6 with summary and discussion of the findings.

2 Data and methods

2.1 SSHA data

The altimetry products used for a comparison with theory were produced by Ssalto/Duacs and distributed by Aviso, with support from CNES. The data we used are the multimission (i.e., up to four satellites at a given time, e.g., TOPEX/Poseidon, Jason 1, Jason 2, Envisat) gridded sea surface heights, sampled on a $1/4° \times 1/4°$ Cartesian grid once a week from 1 January 1993 to 31 December 2012. These data are improved compared to those used in previous studies since the combination of data from several, present-day satellites enables high-precision altimetry in both time and space at finer resolutions. More details on the way the

SSHA data are produced by Aviso can be found at http://www.aviso.altimetry.fr/duacs/.

The SSHA time series of each grid point in this region were low-pass filtered in the present study by performing a 5-week-running average to eliminate short-term variability such as storms, tides (including the fortnightly component) and other variations of periods less than 1 month. Though this filtering leaves parts of the high-frequency signals in the averaged signal, these parts are minute since the window contains many cycles of the high-amplitude signals such as the M2 tides. Calculations with wider windows of 27 and 53 weeks (done to examine the possible contribution of longer-term variability such as seasonal winds) yielded qualitatively identical results (see details in Sect. 3 below).

2.2 Methods of estimating observed phase speed of SSHAs

The basis for estimating the speed of westward propagation of SSHAs is time–longitude (Hovmöller) diagrams of the SSHA field at fixed latitude. In this diagram the westward propagation is evident from the left-upward tilt of constant SSHA values; i.e., same color contours, and the angle between this tilt and the ordinate is directly proportional to the speed of westward propagation. The diagram provides a time series of the SSHA changes at fixed longitude and a longitude variation series at any particular time so fast Fourier transforms can be easily calculated in time and longitude to yield the frequency and zonal wavenumber spectra of observed SSHAs.

Three objective methods are employed in the literature for calculating the phase speed of waves from time–longitude diagrams. The first method is the frequently used (e.g., Chelton and Schlax, 1996; Chelton et al., 2003; Tulloch et al., 2009) Radon transform that is used in image processing for detecting structures on any digital image (see details in, e.g., Jain, 1989). The Radon transform of a two-dimensional function $f(x, y)$ that describes the intensity of an image at (x, y), such as SSHA values in a given (longitude, time) domain, is the integral of $f(x, y)$ along a line L inclined at an angle θ relative to the ordinate (i.e., $\theta \pm 90°$ relative to the abscissa) and displaced a distance s from the origin. For each angle θ we sum the squares of the values of the integrals along all lines having the same θ (i.e., having different distance s). The angle at which this sum of squares attains its maximum is the most accurate estimate for the orientation of structures with the same SSHA value on the time–longitude diagram. The tangent of this preferred θ is proportional to the sought westward propagation speed. Note that in order to minimize the effect of few very high entries on the sum of squares, we apply the Radon transform to a modified time–longitude diagram where the signal is scaled on the [0,1] interval and the mean of the scaled signal is subtracted.

The second method is a relatively new algorithm (Polito and Liu, 2003; Barron et al., 2009) that constitutes an adaptation of Radon transform to a propagating wave. In this method the **variance** of amplitude values is calculated along the same lines. For each angle θ we average the variances along all lines at different distances s and the westward propagation speed is then determined by the tangent of the angle θ at which the mean of variances is minimal. The third method commonly used (e.g., Zang and Wunsch, 1999; Osychny and Cornillon, 2004) to obtain the observed phase speed is the application of the 2D FFT to the time–longitude diagram to get a frequency–wavenumber (i.e., ω, k) diagram of the signal's amplitude. The phase speed is obtained by locating the values of ω and k where the amplitude is maximal (i.e., maximum spectral coefficient) and calculating $C = \omega/k$ at this point of maximum spectral coefficient. Alternatively, the directionality of the spectral coefficients in the (ω, k) diagram can be found by sweeping over all lines that pass through the origin and inclined at angles ranging from 0° to 180° relative to the abscissa (i.e., ω/k lines). The value of C is then determined as the slope of the line of maximal sum of squares of all spectral coefficients ("total energy").

A comparison between the three methods was made using synthetic signals (De-Leon and Paldor, 2017). Based on the insight gained from the study of synthetic signals, an estimation of the observed phase speed is accepted here only when an isolated peak (characterized by the point at which the derivative changes sign at a clearly defined sharp peak and maintains the same sign in bands that are at least 3° wide on either side of the peak) is evident in at least two of the three methods and the phase speeds that correspond to these peaks agree by better than 10 %.

Note that the observed phase speed is obtained from the Hovmöller diagrams in units of $1/4°$ longitude per week, which is converted to units of 1 cm per second by multiplying the observed phase speed by $4.6 \cos \phi_m$ (where ϕ_m is the latitude of observation).

2.3 The study domain in the Indian Ocean

The trapped wave theory in mid-latitudes applies without any modification to domains of large meridional extent (so the β-plane approximation applies) that are bounded on their equatorward side by a wide zonal boundary. As shown in Fig. 1 such a nearly zonal boundary exists in the Indian Ocean south of Australia. The domain of study extends from the south coast of Australia at about 31.5° S to only about 45° S since south of this latitude, the SSHA field is strongly affected by the nearly 2000 km wide, fast and strongly meandering Antarctic Circumpolar Current (ACC).

3 The meridional structure of SSHAs

The standard deviation of the temporal changes of SSHA observations in each point of this domain over the entire 20 years is shown in Fig. 2a, which clearly demonstrates an

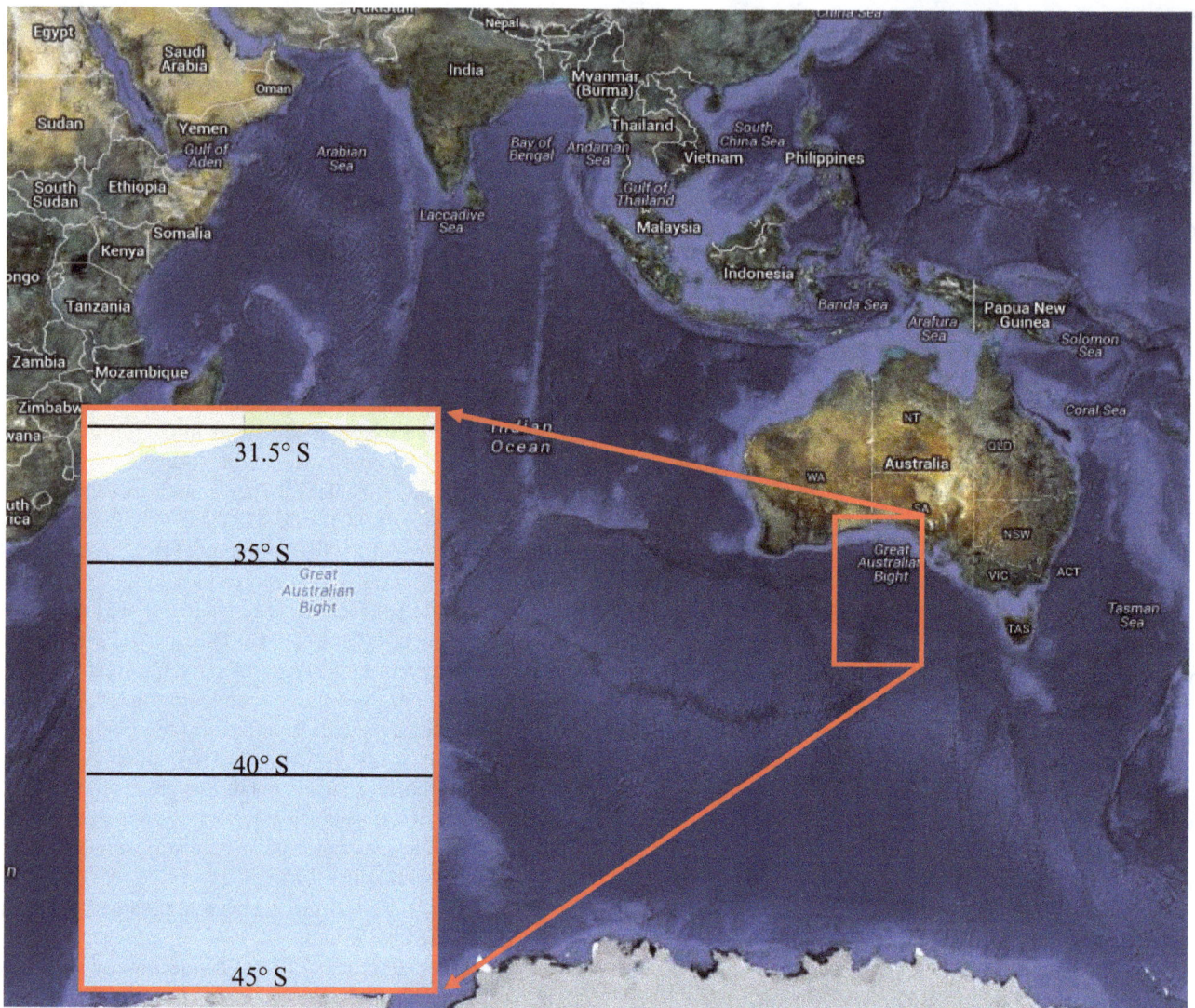

Figure 1. The domain under study in the Indian Ocean with a zoom in on the longitude band of 124.5°–134.5° E, where altimetry data are analyzed (reproduced from Google Maps).

increase in the SSHA signal from 4 to 9 cm over the span of 2–3° from the Australian coast. The meridional structure of the observed SSHAs is clearly non-uniform, while in the harmonic (oscillatory) theory the height field is uniform (i.e., constant) for $l = 0$ and sinusoidal for $l > 0$. Although the ocean depth decreases towards the shore, the observed variability of SSHA signal there cannot be attributed to topography since steady winds affect only the average displacement of the sea surface, which is subtracted from the SSHA signal when the standard deviation is calculated, while the effect of winds of periods shorter than 5 weeks are filtered out by our low-pass filter. In order to examine the possible effect of longer-term winds (seasonal to annual), the calculations were repeated with windows of 27 and 53 weeks. These calculations yielded very similar results to those obtained with the 5 weeks window but with slight decrease in

the amplitude of the main signal near the coast and minute changes in the structure far from it. For the same reason this coastal peak cannot be associated with a mean long-coast current since such a current will not show up on a map of temporal standard deviation. The 5 week filter also eliminates high-frequency waves, such as Kelvin waves and topographic Rossby waves (or continental shelf waves), since in the Great Australian Bight where the slope is 0.01 the period of these waves is O(1 day) (see e.g., Cushman-Roisin, 1994, for harmonic waves and Cohen et al., 2010, for nonharmonic waves).

The mean over all longitudes (Fig. 2b, thick blue line) of the 41 individual latitudinal cross sections (thin light-blue lines) is compared with analytical expressions (described in the next section) for the meridional structure of the height field of both the trapped wave theory (dashed red line) and

(a)

(b)

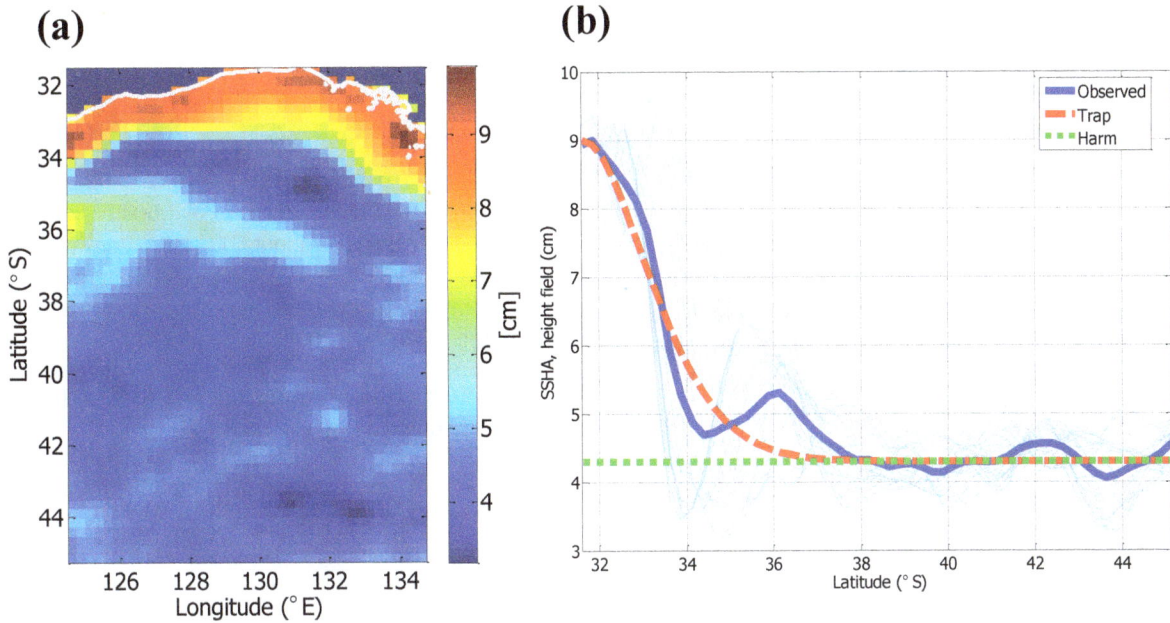

Figure 2. **(a)** The temporal standard deviation of satellite-derived SSHAs over the entire 20-year period poleward of the Great Australian Bight. The coastline is plotted in white. **(b)** Latitudinal cross sections of the data of panel **(a)** every 0.25° longitude (thin light-blue lines), the mean over all longitudes of the latitudinal cross sections (thick blue line) and the analytical expression for the meridional structure of the height field of trapped waves for zero zonal wavenumber and zero meridional mode number (dashed red line; see Eq. 5 below). The maximal trapped wave amplitude is set to match that of the mean observed cross section where the off-shore minimum is about 4 cm since the temporal mean of Aviso's original data is not zero. The analytical expression for the meridional structure of the height field of harmonic waves for $k = 0$ and $l = 0$ is constant, i.e., described by a straight line parallel to the abscissa at arbitrary value of the ordinate; here it is set to match the off-shore minimum of about 4 cm (dotted green line).

the harmonic theory (dotted green line). The decay rates with latitude of both observed and trapped wave theoretical curves are similar in contrast to the flat curve of the harmonic theory. An unexplained minor secondary peak is found near 36° on the observed curve (also evident near 36° S, 125° E in panel a), and upon examining a larger SSHA map it turns out that this secondary peak is an eastward extension of the Leeuwin current that flows poleward along the west coast of Australia between March and July (Godfrey and Ridgway, 1985). Alternatively, this peak can be interpreted as a poleward propagation (into the Indian Ocean) of energy generated in the equatorial Pacific Ocean by the wind and by Ekman pumping, which forms Rossby waves in the study area (Potemra, 2001).

4 Application of wave theories to observations

The relevance of the trapped wave theory to observations can be best assessed by comparing the theoretical phase speed and meridional structure of the waves with observations such as those described above. In addition, it is also natural to compare the observations with phase speed and meridional structure of the harmonic planar theory and use the observations to examine the applicability of each of these theories.

4.1 Explicit expressions for the phase speeds of the two wave types

The Coriolis frequency on the β-plane, expanded linearly about some latitude, ϕ_0, is given by $f(y) = f_0 + \beta y = 2\Omega \sin \phi_0 + \frac{2\Omega}{a} \cos \phi_0 \cdot y$, where Ω is the frequency of Earth's rotation about its polar axis, a is Earth's radius and $y = (\phi - \phi_0) \times a$ (where ϕ is the latitude) is the north coordinate.

In a channel on the mid-latitudes β-plane where the Coriolis frequency is expanded near $\phi_0 = \phi_m$, the latitude of observation, the fastest baroclinic phase speed (in units of meters per second) of harmonic Rossby waves is (see Cushman-Roisin, 1994; Vallis, 2006):

$$C^{\text{harm}} = \frac{-\beta}{k^2 + l^2 + \frac{f_0^2}{g'H'}} = \frac{-\frac{2\Omega}{a} \cos \phi_m}{k^2 + l^2 + \frac{(2\Omega)^2 \sin^2 \phi_m}{g'H'}}, \quad (1)$$

where k and l (the latter is denoted in other studies by n) are the zonal and meridional wavenumbers of the Cartesian coordinates, respectively, g' is the reduced gravity and H' is the weighted depth of the two (or more) layers that make up the baroclinic ocean; therefore, $(g'H')^{\frac{1}{2}}$ is the speed of gravity waves. For sufficiently long waves when both k and l

can be neglected this phase speed reduces to

$$C^{\text{harm}} = \frac{-\beta}{\frac{f_0^2}{g'H'}} = \frac{-g'H'\cos\phi_m}{2\Omega a \sin^2\phi_m}. \tag{2}$$

In contrast to the harmonic wave theory, which is fully described in many textbooks, the application of the trapped wave theory requires some more detailed explanation. In this theory, the waves are trapped next to a single wall that marks the equatorward boundary of the domain and the meridional variation of the wave's amplitude is given by the regular (at infinity) Airy function, Ai, which oscillates (but is not periodic in contrast to harmonic/sinusoidal oscillations) in the $(-\infty, 0)$ interval and decays to zero faster than exponential in the $(0, +\infty)$ interval (see e.g., Abramowitz and Stegun, 1972). The phase speed of trapped waves in a mid-latitude channel is (see Eq. 6 of Gildor et al., 2016)

$$C^{\text{trap}} = \frac{-\beta}{k^2 + \frac{f_0^2}{g'H'} + \zeta_n \times \left(\frac{2f_0\beta}{g'H'}\right)^{\frac{2}{3}} - \frac{2f_0\beta}{g'H'} \times y_w}, \tag{3}$$

where ζ_n is the absolute value of the nth zero of Ai and y_w is the location of the equatorward wall. Following the studies of Paldor and Sigalov (2008) and De-Leon and Paldor (2009), we expand here the Coriolis frequency near $\phi_0 = \phi_w$, where ϕ_w is the latitude of the equatorward boundary of the domain so $y_w = 0$ there, in which case the last term in the denominator of Eq. (3) vanishes (in contrast to Gildor et al., 2016, where the wall was placed at $y_w = -L/2$, where L is the channel width). In addition, the boundary condition at ϕ_w in Gildor et al. (2016) is the vanishing of the meridional velocity, whereas in the present application Fig. 2b implies that the meridional derivative of the height field vanishes at ϕ_w. Thus, ζ_n in Eq. (3) should be replaced in the present application by ξ_n – the absolute value of the nth zero of the derivative of Ai (see the discussion following Eq. 5 below). The resulting expression for the phase speed of the first baroclinic mode of sufficiently long trapped waves (i.e., for zonal wavenumber $k = 0$ and meridional mode number $n = 0$ for which $\xi_0 = 1.0188$, see p. 478 of Abramowitz and Stegun, 1972) is

$$C^{\text{trap}} = \frac{-\beta}{\frac{f_0^2}{g'H'} + 1.0188 \times \left(\frac{2f_0\beta}{g'H'}\right)^{\frac{2}{3}}}$$

$$= \frac{-\frac{2\Omega}{a}\cos\phi_w}{\frac{(2\Omega)^2\sin^2\phi_w}{g'H'} + 1.0188 \times \left(\frac{(2\Omega)^2}{ag'H'}\sin 2\phi_w\right)^{\frac{2}{3}}}. \tag{4}$$

Note that in contrast to the planar harmonic theory where l is a meridional wavenumber (measured in units of m^{-1}), which cannot be determined when no channel exists (and the same is true for the zonal wavenumber, k), in the trapped wave theory n is a non-dimensional mode number that counts the

number of zeros of the eigenfunction inside the meridional domain.

The trapped wave theory is valid when the meridional range is larger than $(2 + \zeta_n)\left(\frac{ag'H'}{4\Omega^2\sin 2\phi_w}\right)^{\frac{1}{3}}$ (see Eq. 7 of Gildor et al., 2016). For $n = 0$, for typical values of $(g'H')^{\frac{1}{2}}$ of 2 to $3\,\text{m s}^{-1}$ and $\phi_w = 30°$ this condition is satisfied when the domain is wider than about 500 km. Accordingly, the harmonic theory applies only in unrealistically narrow channels that are only a few hundred kilometers wide (see also Fig. 3 in Paldor and Sigalov, 2008).

4.2 Explicit expressions for the meridional structure of the two wave types

The meridional structure of the height field of harmonic waves in mid-latitudes varies with y, the meridional coordinate, as $A\cos(ly + \Gamma)$ (where A is an arbitrary amplitude and Γ is a phase angle that guarantees, together with l, that the wave satisfies the boundary conditions), which for $l = 0$ yields height and velocity fields that do not vary with y.

The meridional structure of the height field of trapped waves is (see Eq. 5 in Gildor et al., 2016 with the modifications outlined in Sect. 4.1)

$$\eta(y) = \frac{H'av_0}{g'H' - C^2}\left\{ C \times \left(\frac{4\Omega^2}{ag'H'}\sin 2\phi_w\right)^{\frac{1}{3}} \right.$$

$$\times Ai'\left(\left(\frac{4\Omega^2}{ag'H'}\sin 2\phi_w\right)^{\frac{1}{3}} \times y - \xi_n\right)$$

$$\left. - f(y)Ai\left(\left(\frac{4\Omega^2}{ag'H'}\sin 2\phi_w\right)^{\frac{1}{3}} \times y - \xi_n\right)\right\}, \tag{5}$$

where v_0 is an arbitrary amplitude and the phase speed, C, is given by C^{trap} of Eq. (3) with the modifications outlined in Sect. 4.1. Note that this theoretical expression for $\eta(y)$ consists of two terms: $Ai(y)$ and $Ai'(y)$; therefore, $d\eta/dy$ contains terms proportional to $Ai(y)$, $Ai'(y)$ and $Ai''(y)$. The Airy differential equations relates $Ai''(y)$ to $y \times Ai(y)$; therefore, at $y = 0$ the $Ai''(y)$ term vanishes and the coefficient of $Ai(y)$ is negligible compared to that of $Ai'(y)$, which clarifies why the extremum of η occurs at $y \approx 0$.

5 Results and comparison between observations and theories

5.1 Meridional structure of the height field

The meridional structure of the height field of the trapped waves curve in the area of study is computed from $\eta(y)$ of Eq. (5) with $C = C^{\text{trap}}$ of Eq. (4). In the calculation of these expressions of C^{trap} and $\eta(y)$, the value of ϕ_w was set to 31.5° S, $k = 0 = n$ (so $\xi_n = \xi_0 =$

1.0188), and $(g'H')^{\frac{1}{2}}$, the speed of gravity waves, was set to $2.8\,\mathrm{m\,s^{-1}}$ following Fig. 2 in Chelton et al. (1998) (see also http://www-po.coas.oregonstate.edu/research/po/research/rossby_radius/). The analytical expression for the meridional structure of the height field of harmonic waves for $k = 0$ and $l = 0$ is constant, i.e., described by a straight line parallel to the abscissa at arbitrary value of the ordinate. As shown in Fig. 2b, the curve of the trapped wave theory (dashed red line) fits the observed one (solid blue line) much better than that of the harmonic theory (dotted green line).

5.2 Phase speeds

An estimation of the speed of westward propagation of observed SSHAs is obtained by analyzing time–longitude (Hovmöller) diagrams of the SSHA field at fixed latitude as explained in Sect. 2.2. Figure 3 shows two examples of such diagrams calculated at $36°\,\mathrm{S}$ (panel a) and at $45°\,\mathrm{S}$ (panel b); both are sufficiently far from any major current or continent and sufficiently far from the equatorward boundary so that the condition for the validity of the trapped wave theory derived in the paragraph following Eq. (4) is satisfied (and sufficiently far (at least $200\,\mathrm{km}$) from the ACC). Also plotted on these diagrams are the two lines corresponding to the theoretical phase speeds for $k = 0$ and $n = 0$ of trapped wave theory (Eq. 4, dashed) and the harmonic wave theory (Eq. 2, dotted). A casual visual inspection shows that the line of trapped wave theory fits the observed tilt of SSHA features more closely than the harmonic one (especially at $45°\,\mathrm{S}$; panel b).

The various objective methods for obtaining the phase speed from the Hovmöller diagram are now applied to the diagram in Fig. 3b. The distribution of the sum of squares (or standard deviation) of the Radon transform as a function of the angle θ (for θ values near the peak) is shown in Fig. 4a (solid blue curve) where the maximum is at $\theta \approx 30°$ (i.e., $C \approx 1.9\,\mathrm{cm\,s^{-1}}$, in absolute value, hereafter C is positive although the wave propagates westward). The distribution of the mean of variances as a function of the angle θ is shown in Fig. 4b (solid blue curve) where the mean of variances is minimal at $\theta = 33°$ (i.e., $C \approx 2.1\,\mathrm{cm\,s^{-1}}$). The θ values corresponding to the phase speeds of trapped waves (obtained from Eq. 4, $\theta \approx 37°$, i.e., $C \approx 2.4\,\mathrm{cm\,s^{-1}}$; solid red vertical line) and to the harmonic phase speed (obtained from Eq. 2, $\theta \approx 20°$, i.e., $C \approx 1.2\,\mathrm{cm\,s^{-1}}$; dashed green vertical line) are also shown in panels (a), (b) and (d) of Fig. 4.

The frequency–wavenumber diagram obtained by applying 2D FFT to the time–longitude diagram at this latitude is shown in Fig. 4c in the range of low frequency and low wavenumber (in the rest of the frequency–wavenumber plane the amplitudes vanish). The maximum amplitude (outside $k = 0$ since only $k \neq 0$ values yield finite westward phase speeds by ω/k) of the frequency–wavenumber diagram shown in Fig. 4c occurs at $k = 0.1571$, which is a sufficiently small value that justifies the long-wave approx-

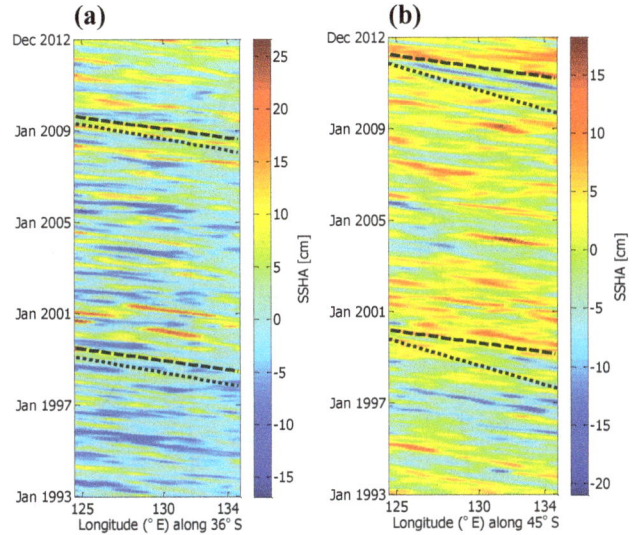

Figure 3. Time–longitude (Hovmöller) diagrams at $\phi_m = 36°\,\mathrm{S}$ (**a**) and at $\phi_m = 45°\,\mathrm{S}$ (**b**), where the abscissa is longitude and the ordinate is the date. The temporal average was subtracted from the record of each grid point. Dashed lines: trapped wave phase speeds (Eq. 4); Dotted lines: harmonic wave phase speeds (Eq. 2).

imation made earlier ($k = 0.1571$ corresponds to wavelength of about $160°$ of longitude). The frequency with maximal spectral amplitude at this wavenumber is -0.09045; therefore, the resulting phase speed of maximal spectral amplitude is $-0.09045/0.1571 = -0.5757$ (in degrees of longitude per 4 weeks, i.e., $C \approx 1.9\,\mathrm{cm\,s^{-1}}$) and this phase speed equals the phase speed obtained independently by the Radon transform. Figure 4c also compares the phase speeds of the two theories with the observed speed and it demonstrates that the phase speed of trapped waves (dashed red line) is slightly (but not significantly) closer to the observed speed (defined by both the maximum amplitudes and the directionality of the band of high amplitudes in frequency–wavenumber plane) than that of the harmonic waves (dotted light-green line). Though this red line (that corresponds to trapped waves) connects the two maximal values of the 2D FFT at the smallest $\pm k \neq 0$ (and passes through the origin as expected), at larger k its fit to the location of maximal amplitude is no better than that of the line corresponding to harmonic waves.

The distribution of the sum of squares of the spectral coefficients along ω/k lines versus the inclination angle, $\arctan(C)$, is shown in Fig. 4d (solid blue curve) where the curve attains its maximum at $\arctan(C) \approx 151°$, i.e., $\theta \approx 29°$ in terms of the Radon transform method ($C \approx 1.8\,\mathrm{cm\,s^{-1}}$).

For this time–longitude diagram the phase speed obtained by the variance method differs by about 11 to 14 % from that obtained by the Radon and 2D FFT methods that yield nearly identical phase speeds; therefore, according to our criteria mentioned in the end of Sect. 2.2, the latter estimate for

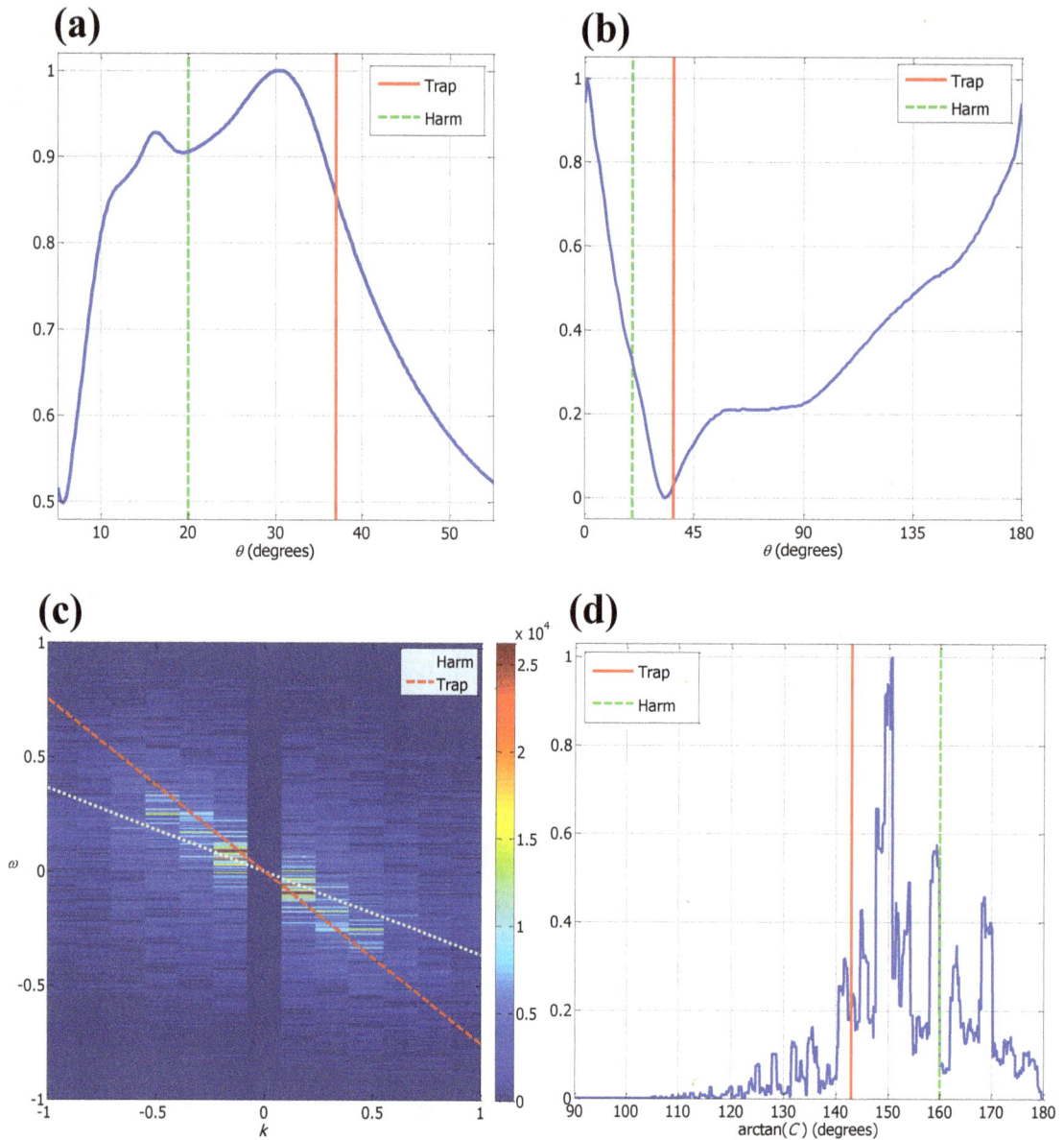

Figure 4. Analyses of the phase speeds of the Hovmöller diagram of Fig. 3b. **(a)** Solid blue curve: the sum of squares of the Radon transform as a function of θ (near the peak) normalized such that the maximum value equals 1; Dashed green vertical line: the angle of the harmonic wave theory (Eq. 2); Solid red vertical line: the angle of the trapped wave theory (Eq. 4). The same two vertical lines appear also in panels **(b)** and **(d)**. **(b)** Solid blue curve: the distribution of the mean of variances versus θ, normalized such that the maximum (minimum) value equals 1 (0). **(c)** The 2D FFT frequency–wavenumber diagram in the low-frequency–low-wavenumber regime (k is measured in units of $(1/4°$ of longitude$)^{-1}$, ω is measured in units of week^{-1} and the amplitude units are arbitrary). Dashed red line: trapped wave's phase speed, Eq. (4); dotted light-green line: harmonic wave's phase speed, Eq. (2). **(d)** The distribution of the sum of squares of the 2D FFT amplitudes along different lines (sweeping) versus arctan(C), normalized such that the maximum value equals 1 (solid blue curve). Only values of $90° < $ arctan(C) $< 180°$ are shown since only these values correspond to westward propagating speeds.

the phase speed is accepted. However, this observed phase speed does not clearly validate any of the two theoretical phase speeds since the corresponding vertical lines in panels (a) and (d) of Fig. 4 are located at nearly the same distance on both sides of the observed peak. In contrast, the estimate of the observed phase speed obtained by the variance method (Fig. 4b) is much closer to that of the trapped wave phase speed than the harmonic one. Thus, the determination of the relevant theory that yields the correct phase speed that matches the propagation rate determined from observations cannot rely solely on the match at any particular latitude and therefore the match over an entire range of latitudes was also examined.

The implications from similar comparisons carried out every 0.5° between 33 and 45.5° S can be summarized as follows: at about a third of the diagrams analyzed the signal was too blurred or the three methods yielded three different phase speed estimates. The application of a single method over the entire range of latitudes yields estimates that occasionally vary by over 50 % between adjacent 0.5° latitudes; therefore, the latitudinal continuity of the phase speed rules out the use of a single method. In only one or two latitudes (out of 22) all three methods have yielded the same (up to 10 %) estimate. Our conclusion from these comparisons bolsters our criteria that only when at least two of the three methods yield phase speed estimates that are closer to one another by less than 10 %, the resulting phase speed estimate can be considered reliable.

Phase speed estimates north of 35° S and between 37 and 39° S have not satisfied the agreement criteria between methods outlined in the end of Sect. 2.2. The lack of reliable phase speed estimates at these latitudes even though the amplitudes of the SSHAs there are higher than in adjacent latitudes in which the phase speed estimates were deemed reliable (and especially north of 35° S) requires an explanation. In linear theories amplitudes can only be determined up to a multiplicative factor while phase speeds are determined completely. Accordingly, the harmonic theory, where the solution is determined by ϕ_m alone, does not provide any information on the variation of the SSHA amplitude with ϕ_m, while in the trapped wave theory the variation of the amplitude with ϕ_m is determined up to an overall multiplicative constant. Regardless of whether the meridional structure of SSHAs is determined or not it should be stressed that higher/lower amplitudes do not necessary imply that the corresponding phase speed estimates are more/less reliable and it is possible for the amplitude to be high while the phase speed estimates are not reliable (using the methods and criteria we apply) or for the phase speed to be significant where the amplitudes are small (e.g., south of 40° S).

Figure 5 shows the observed and the two theoretical speeds as a function of ϕ_m between 35 and 45.5° S, where reliable estimates are obtained. The theoretical trapped speed is calculated using Eq. (4) and the theoretical harmonic speed is calculated using Eq. (2). It is clear that the trapped speeds (solid red line) are closer to the observed speeds (blue dots, squares and triangles) than the harmonic speed (dashed green line). A quantitative confirmation of this qualitative conclusion can be obtained by calculating the sum of squares of the distances between the observed and theoretical speeds. This calculation shows that the trapped speeds with sum of squares that equals 3.5 are much closer to the observed speeds than the harmonic speeds where the sum of squares is 15.3, i.e., more than 4 times that of trapped waves. Since the value of 10 % agreement (shown by blue circular dots) is somewhat arbitrary, we also include in Fig. 5 estimates of 11

and 12 % agreement (light-blue triangles) and estimates that agree by 25 % (light-blue squares).

6 Discussion and summary

The phase speed of harmonic waves decreases monotonically with the latitude of observation, ϕ_m, as is evident from Eqs. (1) and (2). In contrast, the phase speed of trapped waves depends on ϕ_w only (i.e., the latitude of the zonal boundary) and is independent of ϕ_m. Our analyses of the propagation speeds of SSHA signals show that the rate at which the observed speed decreases with ϕ_m (the trend of the data in Fig. 5 is 0.12 cm s^{-1} deg^{-1}) exceeds the rates of decrease of both harmonic (where the trend is 0.09 cm s^{-1} deg^{-1}) and trapped (no trend) phase speeds. In contrast to the meridional trends, the actual values of the observed phase speeds are much closer to the trapped phase speeds than to the harmonic speeds.

Colin de Verdière and Tailleux (2005) argued that the addition of mean flows affects the propagation speed of Rossby waves via its curvature: increase (decrease) of the westward phase speed for eastward (westward) surface mean flow. However, in the domain of the Indian Ocean studied here it is not clear whether or not a mean flow exists (in contrast to west of Australia where a subtropical gyre has been observed, see, e.g., Stramma and Lutjeharms, 1997) or what is its direction (some of the flows vary seasonally, see Wyrtki, 1973); therefore, even if the numerical values of parameters such as Richardson number or buoyancy could be somehow estimated, it is still unclear whether the mean flow increases or decreases the phase speed.

We should note that the simple choice, here and in many other prior studies, to interpret that observed SSHA propagation as that of the first baroclinic mode is not the only possible choice. Other choices of a single mode to fit the observation require detailed analyses of the hydrography while a (linear) combination of several modes (including the fast barotropic mode) with weights that are tuned so as to fit the observed speed can yield a better fit (see Hochet et al., 2015).

Chelton et al. (2007, 2011) argued that most of the observed SSHA features in the global ocean are nonlinear mesoscale eddies, whose propagation speed is close to the phase speed of long harmonic Rossby waves (but linear eddies move much faster). The nonlinearity in those studies is determined by a combination of second-order spatial derivatives of the SSHAs that are used in the calculation of the Okubo–Weiss parameter. Since no derivatives can be computed by any of the methods used in the present study, it is impossible to use these methods to directly determine whether the SSHA features examined in the present study are linear or not. However, all observed propagation speeds calculated here move faster than the phase of long harmonic Rossby waves (see Fig. 5), which implies that only linear (in the sense defined in Chelton et al., 2007) eddies that prop-

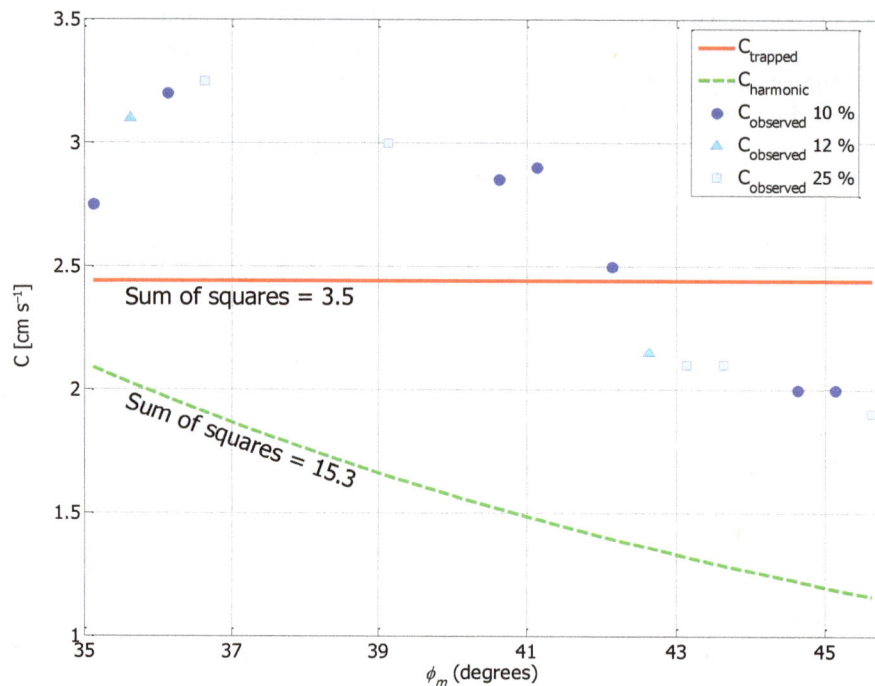

Figure 5. The observed phase speeds and the two theoretical phase speeds (trapped and harmonic) as a function of ϕ_m in intervals of $0.5°$ latitude. Blue dots denote latitudes where the estimates of at least two methods agreed by 10 % or less, triangles denote latitudes where such estimates agreed by 11 to 12 % and squares denote latitudes where the agreement is 25 %. No reliable estimates were obtained north of 35° S and in some more latitudes. The sum of squares of the distances in $(\text{cm s}^{-1})^2$ between trapped wave phase speeds and observed speeds (3.5) is much smaller than that of harmonic phase speeds (15.3).

agate faster than the phase speed of long harmonic Rossby waves exist in the Indian Ocean south of Australia.

As was concluded in De-Leon and Paldor (2017), an estimation of the observed phase speed using one method only is not reliable in most of the observed signals. On the other hand, even when estimates of the observed speed of at least two methods agree with each other, a comparison of the observed speed and the theoretical speeds varies in accordance with the method used for obtaining the observed speed. For example, in Fig. 4 the observed speed obtained by the variance method (panel b) is much closer to the trapped speed than to the harmonic speed while the observed speed obtained by the Radon and 2D FFT methods does not fit either the trapped or the harmonic speed preferentially. These differences between different methods point to the low accuracy/reliability of existing SSHA data.

As mentioned in the introduction, many studies compared observations of Rossby waves in the ocean with the harmonic Rossby waves. However, from a theoretical point of view, the harmonic theory in mid-latitudes is valid only in domains narrower than a few hundred kilometers; therefore, it is not clear why one should expect the harmonic speed to match the observed speed at unbounded domains. The case of Australia is unique since the trapped wave theory applies there while no other place exists that has a sufficiently wide, nearly straight, zonal coast line and meridional extent of the ocean

that spans over $10°$ poleward of the equatorward boundary. In cases of narrower straight zonal coast line such as Puerto Rico the trapped wave theory is inapplicable. In unbounded domains of the world ocean, the trapped wave theory does not apply straightforwardly and additional theoretical considerations have to be developed.

Competing interests. The authors declare that they have no conflict of interest.

Acknowledgements. The authors are grateful to Carl Wunsch of MIT/Harvard University for his helpful and instructive comments on an earlier version of this work. The comments of two anonymous reviewers helped us clarify the focus of the paper and improve its presentation.

Edited by: John M. Huthnance

References

Abramowitz, M. and Stegun, I. A.: Handbook of Mathematical Functions, Dover Publications, New York, USA, 1972.

Barron, C. N., Kara, A. B., and Jacobs, G. A.: Objective estimates of westward Rossby wave and eddy propagation from sea surface height analyses, J. Geophys. Res., 114, C03013, https://doi.org/10.1029/2008JC005044, 2009.

Chelton, D. B. and Schlax, M. G.: Global observations of oceanic Rossby waves, Science, 272, 234–238, https://doi.org/10.1126/science.272.5259.234, 1996.

Chelton, D. B., deSzoeke, R. A., Schlax, M. G., El Naggar, K., and Siwertz, N.: Geographical variability of the first baroclinic Rossby radius of deformation, J. Phys. Oceanogr., 28, 433–460, https://doi.org/10.1175/1520-0485(1998)028<0433:GVOTFB>2.0.CO;2, 1998.

Chelton, D. B., Schlax, M. G., Lyman, J. M., and Johnson, G. C.: Equatorially trapped Rossby waves in the presence of meridionally sheared baroclinic flow in the Pacific Ocean, Prog. Oceanogr., 56, 323–380, https://doi.org/10.1016/S0079-6611(03)00008-9, 2003.

Chelton, D. B., Schlax, M. G., Samelson, R. M., and de Szoeke, R. A.: Global observations of large oceanic eddies, Geophys. Res. Lett., 34, L15606, https://doi.org/10.1029/2007GL030812, 2007.

Chelton, D. B., Schlax, M. G., and Samelson, R. M.: Global observations of nonlinear mesoscale eddies, Prog. Oceanogr., 91, 167–216, https://doi.org/10.1016/j.pocean.2011.01.002, 2011.

Cohen, Y., Paldor, N., and Sommeria, J.: Laboratory experiments and a non-harmonic theory for topographic Rossby waves over a linearly sloping bottom on the f-plane, J. Fluid Mech., 645, 479–496, https://doi.org/10.1017/S0022112009992862, 2010.

Colin de Verdière, A. and Tailleux, R.: The interaction of a baroclinic mean flow with long Rossby waves, J. Phys. Oceanogr., 35, 865–879, https://doi.org/10.1175/JPO2712.1, 2005.

Cushman-Roisin, B.: Introduction to Geophysical Fluid Dynamics, Prentice-Hall, Englewood Cliffs, New Jersey, USA, 1994.

De-Leon, Y. and Paldor, N.: Linear waves in Mid-latitudes on the rotating spherical Earth, J. Phys. Oceanogr., 39, 3204–3215, https://doi.org/10.1175/2009JPO4083.1, 2009.

De-Leon, Y. and Paldor, N.: An accurate procedure for estimating the phase speed of ocean waves from observations by satellite borne altimeters, Acta Astronaut., 137, 504–511 , https://doi.org/10.1016/j.actaastro.2016.11.016, 2017.

Gildor, H., Paldor, N., and Ben-Shushan, S.: Numerical simulation of harmonic, and trapped, Rossby waves in a channel on the midlatitude β-plane, Q. J. Roy. Meteor. Soc., 142, 2292–2299, https://doi.org/10.1002/qj.2820, 2016.

Godfrey, J. S. and Ridgway, K. R.: The Large-Scale Environment of the Poleward-Flowing Leeuwin Current, Western Australia: Longshore Steric Height Gradients, Wind Stresses and Geostrophic Flow, J. Phys. Oceanogr., 15, 481–495, https://doi.org/10.1175/1520-0485(1985)015<0481:TLSEOT>2.0.CO;2, 1985.

Hochet, A., Colin De Verdiere, A., and Scott, R.: The vertical structure of large-scale unsteady currents, J. Phys. Oceanogr., 45, 755–777, https://doi.org/10.1175/JPO-D-14-0077.1, 2015.

Jain, A. K.: Fundamentals of Digital Image Processing, Prentice-Hall, Englewood Cliffs, New Jersey, USA, 1989.

Killworth, P. D. and Blundell, J. R.: The dispersion relation of planetary waves in the presence of mean flow and topography. Part II: Two-dimensional examples and global results, J. Phys. Oceanogr., 35, 2110–2133, https://doi.org/10.1175/JPO2817.1, 2005.

Killworth, P. D., Chelton, D. B., and de Szoeke, R. A.: The speed of observed and theoretical long extratropical planetary waves, J. Phys. Oceanogr., 27, 1946–1966, https://doi.org/10.1175/1520-0485(1997)027<1946:TSOOAT>2.0.CO;2, 1997.

LaCasce, J. H. and Pedlosky, J.: The instability of Rossby basin modes and the oceanic eddy field, J. Phys. Oceanogr., 34, 743–769, 2004.

Nof, D.: On the beta-induced movement of isolated baroclinic eddies, J. Phys. Oceanogr., 11, 1662–1672, 1981.

O'Brien, R. C., Cipollini, P., and Blundell, J. R.: Manifestation of oceanic Rossby waves in long-term multiparametric satellite datasets, Remote Sens. Environ., 129, 111–121, https://doi.org/10.1016/j.rse.2012.10.024, 2013.

Osychny, V. and Cornillon, P.: Properties of Rossby waves in the North Atlantic estimated from satellite data, J. Phys. Oceanogr., 34, 61–76, https://doi.org/10.1175/1520-0485(2004)034<0061:PORWIT>2.0.CO;2, 2004.

Paldor, N. and Sigalov, A.: Trapped waves in Mid-latitudes on the β-plane, Tellus A, 60, 742–748, https://doi.org/10.1111/j.1600-0870.2008.00332.x, 2008.

Paldor, N., Rubin, S., and Mariano, A. J.: A consistent theory for linear waves of the Shallow Water Equations on a rotating plane in mid-latitudes, J. Phys. Oceanogr., 37, 115–128, https://doi.org/10.1175/JPO2986.1, 2007.

Pedlosky, J.: Geophysical Fluid Dynamics, Springer-Verlag, New York, USA, 1982.

Polito, P. S. and Liu, W. T.: Global characterization of Rossby waves at several spectral bands, J. Geophys. Res., 108, 3018, https://doi.org/10.1029/2000JC000607, 2003.

Polito, P. S. and Sato, O. T.: Do eddies ride on Rossby waves?, J. Geophys. Res.-Oceans, 120, 5417–5435, https://doi.org/10.1002/2015JC010737, 2015.

Potemra, J. T.: Contribution of equatorial Pacific winds to southern tropical Indian Ocean Rossby waves, J. Geophys. Res., 106, 2407–2422, https://doi.org/10.1029/1999JC000031, 2001.

Stramma, L. and Lutjeharms, J. R.: The flow field of the subtropical gyre of the South Indian Ocean, J. Geophys. Res., 102, 5513–5530, https://doi.org/10.1029/96JC03455, 1997.

Tailleux, R. and McWilliams, J. C.: The effect of bottom-pressure decoupling on the speed of extratropical baroclinic Rossby waves, J. Phys. Oceanogr., 31, 1461–1476, https://doi.org/10.1175/1520-0485(2001)031<1461:TEOBPD>2.0.CO;2, 2001.

Tulloch, R., Marshall, J., and Smith, K. S.: Interpretation of the propagation of surface altimetric observations in terms of planetary waves and geostrophic turbulence, J. Geophys. Res., 114, C02005, https://doi.org/10.1029/2008JC005055, 2009.

Vallis, G. K.: Atmospheric and Oceanic Fluid Dynamics, Cambridge University Press, Cambridge, UK, 2006.

Watanabe, W. B., Polito, P. S., and da Silveira, I. C. A.: Can a minimalist model of wind forced baroclinic Rossby waves produce reasonable results?, Ocean Dynam., 66, 539–548, https://doi.org/10.1007/s10236-016-0935-1, 2016.

Wyrtki, K.: Physical oceanography of the Indian Ocean, in: The Biology of the Indian Ocean, edited by: Zeitzschel, B. and Gerlach, A., Springer-Verlag, Berlin, Germany, 18–36, 1973.

Zang, X. and Wunsch, C.: The observed dispersion relationship for North Pacific Rossby wave motions, J. Phys. Oceanogr., 29, 2183–2190, https://doi.org/10.1175/1520-0485(1999)029<2183:TODRFN>2.0.CO;2, 1999.

9

Modelling of sediment transport and morphological evolution under the combined action of waves and currents

Guilherme Franz[1], **Matthias T. Delpey**[2], **David Brito**[3], **Lígia Pinto**[1], **Paulo Leitão**[4], and **Ramiro Neves**[1]

[1]MARETEC, Instituto Superior Técnico, Universidade de Lisboa, Av. Rovisco Pais, 1049-001, Lisboa, Portugal
[2]Centre Rivages Pro Tech, SUEZ, 2 allée Théodore Monod, Bidart, France
[3]ACTION MODULERS, Estrada Principal, no. 29, Paz, 2640-583 Mafra, Portugal
[4]HIDROMOD, Rua Rui Teles Palhinha, no. 4, Leião, 2740-278 Porto Salvo, Portugal

Correspondence to: Guilherme Franz (guilherme.franz@tecnico.ulisboa.pt)

Abstract. Coastal defence structures are often constructed to prevent beach erosion. However, poorly designed structures may cause serious erosion problems in the downdrift direction. Morphological models are useful tools to predict such impacts and assess the efficiency of defence structures for different scenarios. Nevertheless, morphological modelling is still a topic under intense research effort. The processes simulated by a morphological model depend on model complexity. For instance, undertow currents are neglected in coastal area models (2DH), which is a limitation for simulating the evolution of beach profiles for long periods. Model limitations are generally overcome by predefining invariant equilibrium profiles that are allowed to shift offshore or onshore. A more flexible approach is described in this paper, which can be generalised to 3-D models. The present work is based on the coupling of the MOHID modelling system and the SWAN wave model. The impacts of different designs of detached breakwaters and groynes were simulated in a schematic beach configuration following a 2DH approach. The results of bathymetry evolution are in agreement with the patterns found in the literature for several existing structures. The model was also tested in a 3-D test case to simulate the formation of sandbars by undertow currents. The findings of this work confirmed the applicability of the MOHID modelling system to study sediment transport and morphological changes in coastal zones under the combined action of waves and currents. The same modelling methodology was applied to a coastal zone (Costa da Caparica) located at the mouth of a mesotidal estuary (Tagus Estuary, Portugal) to evaluate the hydrodynamics and sediment transport both in calm water conditions and during events of highly energetic waves. The MOHID code is available in the GitHub repository.

1 Introduction

The morphological features of the coastal zone depend on the sediment characteristics and the combined action of waves and currents. Wind waves are the main energy source for most beaches. Particularly in the surf zone, waves may induce considerable changes in mean sea level and strong currents (Longuet-Higgins, 1970a, b, 1983). The pattern of surf zone currents varies with the angle of waves approaching the shore as well as with bathymetric heterogeneities, leading to longshore and rip currents. Sediment is usually carried shoreward during low wave conditions, mainly due to the asymmetry of waves in shallow waters (Myrhaug et al., 2004). The sediment accumulated during these periods may be eroded very rapidly under high wave conditions during a major storm. Following these energetic events, the bottom profile may recover its initial shape only if the longshore transport of sediment during the storm is low. Different structures such as breakwaters and groynes can prevent the alongshore movement of sediment (e.g. Dally and Pope, 1986). Consequently, serious erosion problems in the downdrift direction may arise from the construction of these structures. Morphological models are useful tools to assess the impact of protection structures, enabling us to consider different wave conditions and structure designs.

The complexity of morphological models ranges from coastal profile models to 2- or 3-D models. Actually, morphological models are usually a set of different models or modules, depending on the chosen approach. Here the focus is on the coupling of a spectral wave model with a phase-averaged hydrodynamic and sediment transport model. Spectral wave models offer a representation of the physical processes related to the generation, propagation, and dissipation of waves (e.g. Booij et al., 1999). The wave-induced forces computed by a wave model can be provided to a hydrodynamic model in order to simulate wave-related phenomena, such as wave set-up, wave-induced currents, and mixing. On the other hand, the hydrodynamic model can return water levels and currents to the wave model (e.g. Warner et al., 2008). Additional processes can be considered in the hydrodynamic model, such as wind action, tidal motion, and river discharges. The transport of suspended sediment may be simulated by an advection/diffusion model. The mechanism of erosion/deposition of sediments is controlled by the bed shear stress induced by currents and waves. The bathymetry evolution resulting from the total sediment transport (suspended load and bed load) affects the patterns of currents and waves. Due to the interdependence of the physical processes involved in sediment dynamics, all of these models must be coupled.

The hydrodynamics inside the surf zone is influenced by important 3-D effects. Due to the absence of 3-D processes (e.g. undertow), coastal area models (2DH) fail to reproduce a consistent evolution of beach profiles for long periods. This shortcoming can be overcome by predefining invariant equilibrium profiles. In this case, the equilibrium profile only shifts offshore or onshore depending on the overall sediment balance along the profile, similarly to a coastal profile model (e.g. Kriebel and Dean, 1985; Kristensen et al., 2013). As the sediment transport in the swash zone is usually neglected in large-scale 2DH models, this approach also has the advantage of updating the shoreline position. However, limitations arise when a structure is present in the surf zone. Moreover, the processes responsible for sandbar evolution are not considered. Attempts to simulate sandbar dynamics have been performed generally with cross-shore 2-D (2DV) and quasi-3-D models (e.g. Drønen and Deigaard, 2007; Ruessink et al., 2007). Nevertheless, the proper reproduction of sandbar migration is still an active topic of research (e.g. Dubarbier et al., 2015). Inaccuracies in the cross-shore sediment transport may degrade the coastal profile, which is a restriction for the simulation of long-term morphological evolution.

In this work, we test a more flexible approach to overcoming the 2DH model limitations in order to simulate morphological evolution for long periods. Instead of fixing an equilibrium profile to update the bathymetry and shoreline position, we defined a maximum slope that when surpassed generates sediment transport in the downslope direction. Thus, rather than extrapolating erosion or deposition fluxes over the entire profile, only individual grid cells are affected. This approach may be more appropriate in order to consider the effect of non-uniform grain-size distributions on the overall sediment transport, through morphological models that account for multiple sediment fractions. Grain-size sorting is generally observed along the cross-shore beach profile, as well as in the longshore beach direction (Komar, 1998). Furthermore, the method can be generalised to 3-D models for a better representation of sandbar slopes and to update the shoreline position.

This paper is divided into five sections. A brief description of the effect of waves in the nearshore hydrodynamics and sediment transport is given in Sect. 2. The numerical modelling approach is presented in Sect. 3. The methodology was verified for different test cases (Sect. 4). Firstly, the morphological evolution of a schematic beach was evaluated for different designs of coastal defence structures (detached breakwaters and groynes) following a 2DH model configuration. The model was applied later in a 3-D configuration for the same schematic beach to verify the development of sandbars. Finally, the numerical modelling methodology was applied to assess the hydrodynamics and sediment transport under extreme wave conditions in a coastal zone (Costa da Caparica) located at the mouth of a mesotidal estuary (Tagus Estuary, Portugal). The main conclusions found from these test cases are discussed in Sect. 5.

2 Background

The effect of breaking waves on the mean sea level (wave set-up) has been known since the laboratory measurements performed by Saville (1961), confirmed further by Bowen et al. (1968). This tilt of the mean sea level is explained by the horizontal flux of momentum carried by waves or, equivalently, by the radiation stress, a vertically integrated momentum flux whose gradient balances the wave set-up (Longuet-Higgins and Stewart 1962, 1964). In the surf zone, wave heights and orbital velocities decrease towards the shore due to wave breaking. As a consequence, the radiation stress also decreases, resulting in a force directed towards the shore. This force is balanced by a hydrostatic pressure gradient that increases the mean sea level onshore.

Breaking waves can also drive strong currents in the surf zone, which are important for sediment transport and morphological evolution in the coastal zone. The horizontal mass transport associated with waves, or Stokes drift, is oriented shoreward and vertically sheared, being more intense at the surface (e.g. Ardhuin et al., 2008). As a result, mass conservation in the nearshore is satisfied by a seaward transport in the lower part of the water column, called undertow, which has an important role in sandbar formation. The undertow is strongest in steep beaches and may be insignificant for moderate beach slopes, where circulation tends to break up into rip currents (Longuet-Higgins, 1983). Also, obliquely breaking waves generate longshore currents (Longuet-Higgins,

1970a, b) and, consequently, longshore sediment transport. Although nearshore sediment dynamics are dominated by wave action, tidal motion can also play an important role, alternately moving the breaker zone and shoreline position shoreward and seaward, which may prevent the development of longshore bars in the surf zone (e.g. Levoy et al., 2000).

3   Numerical model

The present work is based on the coupling of the MO-HID modelling system (Leitão, 2003; Leitão et al., 2008) and the SWAN wave model (Booij et al., 1999). The MO-HID code organisation follows an object oriented strategy that permits the integration of different scales and processes. Herein, the focus is given to the hydrodynamics and sediment transport in the nearshore area. Thus, we considered the processes related to the hydrodynamics, turbulence, advection/diffusion of suspended sediment, erosion/deposition of sediments, bed load sediment transport, and morphological evolution. A brief description of the most important aspects of the hydrodynamic model for this work is presented in this section, followed by the main novelties implemented in the MOHID code: a new method to calculate the bed load transport under the combined effect of currents and waves; and a bed slope correction considered to overcome the 2DH model limitations and to update the shoreline. This work represents an extension of the development of the morphological model previously described in Franz et al. (2017).

The SWAN wave model represents the processes of wave generation, propagation, refraction, shoaling, non-linear (quadruplet and triad) wave–wave interactions, and dissipation (whitecapping, bottom friction, and depth-induced breaking). More information can be found in the documentation of the SWAN wave model (http://swanmodel. sourceforge.net/). Both MOHID and SWAN can be run with multiple processors using shared (OpenMP) or distributed (OpenMPI) memory architectures. In this work, we implemented MPI directives in the MOHID module responsible for the calculations of bed load transport and morphological evolution (sediment module), which can consider multiple sediment classes. In addition to speeding up morphological changes by an acceleration factor, this new development reduces the computational time required for modelling bed evolution over long time periods.

3.1   Hydrodynamic model

The MOHID hydrodynamic module solves the Navier–Stokes equations, considering the incompressibility, hydrostatic, Boussinesq, and Reynolds approximations (Martins, 2000; Leitão, 2003):

$$\frac{\partial}{\partial t} \int_V \boldsymbol{v}_H \mathrm{d}V = -\oint_A \boldsymbol{v}_H (\boldsymbol{v}.\boldsymbol{n}) \,\mathrm{d}A + \oint_A \upsilon_T (\nabla (\boldsymbol{v}_H).\boldsymbol{n}) \,\mathrm{d}A$$
$$-\frac{1}{\rho} \oint_A p.\boldsymbol{n}_H \mathrm{d}A + \int_V 2\boldsymbol{\Omega} \times \boldsymbol{v}_H \mathrm{d}V + \boldsymbol{F}, \qquad (1)$$

where V represents the control volume, $\boldsymbol{v}_H = (u, v)$ the horizontal velocity vector, $\boldsymbol{v} = (u, v, w)$ the velocity vector, $\boldsymbol{n}$ the normal vector to the bounding surface (A), $\boldsymbol{n}_H$ the normal vector related to the horizontal plane, υ_T the turbulent viscosity, ρ the water density, $p = g \int_z^\eta \rho \mathrm{d}z + p_{atm}$ the water pressure, g the gravitational acceleration, p_{atm} the atmospheric pressure, η the water level, $\boldsymbol{\Omega}$ the earth rotation vector, and $\boldsymbol{F}$ the external forces, which include the wave-induced force (gradient of the radiation stress) computed by the wave model. The wave-induced force was considered in the MOHID hydrodynamic module in previous studies to compute the effects of waves on sea level (Malhadas et al., 2009), water renewal in a coastal lagoon (Malhadas et al., 2010), and coastal water dispersion in an estuarine bay (Delpey et al., 2014).

The spatial discretisation is performed by following the finite-volume method. The water level and vertical velocity are computed through the continuity equation integrated over the entire water column or applied to each control volume, respectively. The equations are solved through the alternating direction implicit (ADI) method in an Arakawa C-grid structure. A generic vertical discretisation allows implementation of different types of vertical coordinates (e.g. Sigma or Cartesian) (Martins, 2001). The turbulent viscosity is computed differently for the horizontal and vertical directions. The horizontal turbulent viscosity is defined as a constant value, based on the grid resolution and a reference velocity, or as a function of horizontal velocity gradients, based on Smagorinsky (1963). The vertical turbulent viscosity is computed by the Global Ocean Turbulence Model (GOTM), which is coupled to MOHID and consists of a set of turbulence-closure models (Buchard et al., 1999; Villarreal et al., 2005). Effects of wave breaking on vertical turbulence can be taken into account through surface boundary conditions (Delpey et al., 2014).

To solve the Navier–Stokes and continuity equation, appropriate boundary conditions are required for the lateral (e.g. land and open sea), surface, and bottom boundaries. MOHID has the option of a great variety of open boundary conditions of several types: Dirichlet, Neumann, radiation, cyclic, relaxation (or nudging), etc. Some boundary conditions can be combinations of the types enumerated, e.g. a combination of radiation with nudging (Blumberg and Kantha, 1985). Open boundary conditions (OBCs) can be imposed by prescribing the values of a specific variable (Dirichlet boundary condition). This condition is commonly applied in coastal models to impose tidal levels when the correspondent barotropic velocities are not available. On the other hand, following a Neumann boundary condition, the gradient of a specific variable is imposed instead of a prescribed value. Assuming a null gradient condition, the value of a variable at a boundary point is equal to the value at an adjacent interior point. When the shoreline location and bathymetry are uniform, e.g. in schematic cases, cyclic boundary conditions can be applied.

A relaxation scheme can be applied as an OBC by assuming a decay time that increases gradually from the boundary to infinite after a defined number of cells (see Martinsen and Engedahl, 1987; Engedahl, 1995):

$$P^{t+\Delta t} = P^* + \left(P^{\text{ext}} - P^* \right) \frac{\Delta t}{T_d}, \tag{2}$$

where P is a generic property, P^* is the property value calculated by the model, P^{ext} is the reference value of the property, Δt is the model time step, and T_d is the relaxation timescale.

Radiation methods can also be used to impose the OBCs, which allow the propagation of internal disturbances on water levels through the open boundaries. These disturbances can be caused, for example, by the wave forces. MOHID has two types of radiation conditions (Leitão, 2003), based on Blumberg and Kantha (1985), Eq. (3), and Flather (1976), Eq. (4):

$$\frac{\partial \eta}{\partial t} + (\boldsymbol{C}_{\text{E}}.\boldsymbol{n}) \Delta \eta = \frac{1}{T_d} (\eta_{\text{ext}} - \eta), \tag{3}$$

$$q - q_{\text{ext}} = (\eta_{\text{ext}} - \eta) (\boldsymbol{C}_{\text{E}}.\boldsymbol{n}), \tag{4}$$

where η and q are the water level and specific flow, respectively; η_{ext} and q_{ext} are the imposed values of η and q at open boundary points; $\boldsymbol{C}_{\text{E}}$ is the celerity of internal water level disturbances or the celerity of external waves ($C_{\text{E}} = \sqrt{gh}$). When T_d is approximated to infinity in the Eq. (3), the OBC becomes totally passive, which means that the water levels at the boundary points are computed only from the internal water levels. An approximated null relaxation timescale means that the water level is imposed as η_{ext}. The Flather radiation condition is mostly used in nested model domains, when velocities and water levels are known at the open boundaries. Many of the OBCs used by MOHID considered the concept of an external (or reference) solution. These solutions can be provided via input file or via a one-way nesting of a chain of models. This last process is used to downscale from large-scale domains to local ones (e.g. Franz et al., 2016).

In the case of land points, the closed boundary condition is imposed as null fluxes of mass and momentum in the perpendicular direction. However, for the parallel direction two boundary conditions can be used in MOHID: no-slip and free-slip (Leitão, 2003). In the no-slip condition, the normal and parallel velocities in the land boundary are assumed to be equal to zero. This generates a persistent gradient in the velocity components parallel to the land boundary and, consequently, a persistent sink associated with the horizontal turbulent diffusion of momentum. The covering and uncovering

of boundary cells can be represented in MOHID by a wetting/drying scheme (Martins et al., 2001).

At the surface, fluxes of momentum from wind action and wave breaking can be considered (Delpey et al., 2014). At the bottom, the method proposed by Soulsby and Clarke (2005) to compute the bed shear stress was implemented in this work, consisting of a steady component due to currents together with an oscillatory component due to waves. In a laminar flow, the combined bed shear stress is a simple linear addition of the laminar current-alone and wave-alone shear stresses. However, in turbulent flows this addition is nonlinear and the mean and oscillatory components of the stress are enhanced beyond the values of the laminar case. The mean bed shear stress is used for determining the friction acting on the current, whereas the maximum shear stress is used to determine the threshold of sediment motion. The turbulence generated by the skin friction acts directly on bottom sediment grains (Einstein, 1950), contrarily to that related to bed forms. Thus, the threshold of sediment motion depends on the grain-related bed shear stress.

3.2 Sediment transport

The transport of sand is divided into suspended and bed load, in which the sand particles are in frequent contact with the bed. The suspended sediment transport is computed by resolving the advection/diffusion equation. This approach is more realistic than considering empirical equations based on the instantaneous bed shear stresses, as the suspended load is not in equilibrium with the instantaneous bed shear stresses in unsteady flows. The net upward flux of suspended sand depends on the equilibrium concentration near the bottom, estimated by empirical equations available in the literature, extrapolated following the Rouse profile to the middle of the near-bed layer, which in 2DH mode means the middle of the water column. The adopted methodology was described previously in Franz et al. (2017), converging for different numbers of vertical layers.

The bed load transport under the combined effect of currents and waves is computed following the semi-empirical formulation of Soulsby and Damgaard (2005). The formulation was derived for current plus sinusoidal and asymmetrical waves, as well as asymmetrical waves alone. Amoudry and Liu (2010) obtained a generally good agreement comparing the results of Soulsby and Damgaard (2005) formulations with a sheet flow model, concluding that it can be implemented in both intrawave and wave-averaged models in order to study sediment transport. The parallel $\Phi_\parallel$ and normal $\Phi_\perp$ components of the non-dimensional bed load transport vector in relation to the current direction are

$$\Phi_\parallel = \max\left(\Phi_{\parallel 1}, \Phi_{\parallel 2}\right); \text{if } \theta_{\max} > \theta_{cr}, \tag{5a}$$

$$\Phi_{\parallel 1} = k_{\Phi 1}\theta_m^{k_{\Phi 2}}\left(\theta_m - \theta_{cr}\right)^{k_{\Phi 3}}, \tag{5b}$$

$$\Phi_{\parallel 2} = k_{\Phi 1}\left(0.9534 + 0.1904\cos 2\varnothing\right)\theta_w^{1/2}\theta_m$$
$$+ k_{\Phi 1}\left(0.229\gamma_w\theta_w^{3/2}\cos\varnothing\right), \tag{5c}$$

$$\Phi_\perp = k_{\Phi 1}\frac{\left(0.1907\theta_w^2\right)}{\theta_w^{3/2} + (3/2)\theta_m^{3/2}}\left(\theta_m\sin 2\varnothing + 1.2\gamma_w\theta_w\sin\varnothing\right);$$
$$\text{if } \theta_{\max} > \theta_{cr}, \tag{6}$$

where θ_m and θ_w are the time-mean and oscillatory part of the bed shear stress (non-dimensional), respectively; $k_{\Phi 1}$, $k_{\Phi 2}$, and $k_{\Phi 3}$ are calibration coefficients that allow one to represent different equations for the bed load transport found in the literature (e.g. Amoudry and Souza, 2011); $\varnothing$ is the angle between the wave propagation and current direction; γ_w is a factor that represents the wave's asymmetry; θ_{cr} is the critical non-dimensional bed shear stress, which depends on sediment diameter and bed-material gradation. The bed load transport is null ($\Phi_\parallel = \Phi_\perp = 0$) if θ_{cr} is greater than or equal to the maximum non-dimensional bed shear stress ($\theta_{\max}$):

$$\theta_{\max} = \max\left(\theta_{\max 1}, \theta_{\max 2}\right), \tag{7a}$$

$$\theta_{\max 1} = \left(\left(\theta_m + \theta_w\left(1 + \gamma_w\right)\cos\varnothing\right)^2\right.$$
$$\left. + \left(\theta_w\left(1 + \gamma_w\right)\sin\varnothing\right)^2\right)^{1/2}, \tag{7b}$$

$$\theta_{\max 2} = \left(\left(\theta_m + \theta_w\left(1 - \gamma_w\right)\cos\left(\varnothing + \pi\right)\right)^2\right.$$
$$\left. + \left(\theta_w\left(1 - \gamma_w\right)\sin\left(\varnothing + \pi\right)\right)^2\right)^{1/2}, \tag{7c}$$

The bed load transport vector, $\Phi = (\Phi_\parallel, \Phi_\perp)$, is enhanced in the presence of waves. Regarding the symmetrical case ($\gamma_w = 0$), the effect of the wave's asymmetry results in an additional increase in the normal component ($\Phi_\perp$), whereas the parallel component ($\Phi_\parallel$) can be increased or reduced, depending on the angle ($\varnothing$) between the wave propagation and current direction. The asymmetry factor ($\gamma_w = \theta_{w,2}/\theta_{w,1}$) is defined as the ratio between the bed shear stress due to the wave's second harmonic ($\theta_{w,2}$) and basic harmonic ($\theta_{w,1}$), set to a maximum value of 0.2 (Soulsby and Damgaard, 2005). Considering the quadratic friction law to determine the magnitude of the wave's bed shear stress, the asymmetry factor is computed as

$$\gamma_w = \left(\frac{U_{w,2}}{U_{w,1}}\right)^2 = \left(\frac{3}{4}\frac{\pi H_w}{L_w\sinh^3(k_w h)}\right)^2, \tag{8}$$

where $U_{w,2}$ and $U_{w,1}$ are the amplitude of near-bed wave-orbital velocity for the second harmonic and basic harmonic of Stokes second-order wave theory (e.g. Greenwood and

Davis, 2011), respectively; H_w is the significant wave height; L_w is the wavelength; $k_w = 2\pi/L_w$ is the wave number; and h is the water depth. When waves propagate into shallow waters, γ_w becomes more significant, as the term $1/\sinh^3(k_w h)$ increases when decreasing h. At a given depth, γ_w tend to be higher for longer incident waves. This effect may be explained by the fact that longer waves propagate over a larger distance in limited water depth ($h < L/2$) than shorter waves. Thus, they are more affected by wave–bottom interaction and, as a consequence, more asymmetric than shorter waves at the same depth.

To compute the sediment fluxes between grid cells, the components of the non-dimensional bed load transport vector are rotated to the grid referential (u-axis and v-axis):

$$\Phi_u = \Phi_\parallel\cos\varnothing_c - \Phi_\perp\sin\varnothing_c, \tag{9a}$$

$$\Phi_v = \Phi_\parallel\sin\varnothing_c + \Phi_\perp\cos\varnothing_c, \tag{9b}$$

where $\varnothing_c$ is the angle between the horizontal velocity vector and the u-axis. Thus, the bed load transport vector, $q = (q_u q_v)$, in mass units (kg m^{-1} s^{-1}) is equal to

$$q_u = \rho_s\Phi_u\left[g\left(\rho_r - 1\right)d^3\right]^{1/2}, \tag{10a}$$

$$q_v = \rho_s\Phi_v\left[g\left(\rho_r - 1\right)d^3\right]^{1/2}, \tag{10b}$$

where ρ_s is the sand particle density (kg m^{-3}); ρ_r is the relative density (ρ_s/ρ); and d is the sand representative diameter (m).

3.3   Bed slope correction

Wave action induces a shoreward sediment transport that has no counterpart in 2DH models, leading to sand accumulation in the nearshore and increasing the steepness of the beach profile. Actually, undertow currents are responsible for a seaward sediment transport, which may generate sandbars. Diverse opposing forces are responsible for creating an equilibrium profile, which depends on sediment characteristics and wave heights (Dean, 1991). To account for the neglected forces in 2DH models, we defined a maximum slope ($\alpha_{\max}$) that when exceeded induces sediment transport in the downslope direction. This artificial sediment transport may act as the undertow, transporting sediment seaward. The mass of sand (M) in the sediment column and, consequently, the bathymetry are updated when the bottom slope (α) is larger than $\alpha_{\max}$:

$$M_{i,j}^{t+1} = M_{i,j}^t - \Delta M, \tag{11a}$$

$$M_{i,j+1}^{t+1} = M_{i,j+1}^t + \Delta M, \tag{11b}$$

where

$$\Delta M = \Delta z_b A\rho_s\left(1 - n\right), \quad \alpha > \alpha_{\max} \tag{12a}$$

$$\Delta M = 0, \quad \alpha \leq \alpha_{\max} \tag{12b}$$

in which n is the sediment porosity, A is the grid cell area, t is an index symbol for time, and i, j are index symbols to identify the grid cell (i – line number; j – column number). Considering the u-direction, the bed change in one time step (Δz_b) is computed as

$$\Delta z_b = \min\left((|\alpha| - \alpha_{\max})\,\Delta x, \Delta z_{b_{\max}}\right), \alpha > 0 \qquad (13a)$$

$$\Delta z_b = \min\left(-(|\alpha| - \alpha_{\max})\,\Delta x, -\Delta z_{b_{\max}}\right), \alpha < 0 \qquad (13b)$$

where $\alpha = \left(z_{b_{i,j+1}} - z_{b_{i,j}}\right)/\Delta x$, Δx is the cell width, z_b is the distance from the bed to a reference height (e.g. the hydrographic zero), and $\Delta z_{b_{\max}}$ is a threshold to avoid numerical instabilities due to large shockwaves. Similar equations are used in the v-direction. Different values of the maximum slope ($\alpha_{\max}$) can be defined in wet and dry cells. This method is based on that of Roelvink et al. (2009), previously applied to simulate dune erosion. The method was implemented in the MOHID code in terms of mass evolution, in order to be applied in the future considering multiple sediment fractions. The shoreline position is also updated following this approach. Other authors have used the same approach to simulate the migration of a tidal inlet (Nahon et al., 2012; Fortunato et al., 2014).

3.4 Model coupling

The wave induced force (radiation stress) computed by the SWAN wave model is provided to the MOHID hydrodynamic model in order to simulate the wave-induced currents and wave set-up. Fields of significant wave height, wave period, wavelength, wave direction, and maximal orbital velocity near the bottom are also provided by SWAN to MOHID in order to compute the bed shear stress and sediment transport. On the other hand, the MOHID hydrodynamic model returns fields of water level and current to SWAN. The water level variation caused by the tidal motion changes the breaker zone and shoreline position, affecting waves and sediment transport. The morphological evolution modifies the currents and waves. Thus, the bathymetry changes computed by MOHID are also updated in SWAN. The different fields computed by SWAN must be updated in MOHID, and the different fields computed by MOHID must be updated in SWAN, with an adequate frequency for each application depending on the variability of forcing conditions and the speed of morphological changes.

The coupling between the MOHID modelling system and the SWAN wave model was performed by files transferring through tools developed in the Fortran language to convert the results to the appropriate format and another tool developed in the Python language to automatically manage the runs of the tools and models. At this time, we have focused on model results instead of numerical efficiency. However, considering the domain decomposition parallelisation approach implemented using MPI directives and the morphological acceleration factor, the computational time required to simulate the presented test cases was feasible through the use of a regular computer with six cores.

4 Test cases

4.1 Coastal defence structures

The morphological evolution of a schematic beach was simulated to assess model results, considering the beach response for different designs of detached breakwaters and groynes. Constant wave conditions were defined along the offshore boundary (1.5 m of wave height, 8 s of peak wave period, and 15° of peak wave direction), following the JONSWAP spectrum. The hydrodynamic model was applied in 2DH mode, considering the vertically integrated wave-induced forces. The MOHID domain was defined as 2 km cross-shore by 3 km alongshore, whereas the SWAN domain was defined as 2 km cross-shore by 9 km alongshore (3 km larger on each side of the MOHID domain). The grid resolution was equal in both models, ranging from 50 m × 50 m to 10 m × 10 m. A larger domain for the wave model was considered to avoid inaccuracies in the lateral boundaries (shadow zones), as incident wave energy was imposed only along the offshore boundary. To prevent discontinuities in the SWAN bathymetry, the part of the domain not covered by the MOHID domain was updated with the depths of the MOHID cross-shore boundaries. The bathymetry evolution was allowed only after a warm up period, considering a morphological acceleration factor of 365. This means that 1 day of simulation time is equivalent to 1 year of morphological changes. The wave forcing was updated in MOHID, as the bathymetry and water levels were updated in SWAN, with a constant frequency of 5 min (or 30 h of morphological evolution).

The open boundary condition was defined as a null gradient for the sediment concentrations in the water column, as well as for the sediment mass evolution at the bed (or, equivalently, for bathymetry). A null gradient condition was also imposed at the open boundaries for the normal and tangential current velocities. The radiation condition of Blumberg and Kantha (1985) was imposed for water level, Eq. (3), assuming a passive condition at the cross-shore boundaries ($T_d = 1e^{32}$ s) and an active condition at the offshore boundary ($T_d = 1e^{-12}$ s). Thus, the water level was imposed at the offshore boundary as equal to the initial condition (zero in this case) to maintain the average water level inside the model domain; otherwise, it would continuously decrease. The effect of lateral friction in land boundaries was considered for a better representation of the flow around the groynes. The horizontal viscosity was set as $1.0\,\text{m}^2\,\text{s}^{-1}$. The sand granulometry was uniform, with a diameter of 0.2 mm.

The initial bathymetry was defined by considering an equilibrium profile of the form $h = \beta y^{2/3}$ (Dean, 1991), where β is a constant set to $0.12\,\text{m}^{1/3}$, and y is the distance to the shoreline. The average slope is approximately 1 : 60 in the first 200 m from the shoreline, decreasing seaward. The maximum slope ($\alpha_{\max}$) was defined as 1 : 50 for the bed slope corrections. The effectiveness of the bed slope correction

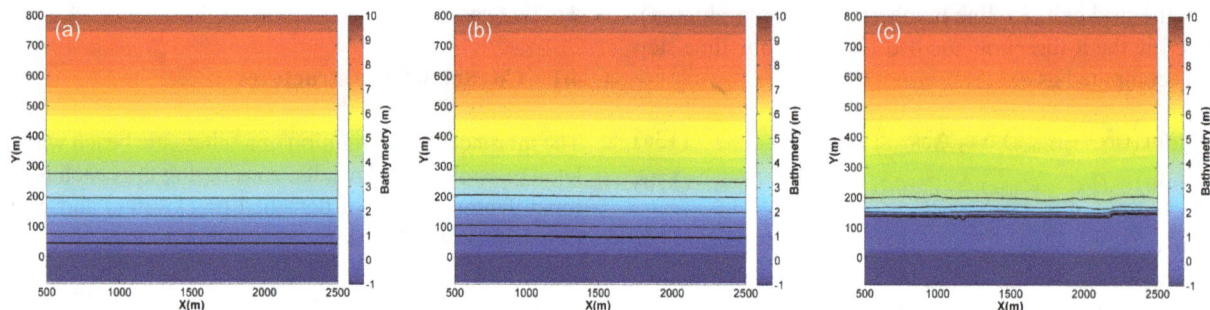

Figure 1. Initial bathymetry **(a)** and bathymetries after 9 years of morphological evolution considering the bed slope correction **(b)** and without the bed slope correction **(c)**. The thicker isoline in the bathymetry represents the shoreline, whereas the remaining ones represent the 1, 2, 3, and 4 m isobaths.

Table 1. Detached breakwaters considered for test case scenarios.

Length	Distance	Ratio (r)	Beach response*
100	500	0.2	Salient
100	200	0.5	Salient
200	200	1.0	Salient/tombolo
200	100	2.0	Tombolo

* Beach response according to Dally and Pope (1986).

was first assessed neglecting coastal defence structures. The bathymetry of the schematic beach shows minor adaptations to the wave regime after 9 years of morphological evolution considering the maximum bed slope, whereas without the bed slope correction the bathymetry reaches unrealistically steep slopes and sand accumulation on the beach (Fig. 1).

4.1.1 Detached breakwaters

Detached breakwaters generate sediment transport from the adjacent coast to the lee side of the structure, leading to the formation of a bulge or salient in the beach planform. Depending on geometrical features of the breakwater, wave climate, and sediment availability, the salient may become attached to the breakwater, forming a tombolo. Based on the analysis of several existing breakwater projects, Dally and Pope (1986) found that a ratio (r) between the breakwater's length and a distance to the shoreline of less than 0.5 prevents the development of a tombolo. In contrast, the development of a tombolo is assured if r is larger than 1.5, assuming sufficient sediment supply. Taking these values into account, we tested the model response for the four different detached breakwater designs described in Table 1.

Model results for a near-equilibrium planform of the shoreline agree with the analysis of Dally and Pope (1986), demonstrating the development of a salient for r equal to 0.2, 0.5, and 1.0 (Figs. 2, 3, and 4), which become attached to the breakwater, forming a tombolo only for r equal to 2.0 (Fig. 5). The obliquity of waves generates a longshore

current and, consequently, longshore sediment transport in the nearshore zone. The shoreline tends to be parallel to the wave crests, creating asymmetric bulges. For r equal to 0.5 and 1.0, longshore currents restricted the size of the salients, preventing the connection with the breakwater. The shoreline advances more on the updrift side for larger values of r, trapping sediment from the littoral drift. On the other hand, the downdrift beach erosion increases. When a tombolo is formed, the detached breakwater affects the shoreline similarly to a groyne.

4.1.2 Groynes

Groynes are applied to reduce the littoral drift in the surf zone, trapping sediment on the updrift side of the structure, which may cause erosion problems on the downdrift side. Moreover, the longshore currents are forced to deviate into deeper water around groynes, causing sediment losses from nearshore to offshore. The morphological impacts of the groynes are a function of their length from the shoreline. Model results were assessed for two designs of groynes, with lengths of 100 and 200 m (Figs. 6 and 7). As expected, greater erosion occurs on the downdrift side for a longer groyne, as more sediment from the littoral drift gets trapped on the updrift side. Furthermore, offshore sediment transport becomes intensified in the 200 m length groyne design, as the deviation of longshore currents is more important. In this case, the retrogradation of the shoreline is similar to the case of a detached breakwater in which a tombolo was formed.

4.2 Sandbars formation

In this test case, we verified the model capacity to generate sandbars in a 3-D approach. The same domain and sand granulometry (0.2 mm) as in the previous 2DH test cases were considered, but without protection structures. Two wave heights (1.5 and 1.0 m) were defined in sequence along the offshore boundary during periods of 45 days of morphological evolution (3 h of simulation time with a morphological acceleration factor of 365). The peak wave period and

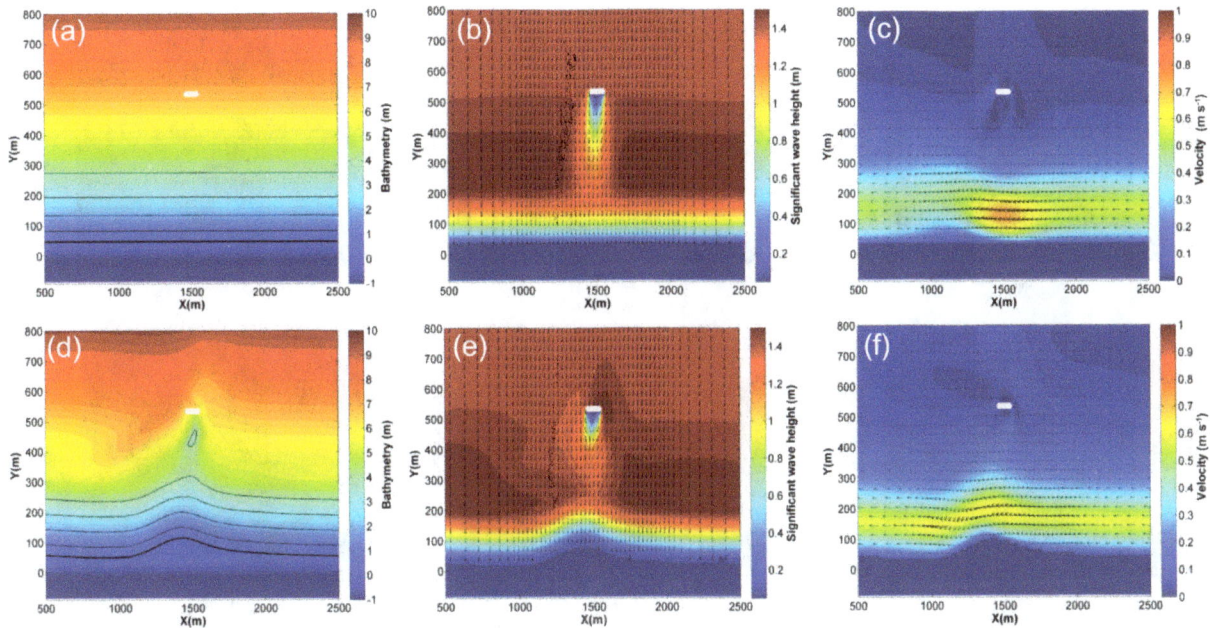

Figure 2. Model results for the breakwater's length-to-distance ratio of 0.2. Bathymetry (**a, d**), waves (**b, e**), and currents (**c, f**) for the initial condition (**a, b, c**) and near equilibrium (**d, e, f**) after 9 years. The thicker isoline in the bathymetry represents the shoreline, whereas the remaining ones represent the 1, 2, 3, and 4 m isobaths.

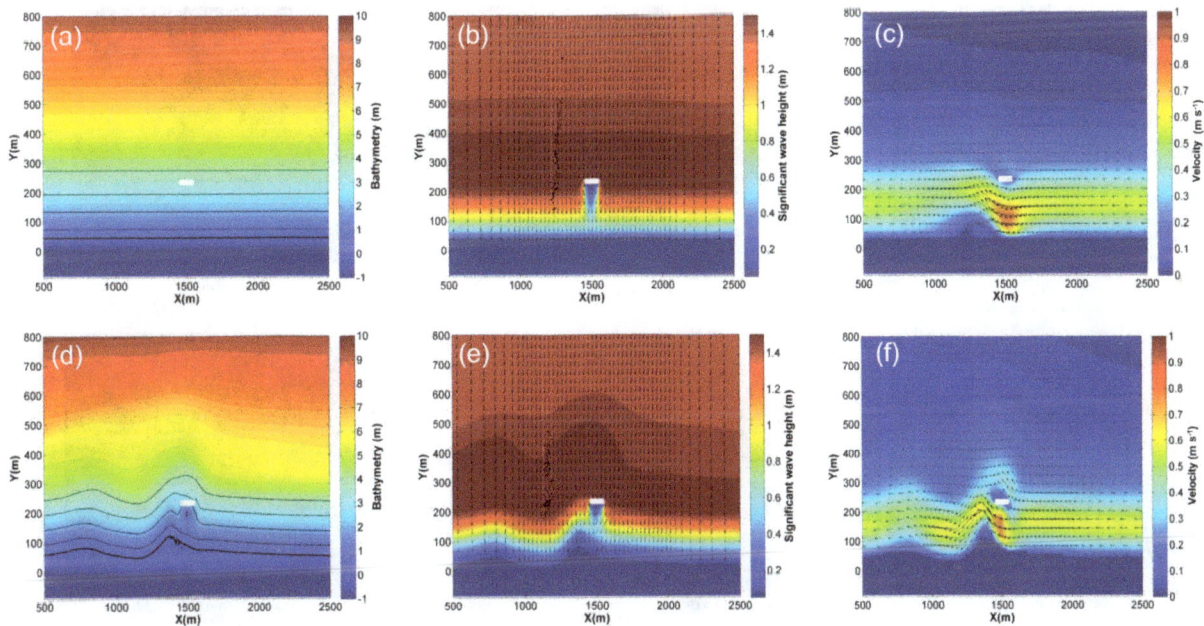

Figure 3. Model results for the breakwater's length-to-distance ratio of 0.5. Bathymetry (**a, d**), waves (**b, e**), and currents (**c, f**) for the initial condition (**a, b, c**) and near equilibrium (**d, e, f**) after 18 years. The thicker isoline in the bathymetry represents the shoreline, whereas the others represent the 1, 2, 3, and 4 m isobaths.

peak wave direction were maintained constant (8 s and 15°). For the 3-D case, a larger maximum slope was defined as 1 : 10 for the bed slope corrections, considering that the seaward sediment transport due to undertow currents can now be represented. The maximum slope in 3-D simulations is use-ful for representing the sand motion induced by excessively steep slopes.

The water column was divided into five layers of a sigma vertical coordinate and a simple exponential approach was followed to consider the vertical variation of the wave-

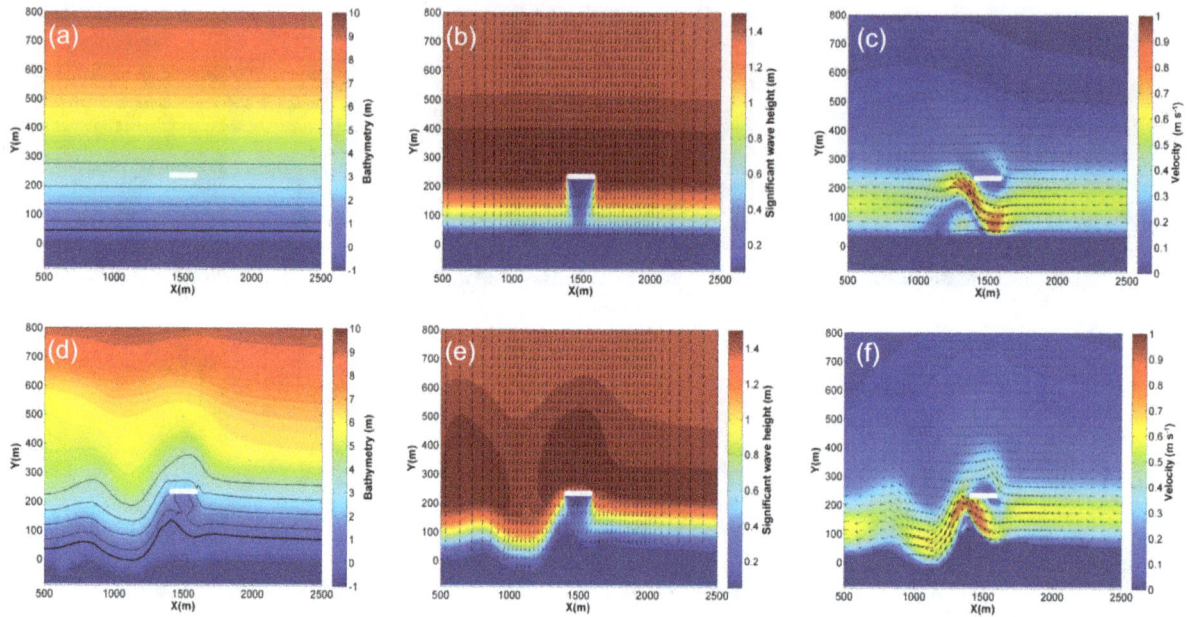

Figure 4. Model results for the breakwater's length-to-distance ratio of 1.0. Bathymetry (**a, d**), waves (**b, e**), and currents (**c, f**) for the initial condition (**a, b, c**) and near equilibrium (**d, e, f**) after 27 years. The thicker isoline in the bathymetry represents the shoreline, whereas the others represent the 1, 2, 3, and 4 m isobaths.

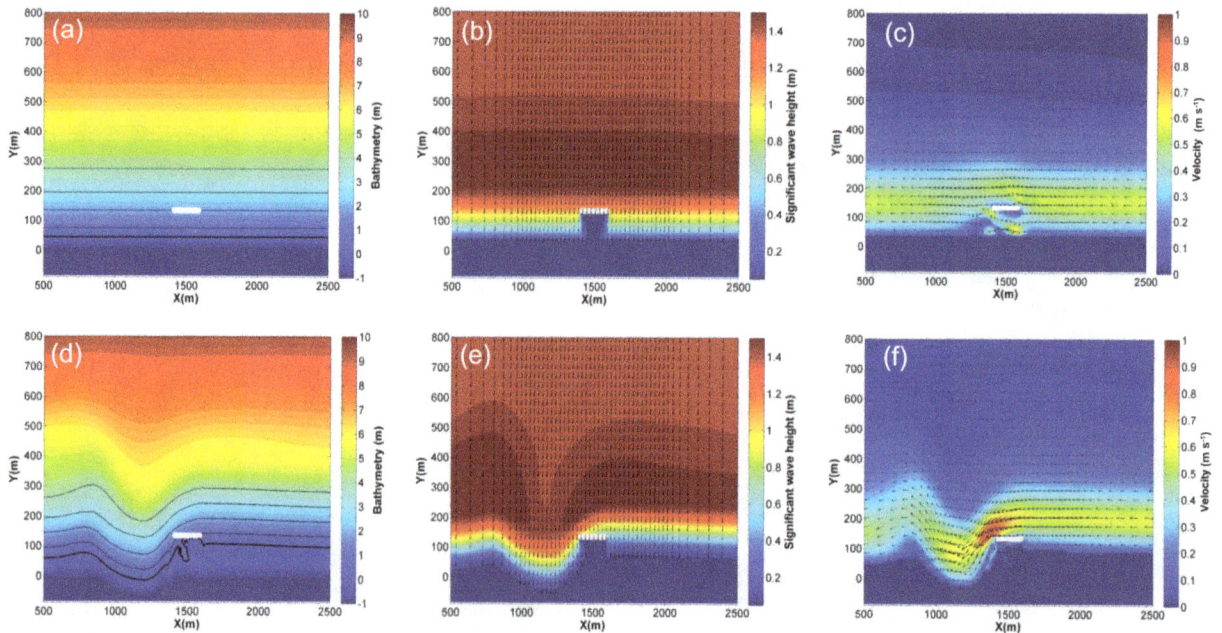

Figure 5. Model results for the breakwater's length-to-distance ratio of 2.0. Bathymetry (**a, d**), waves (**b, e**), and currents (**c, f**) for the initial condition (**a, b, c**) and near equilibrium (**d, e, f**) after 18 years. The thicker isoline in the bathymetry represents the shoreline, whereas the others represent the 1, 2, 3, and 4 m isobaths.

induced forces: an exponential decrease in the radiation stress is imposed from the surface to the bottom, following the same shape as the profile of the orbital velocities, provided by the linear wave theory. The vertical radiation stress profile is designed to conserve the vertically integrated flux of momentum, which remains equal to the flux given by SWAN. The effects of wave breaking on vertical turbulence were disregarded in this work. The idea here is only to provide an approximate representation of the vertical distribution of wave momentum, in order to generate a general

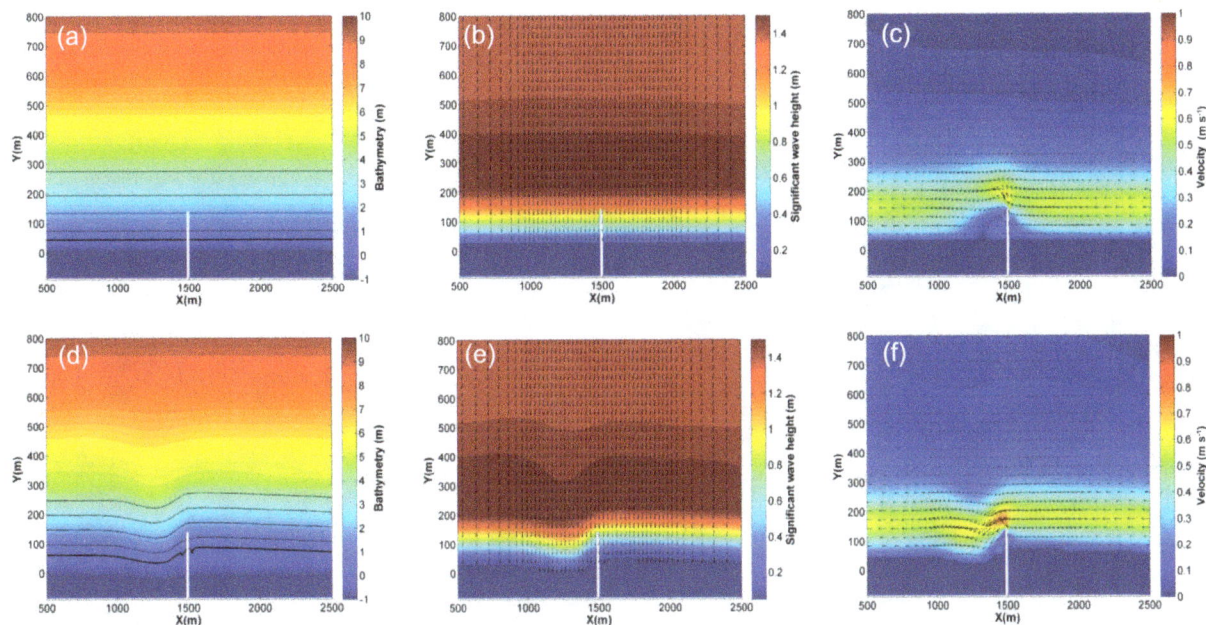

Figure 6. Groyne with a length of 100 m from the shoreline. Bathymetry (**a, d**), waves (**b, e**), and currents (**c, f**) for the initial condition (**a, b, c**) and near equilibrium (**d, e, f**) after 9 years. The thicker isoline in the bathymetry represents the shoreline, whereas the others represent the 1, 2, 3, and 4 m isobaths.

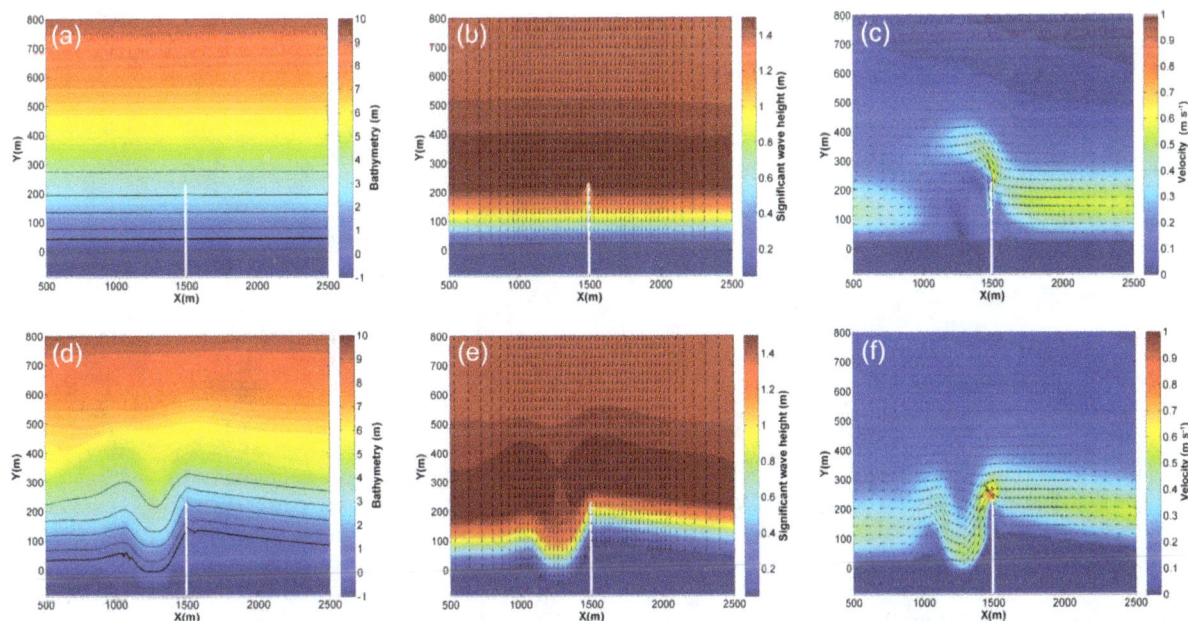

Figure 7. Groyne with a length of 200 m from the shoreline. Bathymetry (**a, d**), waves (**b, e**), and currents (**c, f**) for the initial condition (**a, b, c**) and near equilibrium (**d, e, f**) after 9 years. The thicker isoline in the bathymetry represents the shoreline, whereas the others represent the 1, 2, 3, and 4 m isobaths.

undertow pattern. Thus, the corresponding results should be considered as a first qualitative evaluation of the effect of such an undertow in our morphological module, the latter being our focus here. It is left for further work to use a more advanced formulation of 3-D wave–current interactions for more quantitative investigations. The k-ε turbulence-closure model was used to compute the vertical viscosity, with the MOHID default parameterisation, whereas the horizontal viscosity was set to $1.0 \, \mathrm{m^2 \, s^{-1}}$.

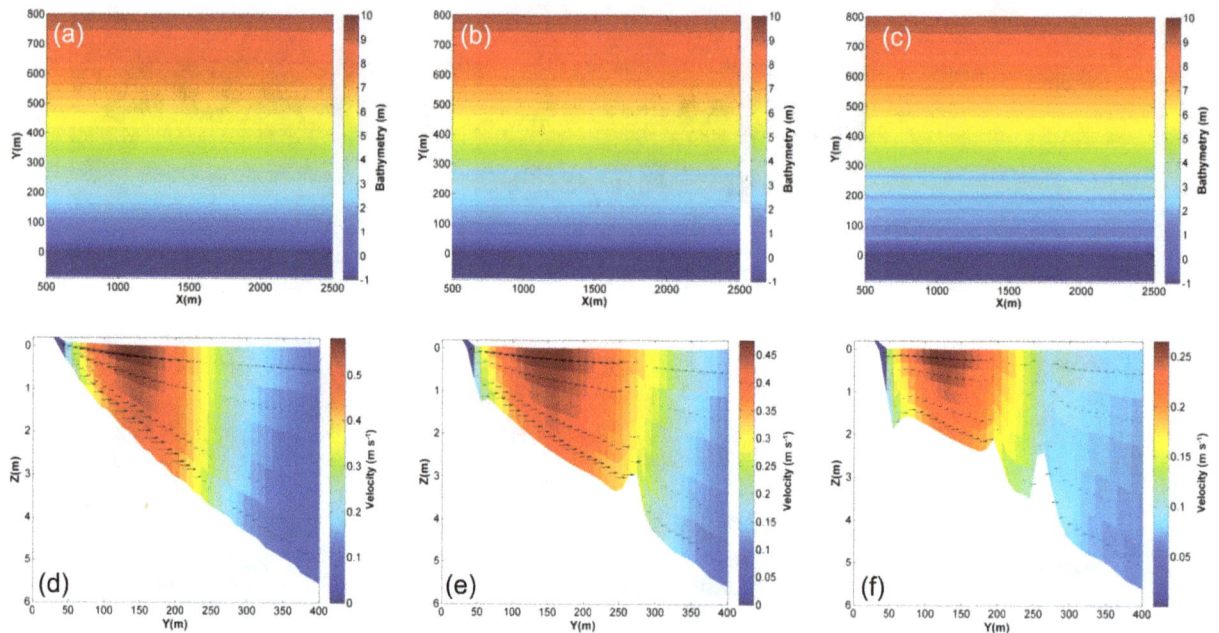

Figure 8. Simulated sandbars for different wave heights. The horizontal plane of the bathymetry (**a, b, c**) and a vertical cut with velocity modulus and vectors (**d, e, f**). (**a, d**) Initial condition. (**b, e**) After 45 days of morphological evolution with 1.5 m wave height. (**c, f**) After an additional 45 days of morphological evolution with 1.0 m wave height.

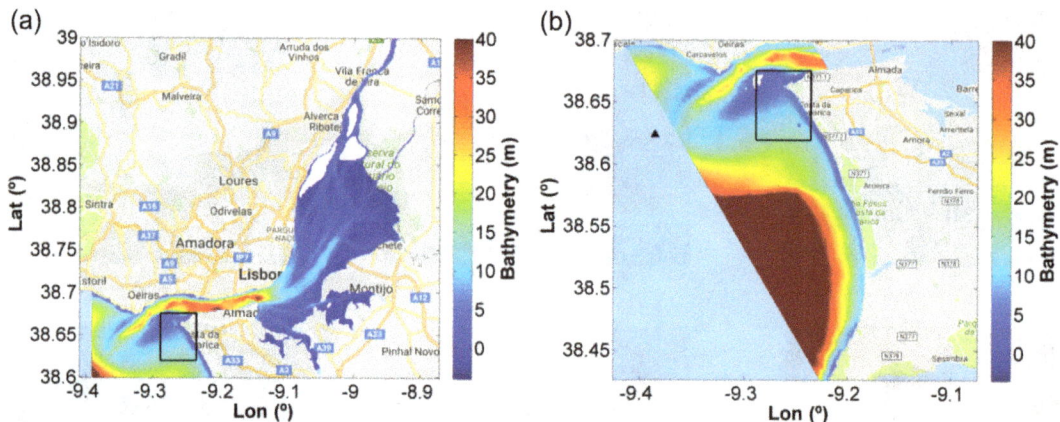

Figure 9. Bathymetries of the Tagus Estuary hydrodynamic model (**a**) and the wave model (**b**) created to propagate the waves until the Tagus Estuary mouth. The domain of the Costa da Caparica model is presented by the black rectangle and the location of the port of Lisbon wave buoy is indicated by the black triangle.

The open boundary conditions for sediment concentrations in the water column and sediment mass evolution at the bed column were defined as a null gradient, as well as the boundary conditions for normal and tangential velocities, as in the previous 2DH test cases. Considering that no structures were included in this test case, the bathymetry evolution was expected to be nearly uniform along the beach. Thus, a cyclic boundary condition was imposed at the cross-shore boundaries together with a Flather radiation condition at the offshore boundary (Eq. 4).

The model was capable of representing an undertow pattern and associated sediment transport that induce the formation of longshore sandbars (Fig. 8). A longitudinal current is presented in the surf zone, similarly to that observed in 2DH. However, a cross-shore velocity component is now represented by the 3-D model. Inside the surf zone, this component is shoreward near the surface and seaward near the bottom. As expected, the cross-shore component has opposite directions before and after the breaking zone. The sandbar migrated seaward, changing the location of the breaking zone until wave heights decreased to 1.0 m after 45 days.

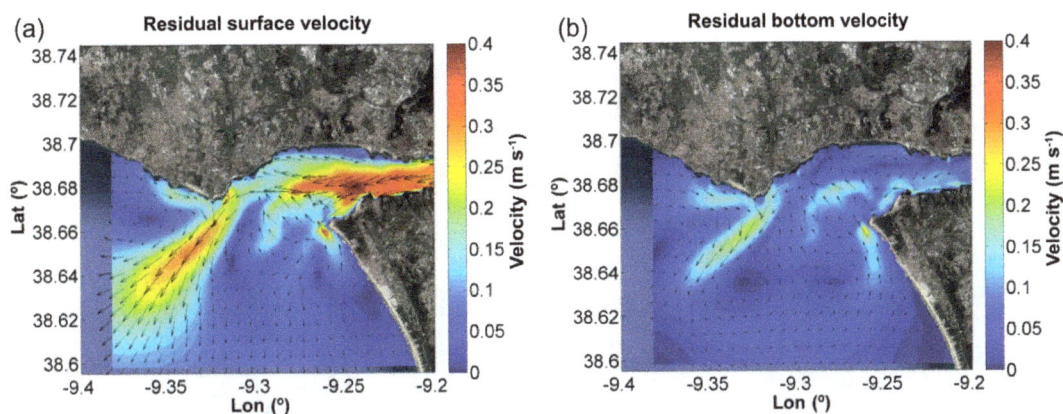

Figure 10. Residual velocity in the Tagus Estuary mouth at the surface **(a)** and near the bottom **(b)** obtained from the 3-D baroclinic model for the Tagus Estuary.

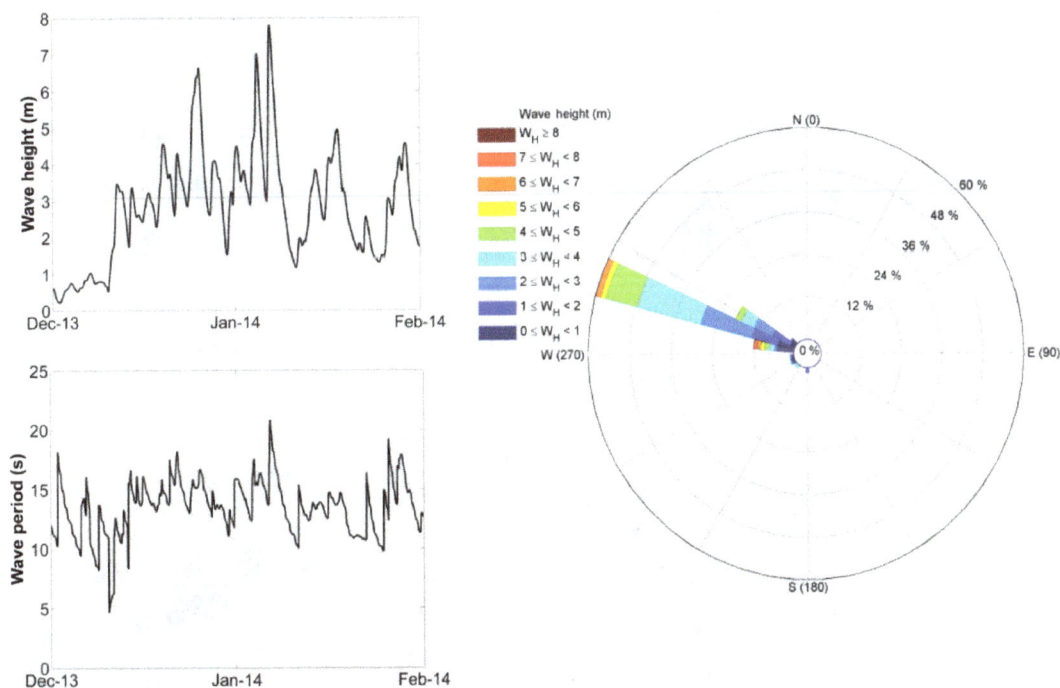

Figure 11. Wave conditions for the winter of 2013/2014 obtained from the Portuguese coast wave model in the location of the port of Lisbon wave buoy.

At this time, waves were able to propagate further without breaking, creating a new sandbar nearer the coast. Finally, the results demonstrate the model's potential to represent the formation of multiple sandbars, which are observed in many places (e.g. Dolan and Dean, 1985; Ruessink et al., 2009).

4.3   Costa da Caparica

The hydrodynamics and sediment transport on the southern coast of the Tagus Estuary mouth (Costa da Caparica) are evaluated under extreme wave conditions by coupling the MOHID modelling system and the SWAN wave model. A

significant coastline retreat was observed in the Costa da Caparica in the last century. Defence structures (groynes) were built around the 1960s to reduce coastal erosion, resulting in some stability until the 2000/2001 winter, when this issue started to receive more attention from the local authorities (Veloso-Gomes et al., 2009). The importance of the problem has augmented due to urbanisation and tourism development. The location near to the Tagus Estuary inlet increases the complexity of sediment dynamics in this zone.

A downscaling approach was followed to provide appropriate boundary conditions for the Costa da Caparica model, considering previous results of the wave and hydrodynamic

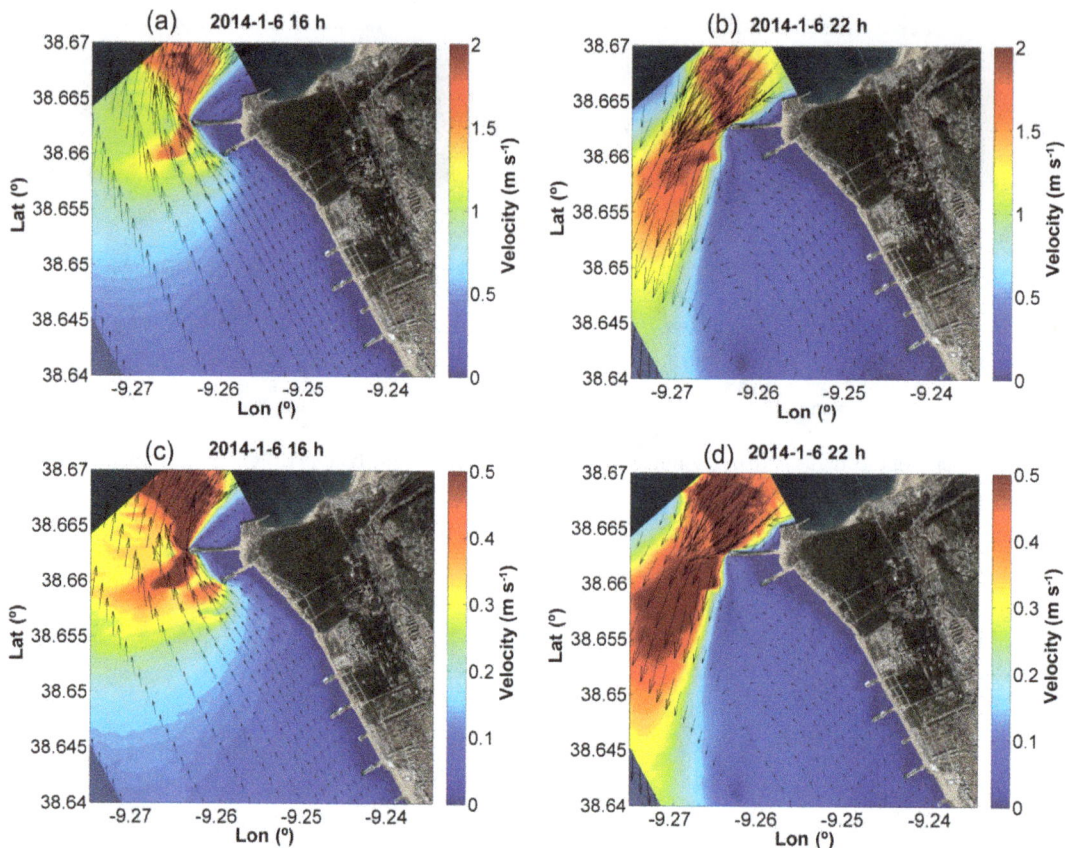

Figure 12. Velocity results of the Costa da Caparica model during flood (**a, c**) and ebb (**b, d**) tides at the surface (**a, b**) and near the bottom (**c, d**) without considering the wave's effect on the hydrodynamics.

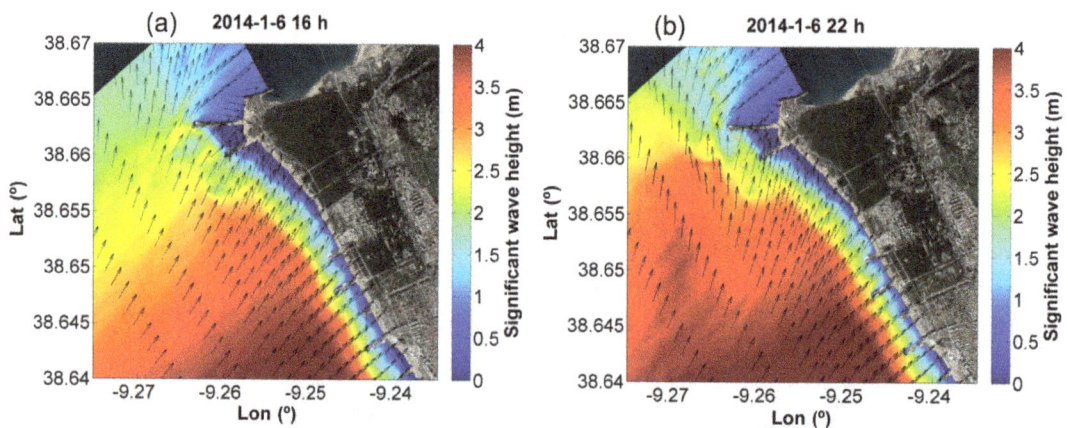

Figure 13. Wave results of the Costa da Caparica model.

modelling system for the Portuguese coast developed by the MARETEC research group (http://forecast.maretec.org). A new domain was created with 100 m × 100 m of grid resolution to propagate the waves until the Tagus Estuary mouth. The bathymetries of the father hydrodynamic model and father wave model are presented in Fig. 9, showing the domain of the Costa da Caparica model. The hydrodynamic boundary conditions (water level, current velocities, salinity, and temperature) for the Costa da Caparica model were provided by the 3-D baroclinic model for the Tagus Estuary (see Franz et al., 2014a, b), through the application of the relaxation scheme together with the Flather radiation condition. The wave model for the Portuguese coast was validated previously considering the data of the port of Lisbon wave buoy,

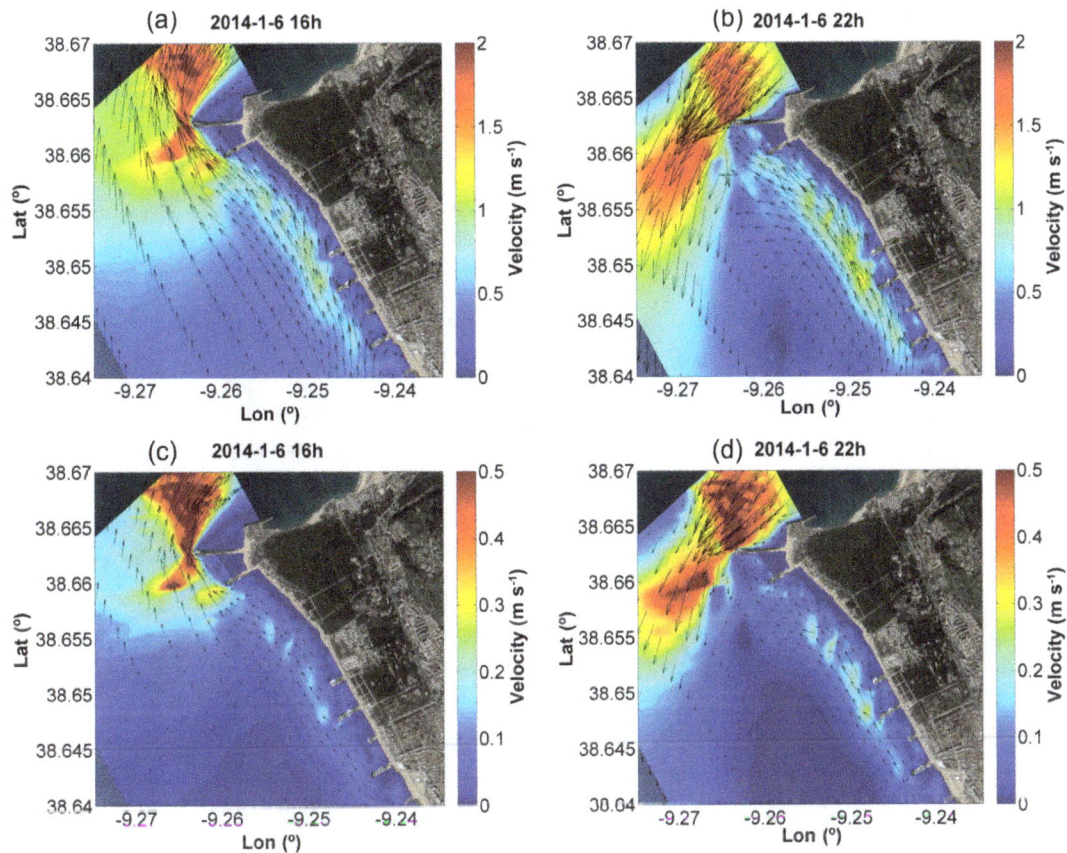

Figure 14. Velocity results of the Costa da Caparica model during flood (**a, c**) and ebb (**b, d**) tides at the surface (**a, b**) and near the bottom (**c, d**) considering the wave's effect on the hydrodynamics.

among other buoys located along the Portuguese coast (see Franz et al., 2014c).

The Tagus Estuary is classified as mesotidal, with an average tidal height of 2.0 m in the mouth (Lemos, 1972). The tide is the main mechanism forcing the flow in the estuary, determining current directions and water level variations (Franz et al., 2014d). The maximum velocities reach 2 m s^{-1} in the estuary mouth. The Tagus River is the estuary's main freshwater source, with an annual average flow of about 300 m^3 s^{-1}. The estuary stratification is strongly related to the river flow and tidal cycle. The residual currents pattern in the Tagus Estuary mouth is characterised by a jet in the inlet channel and two adjacent vortices, causing a residual recirculation into the estuary in front of the Costa da Caparica (Fig. 10).

The effects of the waves on the currents and sediment transport were investigated during a high-energy event in the winter of 2013/2014 caused by the Hercules storm. The wave conditions for the period of study in the location of the port of Lisbon wave buoy reached wave heights higher than 7 m and wave periods of up to 20 s (Fig. 11). A variable grid resolution was defined for the Costa da Caparica domain, ranging from 50 to 10 m near the coast. The water column was

divided into 10 layers, including five layers in the first metre above the bottom with fixed thicknesses ranging from 0.1 to 0.3 m, and five sigma layers on top. The exponential approach described in the last section was followed to represent the vertical variation of the wave-induced forces. The vertical viscosity was computed by the k-ε turbulence-closure model and the horizontal viscosity was set to 5.0 m^2 s^{-1}. The bathymetry data for the coast of the Costa da Caparica were provided by the Portuguese Environment Agency (APA).

In a first scenario, the effects of the waves on the currents and sediment transport were neglected. Therefore, just the influence of hydrodynamic boundary conditions from the Tagus Estuary model was taken into account. The results of the Costa da Caparica model without the wave action demonstrate strong velocities of up to 2 m s^{-1} at the surface and 0.5 m s^{-1} near the bed in the northern zone, with opposite directions on the flood and ebb tides, whereas weak velocities are found in the remaining coast directed to the estuary inlet in both situations (Fig. 12). The strong tidal currents are deflected away from the shoreline by the presence of a long groyne.

During the period of study, the waves propagated from offshore, mainly with a west-northwesterly (WNW) direction

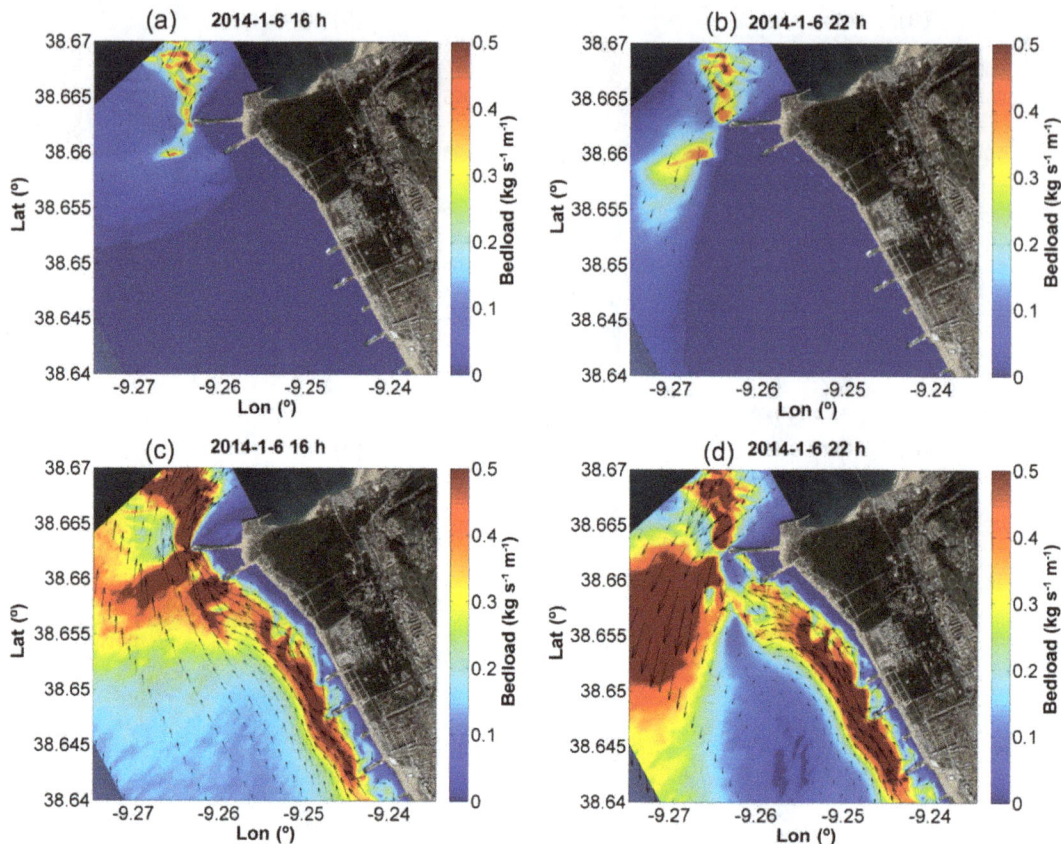

Figure 15. Results of bed load sediment transport for the scenarios without waves **(a, b)** and under extreme wave conditions **(c, d)**.

(Fig. 11). The bathymetric features (Fig. 9) cause the modification of the wave propagation direction through refraction to a west-southwesterly (WSW) or southwesterly (SW) direction in the nearshore (Fig. 13). The effect of the currents and water level variations on wave propagation was also considered, with an update frequency of 1 h, the same frequency at which the wave forcing was updated in MO-HID. The oblique angle of the waves' incidence generates a nearshore longitudinal current oriented to the estuary inlet (Fig. 14), reinforcing the velocities observed in the scenario in which the wave action was neglected. The velocity vectors have a shoreward component at the surface, whereas near the bottom a seaward component is observed, caused by the vertical variation of the wave-induced forces. Although the currents were intensified along the coast of the Costa da Caparica, a small reduction of velocities can be observed in the northern part of the model domain.

Based on a few existing data in the literature, a uniform granulometry was assumed for the Costa da Caparica model with a diameter of 0.3 mm (Freire et al., 2006). A more representative set of granulometry data is necessary to better characterise the sediment distribution in the model domain, considering that grain-size sorting is expected to occur due to the complex hydrodynamics, wave variability and bathymetric heterogeneities. Thus, the results of sediment transport should be seen as a first qualitative assessment. Moreover, the morphological evolution was ignored at this stage.

The patterns of the bed load sediment transport for the scenarios without waves and under extreme wave conditions are presented in Fig. 15. Along the coast of the Costa da Caparica, the results of the bed load sediment transport are practically null when the wave action was disregarded, suggesting that the tidal currents are irrelevant for the sediment transport in this area. By contrast, simulations including wave forcing show a very strong bed load sediment transport due to waves. However, the importance of the tidal currents for the sediment transport in the inlet of the Tagus Estuary is noticeable. The littoral drift caused by the waves is deflected seaward by the tidal currents during the ebb tide and by the longer groynes present near the estuary inlet.

5 Conclusions

The potential of a new modelling approach to simulate the impact of different designs of coastal defence structures was demonstrated in this paper. The coupling between the MO-HID modelling system and the SWAN wave model can be useful for engineering studies in order to evaluate the best solution to protect the coast against erosion. The speed-up of morphological changes, along with the multiprocessing architecture, allows for the modelling of bed evolution for long periods and for the study of several scenarios. The acceleration factor of 365 allowed us to simulate many years of morphological evolution of a schematic beach with constant wave conditions in a feasible computational time. The bathymetry results reached an equilibrium condition, demonstrating the model stability. The acceleration factor is user-defined, depending on the variability of forcing conditions and speed of morphological changes. The morphological changes should not be speeded up (acceleration factor $= 1$) to simulate extreme events that occur within few days, as for the case presented for the Costa da Caparica. A more efficient coupling method is currently being developed inside the MOHID code to further reduce the computational time. Moreover, the interface of the MOHID modelling system (MOHID Studio) was developed to make the coupling with SWAN straightforward for all users.

The potential for modelling the evolution of sandbars was also demonstrated in this paper. In the future, an up-to-date methodology can be applied to resolve the vertical variation of wave-induced forces, as well as wave-induced vertical mixing, based on e.g. the generalised Lagrangian mean (GLM) theory implemented in the MOHID code (see Delpey et al., 2014). The test cases shown in this paper are only a preliminary demonstration of model potential, which is thought to be encouraging. The findings of this work confirmed the applicability of the MOHID modelling system in studying sediment transport and morphological changes in coastal systems under the combined action of waves and currents.

The application of the described modelling methodology to a coastal zone located near the inlet of a mesotidal estuary with strong tidal currents allowed for an assessment of the hydrodynamics and sediment transport in situations of calm water conditions (no waves) and under extreme wave conditions. Although these initial results are just a qualitative assessment of sediment transport, the applicability of the modelling methodology to complex cases was demonstrated. In the future, with a more representative set of data, quantitative studies could be performed, taking into account the morphological evolution. Furthermore, the methodology can be used to evaluate different designs of defence structures in order to propose a more efficient solution for the coastline retreat and intense erosion observed in recent years on the coast of the Costa da Caparica.

Code availability. The MOHID code updated with the developments performed in this work is available in the repository: //github.com/Mohid-Water-Modelling-System/Mohid.

Competing interests. The authors declare that they have no conflict of interest.

Acknowledgements. The authors are grateful to the Portuguese Environment Agency (APA) for providing the bathymetry data for the coast of the Costa da Caparica. The first author was financed by the Brazilian National Council for Scientific and Technological Development (CNPq) under the Ciências Sem Fronteiras programme (research grant no. 237448/2012-2). MARETEC acknowledges the ERDF funds of the Competitiveness Factors Operational Programme (COMPETE), and national funds from the Foundation for Science and Technology (FCT) (project UID/EEA/50009/2013).

Edited by: John M. Huthnance

References

Amoudry, L. O. and Liu, P. L. F.: Parameterization of near-bed processes under collinear wave and current flows from a two-phase sheet flow model, Cont. Shelf Res., 30, 1403–1416, 10.1016/j.csr.2010.04.009, 2010.

Amoudry, L. O. and Souza, A. J.: Impact of sediment-induced stratification and turbulence closures on sediment transport and morphological modelling, Cont. Shelf Res., 31, 912–928, https://doi.org/10.1016/j.csr.2011.02.014, 2011.

Ardhuin, F., Rascle, N., and Belibassakis, K. A.: Explicit wave-averaged primitive equations using a generalized Lagrangian mean, Ocean Model., 20, 35–60, https://doi.org/10.1016/j.ocemod.2007.07.001, 2008.

Blumberg, A. F. and Kantha, L. H.: Open boundary condition for circulation models, J. Hydraul. Eng., 111, 237–255, 1985.

Booij, N., Ris, R., and Holthuijsen, L. H.: A third-generation wave model for coastal regions: 1. Model description and validation, J. Geophys. Res.-Oceans, 104, 7649–7666, 1999.

Bowen, A. J., Inman, D., and Simmons, V.: Wave "set-down" and set-Up, J. Geophys. Res., 73, 2569–2577, 1968.

Burchard, H., Bolding, K., and Villarreal, M. R.: GOTM, a general ocean turbulence model: theory, implementation and test cases, Space Applications Institute, 1999.

Dally, W. R. and Pope, J.: Detached breakwaters for shore protection, DTIC Document, 1986.

Dean, R. G.: Equilibrium beach profiles: characteristics and applications, J. Coastal Res., 7, 53–84, 1991.

Delpey, M., Ardhuin, F., Otheguy, P., and Jouon, A.: Effects of waves on coastal water dispersion in a small estuarine bay, J. Geophys. Res.-Oceans, 119, 70–86, 2014.

Dolan, T. J. and Dean, R. G.: Multiple longshore sand bars in the upper Chesapeake Bay, Estuar. Coast. Shelf Sci., 21, 727–743, 1985.

Drønen, N. and Deigaard, R.: Quasi-three-dimensional modelling of the morphology of longshore bars, Coast. Eng., 54, 197–215, 2007.

Dubarbier, B., Castelle, B., Marieu, V., and Ruessink, G.: Process-based modeling of cross-shore sandbar behavior, Coast. Eng., 95, 35–50, 2015.

Einstein, H. A.: The bed-load function for sediment transportation in open channel flows, Citeseer, 1950.

Engedahl, H.: Use of the flow relaxation scheme in a three-dimensional baroclinic ocean model with realistic topography, Tellus A, 47, 365–382, 1995.

Flather, R.: A tidal model of the northwest European continental shelf, Mem. Soc. R. Sci. Liege, 10, 141–164, 1976.

Franz, G., Fernandes, R., de Pablo, H., Viegas, C., Pinto, L., Campuzano, F., Ascione, I., Leitão, P., and Neves, R.: Tagus Estuary hydro-biogeochemical model: Inter-annual validation and operational model update. 3.as Jornadas de Engenharia Hidrográfica, Lisbon, Portugal, Extended abstracts, 103–106, 2014a.

Franz, G., Campuzano, F., Fernandes, R., Pinto, L., de Pablo, H., Ascione, I., and Neves, R.: An integrated forecasting system of hydro-biogeochemical and waves in the Tagus estuary, Proceedings of the Seventh EuroGOOS International Conference, Lisbon, 2014b.

Franz, G., Campuzano, F., Pinto, L., Fernandes, R., Sobrinho, J., Simões, A., Juliano, M., and Neves, R.: Implementation and validation of a wave forecasting system for the Portuguese Coast, Proceedings of the Seventh EuroGOOS International Conference, Lisbon, 2014c.

Franz, G., Pinto, L., Ascione, I., Mateus, M., Fernandes, R., Leitao, P., and Neves, R.: Modelling of cohesive sediment dynamics in tidal estuarine systems: Case study of Tagus estuary, Portugal, Estuarine, Coast. Shelf Sci., 151, 34–44, 2014d.

Franz, G. A. S., Leitão, P., Santos, A. D., Juliano, M., and Neves, R.: From regional to local scale modelling on the south-eastern Brazilian shelf: case study of Paranaguá estuarine system, Braz. J. Oceanogr., 64, 277–294, 2016.

Franz, G., Leitão, P., Pinto, L., Jauch, E., Fernandes, L., and Neves, R.: Development and validation of a morphological model for multiple sediment classes, Int. J. Sediment Res., https://doi.org/10.1016/j.ijsrc.2017.05.002, in press, 2017.

Freire, P., Taborda, R., and Andrade, C.: Caracterização das praias estuarinas do Tejo, APRH, Congresso da Água, vol. 8, 2006.

Fortunato, A. B., Nahon, A., Dodet, G., Pires, A. R., Freitas, M. C., Bruneau, N., Azevedo, A., Bertin, X., Benevides, P., Andrade, C., and Oliveira, A.: Morphological evolution of an ephemeral tidal inlet from opening to closure: The Albufeira inlet, Portugal, Cont. Shelf Res., 73, 49–63, 2014.

Greenwood, R. and Davis, R. A.: Hydrodynamics and sedimentation in wave-dominated coastal environments, Elsevier, 2011.

Komar, P. D.: Beach Processes and Sedimentation. Upper Saddle River, N.J, Prentice Hall, 1998.

Kriebel, D. L. and Dean, R. G.: Numerical simulation of time-dependent beach and dune erosion, Coast. Eng., 9, 221–245, 1985.

Kristensen, S. E., Drønen, N., Deigaard, R., and Fredsoe, J.: Hybrid morphological modelling of shoreline response to a detached breakwater, Coast. Eng., 71, 13-27, 2013.

Leitao, P., Mateus, M., Braunschweig, L., Fernandes, L., and Neves, R.: Modelling coastal systems: the MOHID Water numerical lab, Perspectives on integrated coastal zone management in South America, 77–88, 2008.

Leitão, P. C.: Integração de Escalas e Processos na Modelação do Ambiente Marinho, 2003.

Lemos, P. A. F.: Estuário do Tejo. Administração Geral do Porto de Lisboa, Lisbon, 1972.

Levoy, F., Anthony, E., Monfort, O., and Larsonneur, C.: The morphodynamics of megatidal beaches in Normandy, France, Mar. Geol., 171, 39–59, 2000.

Longuet-Higgins, M. S. and Stewart, R.: Radiation stress and mass transport in gravity waves, with application to "surf beats", J. Fluid Mech., 13, 481–504, 1962.

Longuet-Higgins, M. S. and Stewart, R.: Radiation stresses in water waves; a physical discussion, with applications, Deep Sea Research and Oceanographic Abstracts, Vol. 11, No. 4, Elsevier, 1964.

Longuet-Higgins, M. S.: Longshore currents generated by obliquely incident sea waves: 1, J. Geophys. Res., 75, 6778–6789, 1970a.

Longuet-Higgins, M. S.: Longshore currents generated by obliquely incident sea waves: 2, J. Geophys. Res., 75, 6790–6801, https://doi.org/10.1029/JC075i033p06790, 1970b.

Longuet-Higgins, M. S.: Wave set-up, percolation and undertow in the surf zone, P. Roy. Soc. Lond. A, 390, 283–291, 1983.

Malhadas, M. S., Leitão, P. C., Silva, A., and Neves, R.: Effect of coastal waves on sea level in Óbidos Lagoon, Portugal, Cont. Shelf Res., 29, 1240–1250, 2009.

Malhadas, M. S., Neves, R. J., Leitão, P. C., and Silva, A.: Influence of tide and waves on water renewal in Óbidos Lagoon, Portugal, Ocean Dynam., 60, 41–55, 2010.

Martins, F.: Modelação Matemática Tridimensional de escoamentos costeiros e estuarinos usando uma abordagem de coordenada vertical genérica, PhD thesis, Universidade Técnica de Lisboa, Instituto Superior Técnico, Lisbon, Portugal, 2000.

Martins, F., Leitão, P. C., Silva, A., and Neves, R.: 3D modelling in the Sado estuary using a new generic vertical discretization approach, Oceanol. Acta, 24, 551–562, 2001.

Martinsen, E. A. and Engedahl, H.: Implementation and testing of a lateral boundary scheme as an open boundary condition in a barotropic ocean model, Coast. Eng., 11, 603–627, 1987.

Myrhaug, D., Holmedal, L. E., and Rue, H.: Bottom friction and bedload sediment transport caused by boundary layer streaming beneath random waves, Appl. Ocean Res., 26, 183–197, 2004.

Nahon, A., Bertin, X., Fortunato, A. B., and Oliveira, A.: Process-based 2DH morphodynamic modeling of tidal inlets: a comparison with empirical classifications and theories, Mar. Geol., 291, 1–11, 2012.

Roelvink, D., Reniers, A., Van Dongeren, A., de Vries, J. v. T., McCall, R., and Lescinski, J.: Modelling storm impacts on beaches, dunes and barrier islands, Coast. Eng., 56, 1133–1152, 2009.

Ruessink, B., Kuriyama, Y., Reniers, A., Roelvink, J., and Walstra, D.: Modeling cross-shore sandbar behavior on the timescale of weeks, J. Geophys. Res.-Earth, 112, F03010, https://doi.org/10.1029/2006JF000730, 2007.

Ruessink, B., Pape, L., and Turner, I.: Daily to interannual cross-shore sandbar migration: observations from a multiple sandbar system, Cont. Shelf Res., 29, 1663–1677, 2009.

Saville, T.: Experimental determination of wave set-up, Proceedings of the Second Technical Conference on Hurricanes, Report No. 50, 242–252, 1961.

Smagorinsky, J.: General circulation experiments with the primitive equations: I. The basic experiment, Mon. Weather Rev., 91, 99–164, 1963.

Soulsby, R. and Clarke, S.: Bed shear-stresses under combined waves and currents on smooth and rough beds, HR Wallingford, Report TR137, 2005.

Soulsby, R. L. and Damgaard, J. S.: Bedload sediment transport in coastal waters, Coast. Eng., 52, 673–689, 2005.

Veloso-Gomes, F., Costa, J., Rodrigues, A., Taveira-Pinto, F., Pais-Barbosa, J., and Neves, L. D.: Costa da Caparica artificial sand nourishment and coastal dynamics, J. Coast. Res., 1, 678–682, 2009.

Villarreal, M. R., Bolding, K., Burchard, H., and Demirov, E.: Coupling of the GOTM turbulence module to some three-dimensional ocean models, in: Marine Turbulence: Theories, Observations and Models, edited by: Baumert, H. Z., Simpson, J. H., and Sündermann, J., Cambridge, Cambridge University Press, 225–237, 2005.

Warner, J. C., Sherwood, C. R., Signell, R. P., Harris, C. K., and Arango, H. G.: Development of a three-dimensional, regional, coupled wave, current, and sediment-transport model, Comput. Geosci., 34, 1284–1306, 2008.

Wave spectral shapes in the coastal waters based on measured data off Karwar on the western coast of India

M. Anjali Nair and V. Sanil Kumar

Ocean Engineering Division, Council of Scientific & Industrial Research-National Institute of Oceanography, Dona Paula, 403 004 Goa, India

Correspondence to: V. Sanil Kumar (sanil@nio.org)

Abstract. An understanding of the wave spectral shapes is of primary importance for the design of marine facilities. In this paper, the wave spectra collected from January 2011 to December 2015 in the coastal waters of the eastern Arabian Sea using the moored directional waverider buoy are examined to determine the temporal variations in the wave spectral shape. Over an annual cycle for 31.15 % of the time, the peak frequency is between 0.08 and 0.10 Hz; the significant wave height is also relatively high (~ 1.55 m) for waves in this class. The slope of the high-frequency tail of the monthly average wave spectra is high during the Indian summer monsoon period (June–September) compared to other months, and it increases with an increase in significant wave height. There is not much interannual variation in the slope for swell-dominated spectra during the monsoon, while in the non-monsoon period when wind-seas have a high level of influence, the slope varies significantly. Since the exponent of the high-frequency part of the wave spectrum is within the range of -4 to -3 during the monsoon period, the Donelan spectrum shows a better fit for the high-frequency part of the wave spectra in monsoon months compared to other months.

1 Introduction

Information on wave spectral shapes is required for designing marine structures (Chakrabarti, 2005), and almost all of the wave parameter computations are based on the wave spectral function (Yuan and Huang, 2012). The growth of waves and the corresponding spectral shape is due to the complex ocean–atmosphere interactions, while the physics of the air–sea interaction is not completely understood (Cav-

aleri et al., 2012). The shape of the wave spectrum depends on the factors governing the wave growth and decay, and a number of spectral shapes have been proposed in the past for different sea states (see Chakrabarti, 2005 for a review). The spectral shape is maintained by the nonlinear transfer of energy through nonlinear four-wave interactions (quadruplet interactions) and whitecapping (Gunson and Symonds, 2014). The momentum flux between the ocean and the atmosphere govern the high-frequency wave components (Cavaleri et al., 2012). According to Phillips (1985) the equilibrium ranges for low-frequency and high-frequency regions are proportional to f^{-5} and f^{-4} (where f is the frequency), respectively. Several field studies conducted since the JONSWAP (Joint North Sea Wave Project) field campaign reveal an analytical form for wave spectra with the spectral tail proportional to f^{-4} (Toba, 1973; Kawai et al., 1977; Kahma, 1981; Forristall, 1981; Donelan et al., 1985). Usually, there is a predominance of swell fields in large oceanic areas, which is due to remote storms (Chen et al., 2002; Hwang et al., 2011; Semedo et al., 2011). The exponent used in the expression for the frequency tail has different values (see Siadatmousavi et al., 2012 for a brief review). For shallow water, Kitaigordskii et al. (1975) suggested an f^{-3} tail and Liu et al. (1989) suggested f^{-4} for growing young wind-seas and f^{-3} for fully developed wave spectra. Badulin et al. (2007) suggested f^{-4} for frequencies with dominant nonlinear interactions. The study carried out at Lake George by Young and Babanin (2006) revealed that in the frequency range $5f_\mathrm{p} < f < 10f_\mathrm{p}$, the average value of the exponent "n" of f^{-n} is close to 4. Other studies in real sea conditions indicate that the high-frequency shape of f^{-4} applies up to a few times the peak frequency (f_p) and then decays faster with frequency. The spectra for coastlines in Currituck Sound with short fetch conditions showed a decay closer to f^{-5} when f is greater than 2 or 3 times the peak frequency (Long and Resio, 2007). Gagnaire-Renou et al. (2010) found that the energy input from wind and the dissipation due to whitecapping have a significant influence on the high-frequency tail of the spectrum.

The physical processes in the northern Indian Ocean have a distinct seasonal cycle (Shetye et al., 1985; Ranjha et al., 2015), and the surface wind–wave field is no exception (Sanil Kumar et al., 2012). In the eastern Arabian Sea (AS), the significant wave height (H_{m0}) up to 6 m is measured in the monsoon period (June to September). During the rest of the periods, H_{m0} is normally less than 1.5 m (Sanil Kumar and Anand, 2004). Sanil Kumar et al. (2014) observed that in the eastern AS, the wave spectral shapes are different at two locations within a 350 km distance, even though the difference in the integrated parameters like H_{m0} is marginal. Dora and Sanil Kumar (2015) observed that waves at 7 m of water depth in the nearshore zone off Karwar are high-energy waves in the monsoon period and low to moderate waves in the non-monsoon period (January to May and October to December). The Dora and Sanil Kumar (2015) study shows a similar contribution of wind-seas and swells during the pre-monsoon period (February to May), while swells dominate the wind-sea in the post-monsoon period (October to January) and the monsoon period. A study was carried out by Glejin et al. (2012) to find the variation in wave characteristics along the eastern AS and the influence of swells in the nearshore waves at three locations during the monsoon period in 2010. This study shows that the percentage of swells in the measured waves was 75 % at the southern part of the AS and 79 % at the northern part of the AS. Wind and wave data measured at a few locations along the western coast of India for a short period of 1 to 2 months as well as the wave model results were analyzed to study the wave characteristics in the deep and nearshore regions during different seasons (Vethamony et al., 2013). From the wave data collected for a 2-year period (2011 and 2012) along the eastern AS, swells of more than 18 s and significant wave heights of less than 1 m, which occur 1.4 to 3.6 % of the time, were separated and their characteristics were studied by Glejin et al. (2016). Anjali Nair and Sanil Kumar (2016) presented the daily, monthly, seasonal and annual variations in the wave spectral characteristics for a location in the eastern AS and reported that over an annual cycle, 29 % of the wave spectra are single-peaked spectra and 71 % are multi-peaked spectra. Recently, Amrutha et al. (2017) analyzed the measured wave data in October and reported that the high waves (significant wave height > 4 m) generated in an area bounded by 40–60° S and 20–40° E in the southern Indian Ocean reached the eastern AS in 5–6 days and resulted in the long-period waves. Earlier studies indicate that the spectral tail of the high-frequency part shows large variation and that its variation with seasons is not known. Similarly, the shape of the

parametric spectra are also different, and hence it is important to identify the spectral shapes based on the measured data covering all seasons and different years.

The discussion above shows that there is a strong motivation to study the high-frequency tail of the wave spectrum. For the present study, we used the directional waverider buoy with measured wave spectral data at 15 m of water depth off Karwar on the western coast of India over 5 years from 2011 to 2015 and evaluated the nearshore wave spectral shapes in different months. This study addresses two main issues: (1) how the high-frequency tail of the wave spectrum varies in different months and (2) the spectral parameters for the best-fit theoretical spectra. This paper is organized as follows: the study area is introduced in Sect. 2, and the details of data used and the methodology are presented in Sect. 3. Section 4 presents the results of the study, and the conclusions are given in Sect. 5.

2 Study area

The coastline at Karwar is 24° inclined to the west from the north, and the 20 m depth contour is inclined 29° to the west. Hence, large waves in the nearshore will have an incoming direction close to 241°, since waves become aligned with the depth contour due to refraction. At 10, 30 and 75 km of distance from Karwar, depth contours are present at 20, 50 and 100 m (Fig. 1). The study region is under the seasonally reversing monsoon winds, with winds from the northeast during the post-monsoon period and from the southwest during the monsoon period. The monsoon winds are strong, and the total seasonal rainfall is 280 cm. There is a 0.24 m annual cycle in the mean sea level from September to January. The average tidal range is 1.58 m during spring tides and 0.72 m during neap tides (Sanil Kumar et al., 2012).

3 Data and methods

The waves off Karwar (14°49′56″ N and 74°6′4″ E) were measured using the directional waverider buoy (DWR-MKIII) . The measurements were carried out from 1 January 2011 to 31 December 2015. The heave data and the two-translational motion of the buoy are sampled at 3.84 Hz. A digital high-pass filter with a cutoff at 30 s is applied to the 3.84 Hz samples. At the same time, it converts the sampling rate to 1.28 Hz and stores the time series data at 1.28 Hz. From the time series data for 200 s, the wave spectrum is obtained through a fast Fourier transform (FFT). During half an hour, eight wave spectra with a 200 s data interval are collected and averaged to get a representative wave spectrum for half an hour (Datawell, 2009). The wave spectrum has a resolution of 0.005 Hz from 0.025 to 0.1 Hz and 0.01 Hz from 0.1 to 0.58 Hz. The bulk wave parameters are the significant wave height (H_{m0}), which equals $4\sqrt{m_0}$, and the mean wave period (T_{m02}) based on second-order moment,

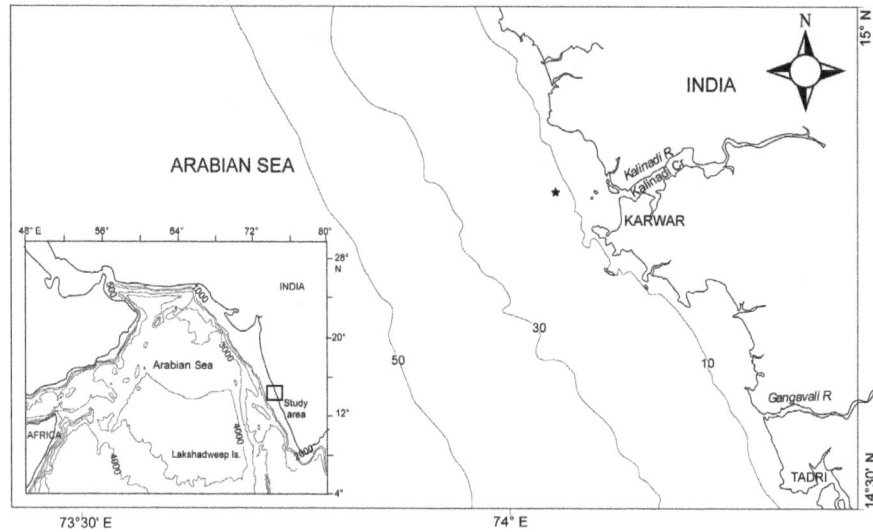

Figure 1. The study area along with the wave measurement location in the eastern Arabian Sea.

which equals $\sqrt{m_0/m_2}$); these are obtained from the spectral moments where m_n is the nth-order spectral moment $(m_n = \int_0^\infty f^n S(f)\,df$, $n = 0$ and 2), $S(f)$ is the spectral energy density and f is the frequency. The spectral peak period (T_p) is estimated from the wave spectrum, and the peak wave direction (D_p) is estimated based on circular moments (Kuik et al., 1988). The wind-seas and swells are separated through the method described by Portilla et al. (2009), and the wind-sea and the swell parameters are computed by integrating over the respective spectral parts. The measurements reported here are in Coordinated Universal Time (UTC), which is 5 h 30 min behind the local time. U_{10} is the wind speed at 10 m of height obtained from the reanalysis data of the zonal and meridional components at 6-hourly intervals from NCEP/NCAR (Kalnay et al., 1996). It is used to study the influence of wind speed on the spectral shape.

Since the frequency bins over which the wave spectrum is estimated are the same in all years, the monthly and seasonally averaged wave spectrum is computed by taking the average of the spectral energy density at the respective frequencies of each spectrum over the specified time.

The wave spectrum continues to develop through nonlinear wave–wave interactions, even for very long times and distances. Hence, most of the wave spectrum is not fully developed and cannot be represented by the Pierson–Moskowitz (PM) spectrum (Pierson and Moskowitz, 1964). Accordingly, an additional factor was added to the PM spectrum in order to improve the fit to the measured spectrum. The JONSWAP spectrum (Hasselmann et al., 1973) is thus a PM spectrum multiplied by an extra peak enhancement factor γ. The high-frequency tail of the JONSWAP spectrum decays in a form proportional to f^{-5}. A number of studies reported that high-frequency decay is by a form propor-

tional to f^{-4}. A modified JONSWAP spectrum that includes Toba's formulation of the saturation range was proposed by Donelan et al. (1985). The JONSWAP and Donelan spectra used in the study are given in Eqs. (1) and (2):

$$S(f) = \frac{\alpha g^2}{(2\pi)^4 f^5} \exp\left[-\frac{5}{4}\left(\frac{f}{f_p}\right)^{-4}\right] \gamma^{\exp\left[-(f-f_p)^2/2\sigma^2 f_p\right]}, \quad (1)$$

$$S(f) = \frac{\alpha g^2}{(2\pi)^4 f^4} f_p \exp\left[-\left(\frac{f}{f_p}\right)^{-4}\right] \gamma^{\exp\left[-(f-f_p)^2/2\sigma^2 f_p^2\right]}. \quad (2)$$

Here, γ is the peak enhancement parameter, α is the Phillips constant, f is the wave frequency, g is the gravitational acceleration and σ is the width parameter:

$$\sigma = \left\{ \begin{array}{ll} 0.07, & f < f_p \\ 0.09, & f \geq f_p \end{array} \right\}.$$

An exponential curve $y = k f^b$ is fitted for the high-frequency part of the spectrum and the exponent (the value of b) and the coefficient k are estimated for the best-fitting curve based on statistical measures such as the least-squares error and the bias. The slope of the high-frequency part of the wave spectrum is represented by the exponent of the high-frequency tail.

For the present study, the JONSWAP spectrum is tested by fitting for the whole frequency range of the measured wave spectrum. It is found out that the JONSWAP spectra do not show a good fit for higher-frequency ranges, whereas the Donelan spectrum shows a better fit for the high-frequency range. Hence, the JONSWAP spectrum is used for the lower-frequency range up to the spectral peak, and the Donelan spectrum is used for the higher-frequency range from the spectral peak for the single-peaked wave spectrum. The theoretical wave spectra are not fitted to the double-peaked wave spectra.

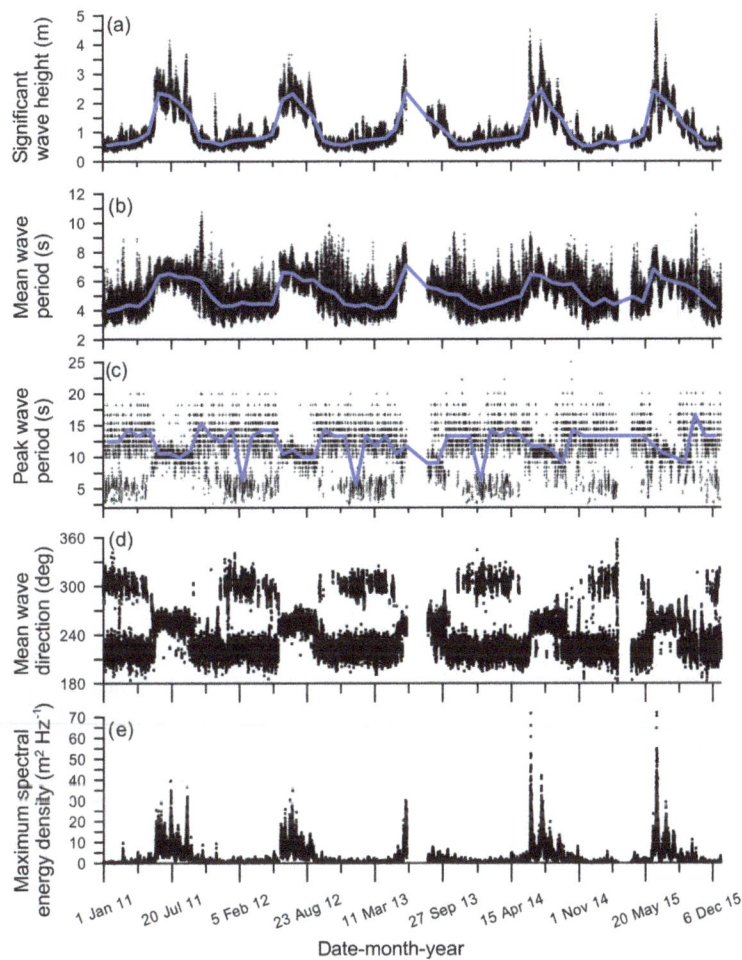

Figure 2. A time series plot of (**a**) the significant wave height, (**b**) mean wave period, (**c**) peak wave period, (**d**) mean wave direction and (**e**) maximum spectral energy density from 1 January 2011 to 31 December 2015. The thick blue line indicates the monthly average values.

4 Results and discussions

4.1 Bulk wave parameters

The wave conditions ($\sim 75\%$) at the buoy location are mostly intermediate- and shallow-water waves (where the water depth is less than half the wavelength, $d < L/2$). This condition is not satisfied during $\sim 25\%$ of the time due to waves with mean periods of 4.4 s or less. This study therefore deals with shallow, intermediate and deepwater wave climatology. Hence, bathymetry will significantly influence the wave characteristics.

The persistent monsoon winds generate choppy seas with average wave heights of 2 m and mean wave periods of 6.5 s. Figure 2 shows that in the monsoon, the observed waves had a maximum H_{m0} of about 5 m, with H_{m0} of 2–2.5 m more common during this period. The maximum H_{m0} measured during the study period is on 21 June 2015 at 17:30 UTC (Fig. 2a). The mean wave periods (T_{m02}) at the measurement location ranged from 4–8 s (Fig. 2b). The wave direction dur-

ing the monsoon is predominantly from the west due to refraction towards the coast. The fluctuation in H_{mo} due to the southwest monsoon is seen in all the years (Fig. 2a). High waves ($H_{m0} > 2$ m) during 27–29 November 2011 are due to the deep depression ARB 04 formed in the AS. During the study period, the annual average H_{m0} is the same (~ 1.1 m) in all the years (Table 1). In 2013, the data for July could not be collected, hence resulting in a lower annual average H_{m0}. Over the 5 years, small waves ($H_{m0} < 1$ m) account for a large proportion (63.94 %) of the measured data and only during 0.16 % of the time did H_{m0} exceed 4 m (Table 2). The 25th and 75th percentiles of the H_{m0} distribution over the entire analysis period are 0.6 and 1.4 m.

The waves with low heights ($H_{m0} < 1$ m) have mean periods in a large range (2.7–10.5 s), whereas the high waves ($H_{m0} > 3$ m) have a mean wave period in a narrow range (6.1–9.3 s) (Table 2). For waves with H_{m0} higher than 3 m, the T_p never exceeded 14.3 s, and for waves with H_{m0} less than 1 m, T_p up to 22.2 s are observed (Fig. 2c). The long-period swells (14–20 s) have $H_{m0} < 2.5$ m. Around 7 % of

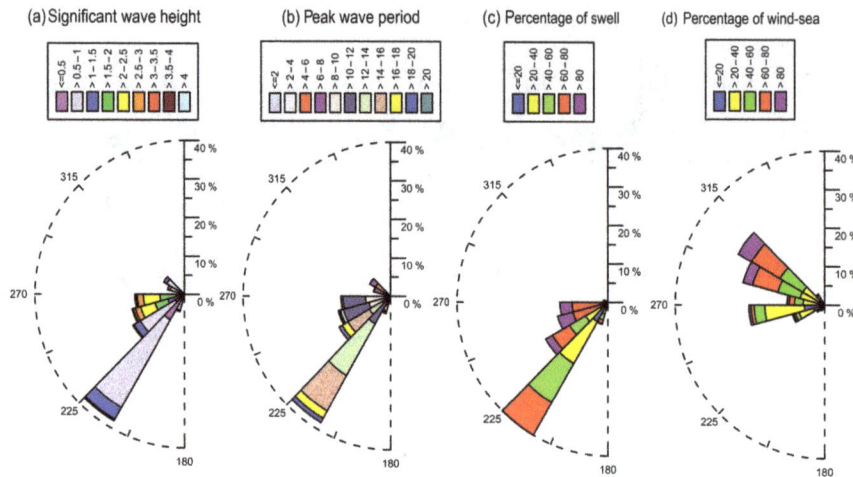

Figure 3. Wave roses during 2011–2015 for **(a)** the significant wave height and mean wave direction, **(b)** the peak wave period and mean wave direction, **(c)** the percentage of swell, **(d)** the percentage of wind-sea and mean wave direction.

Table 1. The amount of data used in the study in different years along with the range of significant wave heights and average values.

Year	Significant wave height (m)		Amount	% of
	Range	Average	of data	data
2011	0.3–4.4	1.1	17 517	99.98
2012	0.3–3.7	1.1	17 323	98.61
2013	0.3–3.6	0.9*	14 531	82.94
2014	0.3–4.5	1.1	17 284	98.65
2015	0.3–5.0	1.1	14 772	84.32

* The average value is estimated excluding the July data.

the time during 2011–2015, the waves had peak periods of more than 16.7 s (Table 3). Peak frequencies between 0.08 and 0.10 Hz, equivalent to a peak wave period of 10–12.5 s, are observed 31.15 % of the time, and the H_{m0} is also relatively high (~ 1.55 m) for waves in this class. During the annual cycle, the wave climate is dominated by low- ($0.5 > H_{m0} > 1$ m) and intermediate-period ($T_p \sim 10$–16 s) southwesterly swells. Waves from the northwest have a T_p less than 8 s (Fig. 3).

The wave roses during 2011–2015 indicate that around 38 % of the time, the predominant wave direction is SSW (225°) with long-period (14–18 s) and intermediate-period (10–14 s) waves (Fig. 3). A small percentage of long-period waves having H_{m0} more than 1 m are observed from the same direction, for which more than 80 % are swells (Fig. 3c). Intermediate-period waves observed with H_{m0} less than 1 m contain 20–60 % swells. Around 10–15 % of the waves observed during the period are from the west, which includes intermediate- and short-period waves with H_{m0} varying from 1.5 to 3 m. These intermediate-period waves from the west with H_{m0} between 2.5 and 3 m contain more than 80 %

swells. Waves from the NW are short-period waves with H_{m0} between 0.5 and 1.5; the swell percentage is very low, showing the influence of the wind-sea (Fig. 3d). The high waves observed in the study area consist of more than 80 % swells.

The date-versus-year plots of the significant wave height (Fig. 4) show that H_{m0} has its maximum values ($H_{m0} > 3$ m) during the monsoon period with a wave direction of WSW and a peak wave period of 10–12 s (the intermediate period). The mean wave period shows its maximum values (6–8 s) during the monsoon period. During January–May in all the years, H_{m0} is low ($H_{m0} < 1$ m) with waves from the SW, W and NW directions. The NW waves observed are the result of strong sea breezes during this period. Long-period ($T_p > 14$ s), intermediate-period ($10 < T_p < 14$ s) and short-period ($T_p < 8$ s) waves are observed during this period; hence, the mean wave period observed is low compared to the monsoon (Fig. 4d). From October to December, similar to the pre-monsoon period, the H_{m0} observed is less than 1 m, but the wave direction is predominantly from the SW and W with the least NW waves. Short-period waves are almost absent during this period, and the condition is similar for all the years. The interannual variations in H_{m0} are less than 15 % (Fig. 4). The primary seasonal variability in the waves is due to the monsoonal wind reversal. During January–March, there is a shift in the occurrences of northwest swells.

4.2 Wave spectrum

The normalized wave spectral energy density contours are presented for different years to identify the wind-sea or swell predominance (Fig. 5). The normalization of the wave spectrum is done to determine the spread of energy in different frequencies. Since the range of the maximum spectral energy density in a year is large (~ 60 m^2 Hz^{-1}), each wave spectrum is normalized by dividing the spectral energy density

Table 2. The characteristics of waves in different ranges of significant wave height.

Significant wave height range	Number (percentage)	Range of T_p (s)	Mean T_p (s)	Range of T_{m02} (s)	Mean T_{m02} (s)
$H_{m0} < 1$ m	52 062 (63.94)	2.6–22.2	12.2	2.7–10.5	4.9
$1 \leq H_{m0} < 2$ m	18 297 (22.47)	3.6–22.2	10.5	3.4–10.7	5.7
$2 \leq H_{m0} < 3$ m	9839 (12.08)	6.2-18.0	10.8	5.0-8.9	6.5
$3 \leq H_{m0} < 4$ m	1096 (1.35)	10.0–14.3	11.8	6.1–9.1	7.2
4 m $\leq H_{m0}$	133 (0.16)	10.5–14.3	12.6	7.2–9.3	7.8

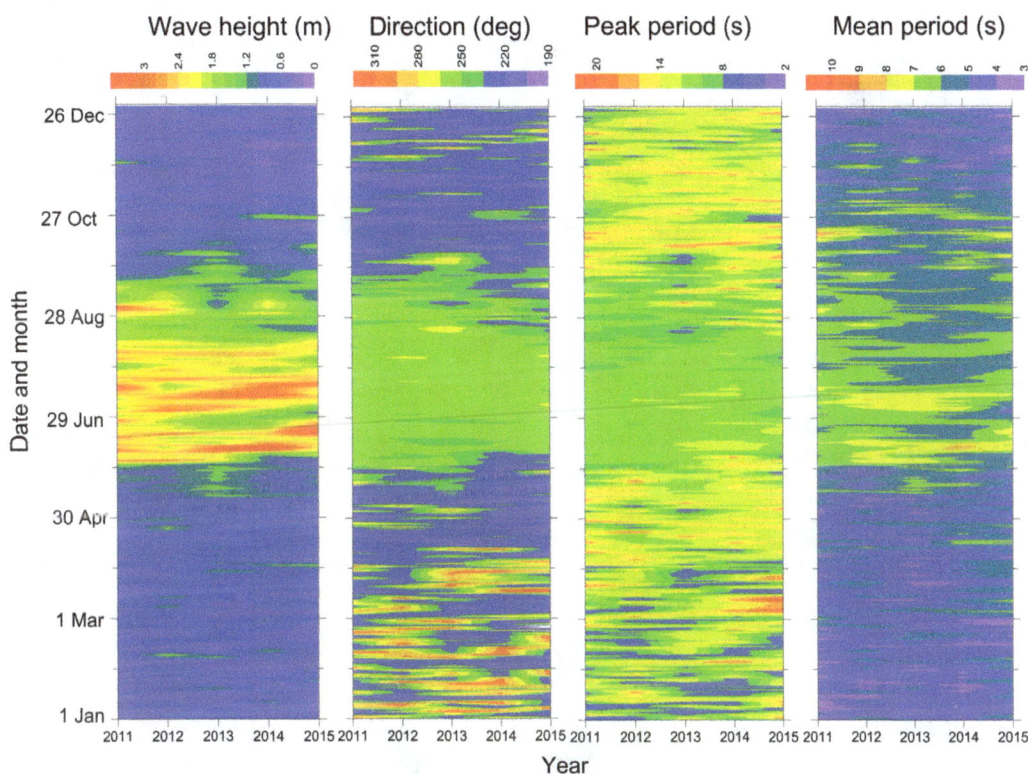

Figure 4. A date-versus-year plot of **(a)** the significant wave height, **(b)** mean wave direction, **(c)** peak wave period and **(d)** mean wave period.

by the maximum spectral energy density of that spectrum. The predominance of both the wind-seas and swells is observed in the non-monsoon period, whereas in the monsoon period only swells are predominant (Fig. 5). The separation of swells and wind-seas indicates that over an annual cycle, around 54 % of the waves are swells. Glejin et al. (2012) reported that the dominance of swells during the monsoon is due to the fact that even though the wind in the study region is strong during the monsoon, the wind over the entire AS will also be strong. When these swells are added to the wave system at the buoy location, the energy of the swell increases (Donelan, 1987) and will result in the dominance of swells. The spread of spectral energy to higher frequencies (0.15 to 0.25 Hz) is predominant during January–May (Fig. 5) due

to the sea breeze in the pre-monsoon period (Neetu et al., 2006; Dora and Sanil Kumar, 2015). In the monsoon during the wave growth period, the spectral peak shifts from 0.12–0.13 to 0.07–0.09 Hz (lower frequencies).

An interesting phenomenon is that the long-period (> 18 s) swells are present for 2.5 % of the time during the study period. The buoy location at 15 m of water depth is exposed to waves from the northwest to the south with the nearest landmass at ∼ 1500 km to the northwest (Asia), ∼ 2500 km to the west (Africa), ∼ 4000 km to the southwest (Africa) and ∼ 9000 km to the south (Antarctica) (Amrutha et al., 2017). Due to its exposure to the southern oceans and the large fetch available, swells are present all year round in the study area, and the swells are dominant in the non-monsoon pe-

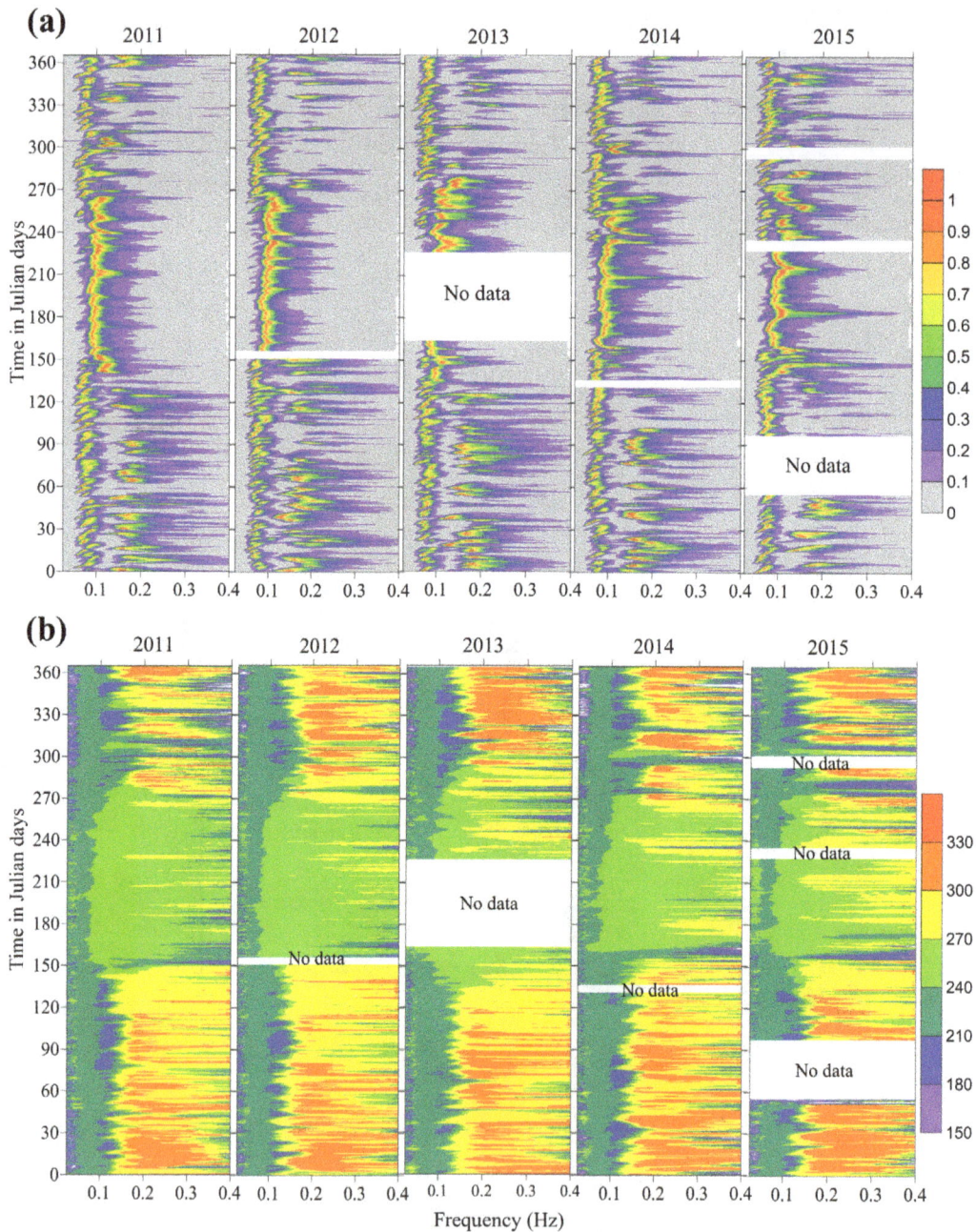

Figure 5. The temporal variation in the normalized spectral energy density **(a)** and the mean wave direction **(b)** with frequency in different years. The value used for normalizing the spectral energy density is presented in Fig. 2e.

riod (Glejin et al., 2013). Throughout the year, waves with periods of more than 10 s (low-frequency < 0.1 Hz waves) are the southwest swells, whereas the direction of short-period waves changes with the seasons (Fig. 5). Amrutha et al. (2017) reported that the long-period waves observed in the eastern AS are the swells generated in the southern Indian Ocean. In the monsoon season, the waves with high frequencies are predominantly from the west-southwest, whereas in the non-monsoon period they are from the north-west. In the non-monsoon period, the predominance of wind-

seas and swells fluctuated, and hence the mean wave direction also changed frequently (Fig. 5). The average direction of waves with $H_{m0} < 1$ m shows the northwest wind-seas and the southwest swells, whereas for the high waves ($H_{m0} > 3$ m), the difference between the swell and wind-sea direction decreases. This is because the high waves become aligned with the bottom contour before 15 m of water depth on their approach to the shallow water.

The interannual changes in the wave spectral energy density for different months in the period 2011–2015 are stud-

Table 3. The average wave parameters and the amount of data in different spectral peak frequencies.

Frequency (f_p) range (Hz)	Amount of data and %	H_{m0} (m)	T_{m02} (s)	Peak wave period (s)
$0.04 < f_p \le 0.05$	318 (0.39)	0.73	5.24	20.19
$0.05 < f_p \le 0.06$	5341 (6.56)	0.82	5.48	17.16
$0.06 < f_p \le 0.07$	14764 (18.13)	0.75	5.22	14.73
$0.07 < f_p \le 0.08$	18221 (22.38)	0.80	5.05	12.96
$0.08 < f_p \le 0.10$	25364 (31.15)	1.55	5.76	10.88
$0.10 < f_p \le 0.15$	9459 (11.62)	1.25	5.35	8.07
$0.15 < f_p \le 0.20$	6355 (7.80)	0.76	4.43	5.72
$0.20 < f_p \le 0.30$	1487 (1.83)	0.78	3.86	4.36
$0.30 < f_p \le 0.50$	118 (0.14)	0.66	3.22	3.09

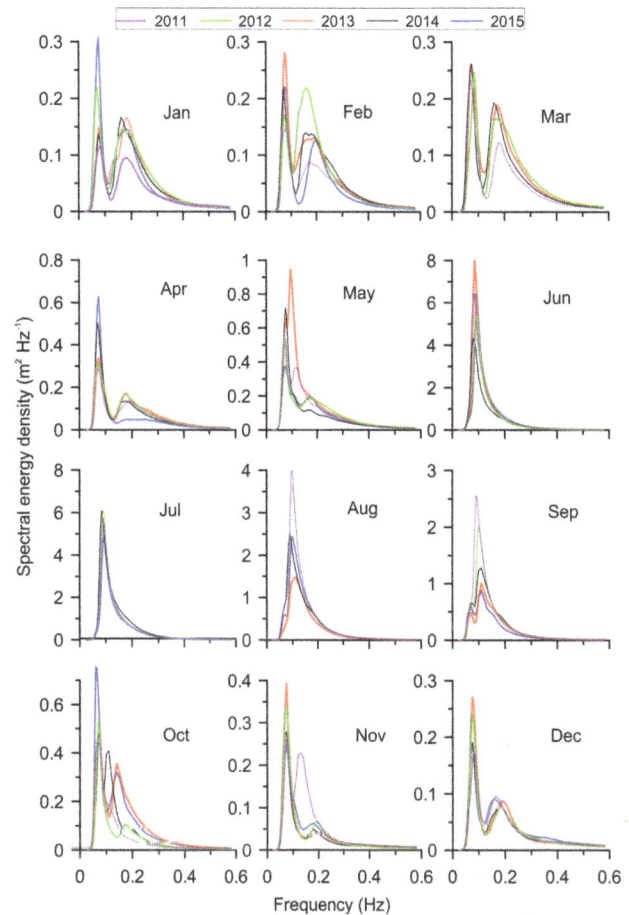

Figure 6. The monthly average wave spectra from 2011 to 2015.

ied by computing the monthly average wave spectra for all the years (Fig. 6). In the non-monsoon period, the wave spectra observed are double peaked, indicating the presence of wind-seas and swells. During the monsoon, due to the strong southwest winds, a single-peaked spectrum is observed, i.e., a swell peak with a low frequency and a high spectral energy density. Along the Indian coast, Harish and Baba (1986), Rao and Baba (1996) and Sanil Kumar et al. (2003) found that wave spectra are generally multi-peaked and that the double-peaked wave spectra are more frequent during low-sea states (Sanilkumar et al., 2004). Sanil Kumar et al. (2014), Sanil Kumar and Anjali (2015) and Anjali and Sanil Kumar (2016) have also observed that double-peaked spectra in the monsoon period in the eastern AS are due to the locally generated wind-seas and the southern Indian Ocean swells. In the study area from January to May and October to December, the swell peak is between the frequencies 0.07 and 0.08 Hz ($12.5 < T_p < 14.3$ s); but in the monsoon period, the swell peak is around 0.10 Hz in all the years studied. This shows the presence of long-period swells ($T_p > 13$ s) in the non-monsoon period and intermediate-period swells ($8 < T_p < 13$ s) in the monsoon. Glejin et al. (2016) also observed the presence of low-amplitude long-period waves in the eastern AS in the non-monsoon period and intermediate-period waves in the monsoon period. This is because the propagation of swells from the Southern Hemisphere is more visible during the non-monsoon period due to the calm conditions (low wind-seas) prevailing in the eastern AS. During the monsoon period, these swells occur less often due to the turbulence in the northern Indian Ocean (Glejin et al., 2013).

Large interannual variations are observed for the monthly average wave spectrum in all months except July. This is because July is known to be the roughest month over the entire annual cycle, and the southwest monsoon reaches its peak during July. Hence, the influence of temporally varying wind-seas on the wave spectrum is the lowest during July compared to other months. Due to the early onset (on 1 June) and advancement of the monsoon during 2013 compared to

other years, the monthly average value of the maximum spectral energy is observed in June 2013 (Fig. 6). The wave spectra of November 2011 are distinct from those of other years, with a low wind-sea peak frequency of 0.13 Hz due to the deep depression ARB 04 that occurred south of India near Cape Comorin from 26 November to 1 December with a sustained wind speed of 55 km h^{-1}. During October 2014, the second peak is observed at 0.11 Hz with comparatively high energy, showing the influence of the cyclonic storm Nilofar. It is an extremely severe cyclonic storm that occurred during the period 25–31 October 2014, originating from a low-pressure area between the Indian and Arabian peninsulas. The highest wind speed was 215 km h^{-1}, and the affected areas included India, Pakistan and Oman. Significant interannual variation is observed in the wind-sea peak frequency. The wave spectra averaged over each season (Fig. 7) show that the interannual variations in the energy spectra averaged over the full-year period almost follow the pattern of the wave spectra averaged over the monsoon period. This indicates the strong influence of monsoon winds on the wave energy spectra in the study area. Interannual variations within the spectrum are higher for the wind-sea region compared

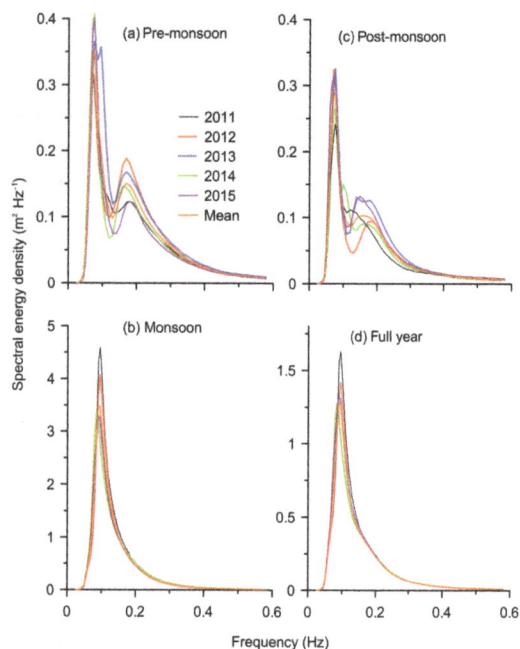

Figure 7. The wave spectra averaged over (**a**) the pre-monsoon period (February–May), (**b**) the monsoon period (June–September), (**c**) the post-monsoon period (October–January) and (**d**) the full year in different years.

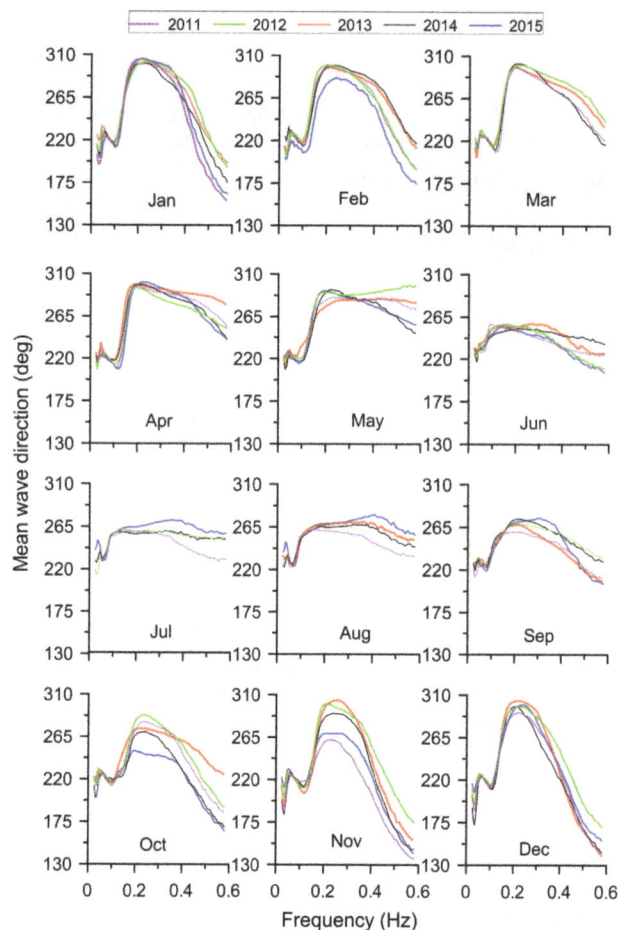

Figure 8. The monthly average wave direction at different frequencies in different months.

to the swell region. During the study period, the maximum spectral energy observed is during the 2011 monsoon.

For different frequencies, the monthly average wave direction is shown in Fig. 8. It is observed that throughout the year, the mean wave direction of the swell peak is southwest (200–250°). In the non-monsoon period, the wind-sea direction is northwest (280–300°), except in October and November. This is due to the wind-seas produced by the sea breeze, which has the maximum intensity during the pre-monsoon season. Interannual variability in the wave direction is the highest during October and November, when the wind-seas from the southwest direction are also observed. This is because during these months, the wind speed and the strength of the monsoon swell decreases, which makes the low-energy wind-seas produced by the withdrawing monsoon winds more visible.

Contour plots of the spectral energy density (normalized) clearly show the predominance of wind-seas and swells during the non-monsoon period (Fig. 9). Only Figs. 5 and 9 present the normalized spectral energy density. In the monsoon period, the spectral energy density is mainly confined to a narrow frequency range (0.07–0.14 Hz) and the wave spectra are mainly single peaked with a maximum energy within the frequency range of 0.08–0.10 Hz and a direction of 240°. Glejin et al. (2012) reported that in the monsoon season, the spectral peak is between 0.08 and 0.10 Hz (12–10 s) for ~72 % of the time in the eastern AS. Earlier studies also reported the dominance of swells in the eastern AS

during the monsoon (Sanil Kumar et al., 2012; Glejin et al., 2012). Above 0.15 Hz, energy gradually decreases with the lowest energy observed between 0.30 and 0.50 Hz. Wind-sea energy is comparatively low during October, November and December and occurs mostly in the frequency range lower than 0.20 Hz; during January–May, the frequency exceeds 0.20 Hz. In the pre-monsoon period, the wind-sea plays a major role in the nearshore wave environment (Rao and Baba, 1996). Wind-sea energy is found to be low during April 2015 (Fig. 6) because of a reduction in local winds. The occurrence of wind-seas is very low during November in most years, except during 2011 due to the deep depression ARB 04.

The behavior of the high-frequency part of the spectrum is governed by the energy balance of the waves generated by the local wind fields. When the wind blows over a long fetch or for a long time, the wave energy for a given frequency reaches the equilibrium range and the energy input from the wind is balanced by energy loss to lower frequencies and by wave breaking (Torsethaugen and Haver, 2004). The high-frequency tail slope of the monthly average wave spectrum in different years shows that the slope is high ($b < -3.1$) during

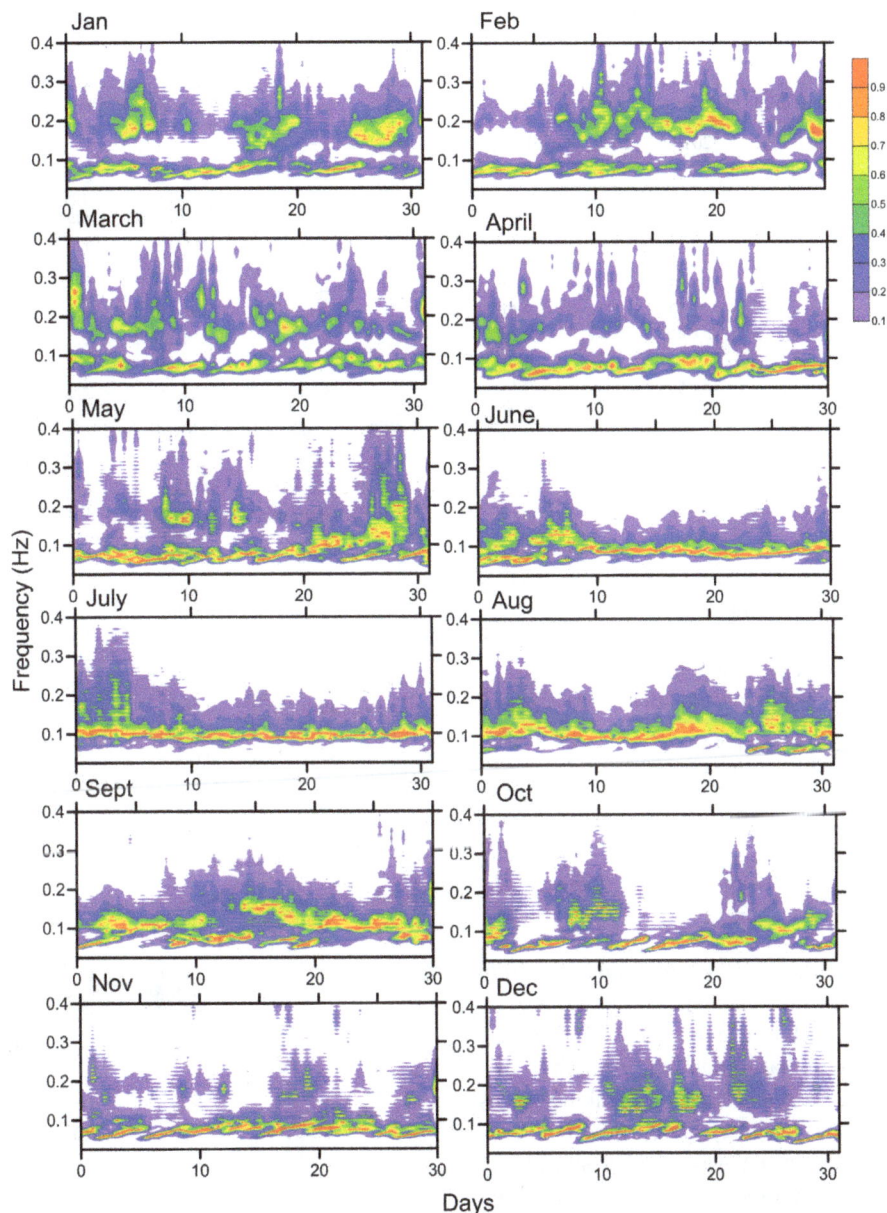

Figure 9. The temporal variation in the normalized spectral energy density in different months (data from 2011 to 2015). The value used for normalizing the spectral energy density is presented in Fig. 2e.

June to September; the case is same for all the years studied (Table 4). During all other months, the exponent in the expression for the frequency tail is within the range of −3.1 to −1.5. The distribution of the exponent values for different significant wave height ranges shows that the slope increases (exponent decrease from −2.44 to −4.20) as the significant wave height increases and reaches a saturation range (Table 5). For frequencies from 0.23 to 0.58 Hz in the eastern AS during January–May, Amrutha et al. (2017) observed that the high-frequency tail has the $f^{-2.5}$ pattern at 15 m of water depth. For frequencies ranging from 0.31 to 0.55 Hz, the high-frequency tail follows f^{-3} at 5 m of water depth. Since

H_{m0} is maximum during the monsoon period, the slope is also maximum from June to September. There is not much interannual variation in the slope for swell-dominated spectra during the monsoon, while in the non-monsoon period when the wind-seas have a high level of influence, the slope varies significantly.

The most obvious manifestations of nonlinearity are the sharpening of the wave crests and the flattening of the wave troughs, and these effects are reflected in the skewness of the sea surface elevation (Toffoli et al., 2006). Zero skewness indicates linear sea states, and a positive skewness value indicates that the wave crests are bigger than the troughs. Fig-

Table 4. The exponent of the high-frequency tail of the monthly average wave spectra in different years.

Months	Exponent of the high-frequency tail					
	2011	2012	2013	2014	2015	2011–2015
January	−2.08	−2.93	−2.97	−2.72	−2.81	−2.72
February	−2.41	−3.02	−2.74	−2.99	−3.06	−2.85
March	−2.75	−2.91	−2.82	−2.76	No data	−2.81
April	−2.56	−2.74	−2.64	−2.71	−2.19	−2.60
May	−2.59	−2.67	−2.63	−2.42	−2.51	−2.56
June	−3.64	−3.53	−3.55	−3.82	−3.58	−3.55
July	−3.76	−3.55	No data	−3.82	−3.63	−3.70
August	−3.63	−3.58	−3.40	−3.52	−3.65	−3.58
September	−3.41	−3.44	−3.16	−3.38	−3.00	−3.30
October	−2.02	−2.77	−3.03	−2.52	−2.61	−2.68
November	−1.78	−2.43	−1.77	−1.55	−1.65	−1.84
December	−1.69	−2.23	−1.95	−2.06	−1.79	−1.94

Table 5. The exponent of the high-frequency tail of the average wave spectra in different wave height ranges.

Range of H_{m0} (m)	Exponent of the high-frequency tail
0–1	−2.44
1–2	−3.26
2–3	−3.67
3–4	−4.21
4–5	−4.21

Table 6. The parameters of the fitted wave spectrum in different years.

Year		JONSWAP spectrum		Donelan spectrum	
		α	Υ	α	Υ
2011	June	0.0013	2.2	0.0028	2.0
	July	0.0016	1.5	0.0021	1.7
	August	0.0013	1.8	0.0029	1.7
	September	0.0004	2.3	0.0021	1.6
2012	June	0.0015	1.6	0.0029	2.0
	July	0.0010	2.1	0.0031	1.9
	August	0.0009	2.2	0.0032	1.7
	September	0.0006	2.0	0.0024	1.8
2013	June	0.0006	3.3	0.0030	1.9
	July		No data		
	August	0.0012	1.1	0.0038	1.4
	September	0.0005	1.9	0.0042	1.4
2014	June	0.0010	1.1	0.0010	1.6
	July	0.0006	2.5	0.0019	1.2
	August	0.0006	1.5	0.0021	1.2
	September	0.0011	1.1	0.0032	1.4
2015	June	0.0011	1.4	0.0023	1.8
	July	0.0011	1.9	0.0024	1.8
	August	0.0008	1.8	0.0024	1.4
	September	0.0006	1.3	0.0043	1.6

ure 10 shows that nonlinearity increases with an increase in H_{m0}. The slope of the high-frequency end of the wave spectrum becomes steeper when the wave nonlinearity increases. Donelan et al. (2012) found that in addition to the k^{-4} dissipation, swells modulate the equilibrium in breaking waves dependent on the mean surface slope, while Melville (1994) also quantified a relation between wave packet slopes and the dissipation rate. These results are specific to breaking waves, but one might expect similar relations between surface dynamics and dissipation rates for non-breaking waves. A function of the form $A \cdot \exp(\lambda H_{m0}) + s0$ with the initial parameters of $A = 8$, $\lambda = -2.4$, $s0 = -3.7$ is found to fit the exponent of the high-frequency tail data with the significant wave height (Fig. 11a). The functional representation of the exponent of the high-frequency tail data with H_{m0} is shown in Fig. 11a and might be useful in revealing the physical connection; at the very least, it could provide a predictive basis for relating spectral slopes with mean significant wave heights as a basis for future research. It is shown in Fig. 11b that the exponent decreases (slope increases) as the mean wave period increases. The study shows that the tail of the spectrum is influenced by the local wind conditions (Fig. 11c), and the influence is higher on the zonal component (u) of the wind than on the meridional component (v) (Fig. 11e and f). The exponent of the high-frequency tail de-

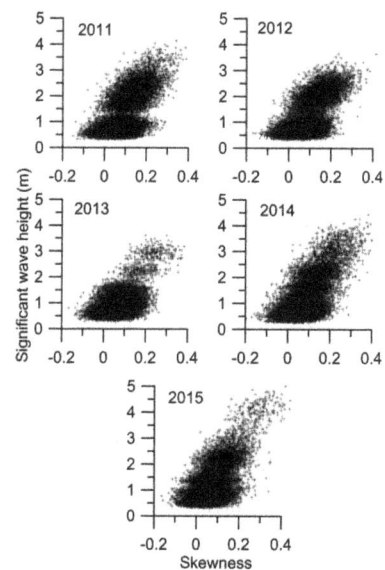

Figure 10. A scatter plot of the significant wave height with the skewness of the sea surface elevation in different years.

creases with the increase in the inverse wave age (U_{10}/c), where c is the celerity of the wave.

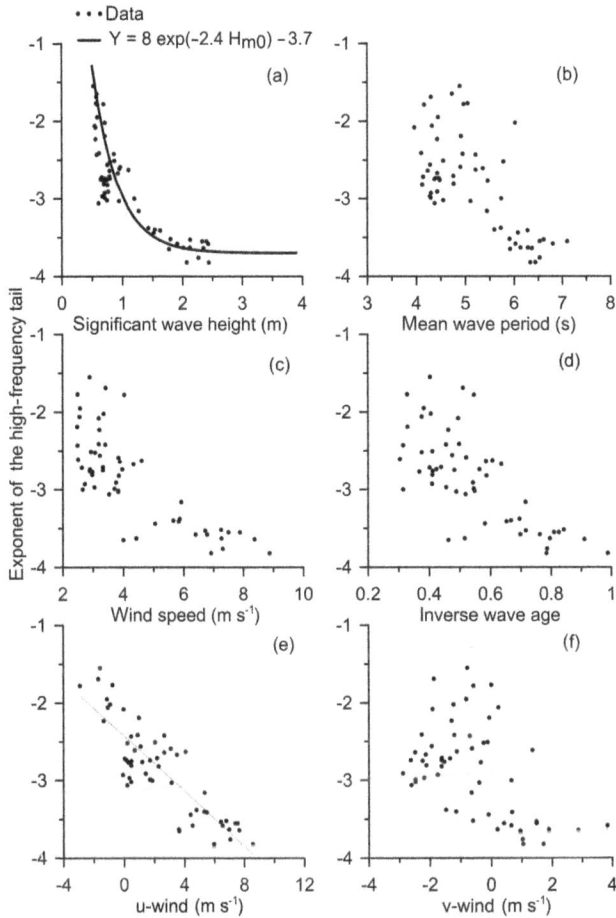

Figure 11. A plot of the exponent of the high-frequency tail with (a) the significant wave height, (b) mean wave period, (c) wind speed, (d) inverse wave age, (e) u-wind and (f) v-wind.

4.3 Comparison with theoretical wave spectra

In the monsoon period, the spectrum is single peaked with a high spectral energy density. During this period, the JONSWAP spectrum is fitted up to the peak frequency; after that, the Donelan spectrum is used. The monthly average wave spectra during the monsoon period for the year 2011 is compared with the JONSWAP and Donelan theoretical wave spectra in Fig. 12. It is found that the JONSWAP and Donelan spectra with modified parameters describe the wave spectra well at low frequencies and high frequencies, respectively. The values for α and Υ were varied from 0.0001 to 0.005 and 1.1 to 3.3, respectively, to find the values for which the theoretical spectrum best fits the measured spectrum; those values were used to plot the theoretical spectrum. The values of α and Υ thus obtained for June, July, August and September are given in Table 6. From the table, the average values of α and Υ for the monsoon months are obtained as 0.0009 and 1.82 for the JONSWAP spectra and 0.0274 and 1.64 for the Donelan spectra. These values are lower than the

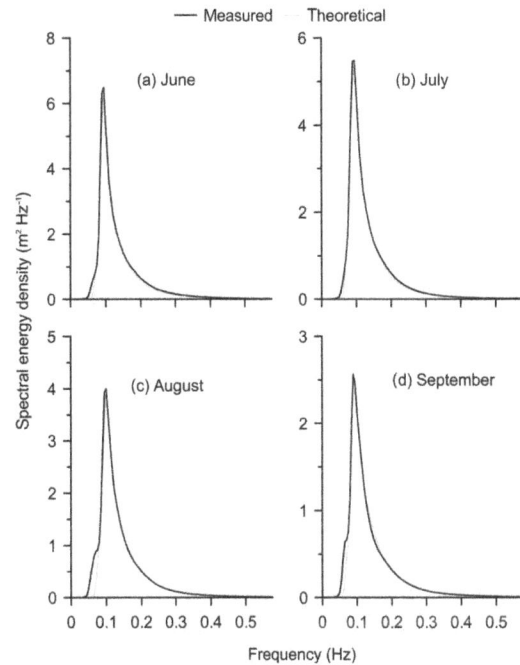

Figure 12. The fitted theoretical spectra along with the monthly average wave spectra for different months.

generally recommended values of α and Υ, which are 0.0081 and 3.3. The α value is a constant that is related to the wind speed and fetch length. For all the data, the fitted Donelan spectrum is proportional to f^{-n}, where n is the exponent value of the high-frequency tail. The theoretical spectrum of JONSWAP and Donelan cannot completely describe the high-frequency tail of the measured spectrum since the high-frequency tail in these spectra decays in the forms of f^{-5} and f^{-4}, respectively. Since the exponent of the high-frequency tail of the wave spectrum is within the range of -4 to -3 during the monsoon period, the Donelan spectrum shows a better fit for the monsoon spectra compared to other months (Fig. 11).

5 Concluding remarks

In this paper, the variations in the wave spectral shapes in different months for a nearshore location are investigated based on in situ wave data obtained from a moored directional waverider buoy. There are more interannual variations within the spectrum for wind-seas compared to swells. The maximum significant wave height measured at 15 m of water depth is 5 m, and the annual average H_{m0} has a similar value (~ 1.1 m) in all the years. Over the 5 years, small waves ($H_{m0} < 1$ m) account for a large proportion of the measured data (63.94 % of the time). The study shows that high waves ($H_{m0} > 2$ m) have a spectral peak period between 8 and 14 s, and the long-period swells (14–20 s) are $H_{m0} < 2.5$ m. The high-frequency slope of the wave spectrum (the exponent decreases from −2.44 to −4.20) increases with an increase in the significant wave height and the mean wave period. During the monsoon period, the Donelan spectrum shows a better fit for the monsoon spectra compared to other months, since the exponent of the high-frequency part of the wave spectrum is within the range of −4 to −3. The decay of the high-frequency waves is the fastest with depth; hence, the high-frequency tail values observed in the study will be different for different water depths.

Competing interests. The authors declare that they have no conflict of interest.

Acknowledgements. The authors acknowledge the Earth System Science Organization, Ministry of Earth Sciences, New Delhi for providing the financial support to conduct part of this research. We thank the following people for their help in the collection of data: T. M. Balakrishnan Nair, Head of OSISG; Arun Nherakkol, scientist at INCOIS, Hyderabad; and Jai Singh, technical assistant, CSIR-NIO. We thank U. G. Bhat and J. L. Rathod, Department of Marine Biology, Karnataka University PG Centre in Karwar for providing the logistics required for wave data collection. This work contributes to the PhD work of the first author (Anjali Nair). This paper is dedicated to the memory of our esteemed colleague, Ashok Kumar, in recognition of his substantial contributions in initiating the long-term wave measurements in the shallow waters around India. We thank the topic editor and both the reviewers for their critical comments and suggestions, which improved the scientific content of the publication. This publication is an NIO contribution 6037.

Edited by: A. Sterl

References

Amrutha, M. M., Sanil Kumar, V., and George, J.: Observations of long-period waves in the nearshore waters of central west coast of India during the fall inter-monsoon period, Ocean Eng., 131, 244–262, doi:10.1016/j.oceaneng.2017.01.014, 2017.

Anjali, N. M. and Sanil Kumar, V.: Spectral wave climatology off Ratnagiri – northeast Arabian Sea, N. Hazards, 82, 1565–1588, 2016.

Badulin, S. I., Babanin, A. V., Zakharov, V. E., and Resio, D.: Weakly turbulent laws of wind-wave growth, J. Fluid Mech., 591, 339–378, 2007.

Cavaleri, L., Fox-Kemper, B., and Hemer, M.: Wind-waves in the coupled climate system, B. Am. Meteorol. Soc., 93, 1651–1661, 2012.

Chakrabarti, S. K.: Handbook of Offshore Engineering, in: Ocean Engineering Series, Vol. 1, Elsevier, Amsterdam, the Netherlands, p. 661, 2005.

Chen, G., Chapron, B., Ezraty, R., and Vandemark, D.: A global view of swell and wind-sea climate in the ocean by satellite altimeter and scatterometer, J. Atmos. Ocean. Tech., 19, 1849–1859, 2002.

Datawell: Datawell Waverider Reference Manual, Datawell BV oceanographic instruments, Haarlem, the Netherlands, 123 pp., 10 October 2009.

Donelan, M. A.: The effect of swell on the growth of wind waves, Johns Hopkins APL Technical Digest., 8, 18–23, 1987.

Donelan, M. A., Hamilton, H., and Hui, W. H.: Directional spectra of wind-generated waves, Philos. T. Roy. Soc. Lond. A, 315, 509–562, 1985.

Donelan, M. A., Curcic, M., Chen, S. S., and Magnusson, A. K.: Modeling waves and wind stress, J. Geophys. Res.-Oceans., 117, C00J23, doi:10.1029/2011JC007787, 2012.

Dora, G. U. and Sanil Kumar, V.: Sea state observation in island-sheltered nearshore zone based on in situ intermediate-water wave measurements and NCEP/CFSR wind data, Ocean Dynam., 65, 647–663, 2015.

Forristall, G. Z.: Measurements of a saturated range in ocean wave spectra, J. Geophys. Res.-Oceans, 86, 8075–8084, 1981.

Gagnaire-Renou, E., Benoit, M., and Forget, P.: Ocean wave spectrum properties as derived from quasi-exact computations of nonlinear wave-wave interactions, J. Geophys. Res.-Oceans, 115, C12058, doi:10.1029/2009JC005665, 2010.

Glejin, J., Sanil Kumar, V., Sajiv, P. C., Singh, J., Pednekar, P., Ashok Kumar, K., Dora, G. U., and Gowthaman, R.: Variations in swells along eastern Arabian Sea during the summer monsoon, Open J. Mar. Sci., 2, 43–50, 2012.

Glejin, J., Sanil Kumar, V., Balakrishnan Nair, T. M., and Singh, J.: Influence of winds on temporally varying short and long period gravity waves in the near shore regions of the eastern Arabian Sea, Ocean Sci., 9, 343–353, doi:10.5194/os-9-343-2013, 2013.

Glejin, J., Sanil Kumar, V., Amrutha, M. M., and Singh, J.: Characteristics of long-period swells measured in the in the near shore regions of eastern Arabian Sea, Int. J. Nav. Arch. Ocean Eng., 8, 312–319, 2016.

Gunson, J. and Symonds, G.: Spectral Evolution of Nearshore Wave Energy during a Sea-Breeze Cycle, J. Phys. Oceanogr., 44, 3195–3208, 2014.

Harish, C. M. and Baba, M.: On spectral and statistical characteristics of shallow water waves, Ocean Eng., 13, 239–248, 1986.

Hasselmann, K., Barnett, T. P., Bouws, F., Carlson, H., Cartwright, D. E., Enke, K., Ewing, J. A., Gienapp, H., Hasselmann, D. E., Krusemann, P., Meerburg, A., Muller, P., Olbers, D. J., Richter, K., Sell, W., and Walden, H.: Measurements of wind-wave growth and swell decay during the Joint North Sea Wave Project (JONSWAP), Deutches Hydrographisches Institut, A8, 1–95, 1973.

Hwang, P. A., Garcia-Nava, H., and Ocampo-Torres, F. J.: Dimensionally Consistent Similarity Relation of Ocean Surface Friction Coefficient in Mixed Seas, J. Phys. Oceanogr., 41, 1227–1238, 2011.

Kahma, K. K.: A study of the growth of the wave spectrum with fetch, J. Phys. Oceanogr., 11, 1503–1515, 1981.

Kalnay, E., Kanamitsu, M., Kistler, R., Collins, W., Deaven, D., Gandin, L., Iredell, M., Saha, S., White, G., Woollen, J., and Zhu, Y.: The NCEP/NCAR 40-year reanalysis project, B. Am. Meteorol. Soc., 77, 437–471, doi:10.1175/1520-0477(1996)077<0437:TNYRP>2.0.CO;2, 1996.

Kawai, S., Okada, K., and Toba, Y.: Field data support of three-seconds power law and $gu^*\sigma-4$-spectral form for growing wind waves, J. Oceanogr., 33, 137–150, 1977.

Kitaigordskii, S. A., Krasitskii, V. P. and Zaslavskii, M. M.: On Phillips' theory of equilibrium range in the spectra of wind-generated gravity waves, J. Phys. Oceanogr., 5, 410–420, 1975.

Kuik, A. J., Vledder, G., and Holthuijsen, L. H.: A method for the routine analysis of pitch and roll buoy wave data, J. Phys. Oceanogr., 18, 1020–1034, 1988.

Liu, A. K., Jackson, F. C., Walsh, E. J., and Peng, C. Y.: A case study of wave-current interaction near an oceanic front, J. Geophys. Res.-Oceans, 94, 16189–16200, 1989.

Long, C. E. and Resio, D. T.: Wind wave spectral observations in currituck sound, north Carolina, J. Geophys. Res.-Oceans, 112, C05001, doi:10.1029/2006JC003835, 2007.

Melville, W. K.: Energy dissipation by breaking waves, J. Phys. Oceanogr., 24, 2041–2049, 1994.

Neetu, S., Satish, S., and Chandramohan, P.: Impact of sea breeze on wind-seas off Goa, west coast of India, J. Earth Syst. Sci., 115, 229–234, 2006.

Phillips, O. M.: Spectral and statistical properties of the equilibrium range in wind-generated waves, J. Fluid Mech., 156, 505–531, 1985.

Pierson, W. J. and Moskowitz, L.: A proposed form for fully developed seas based on the similarity theory of S. A. Kitaigorodski, J. Geophys. Res.-Oceans, 69, 5181–5190, 1964.

Portilla, J., Ocampo-Torres, F. J., and Monbaliu, J.: Spectral Partitioning and Identification of Wind-sea and Swell, J. Atmos. Ocean. Tech., 26, 117–122, 2009.

Ranjha, R., Tjernström, M., Semedo, A., and Svensson, G.: Structure and variability of the Oman Coastal Low-Level Jet, Tellus A, 67, 25285, doi:10.3402/tellusa.v67.25285, 2015.

Rao, C. P. and Baba, M.: Observed wave characteristics during growth and decay: a case study, Cont. Shelf Res., 16, 1509–1520, 1996.

Sanilkumar, V., Ashokkumar, K., and Raju, N. S. N.: Wave characteristics off Visakhapatnam coast during a cyclone, Current Science, 86, 1524–1529, 2004.

Sanil Kumar, V. and Anand, N. M.: Variation in wave direction estimated using first and second order Fourier coefficients, Ocean Eng., 31, 2105–2119, 2004.

Sanil Kumar, V. and Anjali Nair, M.: Inter-annual variations in wave spectral characteristics at a location off the central west coast of India, Ann. Geophys., 33, 159–167, doi:10.5194/angeo-33-159-2015, 2015.

Sanil Kumar, V., Anand, N. M., Kumar, K. A., and Mandal, S.: Multipeakedness and groupiness of shallow water waves along Indian coast, J. Coast. Res., 19, 1052–1065, 2003.

Sanil Kumar, V., Johnson, G., Dora, G. U., Chempalayil, S. P., Singh, J., and Pednekar, P.: Variations in nearshore waves along Karnataka, west coast of India, J. Earth Syst. Sci., 121, 393–403, 2012.

Sanil Kumar, V., Shanas, P. R., and Dubhashi, K. K.: Shallow water wave spectral characteristics along the eastern Arabian Sea, Nat. Hazards, 70, 377–394, 2014.

Semedo, A., Sušelj, K., Rutgersson, A., and Sterl, A.: A global view on the wind-sea and swell climate and variability from ERA-40, J. Climate, 24, 1461–1479, 2011.

Shetye, S. R., Shenoi, S. S. C., Antony, A. K., and Kumar, V. K.: Monthly-mean wind stress along the coast of the north Indian Ocean, J. Earth Syst. Sci., 94, 129–137, doi:10.1007/BF02871945, 1985.

Siadatmousavi, S. M., Jose, F., and Stone, G. W.: On the importance of high frequency tail in third generation wave models, Coast. Eng., 60, 248–260, 2012.

Toba, Y.: Local balance in the air-sea boundary processes, J. Oceanogr., 29, 209–220, 1973.

Toffoli, A., Onorato, M., and Monbaliu, J.: Wave statistics in unimodal and bimodal seas from a second-order model, Eur. J. Mech. Fluids B, 25, 649–661, 2006.

Torsethaugen, K. and Haver, S.: Simplified double peak spectral model for ocean waves, in: Proceeding of the 14th International Offshore and Polar Engineering Conference, 23–28 May 2004, Toulon, France, 2004.

Vethamony, P., Rashmi, R., Samiksha, S. V., and Aboobacker, M.: Recent Studies on Wind Seas and Swells in the Indian Ocean: A Review, Int. J. Ocean Clim. Syst., 4, 63–73, 2013.

Young, I. R. and Babanin, A. V.: Spectral distribution of energy dissipation of wind-generated waves due to dominant wave breaking, J. Phys. Oceanogr., 36, 376–394, 2006.

Yuan, Y. and Huang, N. E.: A reappraisal of ocean wave studies, J. Geophys. Res.-Oceans., 117, C00J27, doi:10.1029/2011JC007768, 2012.

11

Different approaches to model the nearshore circulation in the south shore of O'ahu, Hawaii

Joao Marcos Azevedo Correia de Souza[1,2] **and Brian Powell**[2]

[1]Centro de Investigación Científica y de Educación Superior de Ensenada, Baja California (CICESE), Carretera Ensenada-Tijuana No. 3918, Zona Playitas, C.P. 22860, Ensenada, B.C., Mexico
[2]Department of Oceanography, University of Hawaii, 1000 Pope Rd., MSB, Honolulu, 96822 HI, USA

Correspondence to: Joao Marcos Azevedo Correia de Souza (jazevedo@cicese.mx)

Abstract. The dynamical interaction between currents, bathymetry, waves, and estuarine outflow has significant impacts on the surf zone. We investigate the impacts of two strategies to include the effect of surface gravity waves on an ocean circulation model of the south shore of O'ahu, Hawaii. This area provides an ideal laboratory for the development of nearshore circulation modeling systems for reef-protected coastlines. We use two numerical models for circulation and waves: Regional Ocean Modeling System (ROMS) and Simulating Waves Nearshore (SWAN) model, respectively. The circulation model is nested within larger-scale models that capture the tidal, regional, and wind-forced circulation of the Hawaiian archipelago. Two strategies are explored for circulation modeling: forcing by the output of the wave model and online, two-way coupling of the circulation and wave models. In addition, the circulation model alone provides the reference for the circulation without the effect of the waves. These strategies are applied to two experiments: (1) typical trade-wind conditions that are frequent during summer months, and (2) the arrival of a large winter swell that wraps around the island. The results show the importance of considering the effect of the waves on the circulation and, particularly, the circulation–wave coupled processes. Both approaches show a similar nearshore circulation pattern, with the presence of an offshore current in the middle beaches of Waikiki. Although the pattern of the offshore circulation remains the same, the coupled waves and circulation produce larger significant wave heights ($\approx 10\%$) and the formation of strong alongshore and cross-shore currents ($\approx 1\,\mathrm{m\,s^{-1}}$).

1 Introduction

Our objective is to describe how ocean waves and currents interact in the south shore of the island of O'ahu, Hawaii, with a goal towards the development of an operational ocean forecast system. Much of the focus on ocean predictability has been at the larger scales within the ocean basins or on the continental slopes; however, human–ocean interaction is primarily within the nearshore surf zone. Dynamical interplay between currents and bathymetry, currents and waves, ocean waters and estuaries, breaking waves, etc. may all significantly influence the predictability in the nearshore regions. In this paper, we investigate the impacts of surface gravity waves on the nearshore circulation in a high-resolution regional ocean model for the coast of Honolulu, Hawaii (Fig. 1). This work was developed under the umbrella of the Pacific Islands Ocean Observing System (PacIOOS) project (http://oos.soest.hawaii.edu/pacioos/), aiming to improve an operational coastal ocean forecast system for the island of O'ahu. The south shore of O'ahu is mostly contained within Mamala Bay including Waikiki beach. The challenge is to provide useful forecasts of the nearshore circulation in this region that includes the primary dynamical processes while remaining feasible in computational cost. In the present work, alternatives are sought and tested for the development of such a system.

In Mamala Bay, all scales of ocean dynamics are present from strong planetary mean flows, aperiodic mesoscale eddies impinging on the coastal region, internal tides, barotropic tides, and strong and variable trade winds. The south shore of O'ahu within Mamala Bay is an ideal site for a case study to understand the primary drivers of the nearshore

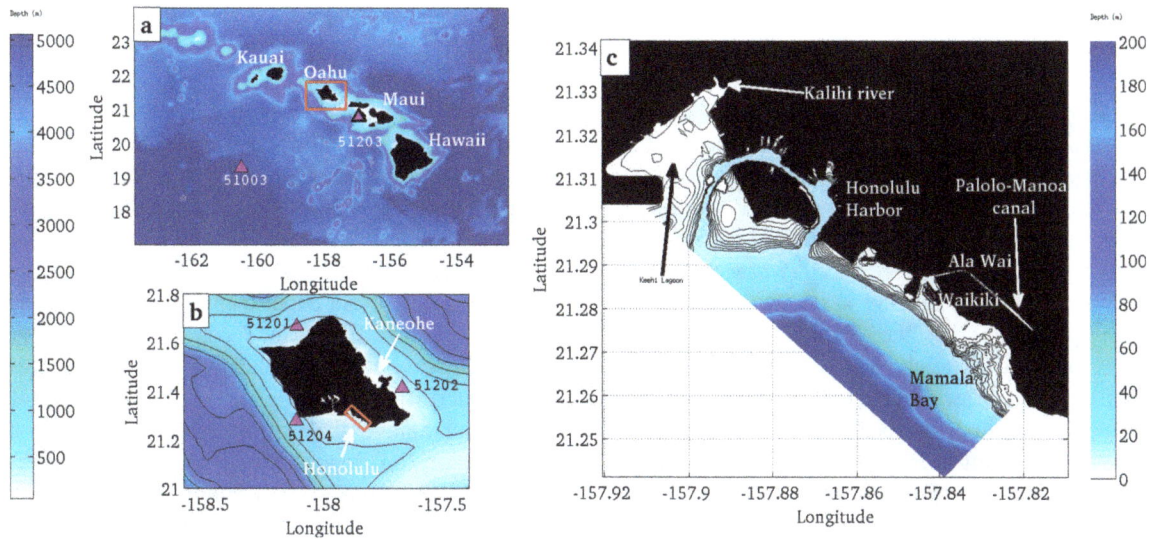

Figure 1. Bathymetry map of the (**a**) Hawaiian Islands, with the island of O'ahu highlighted by the red rectangle and shown in detail in the map (**b**) (contour interval 500 m). The numerical grid used in the present work (≈ 50 m resolution) is highlighted by the red rectangle in (**b**) and expanded in (**c**), with the black contours corresponding to the isobathymetric lines every 1 m from the coast to the depth of 10 m. It is possible to observe the intricate bathymetry near the coast of Honolulu associated with the coral reefs. The purple triangles indicate the positions of the NDBC buoys around Hawaii, which are used to validate the wave numerical model system.

circulation, quantify the particular contribution of the surface gravity waves to the circulation under different conditions, and determine what the best options are to represent such processes in operational forecast systems for exposed coral reef areas.

Among these processes influencing the nearshore circulation, Benetazzo et al. (2013) emphasize the interaction between oceanic waves and currents as one of the main driving mechanisms for coastal regions. The authors show that wave–current interactions lead to important modifications of both the wave parameters – mainly wave significant height (H_S) and wave period – and the ocean currents in the Gulf of Venice. However, their bathymetry presents a gentle bottom slope. In the case of oceanic islands such as O'ahu, the steep slope and intricate bathymetry associated with coral reefs present an additional challenge for forecast systems.

The intermittent discharge of freshwaters from the Ala Wai Canal (Fig. 1c) can impact the nearshore density structure and ocean currents. On average, the influence of the canal on the coastal dynamics is small; however, sporadic, heavy rainfall events can force an outflow of freshwater into the coastal zone that alters the stratification. Although numerous modeling studies on large discharge rivers exist (Gracia Berdeal et al., 2002; Pan et al., 2014), the same is not true for rivers with small volume and aperiodic discharge.

In addition to the complicated bathymetry and the freshwater input from the Ala Wai Canal, there are a number of processes that impact the nearshore currents near Waikiki. There are highly variable winds as the island mountains serve to create a wake region of lower but variable winds

(Souza et al., 2015). Volcanic islands are not protected by wide shelves, which subject the nearshore to potential open-ocean variability such as mesoscale eddies (Chavanne et al., 2010). Although lacking wide shelves, the islands are often protected by coral reefs that significantly alter the surface wave conditions and their interaction with ocean currents. Each of these phenomena can be significant in nearshore environments around the world, but Waikiki is an ideal laboratory for examining the influence since each of the varying dynamical scales are present. The implementation of a coupled circulation–wave model provides both the framework for a useful study on the theme and a forecast tool for operational purposes.

Hoeke et al. (2013) implemented a wave–circulation coupled model for the Hanalei Bay in the north shore of the island of Kaua'i, Hawaii (Fig. 1a). Similar to the south shore of O'ahu, this region is characterized by a complex bathymetry, freshwater discharge, and surface waves that can dominate the dynamics. The authors used the Delft3D modeling system that combines the D-Flow circulation component with a wave component based on the Simulating Waves Nearshore (SWAN) model. The impact of waves on the circulation is calculated in the depth-averaged D-Flow momentum equations by including the wave-induced forces as a source term. The enhanced bed shear stresses caused by waves are computed based on the Soulsby et al. (1993) formulation, and the wave forces are interpolated to the velocity points and substituted as explicit radiation stresses in the momentum equation. However, an explicit description of the complex vertical fluxes of wave momentum is required to properly resolve the 3-D circulation. This is particularly important for the wave-induced mixing and the surf zone circulation. Lane et al. (2007) and Uchiyama et al. (2010) showed that the radiation stress approach used in the Delft3D system does not properly decompose the wave effects, and it obscures their underlying impact on the long (infragravity) waves and currents. From the point of view of the wave field, Edwards et al. (2009) show the Delft3D system tends to underestimate the wave height.

Lowe et al. (2009) applied the same model system as Hoeke et al. (2013) to study the circulation in the coastal reef–lagoon system of Kāne'ohe Bay on the northeast coast of the island of O'ahu, Hawaii (Fig. 1b). Most previous studies assume the wave setup – the local increase in the sea level due to the wave breaking – in the reef lagoon to be negligible, as this is common for ocean atolls and barrier reefs. However, the authors emphasize the fact that, if the water exchange is restricted to relatively narrow channels in the reef as in Kāne'ohe Bay and Waikiki, a water level difference between the lagoon and the open ocean will be present and will establish a pressure gradient impacting the local circulation. The authors did not consider the effect of freshwater input from river discharges that influence the lagoon circulation.

While focusing on the fate of harmful bacteria from the Ala Wai plume, Johnson et al. (2013) developed a Regional Ocean Modeling System (ROMS) simulation for a similar region as this study. They examined sporadic events of large flux from the Ala Wai Canal using a relatively coarse horizontal resolution (250 m) model incapable of resolving the several channels in the reef banks. As shown by Plant et al. (2009) the resolved wave height is very sensitive to the cross-shore bathymetry resolution, while the resolved currents are more sensitive to the alongshore resolution. Comparing model results to observations, these authors demonstrate how the errors in the modeled currents increase if one uses smoothing scales larger than 100 m for the bathymetry. To examine the influence of waves, Johnson et al. (2013) used a prescribed wave field as additional forcing to the ROMS model that did not consider the influence of the currents on the waves. Moreover, the authors used a parameterization based on the Mellor (2003, 2005) approach to estimate the modification of the currents by the wave field. In addition to the radiation stress approach problems discussed, this particular method is known to generate inconsistent pressure fields as demonstrated by Ardhuin et al. (2008).

Therefore, a different approach to the wave–current interaction off the south shore of O'ahu is necessary. This paper aims to clarify (i) the effective contribution from the surface waves under different conditions, (ii) the importance of coupled ocean currents and surface wave processes for the local dynamics, and (iii) the influence of different approaches for forecasting the nearshore current field.

This research uses a suite of numerical models to examine these main questions, as there is little observational data available. Although the Kilo Nalu cabled reef observatory once provided real-time observations of several physical and biogeochemical parameters (Samsone et al., 2008), the lack of continuous measurements of the nearshore currents in Honolulu during the modeled period (and in general) makes it impossible to properly validate the results and quantify the performance of each modeling strategy. Unfortunately, there are no results from the atmospheric model and the ocean model system used to provide surface forcing and boundary conditions to the nearshore domain during the period of the Kilo Nalu experiment. However, contrasting the results from the different model strategies helps to reveal robust circulation patterns and clarify what differences should be expected when adopting each method on a modeling system. A qualitative analysis is performed to understand the modifications on the nearshore circulation. This can assist in the future development of nearshore forecasting systems. It is important to note that this study was part of the design of the operational nearshore forecast system for the south shore of O'ahu. This system uses the model setup presented here to provide daily forecasts of waves and currents for the Honolulu shore.

2   Methods

2.1   The ROMS model

The ROMS is a 3-D primitive equation ocean model using hydrostatic and Boussinesq approximations. A full description of the model can be found in Shchepetkin and McWilliams (2005), McWilliams (2009), and at the ROMS website (www.myroms.org). We make use of the Coupled-Ocean-Atmosphere-Wave-Sediment Transport (COAWST) Modeling System described by Warner et al. (2010), which provides online, two-way coupling between ROMS, SWAN, and Weather Regional Forecast (WRF) models through the Model Coupling Toolkit (MCT). We implemented the coupled ROMS–SWAN simulations for the south shore of Honolulu, Hawaii, using a vortex force formalism to account for the wave–current interaction described by Uchiyama et al. (2010) and Kumar et al. (2012), which gives better performance than the traditional (Mellor, 2005, 2008) radiation stress approach (Lane et al., 2007). All simulations use the Mellor and Yamada (1982) turbulence closure model to account for the vertical mixing.

We utilize a horizontally variable grid with ≈ 50 m resolution in the region between Waikiki and the Honolulu Harbor, gradually decreasing to ≈ 100 m at the boundaries (Fig. 1c), and it covers a total length of 11.5 km alongshore and 4.5 km offshore. Therefore the grid is rectangular, with a variable horizontal resolution and rotated to fit O'ahu's south shore orientation. Since the grid is oriented with the mean coastline, the model directly outputs alongshore and cross-shore velocities. Although these velocities are not perfectly oriented parallel and perpendicular to the coast at every individual point, it provides the correct components when considering the regional circulation pattern. There are 10 vertical layers. Since much of the domain is less than 10 m deep, in most areas the vertical resolution is less than 1 m. It is interesting to note that only 18 % of the model water grid cells are over 50 m deep (4 % over 100 m with all deep cells concentrated at the southern boundary). The maximum depth of the domain is 300 m in the southeast corner of the model grid. The minimum depth of the grid water points is of 0.5 m to account for the shallow reef areas. This grid is sequentially nested within three ROMS circulation models of approximately 250 m, 1 km, and 4 km resolutions that span from the south shore to the entirety of the Hawaiian Islands (not presented).

The open boundaries of the nearshore domain are forced with barotropic tides and circulation, temperature, and salinity from the coarser 250 m parent grid. The horizontal resolution minimizes the errors in the resolved circulation, as shown in comparisons with observations by Plant et al. (2009). All of the grids are part of an operational forecast system, the Pacific Islands Ocean Observing System (PacIOOS, http://pacioos.org). The outer grids are run using an incremental strong constraint four-dimensional variational data assimilation scheme, as described by Matthews et al. (2012). A description and evaluation of the assimilation system is provided by Souza et al. (2015). The present simulations were nested inside the ≈ 250 m horizontal resolution grid comprehending the south shore of O'ahu, the same models used by Johnson et al. (2013), with boundary conditions provided every 3 h. A buffer zone gradually merges the two grids in terms of resolution and bathymetry. In the PacIOOS project all simulations are nested offline due to operational reasons, such that the small-scale processes in the inner grids have no impact on the outer grids.

The models use surface forcing fields from a locally produced WRF model run performed under the PacIOOS project. Eleven tidal constituents obtained from the Oregon State University TOPEX/Poseidon Global Inverse Solution (TPXO) (Egbert and Erofeeva, 2002) were introduced as a separate spectral forcing in the outer grids (Janekovic and Powell, 2011). Through this method the tides are introduced in the barotropic velocities and elevation at the open boundaries and as tidal potential at every grid point (amplitude and phase of 11 tide constituents provided at each grid point). Since TPXO does not properly represent the tides in nearshore shallow regions with complicated bathymetry and geography, the results from the outer grids were used to generate similar spectral tidal forcing for the nearshore domain. Internal tides generated in the outer grids are directly introduced to the inner grids through the baroclinic fields at the boundaries. Input fluxes from the Kalihi and Palolo–Mānoa channels were obtained from the USGS (http://waterdata.usgs.gov). Similar to Johnson et al. (2013), the Palolo–Mānoa channel flux was multiplied by 1.3 to account

for the contributions from runoff waters and smaller drainage sources into the Ala-Wai Canal.

As described by Kumar et al. (2012), the effect of the waves on the circulation is expressed in the inclusion of new terms in the right-hand side of the model governing equations:

$$\frac{\partial \mathbf{u}}{\partial t} + (\mathbf{u} \cdot \nabla_\perp)\mathbf{u} + w\frac{\partial \mathbf{u}}{\partial z} + f\hat{z} \times \mathbf{u} + \nabla_\perp \varphi - \mathbf{F} - \mathbf{D}$$
$$+ \frac{\partial}{\partial z}\left(\overline{\mathbf{u}'w'} - v\frac{\partial \mathbf{u}}{\partial z}\right) = -\nabla_\perp \mathscr{H} + \mathbf{J} + \mathbf{F}^w$$
$$\frac{\partial \varphi}{\partial z} + \frac{g\rho}{\rho_0} = -\frac{\partial \mathscr{H}}{\partial z} + K$$
$$\nabla_\perp \cdot \mathbf{u} + \frac{\partial w}{\partial z} = 0, \tag{1}$$

where $\mathbf{u}$ and w are the Eulerian mean horizontal and vertical velocities, $\nabla_\perp$ is the horizontal differential operator, φ is the dynamic pressure normalized by the density, $\mathscr{H}$ is the lower-order Bernoulli head, $\mathbf{J}$ and K are the vortex force, and F^w is the sum of the momentum flux due to the non-conservative wave forces. The continuity equation is included for completeness, and the tracer equation is not presented. It is important to note that we will refer to the quasi-Eulerian mean velocities as the horizontal currents. As defined by Kumar et al. (2012), this velocity is the Lagrangian mean velocity minus Stokes drift. This is the velocity output by the COAWST system.

As described by Uchiyama et al. (2010), the Stokes drift velocities are defined by

$$\mathbf{u}^{st} = \frac{A^2\sigma}{2\sinh^2[\mathscr{H}]}\cosh[2\mathscr{H}]\mathbf{k}, \tag{2}$$

$$\omega^{st} = -\nabla_\perp \cdot \int \mathbf{u}^{st}dz', \tag{3}$$

where $\mathbf{u}^{st}$ and ω^{st} are the 3-D non-divergent Stokes velocities, A is the wave amplitude, σ is the intrinsic frequency, and $\mathscr{H}$ and $\mathscr{L}$ are the normalized vertical lengths. The Stokes velocity is proportional to the squared wave amplitude, with a smaller influence of the wave height on $\mathscr{H}$.

2.2 The SWAN model

SWAN is a third-generation spectral wave model developed at the Delft University of Technology. It solves the spectral action density balance to describe the evolution of wave energy over direction and frequency, time, and space. It is able to resolve the wave generation by winds, energy transfer by the wave–wave interactions, shoaling and refraction due to the bathymetry and currents, and wave dissipation by white capping, bottom friction, and breaking in the nearshore area. It has been a proven tool for modeling complex wave fields in coastal regions with the varying bathymetry and in the presence of complex currents. The model was developed by Booij et al. (1999) and provides an efficient solution for modeling nearshore waves. The action balance equation describes the evolution of the wave action spectrum, N, as

$$\frac{\partial N}{\partial t} + \frac{\partial c_x N}{\partial x} + \frac{\partial c_y N}{\partial y} + \frac{\partial c_\sigma N}{\partial \sigma} + \frac{\partial c_\theta N}{\partial \theta} = \frac{S}{\theta}, \qquad (4)$$

where t is time, x and y are Cartesian coordinates, c_x and c_y are the propagation velocities of wave energy in x and y, θ is the wave direction, σ is the wave frequency, c_θ and c_σ are the propagation velocities in spectral space (θ, σ), and S represents the source terms. The parameterizations in the source terms cater to the wave processes from deep to intermediate water depth, which include wind–wave interactions, quadratic wave interaction, dissipation due to white capping, bottom friction as well as the coastal wave processes including refraction due to a current field, triad wave interactions, and depth-induced wave breaking. Due to its ability to account for the wave–current interaction, SWAN coupled with other circulation models is suitable for nearshore hydrodynamic studies.

With the same computational grid as the nearshore ROMS simulations shown in Fig. 1c, the nearshore SWAN wave model was forced by the same high-resolution WRF wind used by the circulation model, available from PacIOOS. The Hawaiian wave forecast system in the same PacIOOS project outputs the 2-D spectra boundary condition for the SWAN domain to calculate the wave transformation on the coast of Honolulu. The SWAN spectrum is discretized by 24 equal directional bins from 0 to 360° and 25 exponentially increasing frequency bins from 0.0418 to 1 Hz on each grid. The spectral density over the domain was updated every 5 min during the wave modeling, and wave parameters were output every hour.

As described by Booij et al. (1999), triad wave–wave interactions and depth-induced wave breaking are parameterized using the Eldeberky and Battjes (1995) and Battjes and Janssen (1978) models respectively. For more details, please see the Appendix in Booij et al. (1999).

2.3 The coupling process

The models were coupled via the MCT, which allows the exchange of information between the ocean and wave models. This exchange of information is independent of each model's grid and time step, and we a use coupling time step of 120 s. The coupling time step is a compromise between the computational cost and the timescale of the variability of the properties exchanged by the models. Although sensitivity tests showed that a 1 h time step would be sufficient, a more conservative approach was adopted.

As described by Warner et al. (2010), at each coupling time step the wave model (SWAN) provides results on wave height, wave length, wave direction, surface and bottom periods, percent of waves breaking, bottom orbital velocity, and wave energy dissipation to the ocean model (ROMS). At the same time, ROMS provides the near-surface currents inte-

grated over a depth of one wave height, free surface elevation, and bathymetry (constant in time for our case). The exchange of mean wave parameters between models assumes the wave field is dominated by a well-defined sea swell, which was found to be a reasonable approach for O'ahu's south shore.

2.4 Model experiments

Three groups of simulations were designed to study the impact of the surface gravity waves on the currents: (1) stand-alone ROMS model without considering the waves (NOWAVE), (2) ROMS model including hourly forcing from SWAN (WAVEFORCE), and (3) two-way coupled ROMS–SWAN simulations (WAVECOUPLE) using the MCT.

Each of these simulations were run for two 5-day experimental periods with different wave conditions:

– Experiment 1 took place on 8–13 September 2013 with moderate south waves (1 m) and evening rains corresponding to typical boreal autumn conditions. The evening rains translate into freshwater pulses in the river fluxes, particularly from the Ala Wai Canal.

– Experiment 2 took place 21–26 January 2014 with a large north swell that wrapping around the island with the presence of a southeast swell, generating waves above 2 m in the surf zone in Honolulu. A relatively large rain event occurred in the evening of the second day of simulation; however, the river fluxes are over 30 times smaller than previously observed extreme events, such as in the case study presented by Johnson et al. (2013). This means the influence of the river discharge on the water column stratification is restricted to the Ala Wai mouth.

The wave conditions were chosen based on 2 years of data from the Kilo Nalu observatory (Samsone et al., 2008) presented in Fig. 2. This data set shows a predominance of small (under 1 m) waves from the south in the south shore of Honolulu. A few larger events are present in the data set, with very low variability in the direction. This is to be expected since the observatory was located very close to the shoreline and the waves stir to a direction perpendicular to the bathymetry as approaching the shore. Unfortunately the Kilo Nalu observatory was installed northwest of the Waikiki beach and was not operational for the period of the present study.

Despite the difference in the time step at which the wave parameters are provided to the circulation model in the WAVEFORCE (1 h) and WAVECOUPLE (2 min) approaches, it does not impact the model solutions. This relates to the slow pace of change of the wave characteristics for both experiments, which are evident in the Stokes drift velocity kinetic energy time series of Fig. 8e.

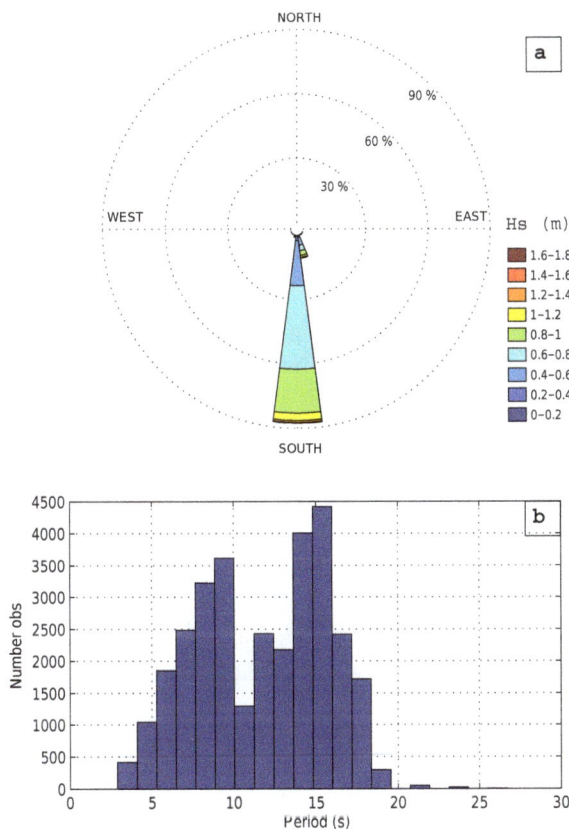

Figure 2. Wave conditions for the south shore of O'ahu as measured by the Kilo Nalu observatory. **(a)** shows the wave rose with frequency of occurrence (%) per wave significant height (H_S) and direction classes – the direction was binned in 10° classes, and **(b)** the histogram of wave peak period. One can see a large prevalence of waves from the south, with H_S usually under 1 m, and two main peak periods around 10 and 15 s.

3 Results and discussion

3.1 Model evaluation

NOAA National Database Buoy Center (NDBC) buoys shown in Fig. 1 provide continuous measurement of the wave parameters around Hawaii. Despite their locations far from the south shore of O'ahu, the wave records at these buoys provide an evaluation for the wave hindcast system. Since no observations are available inside the nearshore domain, we proceed with a comparison of the NDBC buoys with the outer SWAN grid (SWAN only) that provide the boundary conditions to the inner domain. Figure 3 shows good agreement between the measured and modeled significant wave height, peak period, and peak wave direction at offshore buoy 51003 and nearshore buoys 51201, 51202, 51203, and 52104 for the time period of experiment 1. The wave conditions at buoy 51003 indicate dominant trade-wind-generated waves from the east with wave height below 2.5 m. An intermittent north swell with peak period of 15 s is evident at buoys

Table 1. Root-mean-squared deviations for the wave significant height (H_S) and direction, between the model results and the NDBC buoys presented in Fig. 1. All buoys are located in deep water, where the waves do not feel the bathymetry.

Buoy ID	Experiment 1		Experiment 2	
	H_S (m)	Direction (°)	H_S (m)	Direction (°)
51003	0.20	–	–	–
51201	0.18	18	0.61	26
51202	0.16	33	0.45	31
51203	0.11	21	0.29	28
51204	0.22	41	0.64	45

51201 and 51202 for 11 September. Buoys 51203 and 51204 sheltered from east wind waves and north swells show mild south swells with significant wave heights of less than 1 m. The wave hindcast provides a useful tool to reproduce the multi-model wave conditions in Hawaii.

In contrast, both the wave hindcast and measurements in the nearshore buoys 51201, 51202, 51203, and 51204 show large northwest swells with peak period above 15 s for experiment 2 in Fig. 4. The peak wave height decreases from 7 m at buoy 51201 in the north shore to 5 m at buoy 51204 in the south shore of O'ahu. Although no wave data are available in Waikiki for the time periods of experiments 1 and 2, the agreement between the outer wave model and the NDBC buoy records indicate that representative boundary conditions are provided to the O'ahu south shore domain. Therefore, based on the comparisons presented in Figs. 3 and 4, we confirm the skill of the spectral wave model to represent the typical multi-model wave conditions in experiment 1 as well as the arrival of the large swells in experiment 2.

In general, the root-mean-square deviations (RMSDs) show a good agreement between the model results and the buoy data (Table 1). Special attention should be taken on buoy 51204 – the closest buoy to the model domain and the only one on the south side of O'ahu. Further analysis of the model system performance based on 34 years of model hindcasts is provided by Li et al. (2016).

Similar to the waves, there is no data available on the nearshore currents in the south shore of O'ahu during the experiment period. Nevertheless, the model system that provides the boundary conditions of the coastal domain has been validated against satellite sea level anomaly, sea surface temperature, and high-frequency radar surface currents (Souza et al., 2015). The authors calculate root-mean-square errors (RMSEs) for the outer grid of 0.05 ± 0.06 m for sea surface height (SSH), 0.25 ± 0.25 °C for temperature, and 6.5 ± 6 cm s^{-1} for surface velocities, when compared to along-track SSH, temperature from MODIS satellite and Argo profiles, and radial velocities from high-frequency radar (HFR) observations.

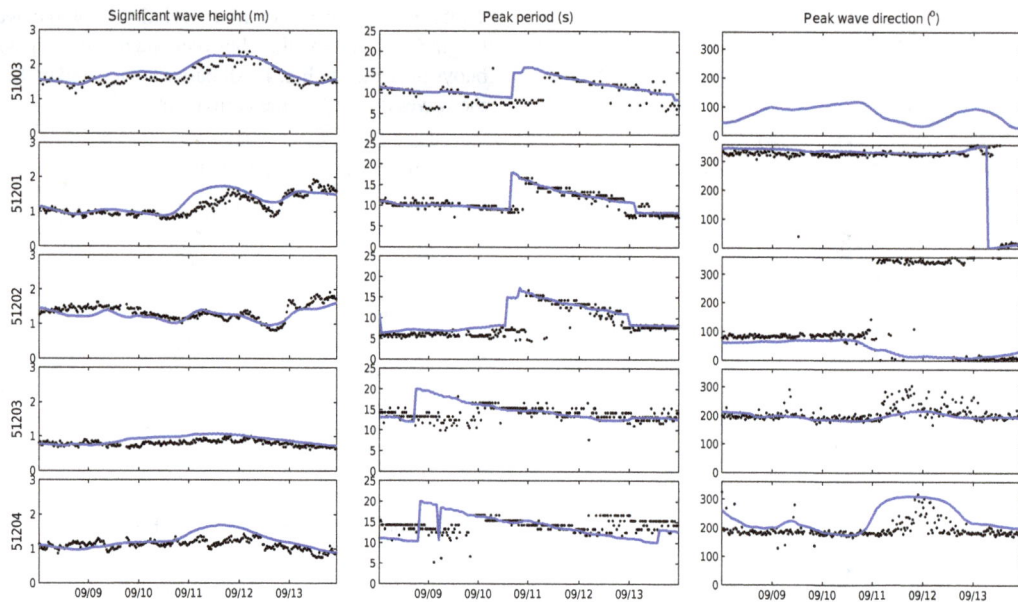

Figure 3. Comparison of the significant wave height, peak period, and direction between the NDBC buoys displayed in Fig. 1 and the wave model hindcast system results for the period of experiment 1. The black dots correspond to the buoy data and the blue line to the model results.

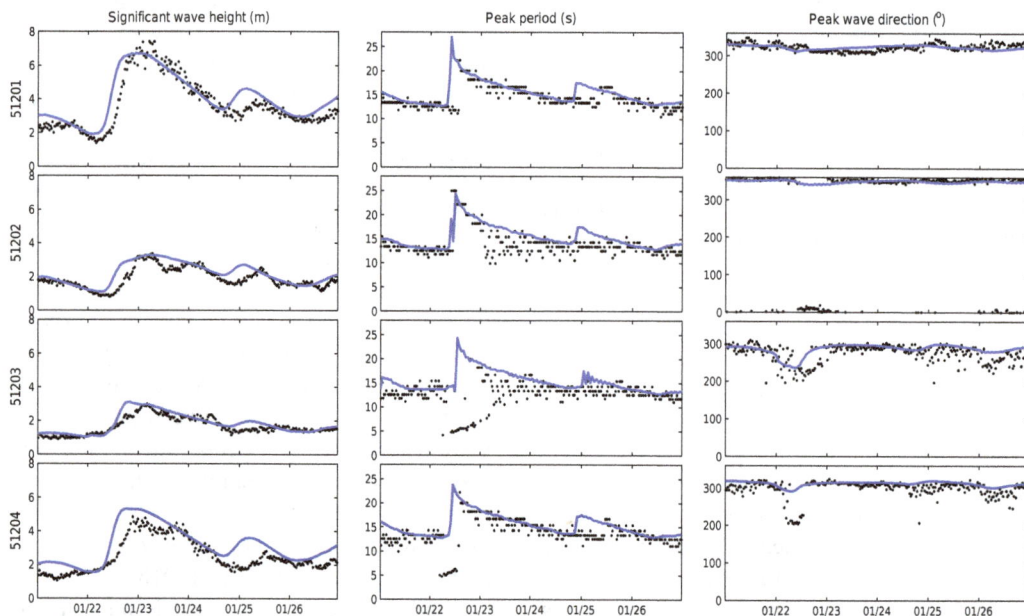

Figure 4. Comparison of the significant wave height, peak period, and direction between the NDBC buoys displayed in Fig. 1 and the wave model hindcast system results for the period of experiment 2. The black dots correspond to the buoy data and the blue line to the model results.

The HFR data give widespread surface current data for the south shore of O'ahu – the region of interest of the present study. Although they do not provide observations of the nearshore circulation, they do provide an important constraint to the outer grid surface velocities used as boundary conditions to the nearshore domain. In their Figs. 2 and 3, Souza et al. (2015) show a good agreement of the modeled radial velocities with the observations for the south shore of O'ahu with RMSEs generally under $10\,\mathrm{cm\,s^{-1}}$ for their experiment A (equivalent to the present model setup). The authors calculate a phase difference of $5.2\,\mathrm{min}$ for the M_2 tide component – the main tide component for this region.

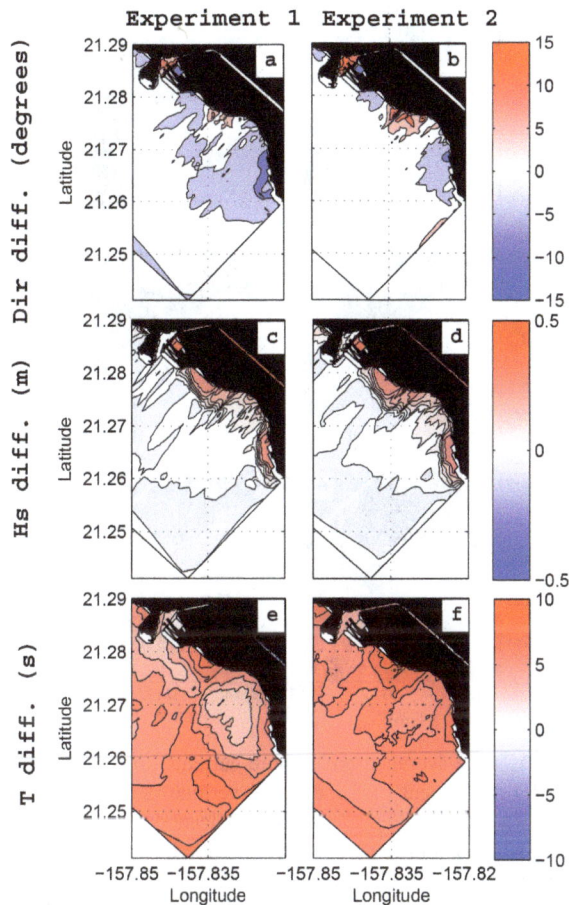

Figure 5. Maps of differences in wave direction (**a, b** – unit: degrees from N – contour interval (c.i.) 5°), wave significant height H_s (**c, d** – unit: meters – c.i. 0.05 m), and wave period T (**e, f** – unit: seconds – c.i. 1 s), between the WAVECOUPLE and WAVEFORCE simulations (WAVECOUPLE–WAVEFORCE). Comparisons of experiment 1 are in the left column and for experiment 2 in the right column. The coupled simulation presents significantly higher H_s near the shore and larger T throughout the model domain for both experiments.

3.2 Effects on the waves

The WAVECOUPLE simulation has longer periods (T) throughout the domain and larger H_s than the WAVEFORCE in the reef region, providing an indication of the effect of the circulation on the waves as described by Lowe et al. (2009) (Fig. 5). According to Warner et al. (2010) the ocean currents affect the modeled waves by modifying the wind stress and the group velocities, $\mathbf{c} = (c_x, c_y)$. Since the model uses bulk formulas to calculate the wind stress, it will reflect the modification of the 10 m winds $\mathbf{u}_{wind}$ by the moving ocean surface $\mathbf{u}_{wind} - \mathbf{u}$, where $\mathbf{u}$ is the surface current velocity.

The ocean currents modify the group velocities $\mathbf{c}$ by adding the current: $\mathbf{c} + \mathbf{u}$. This alters the wave number and allows for current-induced reflection as shown by Fan et al.

(2009). The authors showed that the interaction with the ocean currents causes a Doppler shift for the gravity waves. When the currents are against the waves, the waves are compressed, and when the currents have the same direction as the waves, the waves are elongated. The degree to which this Doppler shift modifies the surface waves depends on the current speed and direction relative to the wave propagation speed and direction; therefore, short waves with slow propagation are most affected by the ocean currents.

Following Oh and Kim (1992) the currents affect the wave's apparent period, ω, through

$$\omega = kU(\mathbf{x}, t) + \sigma(k, h), \qquad (5)$$

where k is the wave number, U is the current intensity along the wave's propagation direction, σ is the intrinsic frequency, and h is the local water depth.

The spatial variability of the differences in H_s, direction, and period (T) between the coupled and forced simulations (WAVECOUPLE–WAVEFORCE) are shown in Fig. 5. Waikiki is characterized by the presence of a coral reef system and complicated bathymetry with strong currents. Due to its complexity and impact for the local community, the following analysis will focus on the region of Waikiki.

WAVECOUPLE exhibits $\approx 10\%$ higher average H_s in the Waikiki area than WAVEFORCE, with differences concentrated in the reef zone. Comparing these difference maps to the bathymetry in Fig. 1c and the time-averaged currents for the NOWAVE cases in Figs. 6a–b and 7a–b, one can note that the differences in H_s are concentrated over shallow reef areas, while differences in wave direction have a more widespread distribution. As expected, the magnification of H_s due to the coupling occurs in the western and eastern extremes of Waikiki, causing increases in the mean water level due to wave setup at the coast.

In the return flow area in the middle of bay formed by the Waikiki beach (see Fig. 1c), there is almost no change in H_s. This area corresponds to a large channel in the reef (see Fig. 1c), where there is very low depth-induced wave breaking, as will be seen in the next section.

To analyze the effect of the circulation on the waves, the differences in wave direction and H_s between the WAVECOUPLE and WAVEFORCE simulations were correlated to the current intensity and direction for both experiments (Table 2). For that, time series of the differences and current direction and intensity in each grid point in the domain covered in Fig. 5 were extracted and the obtained correlations spatially averaged. While experiment 1 exhibits higher correlations with the current direction, experiment 2 was more correlated to the current intensity. Since the mean difference between the waves and currents direction is similar for both experiments ($\approx 60°$), this distinction cannot be explained by the Doppler shift discussed above. It appears the correlation is a function of the relation between the wave and current intensity. Comparing experiments 1 and 2 for both modeling strategies shows that experiment 1 presents stronger ocean

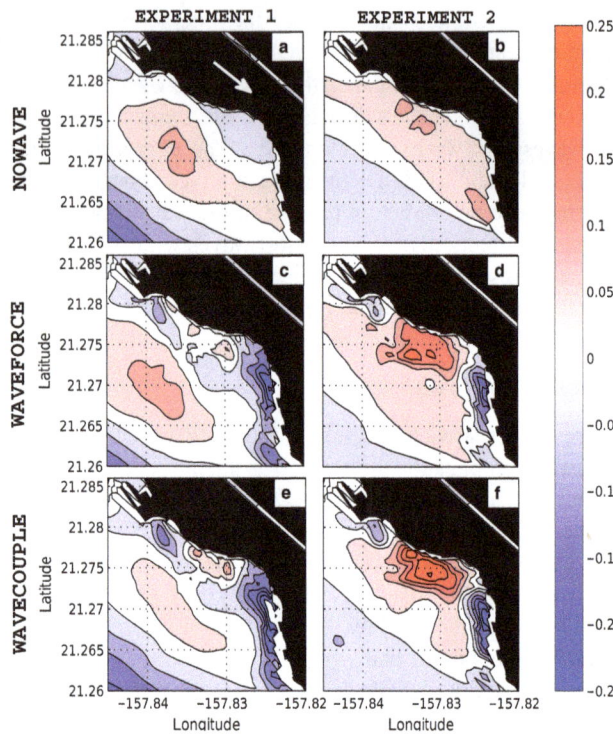

Figure 6. Time-averaged alongshore currents ($m\,s^{-1}$) in Waikiki region for experiment 1 (**a, c, e**) and experiment 2 (**b, d, f**) from the three modeling strategies. The colors indicate the current intensity ($m\,s^{-1}$) with contours every $0.05\,m\,s^{-1}$. It is interesting to observe the appearance of drift currents along the coast in the simulations that consider the effect of the surface waves. Both the WAVE-FORCE and WAVECOUPLE simulations resolve the modification of the nearshore circulation pattern by the waves, with WAVECOUPLE presenting stronger coastal drift currents. The white arrow in panel (**a**) shows the direction of positive alongshore velocities.

Figure 7. Time-averaged cross-shore currents ($m\,s^{-1}$) in Waikiki region for experiment 1 (**a, c, e**) and experiment 2 (**b, d, f**) from the three modeling strategies. The colors indicate the current intensity ($m\,s^{-1}$) with contours every $0.05\,m\,s^{-1}$. It is interesting to observe the appearance of return flow cells perpendicular to the coast in the simulations that consider the effect of the surface waves. Both the WAVEFORCE and WAVECOUPLE simulations resolve the modification of the nearshore circulation pattern by the waves, with WAVECOUPLE presenting stronger currents. The white arrow in panel (**a**) shows the direction of positive cross-shore velocities.

currents (by 90 %), while experiment 2 exhibits larger wave heights (by 15 %). The smaller waves in experiment 1 are subject to the influence of the stronger currents explaining the higher impact of the coupling. The fact the NDBC data show a prevalence of small waves ($H_s < 1\,m$ 94 % of the time) emphasizes the importance of the interaction with the local currents.

3.3 Effects on the currents

Although the effect of the Ala Wai Canal discharge in the nearshore circulation can be significant during large rain events (Johnson et al., 2013), we focus on the effects of the waves in periods when the Ala Wai low-volume flux does not impact the local circulation.

The resulting mean circulations obtained from the three modeling strategies (Figs. 6 and 7) clearly show the effects of the waves and the wave–current interaction on the resolved circulation. The nearshore current pattern drastically changes when introducing the effect of the waves, while the offshore

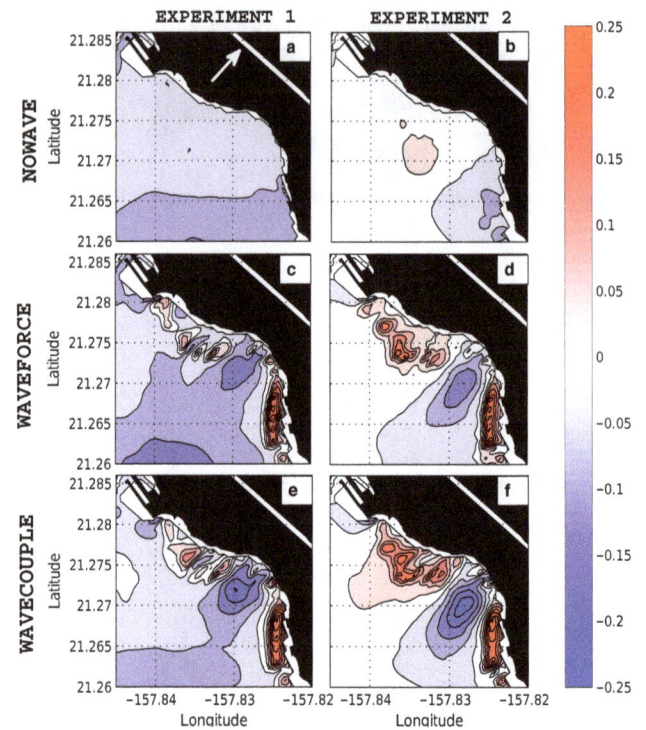

currents keep their general spatial structure (albeit with different intensity). The formation of local circulation cells in the nearshore is related to the presence of return flows that are forced by the pressure gradient due to the wave setup.

The alongshore component of the velocity (Fig. 6) shows the formation of coastal drift currents in both WAVEFORCE and WAVECOUPLE simulations. For the remainder of the text we refer to alongshore as along the mean shoreline orientation. A similar reasoning is used for the cross-shore velocity component. The pattern shows the convergence of this flow in the central area of Waikiki, with the WAVECOUPLE exhibiting larger intensities. Similarly, Fig. 7 presents the cross-shore component of the surface velocity. The formation of onshore/offshore flow cells is evident in the figure, with the WAVECOUPLE exhibiting intensified flow. A strong negative (offshore) current is present in the area of convergence of the coastal drift. This corresponds to a strong nearshore circulation cell with the presence of intensified cross-shore current. In experiment 1, several smaller cross-shore current cells are present in the western portion of Waikiki with onshore currents over the reef heads and offshore currents in the small channels in the reef. Experiment 2 shows a pattern dominated by a unique offshore flow at the convergence area with positive onshore flows both to the east and west. The eastern part of Waikiki is dominated by a strong northwest flow in both experiments that is independent of the modeling strategy. This is related to the fact that – independent of the direction of the incident swell – the nearshore waves have similar direction when approaching the coastline, and they break over the shallow reef close to the beach in the east portion of Waikiki. A direct comparison between both modeling approaches in Figs. 6 and 7 demonstrates that both reproduce the same nearshore circulation pattern but with different intensities.

The modification of the circulation by the waves is expressed by the right-hand-side terms in Eq. (1). It includes the influences of the vortex force, Bernoulli head, and non-conservative wave forces. These wave effects enter the ROMS primitive equations as momentum and tracer fluxes. The vortex force (VF) terms represent the interaction between the Stokes drift and the vorticity of the mean flow. Since this term is not explicitly written in the model output, it is difficult to quantify its contribution to the momentum balance. Nevertheless, it is directly related to the Stokes drift velocities obtained from the resolved wave field, as expressed by Benetazzo et al. (2013):

$$\mathrm{VF_{hor}} = -\hat{z} \times \mathbf{u}^{\mathrm{st}} \left(\hat{z} \cdot \nabla_\perp \times \mathbf{u} + f \right) - w^{\mathrm{st}} \frac{\partial \mathbf{u}}{\partial z}, \qquad (6)$$

where $\mathrm{VF_{hor}}$ is the horizontal component of the vortex force, and $\hat{z}$ is the vertical unit vector.

The maps in Fig. 8a–d show the spatial distribution of the surface Stokes velocities in the Waikiki area. The large signal in the kinetic energy in experiment 2 shows the arrival of the large swell on 22 January. Although the WAVEFORCE simulations exhibit slightly higher Stokes current intensity throughout the domain, the shallow nearshore region over the reef (under 10 m depth) has stronger Stokes currents in the WAVECOUPLE simulations associated with the magnification of the waves in the surf zone due to the interaction with the currents. This is evident in the time series of the kinetic energy associated with the Stokes velocities in the Fig. 8e and f. The Stokes drift velocities are nearly opposite to the mean circulation resolved by the NOWAVE simulations (Figs. 6 and 7a, b). This explains the smaller average velocities and the smaller total kinetic energy resolved by the WAVECOUPLE simulations in the offshore region. While for experiment 1 both modeling strategies show similar Stokes velocity kinetic energy time variability, experiment 2 exhibits a clear peak associated with the arrival of the swell.

This difference in the Stokes drift velocities, however, is not enough to explain the observed differences in the total currents. Taking only the wave effects into consideration, the differences in the total velocity intensities are mainly

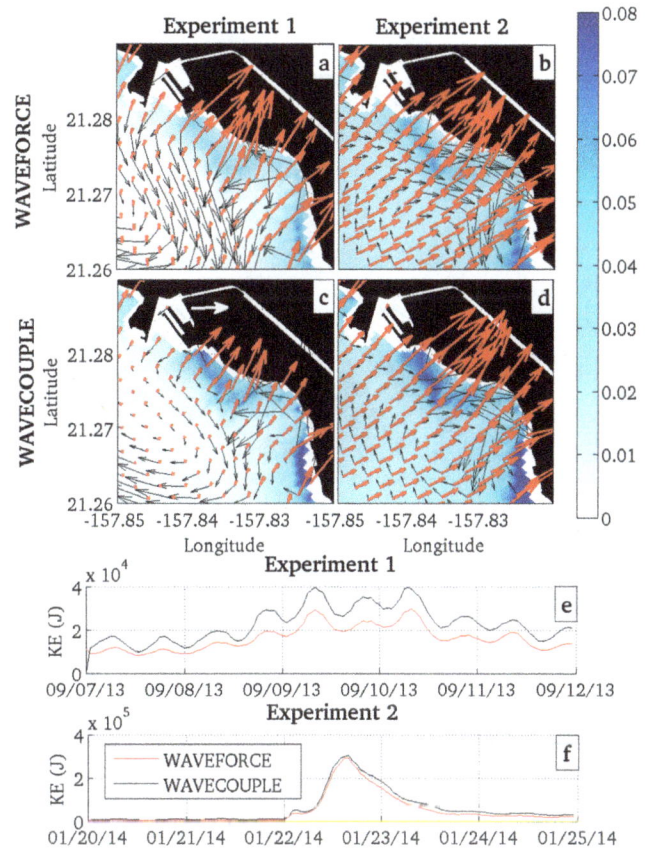

Figure 8. Maps of the time-averaged Stokes drift velocities (red arrows – m s^{-1}) and total surface velocities (black arrows – m s^{-1}) for the experiments 1 (**a, c**) and 2 (**b, d**), and the associated time series of Stokes drift velocities kinetic energy (**e, f** – J). The white arrow in panel (**c**) indicates a 10 cm s^{-1} scale for the velocity vectors. Please note the difference in the vertical scale of panels (**e**) and (**f**).

a consequence of the wave setup/setdown. The presence of cross-shore current cells is the main feature in the velocity maps of Fig. 7. These circulation cells are a consequence of mean sea level increases (wave setup) shoreward of the wave-breaking zone, generating a pressure gradient that balances the radiation stress. As pointed out by Dalrymple et al. (2011), cross-shore currents are usually generated simply by alongshore variations in breaking wave heights. In bays such as Waikiki, cross-shore currents can form in the center of the beach and extend significantly offshore. The model results indicate cross-shore currents exceeding 1 m s^{-1} in the nearshore region in Waikiki for both WAVECOUPLE and WAVEFORCE simulations, as typically observed for atolls and barrier reefs according to Gourlay and Colleter (2005). The differences between the two experiments show that the intensity and duration of the high wave event dominate the circulation response in experiment 2. While in experiment 1 the wave regime is quasi-constant and the nearshore circulation is in balance, the arrival of a large swell in experiment 2 perturbs the balance and causes a stronger circulation

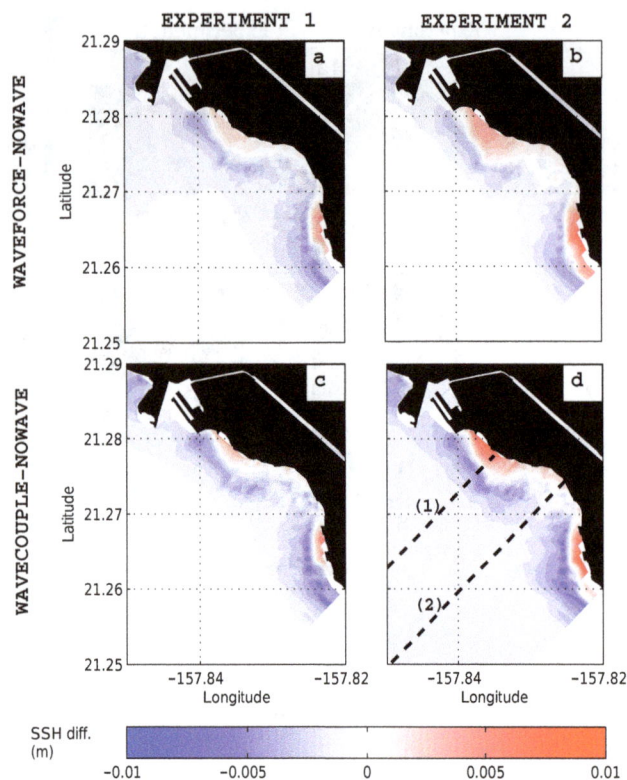

Figure 9. Time-averaged SSH difference (m) between the simulations that include the effect of surface gravity waves (WAVEFORCE and WAVECOUPLE) and the NOWAVE case for experiment 1 (**a, c**) and experiment 2 (**b, d**). The black dashed lines in (**d**) indicate the position of the two cross-shore sections used to analyze the wave setup and the cross-shore balance.

response in the WAVECOUPLE approach. The modification of the wave direction due to the interaction with the currents acts to modify the setup and generate a feedback on the currents.

Therefore, it is necessary to quantify the modification of the sea level by the waves and the balance with the dissipation of wave energy. To achieve this, the SSH differences between the simulations that include the effects of the waves (WAVEFORCE and WAVECOUPLE) and the NOWAVE were calculated and are presented in Fig. 9. Although these differences are small (on the order of 1 cm), the consequences on the nearshore circulation are important (as shown in the nearshore current pattern in Figs. 6 and 7). Figure 10 reveals the differences in the cross-shore pressure gradient. From here, we will present the pressure gradient as a per density unit.

The elevation of the sea surface near the coast due to the waves is observed for each experiment, followed by a lowering of the sea level towards the open ocean. The WAVECOUPLE cases show overall larger magnitude of elevation, both positive and negative, than the WAVEFORCE. These differ-

ences in the sea level impose a cross-shore pressure gradient that affects the local currents.

Taking the cross-shore section (1) shown in Fig. 9d as an example, Fig. 10 shows the pressure gradient obtained by each modeling strategy for both experiments with the associated cross-shore surface velocities. The larger shaded area in Fig. 10b and d in relation to Fig. 10a and c reflects the arrival of the larger swell waves in experiment 2, and the consequent increase of the cross-shore pressure gradient and velocity. There is a difference in the spatial distribution of the pressure gradient between the WAVEFORCE and the WAVECOUPLE simulations. Both maxima are aligned to the reef break, but the WAVECOUPLE simulation shows an overall smoother transition towards the coast and offshore, with smaller gradients in the pressure that reflect the larger and broader wave setup as observed in Fig. 9. The cross-shore velocities follow a similar pattern, mirroring the pressure gradient sections. There is a shift in the maxima between the WAVECOUPLE and WAVEFORCE that reflects the smoother transition of the wave setup observed in the pressure gradient sections. The feedback circle is closed when the Doppler shift by the nearshore currents modify the wave field that generates the pressure gradient against the shoreline. This generates a modified cross-shore pressure gradient affecting the currents and closing the feedback. In the WAVEFORCE experiments this feedback is broken because only half of it is represented by the model dynamics. As explained by Kumar et al. (2012), there is a balance between the wave setup derived pressure gradient and part of the wave energy dissipation that contributes to the momentum flux in the surf zone. This dissipation is part of the nonconservative wave forcing in Eq. (1), which includes depthinduced wave breaking (and white capping) near the surface and frictional wave dissipation near the bottom. The remaining energy from the wave dissipation is involved in the creation of wave rollers, which are related to turbulent mixing in the surf zone and dissipation. Figure 11 shows the dissipation by wave breaking for the WAVECOUPLE simulations, which is $\approx 10^2$ larger than the dissipation by white capping and $\approx 10^6$ greater than the dissipation through bottom friction. The distribution of dissipation by wave breaking is not uniform along the Waikiki beach. There is a region of low dissipation in the middle of the beach, corresponding to the intense return flow observed in Fig. 7. The energy dissipation for experiment 2 is higher than for experiment 1 because it contains larger wave heights.

To analyze how the different phenomena interact to generate the observed nearshore circulation pattern, the pressure gradient (cross- and alongshore), the non-conservative wave forces (sum of depth-induced breaking, white capping, and bottom friction), and the integrated cross-shore transport by the quasi-Eulerian currents and the Stokes drift are plotted for the two sections shown in Fig. 9d. To isolate the contribution of the wave setup to the surface currents, the velocities obtained by the NOWAVE simulations were subtracted from

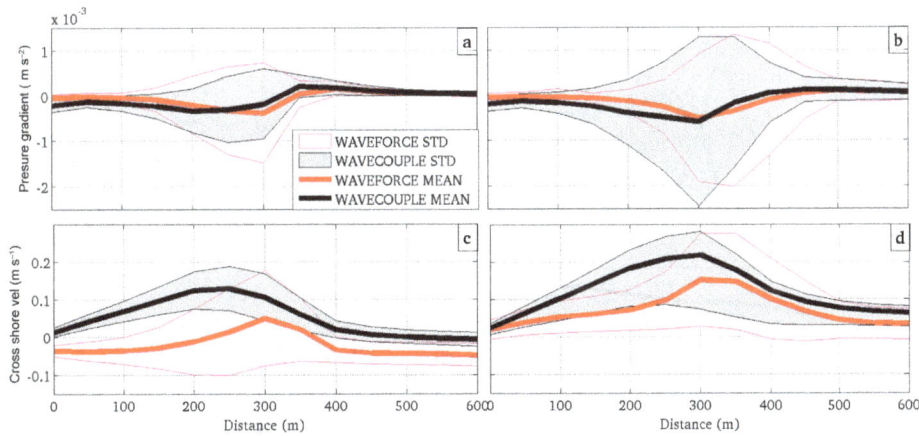

Figure 10. Sections of pressure gradient per density unit and surface cross-shore velocities for experiment 1 (**a, c**) and experiment 2 (**b, d**). The thick lines show the mean while the shaded areas are the standard deviations. Distances are measured from the coastline along the section defined in Fig. 9d. Negative pressure gradient is directed offshore.

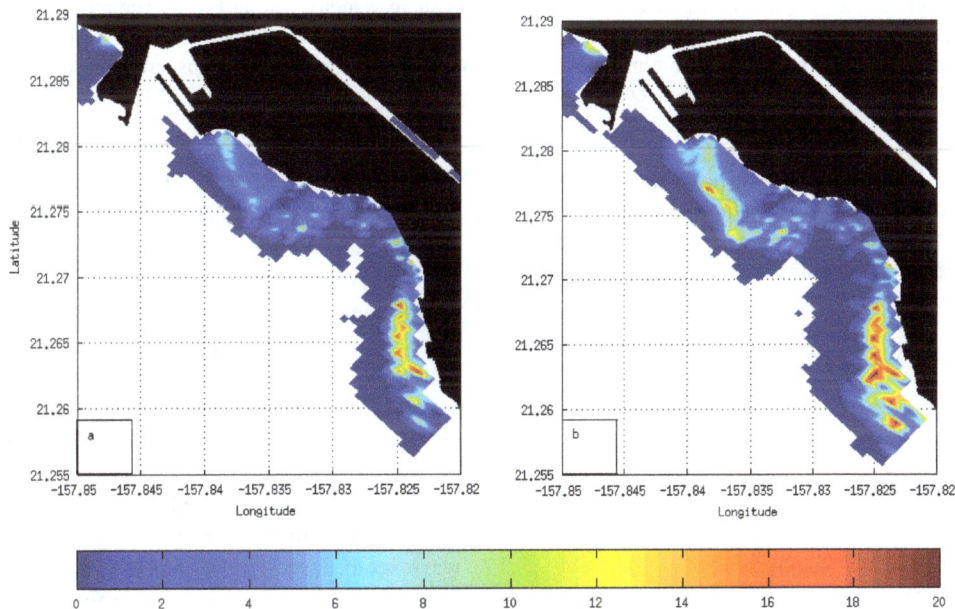

Figure 11. Maps of the time mean wave energy dissipation by depth-induced wave breaking ($W\,m^{-2}$) for the WAVECOUPLE experiment 1 (**a**) and experiment 2 (**b**). The depth-induced wave breaking was the most significant wave dissipation term. The energy dissipation was concentrated in the western and eastern portions of Waikiki due to the reef bathymetry (see Fig. 1). The larger dissipation in (**b**) is due to the large wave heights of experiment 2.

the total cross-shore velocities prior to the transport calculation. The WAVECOUPLE experiment 2 was taken to demonstrate the differences in the balance between the two sections. The results are presented in Fig. 12.

At the section (1) (Fig. 12a) there is a correspondence between the maxima of the non-conservative wave forces on the reef edge, the negative (offshore) pressure gradient, and the onshore water transport due to the Stokes drift currents in the surf zone. The cross-shore transport is insensitive to this balance in the surf zone. The wave energy dissipates

rapidly to zero more than $\approx 400\,m$ offshore as the cross-shore pressure gradient becomes positive (onshore), indicating the wave shoaling region where setdown (reduction of the mean sea level, Fig. 8) takes place. This is a good example of how the energy from the breaking waves drives the currents near the coast. As explained by Dalrymple et al. (2011), at the offshore edge of the surf zone the waves steepen and break, propagate across the surf zone, and run up the beach. Balancing forces are required for the energy loss by wave breaking and consequent change in the cross-shore and/or alongshore

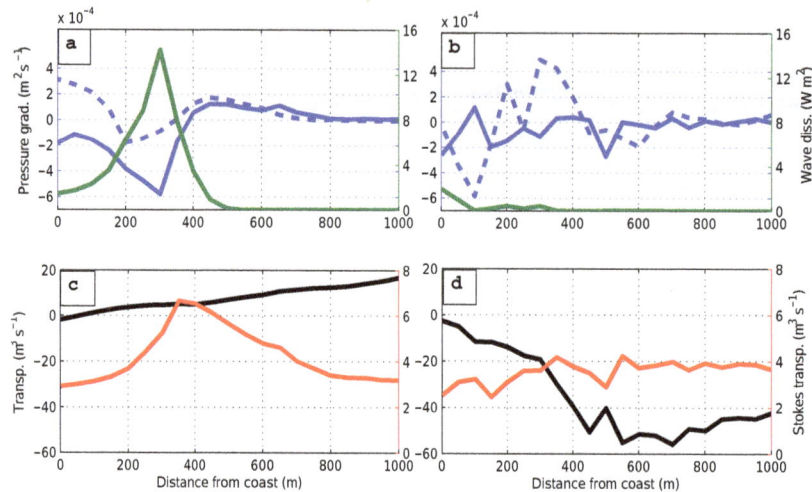

Figure 12. Cross-shore sections 1 **(a, c)** and 2 **(b, d)** of the wave contributions to the nearshore momentum balance for the WAVECOUPLE experiment 2. The section positions are shown in Fig. 9d. While section 1 is over the region of large wave-breaking energy dissipation, the section 2 is in the region of the intense return flow observed in the cross-shore velocity map in Fig. 7. The pressure gradient (continuous line: cross-shore; dashed line: alongshore) due to wave setup is shown in blue, the wave dissipation due to depth-induced wave breaking in green, the vertically integrated Stokes drift transport in red, and the total wave-induced cross-shore transport in black. The total wave-induced cross-shore transport was obtained subtracting the NOWAVE quasi-Eulerian velocities from the WAVECOUPLE velocities.

Table 2. Spatial mean correlation factors between the difference in wave direction and significant height (H_s) for the WAVECOUPLE and WAVEFORCE forced modeling strategies and the surface current intensity and direction. Only correlations significant to the 99 % level were taken into consideration.

	Wave direction difference		Wave H_s difference	
	Exp. 1	Exp. 2	Exp. 1	Exp. 2
Current intensity	0.35	0.36	0.37	0.45
Current direction	0.45	0.26	0.52	0.27

momentum flux. These primarily arise from wave-induced changes in the mean water level at the shoreline (wave setup) that provide a hydrostatic force due to wave-induced currents.

There is a large channel in the reef near the region of section (2). As a consequence of the larger depths close to the coastline, the energy loss by wave breaking is approximately 6 times smaller than in section (1) and concentrated near the coast (shorebreak). The balance explained above for section (1) does not take place, as is evident by the lack of associated maxima in the cross-shore pressure gradient and Stokes drift velocity transport. This section, however, is in the convergence zone of alongshore wave-induced currents observed in Fig. 6. A strong (3 times larger than in section 1) cross-shore transport is generated as the return branch of the nearshore circulation cell. The local alongshore pressure gradient shows large values, ranging from negative close to the

beach to positive offshore. The water transport seems to respond to the larger-scale (order of the beach length) pressure gradient that is evident from the wave setup maps in Fig. 9. Revisiting the cross-shore velocity maps in Fig. 7 and comparing with the non-conservative wave forces of Fig. 10, there is a clear difference between the small wave condition (≈ 1 m) of experiment 1 and the arrival of a large swell in experiment 2. In experiment 1, there is the formation of several small cross-shore cells with a stronger one in the region of alongshore current convergence at the middle of the beach where section (2) is located. For experiment 2, wave breaking dominates the western portion of Waikiki with a stronger wave setup, larger convergence at the middle of the beach, and consequently stronger cross-shore currents in the region of section (2).

With the mechanism of cross-shore circulation cells in mind, it becomes clear how the small observed differences in the pressure gradient associated with the wave setup/setdown resolved by the coupled simulations have important impacts on the nearshore circulation. This demonstrates how significant coupled processes are for the resolved currents and for the skill of nearshore forecast systems.

To quantify these differences, Table 3 presents a comparison of the total velocity, Stokes drift, and setup-associated velocity between the WAVECOUPLE and WAVEFORCE simulations. The setup-associated velocities were calculated by subtracting the NOWAVE velocities from the model quasi-Eulerian velocities. The NOWAVE is taken to represent all of the other contributions not associated with the waves action. The larger intensity of the Stokes drift in the WAVECOUPLE simulations is related to the larger H_s

Table 3. Difference of the total and wave contributions to the surface velocities (%) between the WAVECOUPLE and WAVEFORCE simulations. The results show that while the contribution of the Stokes drift to the total velocities is higher by the same rate for both approaches, the same is not true for the wave setup contribution. A more complex interaction between waves and currents derived from the coupling gives rise to stronger nearshore currents in the WAVECOUPLE simulations (positive values).

	Experiment 1	Experiment 2
Stokes drift	21	21
Wave setup/setdown	5	14
Total	−4	4

($\approx 10\,\%$) in relation to the WAVEFORCE simulations for both experiments.

Therefore, the interaction between the surface waves and nearshore circulation has important impacts on the resolved currents, especially in the nearshore region between the reef crest and the coast. This is an important phenomenon that should be taken into consideration when developing forecast systems that aim to provide a useful description of the nearshore currents.

4 Summary and conclusions

Due to the interaction with the currents that modify the wave number, coupling the circulation and wave models gives rise to larger H_s and slightly different wave directions when compared to the SWAN alone. The smaller waves present in experiment 1 – the type that prevail most the time in the south shore of O'ahu – are overall more sensitive to the local circulation. While the differences in the H_s are concentrated in the reef region, the modification of the directions due to the interaction with the currents is widespread through the domain.

The differences in the resolved wave fields presented feedbacks on the circulation in the coupled simulations (both WAVEFORCE and WAVECOUPLE), since the resolved H_s is reflected in distinct Stokes drift velocities and wave setup against the coast. Such differences are magnified in the reef crest area and in the region between the reef and the coast. The Stokes drift velocities flow opposite to the surface currents resolved by the NOWAVE case, resulting in weaker currents when considering the effect of the waves (both WAVEFORCE and WAVECOUPLE approaches). The wave setup determined the free surface elevation on the reef and dominates the nearshore circulation, similar to the case of Kāne'ohe Bay studied by Lowe et al. (2009). This effect was observed to be stronger in the occurrence of a large swell event in experiment 2. The wave setup was the dominant process determining how the waves affect the nearshore circulation on the Honolulu coast. This conclusion should be trans-ferable to other exposed coral reef coastal areas, where the nearshore circulation is deeply affected by the surface gravity waves particularly in the occurrence of large swell events.

In this paper, we set out to understand (i) the effective contribution from the surface waves under different conditions, (ii) the importance of coupled ocean currents and surface wave processes for the local dynamics, and (iii) the influence of different approaches for forecasting the nearshore current field.

We found that the surface wave field has significant importance to the circulation, even in periods of small (under 1 m) waves. (i) The consequences on the Stokes drift velocities, wave setup, and non-conservative wave forces influence the circulation through the domain, with a particularly important contribution in the surf zone on the reef break. (ii) Although the general nearshore circulation pattern is resolved in both WAVEFORCE and WAVECOUPLE simulations, the inclusion of coupled processes led to differences in the magnitude of the currents and wave parameters. There were important differences in the cross-shore circulation cells resolved by the two approaches, generally with stronger cross-shore currents for the WAVECOUPLE simulations. The inclusion of coupled processes was shown to be important in the representation of both wave and circulation parameters, leading to overall larger H_s ($\approx 10\,\%$), longer wave periods ($\approx 30\,\%$), and stronger associated Stokes drift currents. (iii) Despite the improvements the coupling may bring, one must also consider the computational cost of the numerical simulations – especially for high-resolution coastal grids. Coupled simulations are extremely expensive, and the WAVEFORCE approach can provide an interesting alternative.

The results show the importance of considering coupled processes when aiming to resolve both the nearshore circulation and the wave characteristics in the reef zone. However, the computational cost involved in coupled simulations presents an important obstacle in the use of this approach for operational forecast systems. Once in possession of the wave model results, the WAVEFORCE approach requires one-sixth of the computational time of WAVECOUPLE. The SWAN run alone (uncoupled) takes about one-quarter the computational time of the WAVEFORCE ROMS run. Although this permits a greater malleability in the use of available machine power and time, it does not consider the coupling between the two models and should be viewed as a compromise solution rather than optimal. The ability of resolving the general pattern of nearshore circulation, however, makes the WAVEFORCE an interesting approach for operational purposes.

Competing interests. The authors declare that they have no conflict of interest.

Acknowledgement. Joao Marcos Azevedo Correia de Souza was supported by NOAA grant no. NA07NOS4730207. Brian Powell was supported by ONR grant no. N00014-09-1-0939. We would like to thank Ning Li for providing the SWAN configuration and boundary conditions and helping in the SWAN model description and evaluation. This is SOEST publication no. 9880.

Edited by: A. Sterl

References

Ardhuin, F., Jenkins, A. D., and Belibassakis, K. A.: Comments on the 'The Three-Dimensional Current and Surface Wave Equations', J. Phys. Oceanogr., 38, 1340–1350, doi:10.1175/2007JPO3670.1, 2008.

Battjes, J. A. and Janssen, J. F. M.: Energy loss and set-up due to breaking of random waves, in: Proceedings of the 16th International Conference on Coastal Engineering, 27 August–3 September 1978, Hamburg, Germany, 1978.

Benetazzo, A., Carniel, S., Sclavo, M., and Bergamasco, A.: Wave-current interaction: Effect on the wave field on a semi-enclosed basin, Ocean Model., 70, 152–165, doi:10.1016/j.ocemod.2012.12.009, 2013.

Booij, N., Ris, R. C., and Holthuijsen, L. H.: A third-generation wave model for coastal regions. 1. Model description and validation, J. Geophys. Res., 104, 7649–7666, 1999.

Chavanne, C. P., Flament, P., Luther, D. S., and Gurgel, K. W.: Observations of Vortex Rossby Waves Associated with a Mesoscale Cyclone*, J. Phys. Oceanogr., 40, 2333–2340, doi:10.1175/2010JPO4495.1, 2010.

Dalrymple, R. A., MacMahan, J. H., Reniers, A., and Nelko, V.: Rip Currents, Annu. Rev. Fluid Mech., 43, 551–581, doi:10.1146/annurev-fluid-122109-160733, 2011.

Edwards, K. L., Veeramony, J., Wang, D., Holland, K. T., and Hsu, Y. L.: Sensitivity of Delft3D to input conditions, in: OCEANS 2009 MTS/IEEE Biloxi – Marine Technology for our Future: Global and Local Challenges, 26–29 October 2009, Biloxi, Mississippi, USA, IEEE, 1–8, 2009.

Egbert, G. D. and Erofeeva, S. Y.: Efficient inverse modeling of barotropic ocean tides, J. Atmos. Ocean. Tech., 19, 183–204, 2002.

Eldeberky, Y. and Battjes, J. A.: Parametrization of triad interactions in wave energy models, in: Coastal Dynamics '95, edited by: Dally, W. R. and Zeidler, R. B., 140–148, 1995.

Fan, Y., Ginis, I., and Hara, T.: The effect of wind-wave-current interaction on air-sea momentum fluxes and ocean response in tropical cyclones, J. Phys. Oceanogr., 39, 1019–1034, doi:10.1175/2008JPO4066.1, 2009.

Gourlay, M. R. and Colleter, G.: Wave-generated flow on coral reefs – an analysis for two-dimensional horizontal reef-tops with steep faces, Coast. Eng., 52, 353–387, doi:10.1016/j.coastaleng.2004.11.007, 2005.

Gracia Berdeal, I., Hickey, B. M., and Kawase, M.: Influence of wind stress and ambient flow on a high discharge river plume, J. Geophys. Res., 107, 3130, doi:10.1029/2001JC000932, 2002.

Hoeke, R. K., Storlazzi, C. D., and Ridd, P. V.: Drivers of circulation ina fringing coral reef embayment: A wave-flow coupled numerical study of Hanalei Bay, Hawaii, Cont. Shelf Res., 58, 79–95, doi:10.1016/j.csr.2013.03.007, 2013.

Janekovic, I. and Powell, B.: Analysis of imposing tidal dynamics to nested numerical models, Cont. Shelf Res., 34, 30–40, doi:10.1016/j.csr.2011.11.017, 2011.

Johnson, A. E., Powell, B., and Stewart, G. F.: Characterizing the effluence near Waikiki, Hawaii with a coupled biophysical model, Cont. Shelf Res., 54, 1–13, doi:10.1016/j.csr.2012.12.007, 2013.

Kumar, N., Voulgaris, G., Warner, J., and Olabarrieta, M.: Implementation of a vortex force formalism in the coupled ocean-atmosphere-wave-sediment transport (COAWST) modeling system for the inner-shelf and surf-zone applications, Ocean Model., 47, 65–95, 2012.

Lane, E. M., Restrepo, J. M., and McWilliams, J. C.: Wave-Current Interaction: A Comparison of Radiation-Stress and Vortex-Force Representations, J. Phys. Oceanogr., 37, 1122–1141, doi:10.1175/JPO3043.1, 2007.

Li, N., Cheung, K. F., Stopa, J. E., Hsiao, F., Chen, Y., Vega, L., and Cross, P.: Thirdy-four years of Hawaii wave hindcast from downscaling of climate forecast system reanalysis, Ocean Model., 100, 78–95, doi:10.1016/j.ocemod.2016.02.001, 2016.

Lowe, R. J., Falter, J. L., Monismith, S. G., and Atkinson, M. J.: A numerical study of circulation in a coastal reef-lagoon system, J. Geophys. Res., 114, C06022, doi:10.1029/2008JC005081, 2009.

Matthews, D., Powell, B. S., and Janekovic, I.: Analysis of four-dimensional variational state estimation of the Hawaiian waters, J. Geophys. Res., 117, C03013, doi:10.1029/2011JC007575, 2012.

McWilliams, J. C.: Targeted coastal circulation phenomena in diagnostic analyses and forecast, Dynam. Atmos. Oceans, 49, 3–15, doi:10.1016/j.dynatmoce.2008.12.004, 2009.

Mellor, G.: The Three-Dimensional Current and Surface Wave Equations, J. Phys. Oceanogr., 33, 1978–1989, 2003.

Mellor, G.: Some Consequences of the Three-Dimensional Current and Surface Wave Equations, J. Phys. Oceanogr., 35, 2291–2298, 2005.

Mellor, G.: The Depth-Dependent Current and Wave Interaction Equations: A Revision, J. Phys. Oceanogr., 38, 2587–2596, doi:10.1175/2008JPO3971.1, 2008.

Mellor, G. L. and Yamada, T.: Development of a turbulence closure model for geophysical fluid problems, Rev. Geophys. Space Ge., 20, 851–875, 1982.

Oh, I. S. and Kim, Y. Y.: Wave characteristics changes under a strong tidal current influence, La Mer, 30, 275–285, 1992.

Pan, J., Gu, Y., and Wang, D.: Observations and numerical modelling of the Pearl River plume in summer season, J. Geophys. Res.-Oceans, 119, 2480–2500, doi:10.1002/2013JC009042, 2014.

Plant, N. G., Edwards, K. L., Kaihatu, J. M., Veeramony, J., Hsu, L., and Holland, K. T.: The effect of bathymetric filtering on nearshore process model results, Coast. Eng., 56, 484–493, doi:10.1016/j.coastaleng.2008.10.010, 2009.

Samsone, F. J., Pawlak, G., Stanton, T., McManus, M. A., Glazer, B. T., DeCarlo, E. H., Bandet, M., Sevadjian, J., Stierhoff, K., Colgrove, C., Hebert, A. B., and Chen, I. C.: Kilo Nalu. Physical/biogechemical dynamics above and within permeable sediments, Oceanography, 21, 173–178, 2008.

Shchepetkin, A. F. and McWilliams, J. C.: The regional oceanic modeling system (ROMS): a split-explicit, free-surface, topography-following-coordinate oceanic model, Ocean Model., 9, 347–404, doi:10.1016/j.ocemod.2004.08.002, 2005.

Soulsby, R. L., Hamm, L., Klopmanc, G., Myrhaugd, D., Simonse, R. R., and Thomas, G. P.: Wave-current interaction within and outside the bottom boundary layer, Coast. Eng., 21, 41–69, 1993.

Souza, J. M. A. C., Powell, B., Castillo-Trujillo, A. C., and Flament, P.: The Vorticity Balance of the Ocean Surface in Hawaii from a Regional Reanalysis, J. Phys. Oceanogr., 45, 424–440, doi:10.1175/JPO-D-14-0074.1, 2015.

Uchiyama, Y., McWilliams, J. C., and Shcheptkin, A. F.: Wave-current interaction in an oceanic circulation model with a vortex-force formalism: Application to the surf zone, Ocean Model., 34, 16–35, doi:10.1016/j.ocemod.2010.04.002, 2010.

Warner, J. C., Armstrong, B., He, R., and Zambon, J. B.: Development of a Coupled Ocean-Atmosphere-Wave-Sediment Transport (COAWST) Modeling System, Ocean Model., 35, 230–244, doi:10.1016/j.ocemod.2010.07.010, 2010.

12

A study on some basic features of inertial oscillations and near-inertial internal waves

Shengli Chen, Daoyi Chen, and Jiuxing Xing

Shenzhen Key Laboratory for Coastal Ocean Dynamic and Environment, Graduate School at Shenzhen, Tsinghua University, Shenzhen 518055, China

Correspondence to: Jiuxing Xing (jxx2012@sz.tsinghua.edu.cn)

Abstract. Some basic features of inertial oscillations and near-inertial internal waves are investigated by simulating a two-dimensional $(x - z)$ rectangular basin (300 km × 60 m) driven by a wind pulse. For the homogeneous case, near-inertial motions are pure inertial oscillations. The inertial oscillation shows typical opposite currents between the surface and lower layers, which is formed by the feedback between barotropic waves and inertial currents. For the stratified case, near-inertial internal waves are generated at land boundaries and propagate offshore with higher frequencies, which induce tilting of velocity contours in the thermocline. The inertial oscillation is uniform across the whole basin, except near the coastal boundaries (~ 20 km), where it quickly declines to zero. This boundary effect is related to great enhancement of non-linear terms, especially the vertical non-linear term $(w \partial u / \partial z)$. With the inclusion of near-inertial internal waves, the total near-inertial energy has a slight change, with the occurrence of a small peak at ~ 50 km, which is similar to previous research. We conclude that, for this distribution of near-inertial energy, the boundary effect for inertial oscillations is primary, and the near-inertial internal wave plays a secondary role. Homogeneous cases with various water depths (50, 40, 30, and 20 m) are also simulated. It is found that near-inertial energy monotonously declines with decreasing water depth, because more energy of the initial wind-driven currents is transferred to seiches by barotropic waves. For the case of 20 m, the seiche energy even slightly exceeds the near-inertial energy. We suppose this is an important reason why near-inertial motions are weak and hardly observed in coastal regions.

1 Introduction

Near-inertial motions have been observed and reported in many seas (e.g. Alford et al., 2016; Webster, 1968). They are mainly generated by changing winds at the sea surface (Pollard and Millard, 1970; Chen et al., 2015b). The passage of a cyclone or a front can induce strong near-inertial motions (D'Asaro, 1985), which can last for 1–2 weeks and reach a maximum velocity magnitude of 0.5–1.0 m s^{-1} (Chen et al., 2015a; Zheng et al., 2006; Sun et al., 2011). In deep seas, the near-inertial internal wave propagates downwards to transfer energy to depth (Leaman and Sanford, 1975; Fu, 1981; Gill, 1984; Alford et al., 2012). The strong vertical shear of near-inertial currents may play an important role in inducing mixing across the thermocline (Price, 1981; Burchard and Rippeth, 2009; Chen et al., 2016).

In shelf seas, near-inertial motions exhibit a two-layer structure, with an opposite phase between currents in the surface and lower layers (Malone, 1968; Millot and Crepon, 1981; MacKinnon and Gregg, 2005). By solving a two-layer analytic model using the Laplace transform, Pettigrew (1981) found this "baroclinic" structure can be formed by inertial oscillations without inclusion of near-inertial internal waves. Due to similar vertical structures and frequencies, inertial oscillations and near-inertial internal waves are hardly separable, and could easily be mistaken for each other.

In shelf seas, the near-inertial energy increases gradually offshore, and reaches a maximum near the shelf break, found in both observations (Chen et al., 1996) and model simulations (Xing et al., 2004; Nicholls et al., 2012). Chen and Xie (1997) reproduced this cross-shelf variation in both linear and non-linear simulations, and attribute it to large values of the cross-shelf gradient of surface elevation and the verti-

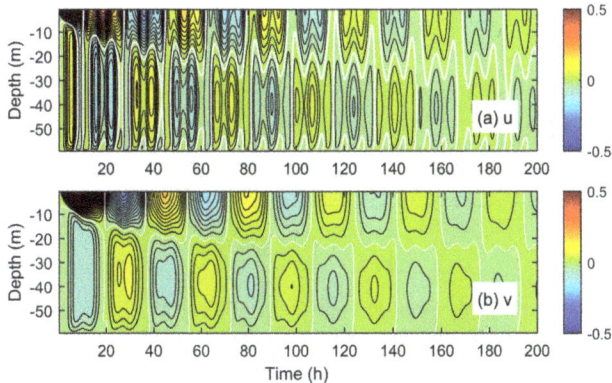

Figure 1. Velocities (u and v, m s^{-1}) at $x = 70$ km. The white lines denote the value of zero. The contour interval is 0.02 m s^{-1} for both panels.

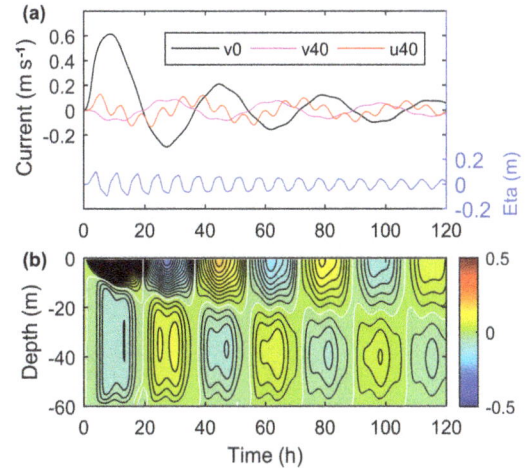

Figure 2. Snapshots of eastward velocity and elevation (η) at $t = 0.5$ and 1 h. The white lines represent the value of zero.

cal gradient of Reynolds stress near the shelf break. By using the analytic model of Pettigrew (1981), Shearman (2005) argued that the cross-shelf variation is controlled by baroclinic waves which emanate from the coast to introduce nullifying effects on the near-inertial energy near the shore. Kundu et al. (1983) found a coastal inhibition of near-inertial energy within the Rossby radius from the coast, which is attributed to leaking of near-inertial energy downward and offshore. As many factors seem to work, the mechanism controlling the cross-shelf variation of near-inertial energy is not clear.

In this paper, simple two-dimensional simulations are used to investigate some basic features of near-inertial motions. Cases with and without vertical stratification are simulated to examine properties and differences between inertial oscillations and near-inertial internal waves. The horizontal distribution of near-inertial energy is discussed in detail. Also, cases with various water depths are simulated to investigate the dependence of near-inertial motions on the water depth.

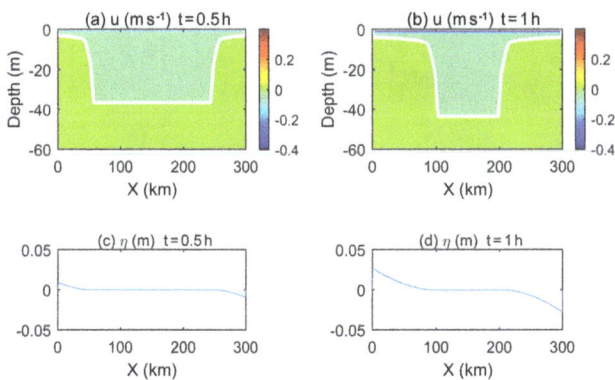

Figure 3. (a) Time series of velocities and elevation at $x = 100$ km. "v0" and "v40" mean the northward velocity (v) at depths of 0 and 40 m, and "u40" is the eastward velocity (u) at 40 m. **(b)** Contours of v at $x = 100$ km. The white lines denote the value of zero, and the contour interval is 0.02 m s^{-1}.

2   Model settings

The simulated region is a two-dimensional shallow rectangular basin (300 km $\times$ 60 m). Numerical simulations are done by the MIT general circulation model (MITgcm; Marshall et al., 1997), which discretizes the primitive equations and can be designed to model a wide range of phenomena. There are 1500 grid points in the horizontal ($\Delta x = 200$ m) and 60 grid points in the vertical ($\Delta z = 1$ m). The water depth is uniform, with the eastern and western sides being land boundaries. The vertical and horizontal eddy viscosities are assumed constant as 5×10^{-4} and 10 m^2 s^{-1}, respectively. The Coriolis parameter is 5×10^{-5} s^{-1} (at a latitude of $20.11°$ N). The bottom boundary is non-slip. The model is forced by a spatially uniform wind which is kept westward and increases from 0 to 0.73 N m^{-2} (corresponding to a wind speed of 20 m s^{-1}) for the first 3 h and then suddenly stops. The model runs for 200 h in total, with a time step of 4 s. The first case is homogeneous, while the second one has a stratification of a two-layer structure initially. For the stratified case, the temperature is $20\,°$C in the upper layer (-30 m $< z < 0$) and $15\,°$C in the lower layer (-60 m $< z < -30$ m). The salinity is constant, and the density is linearly determined by the temperature, with an expansion coefficient of $2 \times 10^{-4}\,°$C^{-1}. The barotropic and baroclinic Rossby radii are 485 and 8 km, respectively.

3   Inertial oscillations

The first case is without the presence of vertical stratification. Thus the near-inertial internal wave is absent, and the near-inertial motion is a pure inertial oscillation.

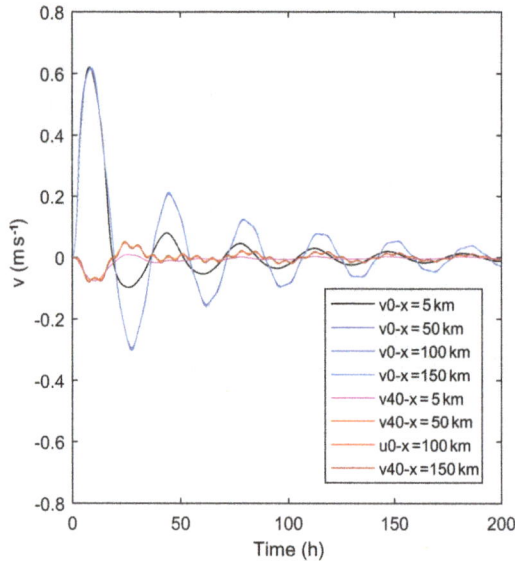

Figure 4. Time series of the northward velocity (v) at different depths and positions. "v0" and "v40" mean v at depths of 0 and 40 m.

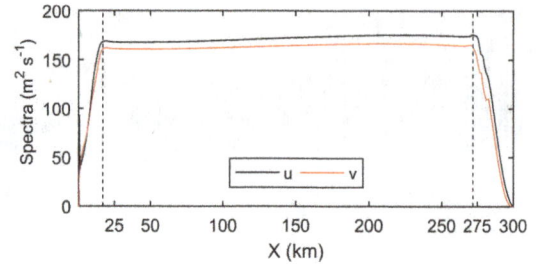

Figure 5. Spatial variation of depth-mean near-inertial spectra of velocities for the homogeneous case.

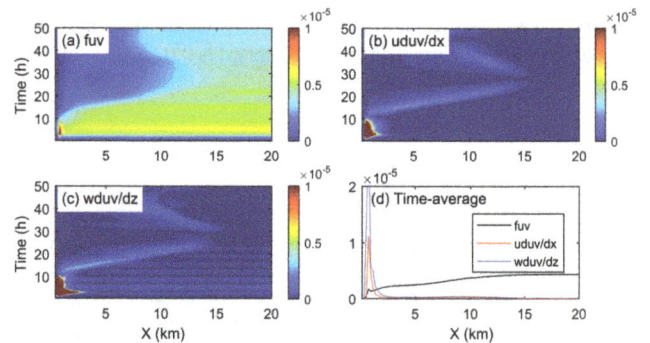

Figure 6. Variation of depth-mean inertial and non-linear terms ($m\,s^{-2}$). The inertial term **(a)** is calculated as $|f(u+iv)|$, the horizontal non-linear term **(b)** is $|u\,(\partial u/\partial x + i\partial v/\partial x)|$, and the vertical non-linear term **(c)** is $|w\,(\partial u/\partial z + i\partial v/\partial z)|$. **(d)** Time-averaged value for the first 50 h.

3.1 Vertical structures

The model simulated velocities (Fig. 1) vary near the inertial period (34.9 h). Spectra of velocities (not shown) indicate maximum peaks located exactly at the inertial period. The spectra of u also have a smaller peak at the period of the first mode seiche (6.9 h). As this simulation is two-dimensional, i.e. the gradient along the y-axis is zero, the u/v of seiches have a value of ω_n/f (equals 5 for the first mode seiche). Thus there is little energy of seiches in v, which shows clearly regular variation at the inertial frequency.

In the vertical direction, currents display a two-layer structure, with their phase being opposite between the surface and lower layers. They are maximum at the surface, and have a weaker maximum in the lower layer (~ 40 m), with a minimum at a depth of ~ 20 m. The velocity gradually diminishes to zero at the bottom due to the bottom friction. This is the typical vertical structure of shelf sea inertial oscillations, which have been frequently observed (Shearman, 2005; MacKinnon and Gregg, 2005). In practice, this vertical distribution can be modified due to the presence of other processes, such as the surface maximum being pushed down to the subsurface (e.g. Chen et al., 2015a). Note that without stratification in this simulation the near-inertial internal wave is absent. However, this two-layer structure of inertial oscillations looks 'baroclinic', which makes it easy to be mistakenly attributed to the near-inertial internal wave (Pettigrew, 1981).

It is interesting that currents of non-baroclinic inertial oscillations reverse between the surface and lower layers. This is usually due to the presence of the coast, which requires the normal-to-coast transport to be zero; thus, currents in the up-

per and lower layers compensate each other (e.g. Millot and Crepon, 1981; Chen et al., 1996). However, it is interesting to see how this vertical structure is established step by step.

As the westward wind blows for the first 3 h, the initial inertial current is also westward and only exists in the surface layer (Fig. 2). In the lower layer there is no movement initially. Thus a westward transport is produced, which generates a rise (in the west) and fall (in the east) in elevation near land boundaries. The elevation slope behaves in the form of a barotropic wave which propagates offshore at a large speed (87 km h^{-1}). The current driven by the barotropic wave is eastward, and uniform vertically. Therefore, with the arrival of the barotropic wave the westward current in the surface is reduced, and the eastward movement in the lower layer commences (Fig. 2). After the passage of the first two barotropic waves (originating from both sides), currents in the lower layer have reached a relatively large value, while currents in the surface layer have largely decreased (Fig. 3a). Accordingly, the depth-integrated transport diminishes significantly. This is a feedback between inertial currents and barotropic waves. If only the depth-integrated transport of currents exists, barotropic waves will be generated, which reduce the surface currents but increase the lower layer currents and thus reduce the current transport. It will end up with iner-

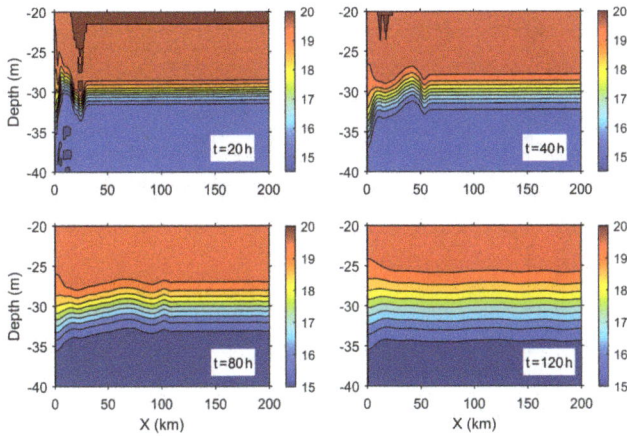

Figure 7. Snapshots of temperature profiles at $t = 20$, 40, 80, and 120 h. The contour interval is 0.5 °C.

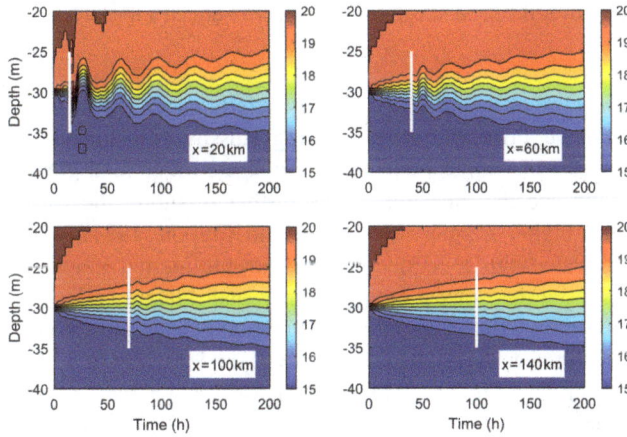

Figure 8. Time series of temperature at $x = 20$, 60, 100, and 140 km. White lines denote the arrival of internal waves. The contour interval is 0.5 °C.

tial currents in the surface and lower layers having opposite directions and comparable amplitudes. As seen from Fig. 1b, the typical vertical structure of inertial currents is established within the first inertial period.

3.2 Horizontal distributions of inertial energy

The inertial velocities are almost entirely the same across the basin (Fig. 4), except near the land boundaries. This indicates that inertial oscillations have a coherence scale of almost the basin width. This is because in our simulation the wind force is spatially uniform, and the bottom is flat. The inertial velocities in the lower layer have slightly more variation across the basin than those in the surface layer, because inertial velocities in the lower layer depend on the propagation of barotropic waves as discussed in Sect. 3.1, while the surface inertial currents are driven by spatially uniform wind. In shelf sea regions, the wind forcing is usually coherent as the synoptic scale is much larger; however, the topography

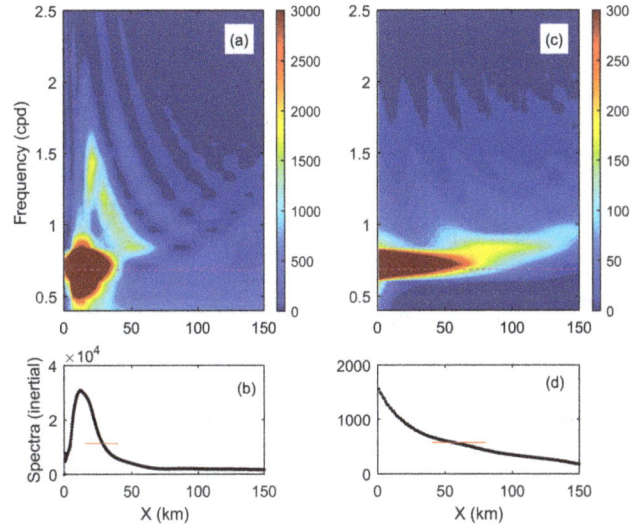

Figure 9. (a) Spectra of the temperature at the mid-depth ($z = -30$ m). The pink dashed line represents the inertial frequency. (b) Sum of spectra in the inertial band with a red line denoting the e-folding value of the peak. (c) Theoretical spectra of mid-depth elevation calculated from the solution in the form of a Bessel function as in Eq. (3.16) of Pettigrew (1981). (d) Same as (b) but for theoretical spectra.

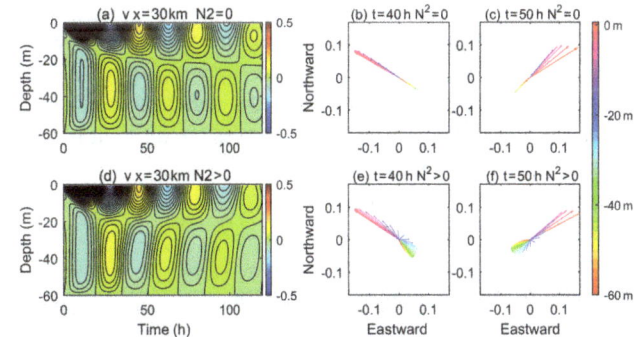

Figure 10. Distribution of near-inertial currents (v, m s^{-1}) and current spirals for the cases without (**a, b, c**) and with (**d, e, f**) stratification at $x = 30$ km. The near-inertial currents are obtained by applying a band-pass filter. The contour interval is 0.02 m s^{-1}.

that is mostly not flat could generate barotropic waves at various places and thus significantly decrease coherence of inertial currents in the lower layer.

The spectra of velocities in the inertial band are almost uniform except near the land boundaries (Fig. 5), consistent with the velocities. Near the boundaries, the inertial energy declines gradually to zero from $x = \sim 20$ km to the land. The eastern side has slightly greater inertial energy and a slightly wider boundary layer compared to the western side.

We calculate the non-linear and inertial terms in the momentum equation and find that non-linear terms are of significantly high value initially within 2 km away from the land

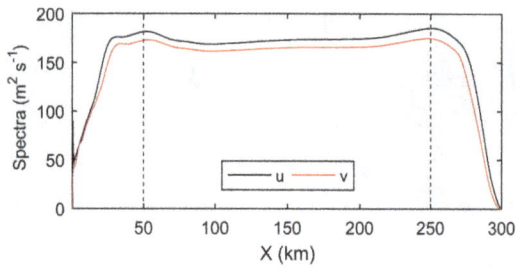

Figure 11. Spatial variation of depth-mean near-inertial spectra of velocities for the stratified case.

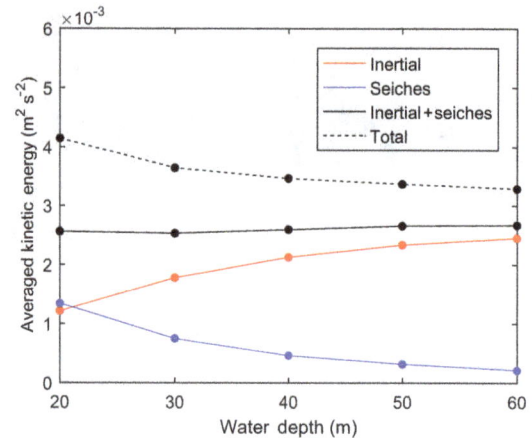

Figure 12. The kinetic energy of near-inertial motions and seiches for different water depths. For each case, the currents are band-pass filtered to get currents for each type of motion which are then averaged over time and integrated over space to obtain a final value.

boundary (Fig. 6bc), where the inertial term is weak (Fig. 6a). For the time-averaged values (Fig. 6d), the vertical non-linear term is 2 times more than the horizontal non-linear term. The inertial term drops sharply near the boundary, and rises gradually with distance away from the boundary. At $x > 15$ km, it keeps an almost constant value which is much greater than non-linear terms. Thus it is concluded that the significant decrease in inertial oscillations near the boundary is due to the influence of non-linear terms, especially the vertical non-linear term.

4 Near-inertial internal waves

In addition to inertial oscillations, near-inertial internal waves are usually generated when the vertical stratification is present. However, due to their close frequencies, inertial oscillations and near-inertial internal waves are difficult to separate. Thus we run a second simulation with the presence of stratification to investigate the differences that near-inertial internal waves introduce.

4.1 Temperature distributions

Figure 7 shows the evolution of temperature profiles with time. One can see an internal wave packet is generated at the western coast and then propagates offshore. The wave phase speed is about 1 km h^{-1}, close to the theoretical value (1.4 km h^{-1}). Before the arrival of internal waves, the temperature at mid-depth diffuses gradually due to vertical diffusion in the model. For a fixed position at $x = 20$ km (Fig. 8), the temperature varies with the inertial period (34.9 h) and the amplitude of fluctuation declines gradually with time. At $x = 60$ km and $x = 100$ km, the strength of internal waves is much reduced, and wave periods are shorter initially, followed by a gradual increase to the inertial period. At $x = 140$ km, the internal wave becomes as weak as the background disturbance.

A spectral analysis of the temperature at mid-depth ($z = -30$ m) is shown in Fig. 9a. The strongest peak is near the inertial frequency (0.69 cpd), but is only confined to the region close to the boundary ($x < 40$ km). In the region 20 km $< x < 70$ km, the energy is also large at higher fre-

quencies of 0.8–1.7 cpd. This generally agrees with properties of Poincaré waves. During Rossby adjustment, the waves with higher frequencies propagate offshore at greater group speeds; thus, for places further offshore, the waves have higher frequencies (Millot and Crepon, 1981), while the wave with a frequency closest to the inertial frequency moves at the slowest group velocity, and it takes a relatively long time to propagate far offshore; thus, it is mostly confined to near the boundary. By solving an idealized two-layer model equation, the response of Rossby adjustment can be expressed in the form of Bessel functions (Millot and Crepon, 1981; Gill, 1982; Pettigrew, 1981), as in Fig. 9cd showing the spectra of mid-depth elevation. The difference from our case is obvious. The frequency of theoretical near-inertial waves increases gradually with distance from the coast, while in our case this property is absent. And the theoretical inertial energy has an e-folding scale of 54 km, while in our case the e-folding scale is much smaller (~ 15 km).

4.2 Velocity distributions

With the presence of near-inertial internal waves, the contours of velocities near the thermocline tilt slightly (Fig. 10d), and indicate an upward propagation of phase and thus a downward energy flux. This can also be seen in vertical spirals of velocities (Fig. 10e and f). With only inertial oscillations, current vectors mostly point toward two opposite directions (Fig. 10b and c). Once the near-inertial wave is included, the current vectors gradually rotate clockwise with depth.

The spatial distribution of the near-inertial energy is also slightly changed compared to the case with only inertial oscillations (Figs. 11 and 5). It is also greatly reduced to zero in the boundary layer (0–20 km) like the case without stratification. But at ~ 50 km away from the boundary the inertial

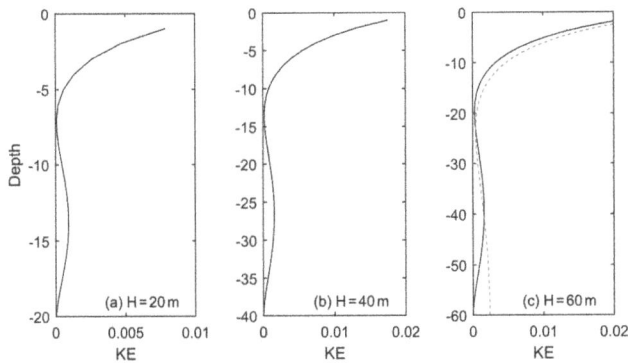

Figure 13. Vertical profile of averaged inertial kinetic energy for the homogeneous cases with water depths of 20, 40, and 60 m. The red dashed line in **(c)** denotes the slip case.

energy reaches a peak. Further away (> 100 km) it becomes a constant. This spatial distribution of inertial energy is similar to that observed in shelf seas, with a maximum near the shelf break (Chen et al., 1996; Shearman, 2005). In our case, the boundary layer effect which induces a sharp decrease to zero makes a major contribution, and near-inertial internal waves which bring a small peak further offshore have a secondary influence.

5 Dependence on the water depth

In coastal regions, near-inertial motions are rarely reported. It is speculated that the strong dissipation and bottom friction in coastal regions suppress the development of near-inertial motions. However, Chen (2014) found the water depth is also a sensitive factor, with significant reduction for the case with smaller water depth. Here we will run cases with different water depths and clarify why the near-inertial energy changes with water depth. Homogeneous cases with water depths of 50, 40, 30, and 20 m are simulated. The vertical model resolution for all cases is 1 m. All the other parameters including viscosities are the same as the homogeneous case of 60 m.

For each case, the currents are band-pass filtered to obtain near-inertial currents. Then near-inertial kinetic energy can be calculated. As seen in Fig. 12, the near-inertial energy gradually declines with decreasing water depth. In this dynamical system, the other dominant process is the seiche induced by barotropic waves. As the elevation induced by seiches is anti-symmetric in such a basin, the potential energy is little. The kinetic energy of seiches can also be calculated by the band-pass filtered currents. We find the energy of seiches, by contrast, increases gradually with decreasing water depth. For the case of 60 m, the near-inertial energy is much greater than the seiche energy. But for the case of 20 m, the energy of seiches has exceeded the near-inertial energy slightly. The total energy of these two processes almost stays constant for all cases. For a shallower water depth, the re-

duction of near-inertial energy equals the increase in seiche energy. The initial current is wind-driven and only distributes in the surface layer. The unbalanced across-shelf flow generates elevation near the land boundary which propagates offshore as barotropic waves and forms seiches. Part of the energy goes to form inertial oscillations. For a shallower water depth, the elevation is enlarged, and more energy is transferred to form seiches and thus with weakened near-inertial motions. Therefore, in coastal regions with water depths less than 30 m, the near-inertial motion is weak, due to the suppression of barotropic waves.

As seen in Sect. 3.1, inertial oscillations behave in a two-layer structure, with currents in the upper layer in opposite phase with those of the lower layer. In terms of kinetic energy, for the case of 60 m (Fig. 13), the near-inertial motion is maximized in the surface, minimized near the depth of 20 m, and then gradually increases with depth to form a much smaller peak at 40 m. Near the bottom, the near-inertial energy gradually reduces to zero due to bottom friction. When we set the bottom boundary condition from non-slip to slip, such a boundary structure vanishes, and near-inertial energy becomes constant in the lower layer. For the other cases of 20 and 40 m, their vertical profiles are almost the same as the 60 m case. The minimum positions are all located at 1/3 of the water depth. This implies the vertical distribution of near-inertial energy is independent of water depth. Note that in our cases the vertical viscosity is set as a constant value. In practice, the viscosity in the thermocline is usually significantly reduced; thus, the minimum position of near-inertial energy is located just below the mixed layer.

6 Summary and discussion

Idealized simple two-dimensional $(x - z)$ simulations are conducted to examine the response of a shallow closed basin to a wind pulse. The first case is homogeneous, in which the near-inertial motion is a pure inertial oscillation. It has a two-layer structure, with currents in the surface and lower layers being opposite in phase, which has been reported frequently in shelf seas. We find that the inertial current is confined in the surface layer initially. The induced depth-integrated transport generates barotropic waves near the boundaries which propagate quickly offshore. The flow driven by the barotropic wave is independent of depth and opposite to the surface flow. Thus the surface flow is reduced but the flow in the lower layer is increased; as a result, the transport diminishes. This feedback between barotropic waves and currents continues and ends up with the depth-integrated transport vanishing, i.e. inertial currents in the upper and lower layers having opposite phases and comparable amplitudes. In our simulation, within just one inertial period the typical structure of inertial currents has been established. By solving a two-layer analytic model using the Laplace transform, Pettigrew (1981) also found the vertical structure of opposite currents to be associated with inertial oscillations. He argued that the arrival of a barotropic wave for a fixed location cancels half of the inertial oscillation in the surface layer, and initiates an equal and opposite oscillation in the lower layer. However, in our simulation the arrival of the first barotropic wave cannot cancel half of the surface flow. The balanced state of upper and lower flows takes more time to reach.

The second case is a set-up with idealized two-layer stratification; thus, near-inertial internal waves are generated. For a fixed position, velocity contours show obvious tilting near the thermocline, and velocity vectors display clearly anticyclonic spirals with depth. These could be useful clues to examine the occurrence of near-inertial internal waves. Near the land boundary the vertical elevation generates fluctuations of the thermocline that propagate offshore. The energy of near-inertial internal waves is confined to near the land boundary ($x < 40$ km). At positions further offshore, the waves have higher frequencies. This is generally consistent with properties of a Rossby adjustment process. However, our simulated results also show evident discrepancies with theoretical values obtained in the classic solutions of the Rossby adjustment problem. These discrepancies are probably due to non-linearity of the model and the changing stratification in the model due to diffusion and mixing, compared to constant density differences between the two layers in theoretical cases.

The inertial oscillation has a very large coherent scale of almost the whole basin scale. It is uniform in both amplitude and phase across the basin, except near the boundary (~ 20 km offshore). The energy of inertial oscillations declines gradually to zero from $x = 20$ km to the coast. This boundary effect is attributed to the influence of nonlinear terms, especially the vertical term ($w\partial u/\partial z$), which is greatly enhanced near the boundary and overweighs the inertial term (fu). When near-inertial internal waves are produced in the stratified case, the distribution of total near-inertial energy is modified slightly near the boundary. A small peak appears at ~ 50 km offshore. This is similar to the cross-shelf distribution of near-inertial energy observed in shelf seas (Chen et al., 1996; Shearman, 2005). This energy distribution has been attributed to downward and offshore leakage of near-inertial energy near the coast (Kundu et al., 1983), the variation of elevation and Reynolds stress terms associated with the topography (Chen and Xie, 1997), and the influence of the baroclinic wave (Shearman, 2005; Nicholls et al., 2012). In our simulations, this horizontal distribution of near-inertial energy is primarily controlled by the boundary effect on inertial oscillations, and the near-inertial internal wave has a secondary effect.

Homogeneous cases with various water depths (50, 40, 30, and 20 m) are also simulated. The inertial energy is reduced with decreasing water depth, while the energy of seiches, by contrast, is increased. For the case of 20 m, the seiche energy slightly exceeds the inertial energy. It is interesting that the reduction of inertial energy just equals the increase in the seiche energy, which implies more energy of initial wind-driven currents is transferred to the seiches for the shallower cases, and thus less energy goes to the inertial process. This is probably an important reason why near-inertial motions are weak and rarely reported in shallow coastal regions.

Competing interests. The authors declare that they have no conflict of interest.

Acknowledgements. We are grateful for discussions with John Huthnance and comments from the editor and reviewers. This study is supported by the National Basic Research Program of China (2014CB745002, 2015CB954004), the Shenzhen government (201510150880, SZHY2014-B01-001), and the Natural Science Foundation of China (U1405233). Shengli Chen is sponsored by the China Postdoctoral Science Foundation (2016M591159).

Edited by: Neil Wells

References

Alford, M. H., Cronin, M. F., and Klymak, J. M.: Annual Cycle and Depth Penetration of Wind-Generated Near-Inertial Internal Waves at Ocean Station Papa in the Northeast Pacific, J. Phys. Oceanogr., 42, 889–909, https://doi.org/10.1175/jpo-d-11-092.1, 2012.

Alford, M. H., MacKinnon, J. A., Simmons, H. L., and Nash, J. D.: Near-Inertial Internal Gravity Waves in the Ocean, Ann. Rev. Mar. Sci., 8, 95–123, https://doi.org/10.1146/annurev-marine-010814-015746, 2016.

Burchard, H. and Rippeth, T. P.: Generation of Bulk Shear Spikes in Shallow Stratified Tidal Seas, J. Phys. Oceanogr., 39, 969–985, 10.1175/2008jpo4074.1, 2009.

Chen, C. and Xie, L.: A numerical study of wind-induced, near-inertial oscillations over the Texas-Louisiana shelf, J. Geophys. Res.-Oceans, 102, 15583–15593, https://doi.org/10.1029/97jc00228, 1997.

Chen, C. S., Reid, R. O., and Nowlin, W. D.: Near-inertial oscillations over the Texas Louisiana shelf, J. Geophys. Res.-Oceans, 101, 3509–3524, https://doi.org/10.1029/95jc03395, 1996.

Chen, S.: Study on Several Features of the Near-inertial Motion, PhD, Xiamen University, 107 pp., 2014.

Chen, S., Hu, J., and Polton, J. A.: Features of near-inertial motions observed on the northern South China Sea shelf during the passage of two typhoons, Acta Oceanol. Sin., 34, 38–43, https://doi.org/10.1007/s13131-015-0594-y, 2015a.

Chen, S., Polton, J. A., Hu, J., and Xing, J.: Local inertial oscillations in the surface ocean generated by time-varying winds, Ocean Dynam., 65, 1633–1641, https://doi.org/10.1007/s10236-015-0899-6, 2015b.

Chen, S., Polton, J. A., Hu, J., and Xing, J.: Thermocline bulk shear analysis in the northern North Sea, Ocean Dynam., 66, 499–508, https://doi.org/10.1007/s10236-016-0933-3, 2016.

D'Asaro, E. A.: The energy flux from the wind to near-inertial motions in the surface mixed layer, J. Phys. Oceanogr., 15, 1043–1059, 1985.

Fu, L. L.: Observations and models of inertial waves in the deep ocean, Rev. Geophys., 19, 141–170, https://doi.org/10.1029/RG019i001p00141, 1981.

Gill, A. E.: Atmosphere-ocean dynamics, Academic Press, 662 pp., 1982.

Gill, A. E.: On the behavior of internal waves in the wakes of storms, J. Phys. Oceanogr., 14, 1129–1151, https://doi.org/10.1175/1520-0485(1984)014<1129:otboiw>2.0.co;2, 1984.

Kundu, P. K., Chao, S. Y., and McCreary, J. P.: Transient coastal currents and inertio-gravity waves, Deep-Sea Res. Pt. I, 30, 1059–1082, https://doi.org/10.1016/0198-0149(83)90061-4, 1983.

Leaman, K. D. and Sanford, T. B.: Vertical energy propagation of inertial waves: a vector spectral analysis of velocity profiles, J. Geophys. Res., 80, 1975–1978, 1975.

MacKinnon, J. A. and Gregg, M. C.: Near-inertial waves on the New England shelf: The role of evolving stratification, turbulent dissipation, and bottom drag, J. Phys. Oceanogr., 35, 2408–2424, https://doi.org/10.1175/jpo2822.1, 2005.

Malone, F. D.: An analysis of current measurements in Lake Michigan, J. Geophys. Res., 73, 7065–7081, https://doi.org/10.1029/JB073i022p07065, 1968.

Marshall, J., Adcroft, A., Hill, C., Perelman, L., and Heisey, C.: A finite-volume, incompressible Navier Stokes model for studies of the ocean on parallel computers, J. Geophys. Res.-Oceans, 102, 5753–5766, https://doi.org/10.1029/96jc02775, 1997.

Millot, C. and Crepon, M.: Inertial oscillations on the continental-shelf of the Gulf of Lions – observations and theory, J. Phys. Oceanogr., 11, 639–657, https://doi.org/10.1175/1520-0485(1981)011<0639:iootcs>2.0.co;2, 1981.

Nicholls, J. F., Toumi, R., and Budgell, W. P.: Inertial currents in the Caspian Sea, Geophys. Res. Lett., 39, L18603, https://doi.org/10.1029/2012gl052989, 2012.

Pettigrew, N. R.: The dynamics and kinematics of the coastal boundary layer off Long Island, 1981.

Pollard, R. T. and Millard, R. C.: Comparison between ovserved and simulated wind-generated inertial oscillations, Deep-Sea Res., 17, 813–821, 1970.

Price, J. F.: Upper ocean response to a hurricane, J. Phys. Oceanogr., 11, 153–175, https://doi.org/10.1175/1520-0485(1981)011<0153:uortah>2.0.co;2, 1981.

Shearman, R. K.: Observations of near-inertial current variability on the New England shelf, J. Geophys. Res., 110, C02012, https://doi.org/10.1029/2004jc002341, 2005.

Sun, Z., Hu, J., Zheng, Q., and Li, C.: Strong near-inertial oscillations in geostrophic shear in the northern South China Sea, J. Oceanogr., 67, 377–384, https://doi.org/10.1007/s10872-011-0038-z, 2011.

Webster, F.: Observation of inertial period motions in the deep sea, Rev. Geophys., 6, 473–490, 1968.

Xing, J. X., Davies, A. M., and Fraunie, P.: Model studies of near-inertial motion on the continental shelf off northeast Spain: A three-dimensional two-dimensional one-dimensional model comparison study, J. Geophys. Res.-Oceans, 109, C01017, https://doi.org/10.1029/2003jc001822, 2004.

Zheng, Q., Lai, R. J., and Huang, N. E.: Observation of ocean current response to 1998 Hurricane Georges in the Gulf of Mexico, Acta Oceanol. Sin., 25, 1–14, 2006.

13

On the meridional ageostrophic transport in the tropical Atlantic

Yao Fu[1], Johannes Karstensen[1], and Peter Brandt[1,2]

[1]GEOMAR Helmholtz Centre for Ocean Research Kiel, Kiel, Germany
[2]Christian-Albrechts-Universität zu Kiel, Kiel, Germany

Correspondence to: Yao Fu (yfu@geomar.de)

Abstract. The meridional Ekman volume, heat, and salt transport across two trans-Atlantic sections near 14.5° N and 11° S were estimated using in situ observations, wind products, and model data. A meridional ageostrophic velocity was obtained as the difference between the directly measured total velocity and the geostrophic velocity derived from observations. Interpreting the section mean ageostrophy to be the result of an Ekman balance, the meridional Ekman transport of 6.2 ± 2.3 Sv northward at 14.5° N and 11.7 ± 2.1 Sv southward at 11° S is estimated. The integration uses the top of the pycnocline as an approximation for the Ekman depth, which is on average about 20 m deeper than the mixed layer depth. The Ekman transport estimated based on the velocity observations agrees well with the predictions from in situ wind stress data of 6.7 ± 3.5 Sv at 14.5° N and 13.6 ± 3.3 Sv at 11° S. The meridional Ekman heat and salt fluxes calculated from sea surface temperature and salinity data or from high-resolution temperature and salinity profile data differ only marginally. The errors in the Ekman heat and salt flux calculation were dominated by the uncertainty of the Ekman volume transport estimates.

1 Introduction

In the tropical Atlantic Ocean, strong and steady easterly trade winds generate a poleward meridional flow in the surface layer. According to the classical linear theory of Ekman (1905), under the momentum balance between steady wind stress and Coriolis force, the wind-driven flow spirals clockwise with depth, the Ekman spiral, while the vertical integration of the spiral results in a net volume transport to the right of the wind direction (Northern Hemisphere), the Ekman transport. A convergence is created in the subtropics, where the poleward Ekman transport induced by the trade winds interacts with the equatorward Ekman transport induced by the mid-latitude westerlies. In simple linear vorticity theory, the Ekman convergence in subtropics drives an equatorward Sverdrup transport that explains many aspects of the wind-driven gyre circulation, such as the subtropical cells (STC). Schott et al. (2004) calculated the Ekman divergence (21–24 Sv, $1\,\mathrm{Sv} = 10^6\,\mathrm{m}^3\,\mathrm{s}^{-1}$) between 10° N and 10° S in the tropical Atlantic using climatological wind to infer the strength of the STC; Rabe et al. (2008) further analysed the variability of the STC using the same sections based on assimilation products, and found that on timescales longer than 5 years to decadal, the variability of poleward Ekman divergence leads the variability of geostrophic convergence in the thermocline.

The meridional Ekman transport is, depending on the latitude, an important upper layer contribution when estimating the strength of the Meridional Overturning Circulation (MOC, Friedrichs and Hall, 1993; Klein et al., 1995; Wijffels et al., 1996). The variations in the meridional Ekman transport have been found to cause barotropic adjustment of the MOC in the ocean interior on different timescales. Cunningham et al. (2007) reported that the upper ocean had an immediate response to the changes in Ekman transport at subseasonal to seasonal timescales, while Kanzow et al. (2010) found that on the seasonal timescale, the Ekman transport was less important than the mid-ocean geostrophic transport, whose seasonal variation was dominated by the seasonal cycle of the wind stress curl. McCarthy et al. (2012) analysed a low MOC case during 2009 and 2010, and also pointed out that on the interannual timescale, although the Ekman transport played a role, its variability was relatively small compared to the variability in mid-ocean geostrophic transport, especially in the upper 1100 m.

Of interest for large-scale overturning studies are also the meridional Ekman-driven heat and freshwater fluxes that provide an important upper layer constraint, for example, for geostrophic end point arrays (McCarthy et al., 2015; McDonagh et al., 2015). In many cases, sea surface temperature (SST) has been found to be a sufficient constraint for the Ekman layer temperature (Wijffels et al., 1994; Chereskin et al., 2002). This probably is not too much of a surprise as the heat flux is primarily determined by the transport and less by the relatively small variability in temperature. However, the unresolved vertical structure of the water column could lead to an unknown bias, for example, due to the difference between the mixed layer depth (MLD) and the depth of the Ekman layer. An extreme case has been reported for the northern Indian Ocean at 8° N at the end of a summer monsoon event (Chereskin et al., 2002), where the direct Ekman temperature transport was 5 % smaller when using the temperature within the top of the pycnocline (TTP, as a proxy of the Ekman layer depth) than using the SST, and the corresponding mean temperature in the Ekman layer was 1.1 °C cooler than the averaged SST. In this case, the mean TTP depth was 92 m deeper than the mean MLD.

Assuming the upper layer ageostrophic flow in Ekman balance, the meridional Ekman transport (M_E^y) can be estimated indirectly from zonal wind stress data or directly from integrating observed ageostrophic Ekman velocity (v_E):

$$M_E^y = \frac{1}{\rho}\frac{\tau_x}{f} = -\int_{-D_E}^{0} v_E \, dz, \quad (1)$$

where τ_x is the zonal wind stress, ρ is the density of seawater, f is the Coriolis parameter of the respective latitude, D_E is the Ekman depth, and z is the upward vertical coordinate. D_E can be defined as the e-folding-scale depth of the Ekman spiral, leading to an analytical solution of $D_E = \sqrt{\frac{2A_v}{f}}$, where A_v is a constant vertical eddy viscosity (Price et al., 1987). Ekman's solution also reveals a surface Ekman velocity $V_0 = \frac{\tau}{\sqrt{\rho^2 f A_v}}$, which is 45° to the right (left) of the wind direction in the Northern (Southern) Hemisphere.

An ageostrophic velocity (v_{ageos}) can be calculated as the difference of the directly observed velocity (v_{obs}) and the geostrophic velocity (v_{geos}). The ageostrophic velocity might consist of an Ekman component (v_E) and components that are not in Ekman balance (e.g. inertial currents). Often the non-Ekman components are assumed to be 0, and v_E is expected to equal v_{ageos}. Under this assumption, the Ekman velocity can be derived as follows:

$$v_E = v_{obs} - v_{geos}. \quad (2)$$

Direct velocity profile data, for example from ADCP, and geostrophic velocities, from hydrographic data, are used in studies comparing direct with indirect Ekman transport estimates (e.g. Chereskin and Roemmich, 1991; Wijffels et al., 1994; Garzoli and Molinari, 2001). The Ekman transport is then derived from vertical integration of the v_E.

For both equations it is relevant to recall that the Ekman balance is derived for an ocean with constant vertical viscosity and infinite depth, forced by a steady wind field (Ekman, 1905). Such conditions are not found in the real ocean; therefore, applications of the indirect (Eq. 1) and direct (Eq. 2) approaches suffer from different kinds of errors. For the indirect approach (Eq. 1) the temporally varying wind field, the momentum flux calculated from the wind speed, and the unknown partitioning of the wind energy input into the Ekman layer at different frequency bands are probably the most important sources of errors introduced into any Ekman current/transport estimate. For the direct approach, unknown lower integration depth, momentum flux variability, errors introduced by the experimental design (e.g. an shipboard ADCP does not resolve the upper 10–20 m of the flow, which is often assumed equal to the values at the first valid bin) or instrument errors can impact obtained results.

Many observational studies on Ekman dynamics that compare indirect and direct approaches have been conducted in the trade wind regions, where at least the wind stress forcing is relatively constant. Using shipboard ADCP data together with conductivity–temperature–depth (CTD) profile data, Chereskin and Roemmich (1991) directly estimated an Ekman transport of 9.3±5.5 Sv at 11° N in the Atlantic by integrating an ageostrophic velocity from the surface to a depth equivalent to TTP. The ageostrophic velocity was obtained by subtracting the geostrophic velocity from the ADCP velocity. Using a similar direct method, Wijffels et al. (1994) estimated an ageostrophic transport of 50.8 ± 10 Sv at 10° N in the Pacific. Chereskin et al. (1997) found Ekman transports of −17.6 ± 2.4 and −7.9 ± 2.7 Sv during and after a southwest monsoon event at 8.5° N in the Indian Ocean, respectively. In all the above studies, the direct estimates agree within 10–20% of the estimates obtained by using the in situ wind data (Eq. 1). Both the direct and indirect approaches also show a consistent transport structure across all the basins, which can be seen from the cumulative meridional Ekman transport curves from one boundary to the other. An indication of the existence of an Ekman balance in the upper ocean is the occurrence of an Ekman spiral. In all the above publications an "Ekman spiral"-like feature has been identified. Because v_{geos} can be estimated only perpendicularly to the CTD stations and all studies are based on more or less zonal CTD sections, the three-dimensional structure of the Ekman spiral can not be obtained. However, the Ekman flow becomes evident by a near-surface maximum of the meridional ageostrophic velocity decreasing smoothly below within the upper 50–100 m to zero.

Despite the fact that the zonal wind in the above studies was predominantly uniform in one direction, their ageostrophic velocity showed a pattern of alternating currents. Also, the section-averaged ageostrophic velocity profiles often exhibited structures that are not a result of an Ekman balance. Chereskin and Roemmich (1991) reported signals of internal wave propagation that was responsible for a peak in their section-integrated ageostrophic transport profile below the Ekman layer. Garzoli and Molinari (2001) also reported on vertically alternating structures in the section-averaged ageostrophic velocity profile at 6° N in the Atlantic. They proposed several possible candidates that could contribute to creating this structure, such as inertial currents within the latitude range of the North Equatorial Counter Current (NECC), and tropical instability waves with northward and southward velocities. Besides, they argued that the advective terms in the momentum equations might also produce a large non-Ekman ageostrophic transport in the presence of large horizontal shears between the NECC and the northern branch of the South Equatorial Current (nSEC).

The appearance of these non-Ekman ageostrophic currents is not surprising, since it has been long recognized that the temporal variability of the wind field leads to wind energy input into the Ekman layer at subinertial and near-inertial frequencies. Wang and Huang (2004) estimated the global wind energy input into the Ekman layer at subinertial frequencies (frequency lower than 0.5 cycles per day) to be 2.4 TW, while Watanabe and Hibiya (2002) and Alford (2003) estimated that at near-inertial frequencies the wind energy input was 0.7 and 0.5 TW, respectively. Elipot and Gille (2009) estimated the wind energy input into the Ekman layer for the frequency range between 0 and 2 cpd at 41° S in the Southern Ocean using surface drifter data. They found that the near-inertial input (between $0.5 f$ and 2 cpd) contributes 8 % of the total wind energy input (here the "total" means the frequency range between 0 and 2 cpd), which may still underestimate the near-inertial contribution due to limitations in their data. All these studies suggest that at least about 10 % of the wind energy (frequency range between 0 and 2 cpd) into the Ekman layer is at near-inertial frequencies, which is used to supply the non-Ekman ageostrophic motions (inertial oscillation, near-inertial internal waves, etc.). Therefore, complicated structures in the directly observed ageostrophic velocity as reported by Chereskin and Roemmich (1991) and Garzoli and Molinari (2001) can be anticipated.

The purpose of the present study is to estimate the Ekman volume, heat, and freshwater transport across two trans-Atlantic sections nominally along 14.5° N and 11° S by using direct and indirect methods, and to analyse the vertical structure of the ageostrophic flow by using high-resolution velocity and hydrographic data. In previous studies, the geostrophic velocity was estimated using CTD profile data with a station spacing of approximately 30–60 nm, and only in situ and climatological wind data were available. In this study, we apply the recently introduced underway-CTD (uCTD), which allows profiling with denser station spacing of about 8–10 nm or less and does not require additional station time by measuring from moving ships (e.g. volunteer commercial and research vessels). We first describe the processing of the uCTD data in detail, and then apply the uCTD data to calculate the Ekman transport. We also test the sensitivity of the Ekman transport estimates with respect to the

CTD profile resolution. We then apply wind data from different sources to indirectly estimate the Ekman transport, including the in situ (ship) wind, satellite-based wind product, and reanalysis wind products. In order to integrate the observation-based Ekman transport estimates into the large-scale tropical Atlantic context, we compared our results with the GECCO2 ocean synthesis data. This work is structured as follows: the processing of the data is described in Sect. 2. The methods used in the calculation of Ekman volume, heat, and salt transport are described in Sect. 3. The vertical and horizontal structures of the ageostrophic velocity, together with the Ekman volume, heat, and salt transport estimated using different datasets and different methods are presented and discussed in Sect. 4, followed by a summary in Sect. 5.

2 Data

Two trans-Atlantic zonal sections near 14.5° N and 11° S were occupied by R/V *Meteor* on three cruise legs (M96, M97, and M98). The 14.5° N section began with cruise M96 off the coast of Trinidad and Tobago on 28 April 2013. The section ended on M96 at about 20° W on 20 May, and was continued to the African coast during M97 from 8 to 9 June (Fig. 1). During these surveys, 64 CTD stations were conducted along the 14.5° N section, with an average spacing of 40 nm (75 km). Parallel to the CTD system, the uCTD system was operated between the adjacent CTD stations when the ship was steaming at 10–12 kn. In total, 317 uCTD profiles were achieved, with an average spacing of 8 nm (15 km). The 11° S section was surveyed during M98 from 6 to 23 July 2013. In this section, the standard CTD was only operated on the shelf and at the shelf break; during the transit across the Atlantic, only the uCTD was in use. All together, 290 uCTD profiles were taken during the survey, with an average spacing of 11 nm (20 km). Shipboard ADCP and anemometer were in continuous operation through the entire cruises.

2.1 CTD and uCTD measurements

The CTD work was carried out with a Sea-Bird Electronic (SBE) 9 plus CTD system. The two temperature sensors were calibrated at the manufacturer just before cruise M96 in March 2013. The conductivity measurements were calibrated by comparing the bottle stop data with salinometer measurements of bottle samples. All CTD system quality control procedures followed the GO-SHIP recommendations (Hood et al., 2010). The accuracy of the CTD data was estimated to be ± 0.001 °C for temperature and ± 0.002 g kg^{-1} for salinity.

The uCTD system used at both zonal sections was an Oceanscience Series II underway-CTD. It consisted of a probe, a tail, and a winch. The probe is equipped with a temperature (SBE-3F), a conductivity (SBE-4) and a pressure sensor from SBE. A tail spool reloading system allows the rope spooled on the tail to be paid out when the probe falls

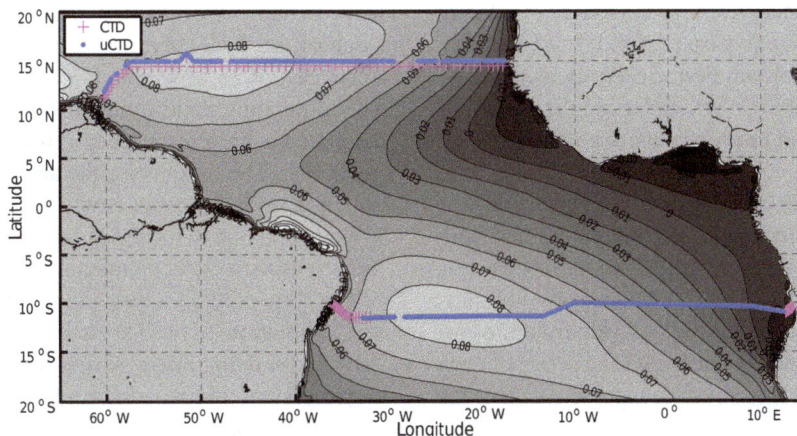

Figure 1. Positions of the CTD (magenta +) and uCTD (blue dots) measurements along the 14.5° N and 11° S sections. The 14.5° N section was completed during RV *Meteor* cruises M96 (28 April to 20 May 2013, west of 20° W) and M97 (8 to 9 June 2013, east of 20° W), the 11° S section during M98 (6 to 23 July 2013). Note that the uCTD position for the 14.5° N section is artificially shifted to the north by 0.5° for visual clarity. The grey shading with contours is the mean zonal wind stress calculated from NCEP/CFSr monthly wind stress between 1979 and 2011 in $N m^{-2}$.

Table 1. Meridional Ekman volume (M^y in Sv), heat (H_e in PW), and salt (S_e in $10^6 kg s^{-1}$) fluxes calculated using different methods, and the transport-weighted temperature Θ_E and salinity S_{AE} in the Ekman layer. Positive and negative fluxes denote northward and southward fluxes, respectively. The uncertainties of the Ekman heat and salt flux are 0.4 PW and $45 \times 10^6 kg s^{-1}$ at 14.5° N, and 0.3 PW and $65 \times 10^6 kg s^{-1}$ at 11° S, respectively. The uncertainties of the transport-weighted Ekman temperature and salinity are 0.20 °C and 0.15 g kg^{-1} at 14.5° N, and 0.11 °C and 0.10 g kg^{-1} at 11° S, respectively.

			Section									
			14.5° N					11° S				
			Θ_E	S_{AE}	M^y	H_E	S_e	Θ_E	S_{AE}	M^y	H_e	S_e
Method	Direct	TTP/profile	25.52	36.33	6.21	0.413	5.40	25.41	36.83	−11.71	−0.842	−17.69
		TTP/surface	25.61	36.34	6.21	0.415	5.49	25.41	36.80	−11.71	−0.842	−17.38
	Indirect	TTP	25.46	36.32	6.68	0.443	5.72	25.13	36.81	−13.64	−0.965	−20.50
		Surface	25.65	36.29	6.68	0.448	5.57	25.20	36.78	−13.64	−0.946	−20.04
		Annual	26.46	36.13	8.31	0.584	5.56	25.53	36.73	−11.02	−0.799	−15.49

freely. The sensors record data at a frequency of 16 Hz. For most of the profiles about 250–300 m of rope were spooled on the tail spool (which set the fall depth) and the recording time length was set to 100 s, and about 1600 data recordings per cast were obtained. From the tail spool the probe sinks freely with a nominal speed of 4 m s^{-1}. However, due to the back-and-forth unspooling of the rope from one end of the tail to the other, the sinking speed typically varies from 3 to 4.5 m s^{-1}. After the rope on the tail is paid out completely, the probe still sinks at speeds less than 2 m s^{-1} in the last tens of metres of its sinking before being winched back to the ship and recovered back to deck. Three probes were used during the two section surveys (nos. 70200126 and 70200068 along the 14.5° N section; nos. 70200068 and 70200138 along the 11° S section). The uCTD winches were out of service several times during the three cruise legs. Although they were

repaired on-board, several measurement gaps were left, for example, between 29 and 27° W (Fig. 1).

The post-calibration of the uCTD data was done in two major steps: the first step is a sensor calibration procedure, which corrects the temperature sensor error due to viscous heating, the conductivity sensor error due to thermal mass delay, and the lag between the conductivity and temperature sensors; the second step is data validation in reference to CTD profile data and to thermosalinograph (TSG) data. The first step was done following Ullman and Hebert (2014) (hereafter UH2014). We will briefly describe the process here; for details, please refer to their work. The uCTD is an unpumped CTD system, the rapid sinking speed of 4 m s^{-1} allowing water to pass through the sensor package at 3.56 m s^{-1} (UH2014). This flow rate is much higher than a pumped CTD system (1 m s^{-1}), which leads to a clear viscous heating effect of the uCTD temperature sensor. This was corrected using a steady-state result of Larson and Pedersen (1996) for the perpendicular flow case (cf. Eq. 8 of UH2014). The thermal mass correction was performed following the algorithm of Lueck and Picklo (1990) and using the mean values of error magnitude and time constant from UH2014 (cf. Table 1 of UH2014).

From the uCTD profiles alone a time lag correction was determined from cross-correlation of temperature and conductivity sensor small-scale variability. The variability was calculated by subtracting a sixth-order Butterworth low-pass filtered profile with a cut-off frequency of 4 Hz from the corresponding temperature and conductivity time series of each profile. The highest correlation was found for a $1/16$ s lag (conductivity leading), which equals the sampling frequency of 16 Hz data. Application of the lag eliminated most of the spikes in the salinity profiles when the sinking speed of the probe was above about $1.5\,\mathrm{m\,s^{-1}}$. However, when the sinking speed was below $1.5\,\mathrm{m\,s^{-1}}$, this correction would cause the spikes pointing in the opposite direction and indicates an inverse dependency of the lag on the sinking speed. This result is consistent with that reported by UH2014, and we corrected the lag following their lag model (cf. Eq. 7 of UH2014), but adjusted their parameters to match our data. The data recorded with a sinking speed smaller than $0.3\,\mathrm{m\,s^{-1}}$ were neglected (including all upcast data).

Validation of the lag corrected uCTD against CTD profile data revealed for the $14.5°$ N section a drift in the conductivity sensors of uCTD probes nos. 70200126 and 70200068. A bias correction in the sense of an absolute salinity offset (uCTD–CTD) was determined based on the temperature–salinity space (Rudnick and Klinke, 2007) by considering the conservative temperature range from 12 to 14 °C and using all uCTDs between adjacent CTD pairs. This particular temperature range was chosen because it belongs to the Atlantic central water, whose T/S relation is nearly linear, which implies that in this temperature range, the spreading of salinity measured during different uCTD casts should be tight. Besides, it was also surveyed by almost all uCTD casts along the section. For probe no. 70200126, the salinity offset fluctuates around a mean value of $0.038\,\mathrm{g\,kg^{-1}}$ west of $39°$ W (CTD station 34), east of which the offset shifts abruptly to around $0.151\,\mathrm{g\,kg^{-1}}$. The calibration was done by applying the mean offset values to the salinity data in the corresponding groups of uCTDs. The salinity data of the last few profiles of probe no. 70200126 (between 30 and $29°$ W) were extremely noisy, and not possible to calibrate. This probe was not further used during the rest of the section due to its poor quality of the salinity data. For probe no. 70200068, the salinity offset remains around 0 west of $36°$ W (CTD station 38), and then abruptly shifts to around $0.295\,\mathrm{g\,kg^{-1}}$ between 36 and $23.5°$ W (CTD station 56). East of $23.5°$ W to the African coast, the offset shows a linear decreasing trend. This is likely due to the increasing portion of South Atlantic central water (SACW) in the central water layer when approaching the coastal region, which is less saline than the North Atlantic central water (NACW), and consequently shifting the slope of the T/S curve. As a result, the linear trend of the offset east of $23.5°$ W should not be due to instrument error. Therefore, only a mean offset was calculated and applied to calibrate each corresponding group of profiles made by no. 70200068. The reasons for the abrupt drift in the salinity (as obtained from the conductivity sensors) are not clear, but it is likely that due to the repeated intensive usage, the conductivity sensors were contaminated or impacted (hit ship hull).

The shipboard TSG provides another source of validation and calibration of uCTD data. On R/V *Meteor*, the TSG (SBE38 for temperature sensor, SBE21 for conductivity sensor) measures temperature and salinity at an intake at approximately 6.5 m depth. For all three legs, the TSG conductivity cell was calibrated from salinity analysis of water samples taken at the water intake, and a comparison with CTD data (if available) was also done. The uCTD salinity calibration was done by calculating the conductivity offset between the uCTD at 6.5 m and the averaged TSG conductivity within 5 min before and after the uCTD downcast. For probe no. 70200126, the drift of its conductivity sensor manifests also east of $39°$ W, the conductivity offset west of $39°$ W is about $-0.022\,\mathrm{S\,m^{-1}}$, and east of that it is about $0.094\,\mathrm{S\,m^{-1}}$. These differences in conductivity correspond to a change in salinity of -0.015 and $0.08\,\mathrm{g\,kg^{-1}}$, respectively. For probe no. 70200068, the conductivity offset west of $36°$ W is indistinguishable from zero, while east of that it is $0.156\,\mathrm{S\,m^{-1}}$, which corresponds to a salinity difference of 0 and $0.15\,\mathrm{g\,kg^{-1}}$. No trend in the offset east of $23.5°$ W is detected. For the $14.5°$ N section, we had uCTD, CTD, and TSG data available and the respective calibrations uCTD/CTD and uCTD/TSG could be compared. This was done in order to see if in case only TSG data are available (as is the case for the $11°$ S section), still reasonable calibration results could be achieved. For both probes, the TSG-derived drifts occurred in the same longitude range as they were detected using the CTD data. However, the magnitude of the offset was generally smaller for the TSG compared to the CTD-based method, especially for probe no. 70200126 in the longitude range west of $39°$ W, where even the signs of the offsets were opposite to each other. Such a difference is likely due to the fact that the CTD-based method employs a specific conservative temperature range where the salinity variation is small, while the TSG-based method focuses only at near-surface values (6.5 m), where the salinity varies in a broad range. Therefore, we would trust more the CTD-based method, and note that if the TSG-based method returns a small conductivity offset ($< 0.03\,\mathrm{S\,m^{-1}}$), one might need more caution to apply this offset to calibrate the uCTD. However, one needs also more caution when applying the CTD-based calibration in regions, where the T/S relation of the central water shows a mixture effect of NACW and SACW. At the $11°$ S section, CTD data were only available at the beginning and end of the section; we could use only the TSG data as the primary source for validation. Fortunately no drift was detected in the uCTD probes' conductivity cell, but a stable offset with a mean value of 0.131 and $0.073\,\mathrm{S\,m^{-1}}$ was detected and applied for probes nos. 70200068 and 70200138, respectively.

After the offset/drift calibration, all the uCTD data were gridded vertically from the original resolution ($\sim 0.25\,\mathrm{m}$ at a nominal sinking speed of $4\,\mathrm{m\,s^{-1}}$) to 1 m for the geostrophic velocity calculation later. Following Rudnick and Klinke (2007), we estimated that the calibrated and gridded uCTD data have an accuracy of $0.02\text{--}0.05\,\mathrm{g\,kg^{-1}}$ in salinity and $0.004\,^{\circ}\mathrm{C}$ in temperature.

All calculations in this study are based on the Thermodynamic Equation of State for seawater 2010 (TEOS-10, McDougall and Barker, 2011). TEOS-10 is introduced to replace the previous Equation of State, EOS-80, and it provides a thermodynamically consistent definition of the equation of state in terms of the Gibbs function for seawater. The most obvious change in TEOS-10 is the adoption of conservative temperature (Θ) and absolute salinity (S_A) to replace the potential temperature and practical salinity. Although the new equation of state has a non-negligible effect on the density field in the deep ocean, its effect in the upper ocean is expected to be small; therefore, our results obtained using TEOS-10 should be comparable with the previous studies.

2.2 ADCP measurements

Direct current velocity profiles were measured continuously during all three cruise legs with vessel-mounted 75 and 38 kHz Teledyne RDI Ocean Surveyors (OS75 and OS38). The OS75 was configured to measure at a rate of 2.2 s and a bin size of 8 m. The measurement range varied between 500 and 700 m. The OS38 was set to measure at a rate of 3.5 s and at 16 m (32 m) bin size during the $14.5^{\circ}\,\mathrm{N}$ ($11^{\circ}\,\mathrm{S}$) section. The measurement range was mostly 1200 m. Ship navigation information was synchronized to the ADCP system. The misalignment angles and amplitude factors were calibrated during post-processing. The processed data contain 10 min averaged absolute velocities in earth coordinates; the first valid bin for OS75 is centred at 18 m at $14.5^{\circ}\,\mathrm{N}$ and 13 m at $11^{\circ}\,\mathrm{S}$, for OS38 is 21 m at both sections. In this study, only the OS75 velocity was used since it has a higher accuracy in upper layers and higher vertical resolution. The uncertainties of 1 h averages were estimated by Fischer et al. (2003) to be $1\text{--}3\,\mathrm{cm\,s^{-1}}$.

Wind data

We used three different wind datasets in our analysis. First, we used the observed wind speed and direction recorded with the R/V *Meteor* anemometer, mounted at a height of 35.3 m. The wind data were stored with a temporal resolution of 1 min. True wind speed and direction were calculated using ship speed and direction from the navigation system. On-station measurements were removed. The reduction from the observation height to 10 m standard height was calculated according to Smith (1988) and wind stress was calculated according to Large and Yeager (2004) assuming neutral stability. The final wind stress used for the Ekman transport calculation was binned in 50 km ensembles to filter out small-scale variability.

The blended Satellite-based level-4 Near-Real-Time wind stress product (hereafter satellite wind stress) from the Copernicus Marine Environment Monitoring Service (CMEMS) was used. The wind speed data are derived from retrievals of scatterometers aboard satellite METOP-A (ASCAT) and Oceansat-2 (OSCAT) and combined with the European Centre for Medium-Range Weather Forecasts (ECMWF) operational wind analysis and gridded to $0.25^{\circ} \times 0.25^{\circ}$ resolution in space and 6 h in time. The wind stress data were estimated using the COARE 3 model (Fairall et al., 2003).

Moreover, the NCEP/NCAR monthly zonal wind stress at $14.5^{\circ}\,\mathrm{N}$ and $11^{\circ}\,\mathrm{S}$ corresponding to the months of the cruises (i.e. May and July 2013) was used to calculate the Ekman transport.

2.3 GECCO2 ocean synthesis data

In order to integrate our local observational results into a large-scale circulation, the GECCO2 ocean synthesis data were used and compared (Köhl, 2015). GECCO2 is a German version of the MIT general circulation model "Estimating the Circulation and Climate of the Ocean system" (ECCO, Wunsch and Heimbach, 2006). It has $1^{\circ} \times \frac{1}{3}^{\circ}$ resolution and 50 vertical levels. GECCO2 includes the Arctic Ocean with roughly 40 km resolution and a dynamic/thermodynamic sea ice model of Zhang and Rothrock (2000). The synthesis uses the adjoint method to bring the model into consistency with available hydrographic and satellite data (Köhl, 2015). The prior estimate of the atmospheric state is included by adjusting the control vector, which consists of the initial conditions for the temperature and salinity, surface air temperature, humidity, precipitation and the 10 m wind speeds from the NCEP RA1 reanalysis 1948–2011 (Köhl, 2015). The surface fluxes are derived by the model via bulk formulae of Large and Yeager (2004). For the study period from May to July 2013 monthly and daily output data were available. It is important to note that the in situ observational data measured during the cruises were not assimilated in the synthesis, while the satellite measured wind speed was assimilated but possibly modified via the synthesis.

3 Methods

According to Eqs. (1) and (2), the meridional Ekman volume transport can be calculated from zonal wind stress data, as well as from observed ageostrophic velocity. Hereafter we refer to the wind-stress-based calculation as the "indirect method", and to the ageostrophic-velocity-based calculation as the "direct method". In this section, we describe some details of the geostrophic and ageostrophic velocity calculation, the definition of the penetration depth of the Ekman flow, the error estimate of the direct Ekman transport calculation, and different methods to derive the Ekman heat and salt fluxes.

3.1   Geostrophic and ageostrophic velocity calculations

According to the thermal wind relation, relative geostrophic velocity referenced to the velocity at the reference depth can be calculated from the density field measured by the CTD and uCTD. At $14.5°$ N, two sets of the relative geostrophic velocity were calculated independently from the CTD and uCTD datasets. For CTDs, the relative geostrophic velocity referenced to 200 m was computed between the adjacent stations (average distance about 75 km). For uCTDs, in order to take advantage of the high spatial resolution, the relative geostrophic velocity to 200 m was calculated between any closest pair of uCTD profiles with a minimum distance of 70 km (roughly the Rossby radius of deformation at this latitude). Along the $11°$ S section, CTD profiles were only taken in the vicinity of the coasts, and over most of the section only uCTD data are available (Fig. 1). Therefore the geostrophic velocity was computed from the combined CTD and uCTD dataset following the methodology applied to uCTD data at the $14.5°$ N section, except that at $11°$ S the minimum distance between the closest profiles was set to 90 km (roughly the Rossby radius of deformation at $11°$ S). Note that the distance between uCTD profiles for geostrophic velocity calculation is an arbitrary choice, and varying the distance from 70 to 110 km has a negligible effect on the total transport (less than 2 %).

To obtain the absolute geostrophic velocity, the reference velocity at 200 m was obtained from the ADCP measurement. The ADCP velocity was projected to the normal direction of the cruise track and then averaged between the corresponding CTD/uCTD pairs. We did not include the ADCP velocity data recorded at the CTD stations, because velocity was repeatedly measured at a CTD station; zonally averaging the ADCP velocity would bias the result towards the on-station velocity. In previous studies (Wijffels et al., 1994; Chereskin et al., 1997; Garzoli and Molinari, 2001), the corresponding ADCP velocity at the reference depth was taken as the reference velocity, assuming that the flow at the reference depth was in geostrophic balance. However, the section-averaged ADCP velocity profile for the $14.5°$ N section shows a complicated vertical structure (Fig. 3a) and it is not obvious at which depth the flow is approximately in geostrophic balance. Thus, referencing the relative geostrophic velocity to the ADCP velocity only at a chosen depth may lead to a biased absolute geostrophic velocity. As a result, the ageostrophic velocity may be sensitive to the choice of the reference level. To overcome this

problem, a reference velocity was calculated as an averaged offset between each relative geostrophic velocity and the corresponding ADCP velocity within a common depth range, over which the ageostrophic components are averaged to about 0. This averaged offset should represent the absolute geostrophic velocity at the reference depth and is roughly independent of the vertical variation due to the ageostrophic components. At $14.5°$ N, the common depth range for the CTD-based calculation is between 70 and 500 m, which is expected to be below the surface Ekman layer and covered by both CTD and ADCP measurement. Due to the limitations in the maximum deployment depth, the uCTD-based calculation covers the depth range between 70 and 250 m. At $11°$ S the depth range is between 100 and 300 m, which should also be below the Ekman layer and was covered by the uCTD and ADCP measurement.

The ageostrophic velocity was then calculated as the difference between the ADCP velocity and the absolute geostrophic velocity. Note that the choice of the depth range still affects the reference velocity due to the vertical variation in the ADCP meridional velocity. For example, using a depth range between 70 and 250 m for the CTD-based calculation (same as the uCTD depth range) would decrease the final ageostrophic velocity by $0.44\,\mathrm{cm\,s^{-1}}$; using another depth range would not result in an absolute difference exceeding this value. This is much smaller compared to the uncertainty caused by using the ADCP velocity at a single depth as the reference velocity (up to $1.75\,\mathrm{cm\,s^{-1}}$), as can be anticipated from the section-averaged meridional ADCP velocity (Fig. 3a). The sensitivity of the absolute geostrophic velocity to the choice of the reference level was also tested at $14.5°$ N. Changing the reference level from 150 to 250 m would make a change in the absolute geostrophic velocity indistinguishable from zero.

3.2   Penetration depth of the Ekman flow

Because the ocean is not homogenous, a control surface must be defined that characterizes the maximum penetration depth of the momentum flux into the upper ocean. One choice would be the MLD, which we defined as the depth where the density increased by $0.01\,\mathrm{kg\,m^{-3}}$ in reference to the value at 10 m (following Wijffels et al., 1994). Along both sections, the MLD is relatively shallow (on average 25.1 m at $14.5°$ N and 32.2 m at $11°$ S), and as such unlikely a representative of D_E (Figs. 3 and 4). According to the Ekman theory, D_E for water at $14.5°$ N with a typical vertical eddy viscosity A_v of $0.02\,\mathrm{m^2\,s^{-1}}$ would be 33.1 m (see the definition of D_E in Eq. 1).

Alternatively a TTP has been defined as the shallowest depth at which the density gradient is larger than $0.01\,\mathrm{kg\,m^{-4}}$ (Wijffels et al., 1994). The TTP is typically deeper than the MLD and better defines the transition depth between well-mixed and stratified ocean, up to which the momentum from the wind is transferred (Chereskin et al., 2002). At some locations along both sections we observed two homogenous layers of slightly different density and possibly a remnant of the seasonal mixed layer cycle. In these cases, the TTP depth was chosen as the deeper one of the depth that satisfies the density gradient criterion. Since TTP was defined based on a gradient criterion, it represents the bottom of a weakly stratified surface layer rather than a specific density surface. Along the 14.5° N section, the mean TTP depth is 45.8 m (Fig. 3a). At both ends of the section, the TTP coincides with the MLD and is relatively shallow, while in the remaining part of the 14.5° N section TTP is deeper than the MLD (Fig. 4c). Along the 11° S section, the mean TTP depth is 56.8 m, and the TTP is deeper than the MLD throughout the section (Figs. 3b and 4d).

3.3 Error estimate of the direct Ekman transport

The errors of the direct Ekman transport were estimated following Chereskin and Roemmich (1991) and Wijffels et al. (1994). Assuming that near-inertial motions are the dominant source of error, decorrelation length scales were calculated as the distance that the ship travelled in a quarter of the inertial period at 14.5° N (47.9 h) and 11° S (62.7 h) resulting in 130 and 230 km, respectively. In total, 38 segments of the 14.5° N section and 25 segments of the 11° S section were obtained by dividing the total distance of each section by the corresponding decorrelation length scale, respectively. The westernmost and easternmost four segments of each section were omitted because of the anomalously weak wind near the eastern boundary and the strong boundary current in the western boundary region. The degree of freedom (DOF) of 30 and 17, respectively, was the number of the remaining segments. The ageostrophic transport within each segment was treated as an independent realization of the Ekman transport. Therefore, standard errors were calculated. Then the final error is given as the standard error times the DOF. Another factor that could lead to an uncertainty is the depth range used to calculate the reference velocity from the ADCP velocity. As discussed above, we argue that the vertical structure of the ageostrophic velocity should arise from the near-inertial motion and therefore, should be included already in this uncertainty estimate.

3.4 Ekman heat and salt flux calculation

The Ekman heat and salt fluxes, H_e and S_e, respectively, were calculated by combining the indirect and direct Ekman volume transport estimates with Θ and S_A from different sources. Note that in order to calculate the Ekman fluxes in the context of mass conservation (Montgomery, 1974), it has to be assumed that the Ekman volume transport in the upper layer is balanced by an equal and opposite geostrophic return flow at depth. This is a reasonable assumption and has been routinely adopted in many inverse studies (Ganachaud and Wunsch, 2003). To account for this return flow, an averaged

conservative temperature, $\overline{\Theta}$, and absolute salinity $\overline{S_A}$ were subtracted from the in situ Θ and S_A at each section. $\overline{\Theta}$ and $\overline{S_A}$ are the zonally and vertically (0–5000 m) averaged conservative temperature and absolute salinity, calculated from the annual climatology of the World Ocean Atlas 2013 v2 (Locarnini et al., 2013; Zweng et al., 2013) at each section. In the following, the calculation details of the Ekman heat flux and transport-weighed Ekman temperature are given; the calculation of the Ekman salt flux and transport-weighted Ekman salinity is an analogue.

3.4.1 Direct methods

By using the in situ Θ together with the ageostrophic velocity within the layer from the sea surface to the TTP (referred to as the TTP layer), the Ekman heat flux H_e was calculated (referred to as the direct TTP/profile).

$$H_e = \rho C_p \int_{x_1}^{x_2} \int_{-TTP}^{0} \left(\Theta(x,z) - \overline{\Theta} \right) v_{ageo}(x,z) dz dx, \quad (3)$$

where C_p is the specific heat capacity of sea water at constant pressure, ρ is the density of sea water, in this study we assumed a constant $C_p = 4000 \, \mathrm{J \, kg^{-1} \, K^{-1}}$ and a constant $\rho = 1025 \, \mathrm{kg \, m^{-3}}$, v_{ageo} is the ageostrophic velocity, Θ is the in situ conservative temperature. $\overline{\Theta}$ is the mean conservative temperature at the corresponding section.

It is useful to consider the Ekman heat flux as the product of the Ekman volume transport, M^y, and the transport-weighted temperature, Θ_E. The transport-weighted temperature then can be calculated as follows:

$$\Theta_E = \frac{\int_{x_1}^{x_2} \int_{-TTP}^{0} \Theta(x,z) v_{ageo}(x,z) dz dx}{M^y_{direct}}. \quad (4)$$

As a comparison to the direct TTP/profile method, the Ekman heat fluxes using only in situ SST from the CTD and uCTD were also calculated (referred to as the direct TTP/surface). This was done by replacing the in situ $\Theta(x,z)$ in Eqs. (3) and (4) with the in situ SST $\Theta(x, z = 0)$.

The uncertainty of the direct Ekman heat and salt fluxes was estimated following Chereskin et al. (2002). Since the wind direction was predominantly uniform and westward, the uncertainty should mainly arise from the ageostrophic velocity that was opposite to the expected Ekman flow direction. Therefore, the uncertainty was calculated still using Eqs. (3) and (4), except that only southward or northward ageostrophic velocity was used in both the numerator and denominator for the 14.5° N or 11° S section, respectively.

3.4.2 Indirect surface method

Often Ekman heat and salt fluxes are estimated by combining the Ekman volume transport inferred from wind stress with the SST and sea surface salinity (SSS) from a climatology or satellite measurements (e.g. McCarthy et al., 2015). Here, we calculated the heat flux using in situ wind and in situ SST data (referred to as the indirect surface) to compare with the direct estimates. Additionally, annual Ekman heat and salt fluxes (referred to as the indirect annual) were calculated using an annual average of the monthly NCEP/NCAR reanalysis wind stress data between 1991 and 2013 and the annual average of SST and SSS from the Roemmich and Gilson (2009) monthly Argo climatology (hereafter RG climatology). Following Levitus (1987), the Ekman heat flux for the indirect surface method was calculated as

$$H_e = C_p \int \left(\Theta(x, z = 0) - \overline{\Theta} \right) \frac{\tau_x}{f} dx, \qquad (5)$$

where τ_x is the in situ wind stress in the tangential direction of the cruise track, f is the Coriolis parameter, and $\Theta(x, z = 0)$ is the in situ SST. The transport-weighted temperature was calculated as follows:

$$\Theta_E = \frac{\int \Theta(x, z = 0) \frac{\tau_x}{\rho f} dx}{M_{indirect}^y}. \qquad (6)$$

The indirect annual method is an analogue to the indirect surface method, except that the Ekman volume transport and SST were derived from the NCEP/NCAR reanalysis wind stress and RG climatology, respectively.

3.4.3 Indirect TTP method

Wijffels et al. (1996) assumed a linear Ekman velocity profile between the surface and TTP and calculated the Ekman heat and salt fluxes using climatological wind stress data, combined with the in situ temperature and salinity. Here we followed their method and used the in situ Θ, S_A, and wind to calculate the Ekman heat and salt fluxes (referred to as the indirect TTP) as a counterpart to the direct TTP method.

$$H_e$$
$$= C_p \int \left[\frac{2}{3} \Theta(x, z = 0) + \frac{1}{3} \Theta(x, z = TTP) - \overline{\Theta} \right] \frac{\tau_x}{f} dx, \qquad (7)$$

where $\Theta(x, z = TTP)$ is the in situ conservative temperature at TTP depth from the CTD/uCTD. The transport-weighted temperature was calculated as follows:

$$\Theta_E = \frac{\int \left[\frac{2}{3} \Theta(x, z = 0) + \frac{1}{3} \Theta(x, z = TTP) \right] \frac{\tau_x}{\rho f} dx}{M_{indirect}^y}. \qquad (8)$$

4 Results and discussion

4.1 Upper layer hydrography at 14.5° N and 11° S

Along both sections (Fig. 2a, b) the typical upward tilting of isotherms towards the east, as a result of the subtropical gyre circulation, can be seen. Along the nominal 14.5° N section, the water in the upper 50 m, compared to that at 11° S, was relatively warm and fresh, with an averaged Θ and S_A of about 26.03 °C and 36.15 g kg^{-1}, respectively. The minimum S_A core near the western boundary probably originates from the freshwater runoff from the Amazon River (Fig. 2c). Together with the warm temperature, it forms the lightest water observed along the section (Fig. 2e). A subsurface salinity maximum layer of Subtropical Underwater (STUW) is centred at 100 m depth with S_A greater than 37.2 g kg^{-1}. STUW is formed in the subtropical Atlantic with a SSS maximum due to excessive evaporation, and is subducted equatorward (Talley et al., 2011). The upward tilt of the isopycnals from west to east is suggestive of a net southward geostrophic transport when excluding the western boundary, where sharp deepening of the isopycnals implies a northward, intensified boundary current (Fig. 2g). At 11° S, the surface water was cooler and more saline than that at 14.5° N, with an averaged Θ and S_A of about 24.52 °C and 36.69 g kg^{-1}. The STUW with maximum salinity larger than 37.3 g kg^{-1} was centred at about 100 m, but was even saltier than that at 14.5° N. Likewise, a net northward geostrophic flow can be anticipated from the displacement of the isopycnals. At the western boundary, the North Brazil Undercurrent (NBUC) is characterized by a narrow and strong northward velocity band west of 35° W (Schott et al., 2005) (Fig. 2h). In the hydrographic data Θ/S_A variability is seen at both sections that are associated with mesoscale eddies. For instance, at 14.5° N/25° W and 11° S/7° E, cyclonic and anticyclonic eddies were characterized by the upward peak of the isotherms, and were clearly visible from the geostrophic velocity sections (Fig. 2g, h).

The daily Θ and S_A data of the GECCO2 synthesis were extracted from the model grid to the nearest time and position of the ship measurement. In general, GECCO2 daily data reproduced the observed hydrographic structure very well (not shown). The upward tilt of the isopycnals from the west to the east and the subsurface salinity maximum with S_A larger than 37.2 g kg^{-1} were clearly captured by GECCO2. However, the most obvious difference was at the western boundary of 11° S, where the surface salinity was not as high as the observed values, and the isopycnals were not tilting in the same direction, indicating that the shallow western boundary current in the GECCO2 flowed in the opposite direction compared to the observation at 11° S. But we expect that this difference should not impact the ageostrophic velocity calculation, since the geostrophic velocity must be removed from the total velocity.

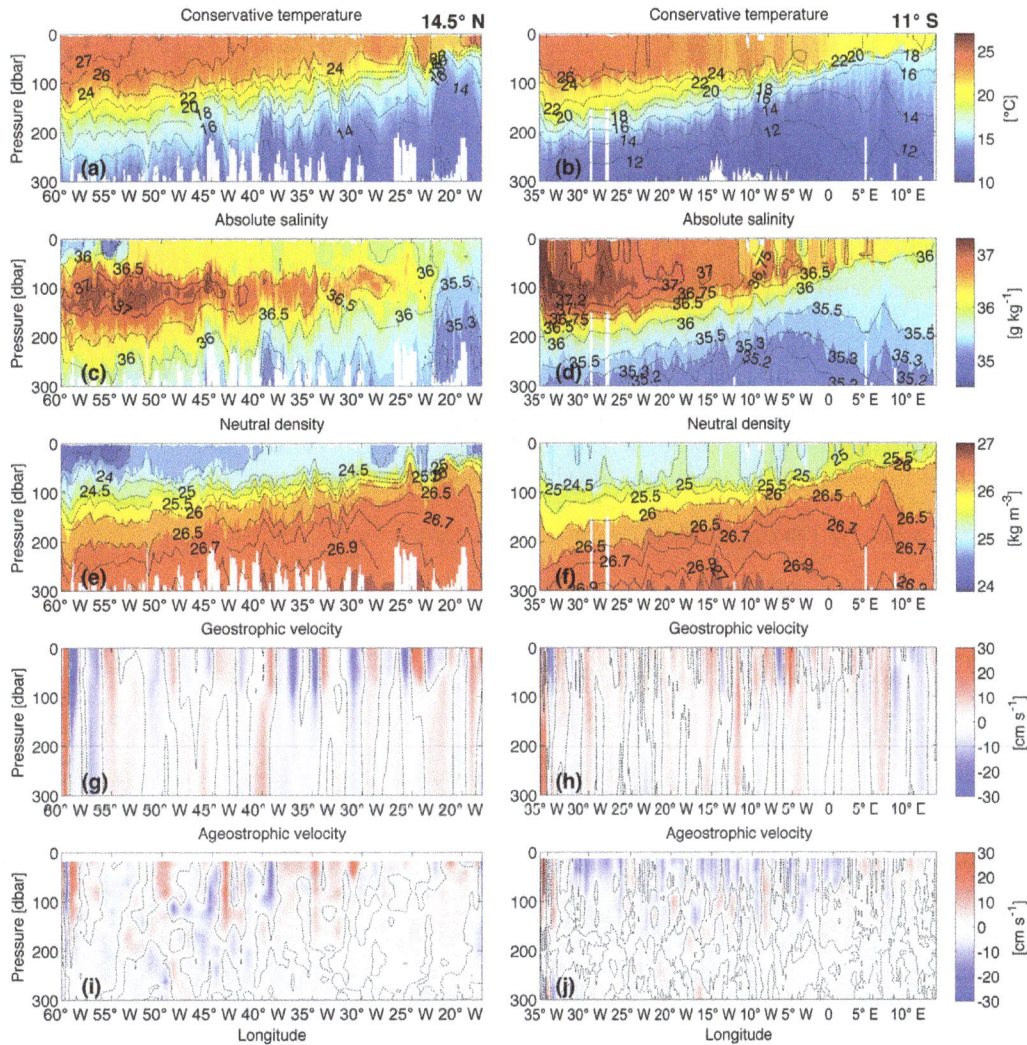

Figure 2. Vertical sections of conservative temperature in °C (**a, b**), absolute salinity in g kg^{-1} (**c, d**), neutral density in kg m^{-3} (**e, f**), geostrophic velocity in cm s^{-1} (**g, h**), and ageostrophic velocity in cm s^{-1} (**i, j**) at 14.5° N (left) and 11° S (right). All the available CTD and uCTD data were used to produce (**a–f**), and the contours were plotted with every fifth value for visual clarity. (**g**) and (**i**) were calculated using only CTD data, while (**h**) and (**j**) were calculated using CTD and uCTD data. The blanks were due to the shallow measurement depth of the uCTD.

4.2 Vertical structure of the ageostrophic flow

Although northward (southward) ageostrophic velocity at 14.5° N (11° S) dominates the upper 50–70 m (Fig. 2i, j), as expected from the persistent westward trade winds, the appearance of southward (northward) velocity at 14.5° N (11° S) in the upper 50–70 m and below indicates the existence of non-Ekman ageostrophic components in the water column. This will be discussed in detail below. The section-averaged ageostrophic velocity based on CTD data at 14.5° N shows a relatively complicated vertical structure with multiple maxima and minima (Fig. 3a). It has a northward maximum velocity of 3.5 cm s^{-1} near the surface, and decreases to about 0.3 cm s^{-1} at about 60 m, followed by a minor peak at about 80 m before approaching 0 at 100 m. Another peak

of 1 cm s^{-1} appears at about 150 m, and below 180 m the velocity changes direction. When the ageostrophic velocity is calculated based on the uCTD data, it has a very consistent structure and strength compared to the CTD-based ageostrophic velocity (Fig. 3a). This is meaningful information as the hydrographic data at 11° S consist primarily of uCTD data. The good agreement between the CTD and uCTD data analysis at 14.5° N justifies the use of either one or the other. At 11° S, the ageostrophic velocity shows a near-surface southward maximum of 4.3 cm s^{-1}, decreases almost linearly in the upper 70 m, and gradually approaches 0 at about 100 m (Fig. 3b). In contrast to the northern section the vertical variations of the ageostrophic velocity profile below 100 m are very small.

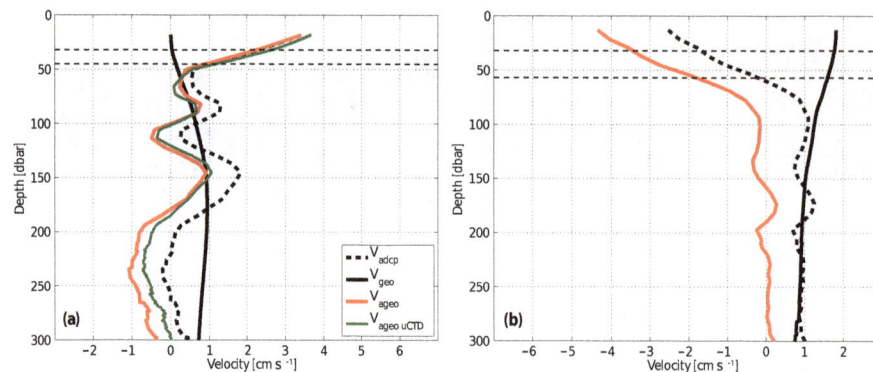

Figure 3. Section-averaged cross-track velocity profiles at **(a)** 14.5° N and **(b)** 11° S. In **(a)**, the solid red curve and the solid black curve are the ageostrophic and geostrophic velocity calculated from the CTD data, respectively. The dark green curve is the ageostrophic velocity profile based only on the uCTD data. In **(b)**, the solid red curve and solid black curve are the ageostrophic and geostrophic velocity calculated from the combination of the CTD/uCTD data, respectively. The dashed black curve is the ADCP velocity. The upper horizontal dashed line denotes the basin-wide averaged MLD and the lower one denotes the basin-wide averaged TTP depth.

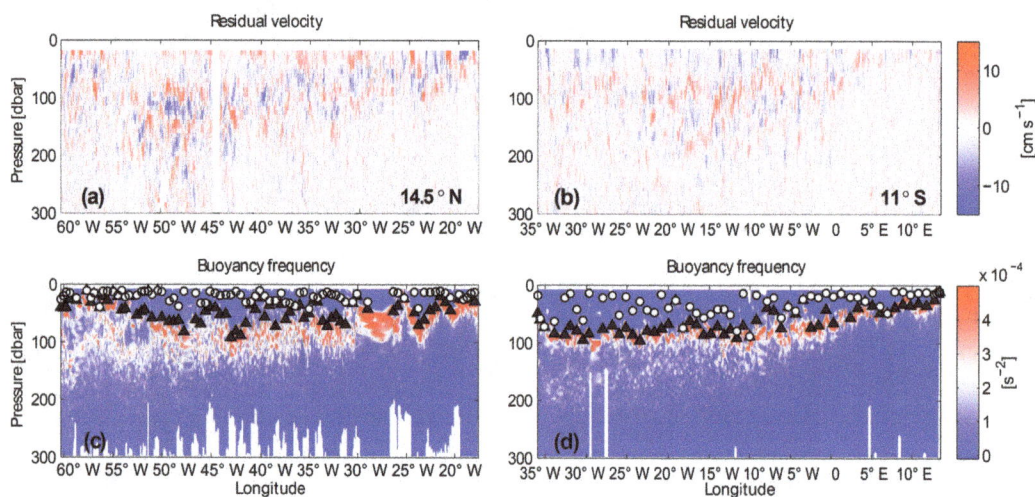

Figure 4. Vertical sections of residual meridional velocity in cm s^{-1} at **(a)** 14.5° N and **(b)** 11° S and of buoyancy frequency calculated from uCTD/CTD at **(c)** 14.5° N and **(d)** 11° S. Northward velocity in **(a, b)** is shaded in red, southward in blue. The residual velocity is calculated by subtracting an 80 m boxcar filtered profile from the original ADCP profile. The white circles in **(c, d)** denote the MLD; the black triangles denote the TTP (see text for details). The MLD and TTP plotted here are subsampled for visual clarity. Note that no uCTD measurements were conducted between 30 and 25° W at 14.5° N.

Assuming that the Ekman balance would hold true along the analysed sections, the ageostrophic velocity would decrease undisturbed from its surface maximum to about 0 at a certain depth (Ekman depth, D_E). However, the observed wave-like structure at 14.5° N indicates that other processes must play a role in setting the section mean ageostrophic flow field. To identify this wave-like structure, we tried to separate the non-Ekman ageostrophic flow from the other components by using the ADCP velocity. A residual velocity was calculated by subtracting an 80 m boxcar filtered velocity profile from the original ADCP meridional velocity (Fig. 4a, b). The 80 m filter window was determined based on the vertical length scale of the wave-like structure in the section-averaged ageostrophic velocity profile by visual inspection.

At 14.5° N, vertically alternating structures with wavelengths of 60–80 m are clearly visible, are coherent and persistent throughout the section, and are most pronounced between 52 and 46° W (Fig. 4a). At 11° S, similar signals are visible for most of the section, but are not as strong as at 14.5° N (Fig. 4b).

Zonally averaging the residual velocity results in a velocity profile with a vertically alternating structure similar to that in the section-averaged ageostrophic velocity in both strength and structure, indicating that the vertical variation in the ageostrophic velocity mainly arises from the presence of high-order baroclinic waves. Figure 4c and d show the buoyancy frequency (N^2) for the two sections, respectively. It appears that the wave-like signals occur mainly in the strongly stratified layer (pycnocline) marked by high N^2 values. N^2 is calculated as follows:

$$N^2 = \frac{g}{\rho_0} \frac{\partial \rho(z)}{\partial z}, \qquad (9)$$

where g is the gravitational acceleration, $\rho_0 = 1025\,\mathrm{kg\,m^{-3}}$ is the reference density, and $\rho(z)$ is the in situ potential density as a function of depth, z. $\rho(z)$ was calculated by using a combination of CTD and uCTD profile data with a re-gridded vertical resolution of 5 m at both sections.

Chereskin and Roemmich (1991) also observed energetic, circularly polarized, relative currents of large horizontal coherence below the base of the mixed layer at 11° N in the Atlantic. They described the signal as the propagation of near-inertial internal waves and argued that the presence of a near-inertial peak in internal wave spectra, together with continuously varying wind forcing, would guarantee the appearance of these waves. Using satellite-based wind stress data, we examined the changes in wind stress at the measurement points within the last 2 weeks before the ship arrived at the measurement points. Although the wind stress strongly changed along the whole section, it is still not indicative why the wave signal is strongest between 52 and 46° W at 14.5° N. It is tempting to believe that these waves are near-inertial internal waves. However, due to the fact that the ship moved nearly constantly except when it was on station, it is extremely difficult to identify what exactly these signals are. More sophisticated methods may be applied to analyse the wave-like signal; for instance, Smyth et al. (2015) took the Doppler shift in the shipboard current measurement into account, and translated observed Yanai wave properties into the reference frame of the mean zonal flow. But this is obviously beyond the scope of this work.

4.3 Ekman transport

4.3.1 Indirect method

According to Eq. (1), the Ekman transport can be calculated from the wind stress data (referred to as the indirect method) by integrating the left-hand side of Eq. (1) zonally. The in situ wind stress data and a satellite-based wind stress product from CMEMS were used. The satellite wind stress data were extracted from the original grid to the nearest time and nearest position of the ship navigation. Both in situ and satellite wind stress were projected to the tangential direction of the cruise track, so that the cross-sectional Ekman transport at each grid point was calculated. Note that both sections were occupied nominally zonally; therefore, we will refer to cross-sectional Ekman transport as meridional Ekman transport for simplicity hereafter.

Overall, the satellite wind stress agrees well with the ship wind stress (Fig. 5) except in the region between 40 and 30° W at 14.5° N, where the zonal ship wind stress is larger than the zonal satellite wind stress, and at 11° S the ship wind

stress is generally smaller than the satellite wind stress. Since the 10 m wind speeds from the ship and satellite are very close to each other at both sections (not shown), the difference in the wind stress may be due to the use of a different drag coefficient formulation (COARE 3 for the CMEMS wind product; Large and Yeager, 2004, for ship wind stress). In comparison to the NCAR/NCEP monthly zonal wind stress, the weaker ship wind stress in the western half of the 14.5° N section indicates that the cruise started with anomalously weak winds, while at 11° S the observed wind stress (both ship and satellite observation) was consistent with the monthly mean wind stress. It is also reported that differences in the different wind stress data may also arise from the unresolved local effect by the satellites and NCEP data (Mason et al., 2011; Pérez-Hernández et al., 2015). For instance, near the Canary Islands, the NCEP monthly data do not resolve the Von Karman structure caused by the interaction of wind with the islands due to its low resolution.

As expected, at 14.5° N, the indirect estimate of the Ekman transport from the in situ wind stress is 6.7 ± 3.5 Sv, only 0.4 Sv larger than that from the satellite wind stress. Using the monthly mean wind stress from NCEP/NCAR during the M96/M97 cruise month (May 2013), the total transport is 8.8 ± 1.4 Sv. The difference between the monthly wind estimate and in situ wind estimate is mainly due to the anomalously weak wind when the cruise started from the western boundary (Fig. 5a). At 11° S, the indirect Ekman transport from the in situ wind stress is 13.6 ± 3.3 Sv, while the transport from the satellite wind stress is 2.0 Sv higher, due to the higher value of the satellite wind stress (Fig. 5b). The NCEP/NCAR monthly wind stress in July 2013 returns a transport of 15.1 ± 1.9 Sv. The errors shown with the indirect ship wind estimates are given by the standard deviation of the long-term Ekman transport calculated using 6 h NCEP/CFSR wind stress between the years 2000 and 2011 at the two latitudes. The errors of the monthly estimates are given by the standard deviation of the monthly mean Ekman transport in May (July) between 1979 and 2013 at 14.5° N (11° S) calculated from the NCEP/NCAR monthly wind stress. Another source of uncertainty may arise from the wind stress calculation using different bulk formulas, which could lead to an uncertainty as large as 20 % (Large and Pond, 1981). This may explain the difference in the indirect estimates between using the in situ wind stress and the satellite wind stress at 11° S.

4.3.2 Direct method

The direct meridional Ekman transport is derived from vertically integrating the ageostrophic velocity profiles (Eq. 1, right-hand side). As already mentioned, one critical assumption is the integration depth (D_E). Applying the TTP as an estimate of D_E, the total Ekman transport at 14.5° N based on CTD data is 6.2 ± 2.3 Sv, while applying a uniform depth of 50 m results in an alternative estimate of

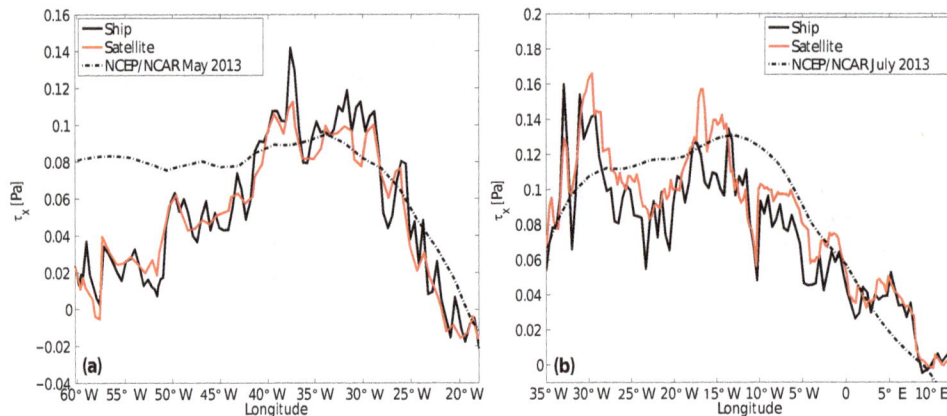

Figure 5. Zonal wind stress along **(a)** 14.5° N and **(b)** 11° S. Ship wind stress (black line) was binned in a 50 km interval. The satellite wind stress data (red) were extracted to the nearest ship time and position. The NCEP reanalysis monthly zonal wind stress (black dashed line) at the same latitude in the cruise month is also plotted.

6.5 ± 1.9 Sv, and applying the local MLD results in a transport of 5.1 ± 1.4 Sv. When integrating the ageostrophic velocity calculated from the uCTD data to the TTP, the Ekman transport is 6.6 ± 2.3 Sv. At 11° S, the direct estimate by applying the TTP, a uniform depth of 80 m, and the MLD is −11.7 ±2.1 Sv, −12.0 ±2.4 Sv, and −8.0 ± 1.4 Sv, respectively ("−" denotes southward transport). The errors given with the transport estimates were calculated by considering the aliasing effect of the inertial waves during the cruises following Chereskin and Roemmich (1991) as described in Sect. 3.3.

Because the shallowest valid bin depth of the ADCP measurement was 18 m (13 m) at 14.5° N (11° S), the ageostrophic velocity was extrapolated linearly to the surface using the value of the first two bins. Note that we did not assume a surface maximum of the ageostrophic velocity everywhere, since for individual profiles the ageostrophic velocity at the first bin depth may be smaller than that at the second bin, which would result in a smaller surface ageostrophic velocity. In previous studies (Chereskin and Roemmich, 1991; Wijffels et al., 1994), the velocity above the first ADCP bin was assumed to equal the value at the first bin. Using this assumption would reduce Ekman transport at 14.5° N by 0.56 Sv (9 % of total northward transport), and at 11° S by 0.14 Sv (1 % of total southward transport). According to the classical Ekman theory, the surface Ekman velocity (V_0) is 45° to the right (left) of the wind blowing direction in the Northern (Southern) Hemisphere and can be derived from the total wind stress (see the definition of V_0 in Eq. 1). As a comparison to the linear extrapolation above the first ADCP bin, we also calculated the meridional Ekman velocity at the surface using the total in situ wind stress and a constant A_v of 0.02 m² s⁻¹. Then the meridional ageostrophic velocity above the first ADCP bin was linearly interpolated using the value at the first bin and the surface meridional Ekman velocity predicted from the in situ wind stress. The resulting

Ekman transport is 1.2 Sv (0.7 Sv) smaller than that using a linear extrapolation method at 14.5° N (11° S). Note that we chose a linear extrapolation method, because it resulted in a better agreement between the indirect and direct estimates, but it may overestimate the total ageostrophic transport.

A question that follows is whether the ageostrophic flow in the mixed layer has shear or is constant with depth referred to as a slab-like shape. Given the large variation of the MLD throughout the sections, basin-wide averages are inconclusive. Chereskin and Roemmich (1991) found shear structure in the mixed layer at 11° N in the Atlantic, while Wijffels et al. (1994) reported a slab-like shape at 10° N in the Pacific, and attributed the shear structure found by Chereskin and Roemmich (1991) to an improper definition of the MLD. Following their method, the depth was normalized by the local MLD before averaging the ageostrophic velocity across the basin (Fig. 6). At 14.5° N, for a slab-like ageostrophic structure, Fig. 6 would show a nearly constant profile from the surface to about the MLD. Instead, it shows strong shear above the MLD. Such strong shear is insensitive to the definition of the MLD. For example, choosing a density threshold of 0.005 kg m⁻³, the shear still exists below 0.4 MLD. At 11° S, no slab-like structure in the ageostrophic velocity was found, either. The constant value above 0.4 MLD is a consequence of using a constant velocity above 18 m, the shallowest ADCP bin. Therefore, we would conclude that ageostrophic shear exists within the mixed layer in our cases, as expected from the classical Ekman theory.

The cumulative Ekman transport from the western to the eastern boundary shows an overall match between the direct and indirect methods (Fig. 7a, b). At 14.5° N, the in situ wind was relatively weak at the beginning and the end of the section. Correspondingly, the increment in transport within these two segments was moderate, while in the central part of the section, where the wind was strong, the rapid accumulation of Ekman transport is directly visible in both indi-

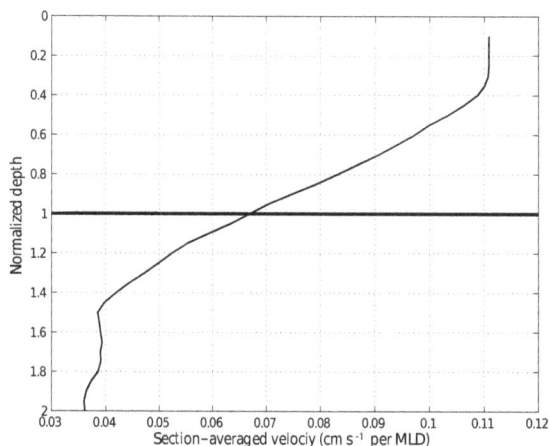

Figure 6. Section-averaged ageostrophic velocity at 14.5° N, normalized in depth by the local MLD. Velocity above 18 m is set equal to the velocity at 18 m. MLD is defined as the depth where the density is $0.01\,\mathrm{kg\,m^{-3}}$ different from the value at 10 m.

rect and direct estimates. The direct estimates using TTP and 50 m depth are very close to the in situ wind estimates. The estimate using 50 m depth tends to overestimate the transport close to both ends of the section. Applying the MLD as integration depth tends to underestimate the total transport by about 1.5 Sv, compared to the ship wind estimate. This is mainly because it fails to capture the increase between 30 and 25° W. Note that the uCTD-based direct estimate is consistent with the CTD-based estimates, though it overestimates the transport in the middle of the section; the total transport as well as the transport structure are similar. This may be a result of the higher spatial resolution of the uCTD measurement, which captures more details in the horizontal features introduced by the wind.

At 11° S, the wind was strong in the western half of the basin and gradually weakened in the eastern half towards the eastern boundary. Correspondingly, the Ekman transport accumulates rapidly to about 12 Sv at 0° E, east of which the increment is very small for both direct and indirect estimates. Among the direct estimates, integrating the ageostrophic velocity to 80 m and TTP returns nearly identical transport in the western half of the section; the difference in the eastern half mainly reflects the shallower TTP towards the eastern boundary, while using the MLD for the integration underestimates the Ekman transport from the very beginning. Note that at both sections, the direct estimate using MLD is about one-fourth smaller than that using TTP depth. This agrees with the findings at 10° N in the Pacific by Wijffels et al. (1994), who reported that the Ekman flow within the mixed layer accounted for about two-thirds of the total Ekman transport, and the in situ wind predicted the Ekman transport down to the TTP. Recalling the definition of D_E in Eq. (1), the vertical eddy viscosity A_v can be estimated by using TTP as a representative of D_E. At 14.5° N, the mean A_v

is $0.038\,\mathrm{m^2\,s^{-1}}$, and at 11° S, the mean A_v is $0.045\,\mathrm{m^2\,s^{-1}}$. These results fall in the range of previous estimates of A_v, which vary by more than 1 order of magnitude (Price et al., 1987; Chereskin, 1995; Lenn and Chereskin, 2009).

4.4 Ekman transport from GECCO2

The daily data of the GECCO2 synthesis (2008–2014) allowed us to estimate the model Ekman transport, inspect the vertical structure of the ageostrophic velocity in the model, and compare these results with the observations for the corresponding cruise time periods. The daily data were first extracted from the model grid to the nearest ship time and position. The Ekman transport in GECCO2 was calculated in a similar manner to the direct method used for the observational data. An ageostrophic velocity was calculated as the difference between the geostrophic velocity and total velocity with a reference depth of 200 m. The geostrophic velocity was computed from the temperature and salinity of GECCO2. The direct estimate of the meridional Ekman transport in GECCO2 was obtained by integrating the meridional ageostrophic velocity vertically and zonally.

The section-averaged ageostrophic velocity at both sections shows a near-surface maximum at about 15 m, and then decreases sharply to 0 at about 50 m; the flow is purely geostrophic below 60 m (not shown). This vertical distribution of ageostrophic velocity indicates that the wind-driven Ekman component is the predominant contributor to the ageostrophic velocity in the GECCO2 model, and that nearly all the wind energy is utilized for the Ekman transport and confined to the upper 50 m at both sections. The total transport by integrating the ageostrophic velocity to 50 m is 7.6 Sv at 14.5° N and 12.0 Sv at 11° S, respectively (Fig. 7), which is close to the indirect Ekman transport estimates based on GECCO2 daily wind stress of 7.4 and 13.4 Sv, respectively.

This result agrees very well with the observed direct Ekman transport, which is likely due to the fact that GECCO2 daily wind stress has a similar magnitude to the in situ wind stress. The observed ageostrophic cumulative transport shows strong mesoscale fluctuations throughout the sections, which are characterized by the presence of northward and southward ageostrophic velocity even though the in situ wind is persistently westward, while the GECCO2 ageostrophic transport accumulates smoothly (Fig. 7).

4.5 Ekman heat and salt fluxes

The Ekman volume, heat, and salt fluxes calculated using different methods are summarized in Table 1. In the previous sections, we have shown that the TTP is a reasonable assumption for the depth of D_E for both sections. Hence, the direct TTP/profile method should give us the best estimate of the heat and salt fluxes. It is clear that the differences in Ekman volume transports dominate the differences in the resulting Ekman heat and salt fluxes. The higher Ekman volume

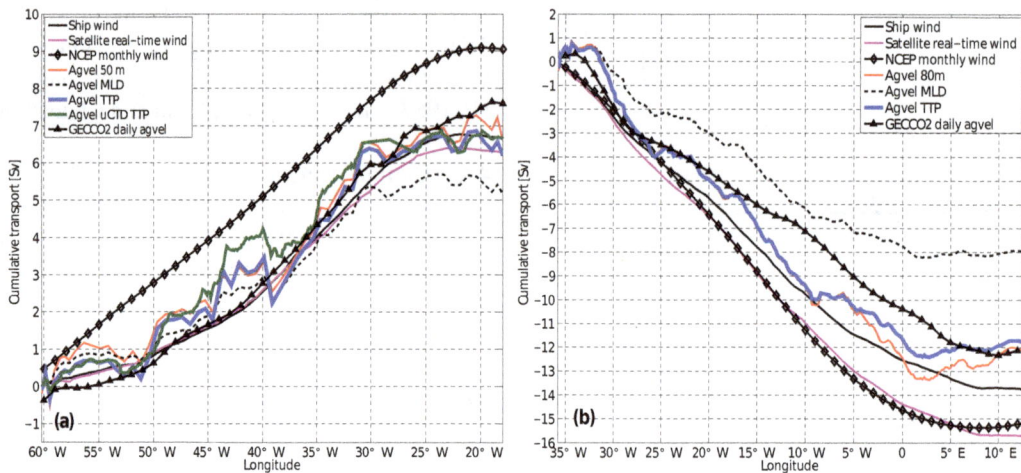

Figure 7. Cumulative meridional Ekman transport from the western to the eastern coast **(a)** at 14.5° N and **(b)** at 11° S. For both sections, the black solid curve marks the indirect Ekman transport estimate from the in situ wind stress; the magenta solid curve denotes that from the satellite wind stress; and the black diamond line denotes that from the NCEP/NCAR monthly wind stress. The solid blue curve denotes the direct estimate by integrating the ageostrophic velocity to the TTP, the red solid to a uniform depth (50 m at 14.5° N and 80 m at 11° S), and the black dashed line to the MLD. The black triangle line represents the direct estimate based on the GECCO2 daily data. The dark green line in **(a)** represents the direct estimate integrated to the TTP based only on the uCTD data.

transport by the indirect methods leads to higher heat and salt fluxes compared to the direct methods at both sections. At 14.5° N, the transport-weighted Ekman temperature from the direct TTP/surface method is 0.10 °C higher than that from the direct TTP/profile method. This temperature difference corresponds to a change in the heat flux of only less than 1 %, which is very small compared to the uncertainty caused by the volume transport uncertainty. The indirect TTP method returns the Ekman temperature and salinity value very close to that of the direct TTP/profile method, indicating that the assumption of a linear Ekman velocity profile between the surface and the TTP depth may be reasonable. This could be potentially interesting, since this method is independent of the ageostrophic velocity.

At 11° S, the difference between the direct TTP/profile and direct TTP/surface methods is negligible. The transport-weighted Ekman temperature from the indirect TTP and surface methods is significantly smaller than that from the direct methods. This may be caused by a combined effect of stronger Ekman volume transport by the indirect method near the eastern boundary (Fig. 7) and the cooler water temperature due to coastal upwelling. In other words, the indirect calculation tends to give excessive weighting to the cooler water, which results in lower values in the transport-weighted Ekman temperature. Such a combined effect has a limited impact on the Ekman salinity.

The difference in the Ekman heat flux when using temperature at the surface or within the TTP layer is much smaller than that for the extreme case at the end of the summer monsoon in the Indian Ocean in September 1995 reported by Chereskin et al. (2002). The choice of Ekman temperature

and salinity has a negligible effect on the resulting heat and salt fluxes across the sections studied here.

Note that at 14.5° N (except for the indirect annual method) the Ekman heat fluxes (0.41–0.44 PW) estimated using direct and indirect methods based on ageostrophic velocity and in situ wind are generally smaller compared with the estimates of 0.7–0.8 PW by Levitus (1987) or 0.6–0.8 PW by Sato and Polito (2005). As described above, both the direct and indirect methods in this study reflect the Ekman transport driven by the in situ wind, which is weak compared to the monthly wind, especially in the western basin (Fig. 5a). By using the annual mean wind stress from NCEP/NCAR reanalysis and SST from RG climatology, the Ekman heat flux is 0.58 PW, which is close to the estimates of Sato and Polito (2005). At 11° S, the direct and indirect Ekman heat fluxes (0.8–0.96 PW) are rather close to the estimate of 1.05 PW by Levitus (1987) or within the range of values (0.7–1.0 PW) estimated by Sato and Polito (2005). Here, the Ekman volume transports estimated from in situ wind and from the annual averaged wind were similar.

It is worth noting that the Ekman salt flux presented in this study may not be representative of an annual or long-term mean Ekman salt flux, but it may provide insight into the sensitivity of the Ekman salt flux to changes in Ekman volume transport and Ekman-layer salinity. This might help in constraining salt conservation and resulting freshwater flux in studies of the meridional overturning circulation in the same region.

5 Summary

The Ekman volume, heat, and salt transport across zonal sections at 14.5° N and 11° S in the Atlantic were calculated by using an ageostrophic-velocity-based method (direct method) and a wind-stress-based method (indirect method). A cross-sectional ageostrophic velocity was calculated at each section following Chereskin and Roemmich (1991) and Wijffels et al. (1994) by subtracting the geostrophic velocity from the cross-sectional component of the ADCP velocity. At both sections, underway-CTD profiles were used for the ageostrophic velocity calculation. A comparison between the results based on standard CTD and uCTD data at 14.5° N revealed a consistent transport estimate with a robust vertical ageostrophic velocity structure and horizontal distribution of the Ekman transport. This has established our confidence in performing a similar calculation for the 11° S section, along which primarily uCTD profiles were taken.

The section-averaged ageostrophic velocity at 14.5° N and 11° S has a near-surface northward and southward maximum of 3.5 and 4.3 cm s^{-1}, and decreases below to reach about zero at 60 and 100 m, respectively. This is an indication of the Ekman spiral, and is consistent with the findings of Chereskin and Roemmich (1991) at 11° N in the Atlantic, Wijffels et al. (1994) at 10° N in the Pacific, and Chereskin et al. (1997) at 8.5° N in the Indian Ocean. Near-inertial oscillations are regarded as a dominant source of ageostrophic noise, which is superimposed upon the Ekman flow, but zonal averaging or integration over several inertial periods should remove most of it. However, below the surface-intensified Ekman flow, the ageostrophic velocity along both sections shows wave-like structures of 50–80 m vertical scale with multiple maxima and minima. By applying a boxcar filter, these wave-like signals were separated from the cross-sectional ADCP velocity (Fig. 4). The appearance of these structures is mainly below the TTP and coincides with the layer of maximum buoyancy frequency. They are characterized by vertically alternating horizontal velocities with large horizontal coherence. Chereskin and Roemmich (1991) also reported the presence of such waves within the main thermocline, which were coherent over large horizontal space scales. These are thought to be near-inertial internal waves.

The section-averaged ageostrophic velocity had its maximum at the depth of the first valid bin of the ADCP (13–18 m), indicating that a shear existed within the ML, despite its homogeneous density. Chereskin and Roemmich (1991) examined this at 11° N in the Atlantic by zonally averaging a MLD-normalized ageostrophic velocity, and concluded that shear existed in the ML. However, Wijffels et al. (1994) applied the same technique and found a slab-like layer of ageostrophic velocity above the MLD at 10° N in the Pacific and attributed the discrepancy to a loose definition of the MLD by Chereskin and Roemmich (1991). Following their methods, we also examined whether there was shear in the ageostrophic velocity within the ML along the two sections.

It appears that at both sections, an ageostrophic shear existed in the ML, and this conclusion does not change if a more rigorous constraint on the MLD is used (Fig. 6).

The Ekman transport estimated by integrating the ageostrophic velocity zonally through the section and vertically to the local TTP depth is 6.2 ± 2.3 and 11.7 ± 2.1 Sv at 14.5° N and 11° S, respectively, which compares reasonably well to the estimates of 6.7 ± 3.5 and 13.6 ± 3.3 Sv by using the in situ wind stress data. By using a fixed integration depth of 50 m at 14.5° N and 80 m at 11° S, the ageostrophic Ekman transport is not significantly different from those calculated using the TTP depth, while using the MLD as the integration depth, the ageostrophic Ekman transport is about one-fourth smaller than using the TTP depth. This is an indication that the wind-driven flow penetrates beyond the ML to the TTP, and it is consistent with the findings of Wijffels et al. (1994), who reported that two-thirds of the wind-driven transport was within the ML and that the TTP is a reasonable choice for the integration depth of the Ekman flow. Note that above the first ADCP bin (13–18 m), the meridional ageostrophic velocity was linearly extrapolated using the values from the first two bins. However, when the surface meridional Ekman velocity is assumed equal to the velocity of the first measured ADCP bin (constant extrapolation), or extrapolated toward the theoretical Ekman solution for the surface velocity, the total ageostrophic transport would be up to 1.2 Sv smaller than the results shown above. Therefore, the linear extrapolation may to some extent overestimate the Ekman transport.

Between the two sections, the poleward Ekman transport divergence is 17.9 Sv, and the equatorward geostrophic convergence in the TTP layer is 6.2 Sv. This result agrees with the conclusion derived from theoretical consideration by Schott et al. (2004), who stated that the poleward Ekman divergence induced by the trade winds in both hemispheres is compensated for by an equatorward convergence due to the geostrophic flow in the upper layer, but the compensation is generally assumed not to be strong enough to reverse the Ekman divergence.

Note that the Ekman volume transport and the Ekman divergence estimated above were obtained by using data sampled during two Atlantic transects. The time series of Ekman transport calculated from the 6 h NCEP/CFSr wind stress from 2000 to 2011 shows a clear seasonal cycle and interannual variability at both latitudes. The transport strength reaches its maximum in the winter months of the respective hemisphere, and its minimum in the summer months of the respective hemisphere. The annual climatology with standard deviation from this time series is 7.9 ± 3.5 Sv at 14.5° N and −10.4 ± 3.3 Sv at 11° S. Our direct Ekman transport estimates agree well with the annual climatology. The uncertainties in the direct Ekman transport estimates are given by considering the aliasing effect of the near-inertial waves during the cruise. However, a larger uncertainty can be expected when the seasonal to interannual variability of the Ekman transport is taken into account.

The cumulative Ekman transport shows a rapid increase in the middle of the section and very little changes in the last quarter near the eastern boundary at both latitudes. This is because the zonal trade winds are generally strong and persistent in the western and middle parts of the basin, while relatively weak and unstable in strength and direction near the eastern boundary. Similar horizontal characteristics of Ekman transport were also seen at $11°$ N and $6°$ S in the Atlantic (Chereskin and Roemmich, 1991; Garzoli and Molinari, 2001) and $10°$ N in the Pacific (Wijffels et al., 1994). The GECCO2 ocean synthesis daily data were also used to calculate the meridional Ekman transport at $14.5°$ N and $11°$ S in the Atlantic by using the direct approach, which agrees very well with the observed results in respect to horizontal transport structure throughout the basin and the total transport amount. This was mostly due to the fact that GECCO2 assimilates the observed wind, and with a daily temporal resolution, it is possible for GECCO2 to reproduce the strength of the in situ wind, hence the Ekman transport. This good agreement has lent us more confidence in using GECCO2 as a reference in further studies on the MOC at the same latitudes.

An Ekman layer temperature and salinity must be assigned when calculating the Ekman heat and salt fluxes. Our results suggest that using the SST and SSS for the meridional Ekman heat and salt flux calculation at the two sections is only marginally different from calculations using the temperature and salinity in the TTP layer. It is rather the uncertainty in the Ekman volume transport estimates that dominates the uncertainties in the Ekman heat and salt fluxes. This is in good agreement with the finding at $10°$ N in the Pacific by Wijffels et al. (1994), while in striking contrast to that at $11°$ N in the Atlantic by Chereskin and Roemmich (1991), who found the transport-weighted Ekman temperature is $1 °C$ cooler than the surface value. The reason for such a contrast is not clear, but it is possible that in their case the TTP was much deeper than the MLD, especially in the western half of the basin.

Since Ekman volume, heat, and salt transport are significant upper layer components of the MOC with respect to the mass, heat, and freshwater conservation, further studies on the vertical and horizontal structures of the Ekman flow, as well as on the Ekman heat and salt fluxes, are expected to deepen our understanding and facilitate the studies on MOC strength and variability. This study would also provide some reference for the follow-up studies on the MOC at the same latitudes.

Competing interests. The authors declare that they have no conflict of interest.

Acknowledgements. We thank Toste Tanhua for closing the M96 section towards the African coast during M97, as well as all the research teams and crews onboard R/V *Meteor* for their dedicated work during the three cruise legs. We thank Armin Köhl for providing the GECCO2 data and the information about the data. We also thank Richard Greatbatch for the comments on a previous version of the manuscript, and Gerd Krahmann for providing the Matlab codes for uCTD sensor calibration. The blended level-4 wind stress data were provided by the Copernicus Marine Environment Monitoring Service (CMEMS). This study is funded by the Deutsche Forschungsgemeinschaft as part of cooperative project FOR1740 and by the German Federal Ministry of Education and Research as part of cooperative projects RACE (03F0605B) and SACUS (03F0751A).

Edited by: David Stevens

References

Alford, M. H.: Improved global maps and 54-year history of wind-work on ocean inertial motions, Geophys. Res. Lett., 30, 1424, https://doi.org/10.1029/2002GL016614, 2003.

Chereskin, T. K.: Direct evidence for an Ekman balance in the California Current, J. Geophys. Res., 100, 18261–18269, https://doi.org/10.1029/95JC02182, 1995.

Chereskin, T. K. and Roemmich, D.: A Comparison of Measured and Wind-derived Ekman Transport at $11°$ N in the Atlantic Ocean, J. Phys. Oceanogr., 21, 869–878, 1991.

Chereskin, T. K., Wilson, W. D., Bryden, H. L., Ffield, A., and Morrison, J.: Observations of the Ekman balance at $8°30'$ N in the Arabian Sea during the 1995 southwest monsoon, Geophys. Res. Lett., 24, 2541, https://doi.org/10.1029/97GL01057, 1997.

Chereskin, T. K., Wilson, W. D., and Beal, L. M.: The Ekman temperature and salt fluxes at $8°30'$ N in the Arabian Sea during the 1995 southwest monsoon, Deep-Sea Res. Pt. II, 49, , 1211–1230, https://doi.org/10.1016/S0967-0645(01)00168-0, 2002.

Cunningham, S. A., Kanzow, T., Rayner, D., Baringer, M. O., Johns, W. E., Marotzke, J., Longworth, H. R., Grant, E. M., Hirschi, J. J., Beal, L. M., Meinen, C. S., and Bryden, H. L.: Temporal Variability of the Atlantic Meridional Overturning Circulation at $26.5°$ N, 317, 935–938, 2007.

Ekman, V. W.: On the influence of the earth's rotation on ocean-currents, Arkiv för matematik, astronomi och fysik, Bd. 2, no. 11., 68 pp., 1905.

Elipot, S. and Gille, S. T.: Estimates of wind energy input to the Ekman layer in the Southern Ocean from surface drifter data, J. Geophys. Res. Ocean., 114, 1–14, https://doi.org/10.1029/2008JC005170, 2009.

EU Copernicus: Wind: The L4 near-real-time wind stress data from Copernicus Marine Environment Monitoring Service, EU Commission, available at: https://marine.copernicus.eu, last access: 20 September 2016.

Fairall, C. W., Bradley, E. F., Hare, J. E., Grachev, A. A., and Edson, J. B.: Bulk parameterization of air-sea fluxes: Updates and verification for the COARE algorithm, J. Climate, 16, 571–591, https://doi.org/10.1175/1520-0442(2003)016<0571:BPOASF>2.0.CO;2, 2003.

Fischer, J., Brandt, P., Dengler, M., Müller, M., and Symonds, D.: Surveying the upper ocean with the Ocean Surveyor: A new phased array Doppler current profiler, J. Atmos. Ocean. Tech., 20, 742–751, https://doi.org/10.1175/1520-0426(2003)20<742:STUOWT>2.0.CO;2, 2003.

Friedrichs, M. A. M. and Hall, M. M.: Deep circulation in the tropical North Atlantic, J. Mar. Res., 51, 697–736, https://doi.org/10.1357/0022240933223909, 1993.

Fu, Y., Karstensen, J., and Brandt, P.: Physical oceanography and meteorology during METEOR cruises M96, M97 and M98, https://doi.org/10.1594/PANGAEA.870516, 2017.

Ganachaud, A. and Wunsch, C.: Large-Scale Ocean Heat and Freshwater Transports during the World Ocean Circulation Experiment, J. Climate, 16, 696–705, https://doi.org/10.1175/1520-0442(2003)016<0696:LSOHAF>2.0.CO;2, 2003.

Garzoli, S. L. and Molinari, R. L.: Ageostrophic transport in the upper layers of the tropical Atlantic Ocean, Geophys. Res. Lett., 28, 4619–4622, https://doi.org/10.1029/2001GL013473, 2001.

Hood, E. M., Sabine, C. L., and Sloyan, B. M.: GO-SHIP Repeat Hydrography Manual: A Collection of Expert Reports and Guidelines, IOCCP Report No 14, ICPO Publication Series No. 134, available at: http://www.go-ship.org/HydroMan.html (last access: 8 August 2016), 2010.

Kanzow, T., Cunningham, S. A., Johns, W. E., Hirschi, J. J.-M., Marotzke, J., Baringer, M. O., Meinen, C. S., Chidichimo, M. P., Atkinson, C., Beal, L. M., Bryden, H. L. and Collins, J.: Seasonal Variability of the Atlantic Meridional Overturning Circulation at 26.5° N, J. Climate, 23, 5678–5698, https://doi.org/10.1175/2010JCLI3389.1, 2010.

Klein, B., Molinari, R. L., Müller, T. J., and Siedler, G.: A transatlantic section at 14.5N: Meridional volume and heat fluxes, J. Mar. Res., 53, 929–957, https://doi.org/10.1357/0022240953212963, 1995.

Köhl, A.: Evaluation of the GECCO2 ocean synthesis: transports of volume, heat and freshwater in the Atlantic, Q. J. Roy. Meteor. Soc., 141, 166–181, https://doi.org/10.1002/qj.2347, 2015.

Large, W. G. and Pond, S.: Open ocean momentum flux measurements in moderate to strong winds, J. Phys. Oceanogr., 11, 324–336, https://doi.org/10.1175/1520-0485(1981)011<0324:OOMFMI>2.0.CO;2, 1981.

Large, W. G. and Yeager, S. G.: Diurnal to decadal global forcing for ocean and sea-ice models: The data sets and flux climatologies, NCAR Technical Note NCAR/TN-460+STR, https://doi.org/10.5065/D6KK98Q6, 2004.

Larson, N. and Pedersen, A. M.: Temperature measurements in flowing water: Viscous heating of sensor tips, Proceedings of 1st IGHEM Meeting, Montreal, QC, Canada, June 1996, available at: http://www.seabird.com/viscous-heating-sensor-tips (last access: 8 August 2016), 1996.

Lenn, Y. and Chereskin, T. K.: Observations of Ekman Currents in the Southern Ocean, J. Phys. Oceanogr., 39, 768–779, https://doi.org/10.1175/2008JPO3943.1, 2009.

Levitus, S.: Meridional Ekman Heat Fluxes for the World Ocean and Individual Ocean Basins, J. Phys. Oceanogr., 17, 1484–1492, https://doi.org/10.1175/1520-0485(1987)017<1484:MEHFFT>2.0.CO;2, 1987.

Locarnini, R. A., Mishonov, A. V., Antonov, J. I., Boyer, T. P., Garcia, H. E., Baranova, O. K., Zweng, M. M., Paver, C. R., Reagan, J. R., Johnson, D. R., Hamilton, M., and Seidov, D.: World Ocean Atlas 2013, Volume 1: Temperature, edited by: Levitus, S., technically edited by: Mishonov, A., NOAA Atlas NESDIS 73, 40 pp., 2013.

Lueck, R. G. and Picklo, J. J.: Thermal Inertia of Conductivity Cells: Observations with a Sea-Bird Cell, J. Atmos. Ocean. Tech., 7, 756–768, https://doi.org/10.1175/1520-0426(1990)007<0756:TIOCCO>2.0.CO;2, 1990.

Mason, E., Colas, F., Molemaker, J., Shchepetkin, A. F., Troupin, C., McWilliams, J. C., and Sangrà, P.: Seasonal variability of the Canary Current: A numerical study, J. Geophys. Res., 116, C06001, https://doi.org/10.1029/2010JC006665, 2011.

McCarthy, G., Frajka-Williams, E., Johns, W. E., Baringer, M. O., Meinen, C. S., Bryden, H. L., Rayner, D., Duchez, A., Roberts, C., and Cunningham, S. A.: Observed interannual variability of the Atlantic meridional overturning circulation at 26.5° N, Geophys. Res. Lett., 39, 1–5, https://doi.org/10.1029/2012GL052933, 2012.

McCarthy, G. D., Smeed, D. A., Johns, W. E., Frajka-Williams, E., Moat, B. I., Rayner, D., Baringer, M. O., Meinen, C. S., Collins, J., and Bryden, H. L.: Measuring the Atlantic Meridional Overturning Circulation at 26° N, Prog. Oceanogr., 130, 91–111, https://doi.org/10.1016/j.pocean.2014.10.006, 2015.

McDonagh, E. L., King, B. A., Bryden, H. L., Courtois, P., Szuts, Z., Baringer, M., Cunningham, S. A., Atkinson, C., and McCarthy, G.: Continuous estimate of Atlantic oceanic freshwater flux at 26.5° N, J. Climate, 28, 8888–8906, https://doi.org/10.1175/JCLI-D-14-00519.1, 2015.

McDougall, T. J. and Barker, P. M.: Getting started with TEOS-10 and the Gibbs Seawater (GSW) Oceanographic Toolbox, SCOR/IAPSO WG127, 28 pp., available at: http://www.teos-10.org/pubs/Getting_Started.pdf (last access: 10 August 2016), 2011.

Montgomery, R. B.: Comments on "Seasonal variability of the Florida Current", by Niiler and Richardson, J. Mar. Res., 32, 533–535, 1974.

National Centers for Environmental Prediction/National Weather Service/NOAA/U.S. Department of Commerce: NCEP/NCAR Reanalysis Monthly Mean Subsets (from DS090.0), 1948-continuing, Research Data Archive at the National Center for Atmospheric Research, Computational and Information Systems Laboratory, available at: http://rda.ucar.edu/datasets/ds090. 2/ (last access: 20 September 2016), 1996 (updated monthly).

Pérez-Hernández, M. D., McCarthy, G. D., Vélez-Belchí, P., Smeed, D. A., Fraile-Nuez, E., and Hernández-Guerra, A.: The Canary Basin contribution to the seasonal cycle of the Atlantic Meridional Overturning Circulation at 26° N, J. Geophys. Res.-Oceans, 120, 7237–7252, https://doi.org/10.1002/2015JC010969, 2015.

Price, J. F., Weller, R. A., and Schudlich, R. R.: Wind-Driven Ocean Currents and Ekman Transport, Science, 238, 1534–1538, https://doi.org/10.1126/science.238.4833.1534, 1987.

Rabe, B., Schott, F. A., and Köhl, A.: Mean Circulation and Variability of the Tropical Atlantic during 1952–2001 in the GECCO Assimilation Fields, J. Phys. Oceanogr., 38, 177–192, https://doi.org/10.1175/2007JPO3541.1, 2008.

Roemmich, D. and Gilson, J.: The 2004–2008 mean and annual cycle of temperature, salinity, and steric height in the global ocean from the Argo Program, Prog. Oceanogr., 82, 81–100, https://doi.org/10.1016/j.pocean.2009.03.004, 2009.

Rudnick, D. L. and Klinke, J.: The underway conductivity-temperature-depth instrument, J. Atmos. Ocean. Tech., 24, 1910–1923, https://doi.org/10.1175/JTECH2100.1, 2007.

Saha, S., Moorthi, S., Pan, H.-L., Wu, X., Wang, J., Nadiga, S., Tripp, P., Kistler, R., Woollen, J., Behringer, D., Liu, H., Stokes, D., Grumbine, R., Gayno, G., Wang, J., Hou, Y.-T., Chuang, H.-Y., Juang, H.-M. H., Sela, J., Iredell, M., Treadon, R., Kleist, D., Van Delst, P., Keyser, D., Derber, J., Ek, M., Meng, J., Wei, H., Yang, R., Lord, S., van den Dool, H., Kumar, A., Wang, W., Long, C., Chelliah, M., Xue, Y., Huang, B., Schemm, J.-K., Ebisuzaki, W., Lin, R., Xie, P., Chen, M., Zhou, S., Higgins, W., Zou, C.-Z., Liu, Q., Chen, Y., Han, Y., Cucurull, L., Reynolds, R. W., Rutledge, G., and Goldberg, M.: NCEP Climate Forecast System Reanalysis (CFSR) 6-hourly Products, January 1979 to December 2010, Research Data Archive at the National Center for Atmospheric Research, Computational and Information Systems Laboratory, https://doi.org/10.5065/D69K487J, 2010.

Sato, O. T. and Polito, P. S.: Comparison of the Global Meridional Ekman Heat Flux Estimated from Four Wind Sources, J. Phys. Oceanogr., 35, 94–108, https://doi.org/10.1175/JPO-2665.1, 2005.

Schott, F. A., Dengler, M., Zantopp, R., Stramma, S., Fischer, J., and Brandt, P.: The Shallow and Deep Western Boundary Circulation of the South Atlantic at 5°–11° S, J. Phys. Oceanogr., 35, 2031–2053, https://doi.org/10.1175/JPO2813.1, 2005.

Schott, F. A., Mccreary, J. P., and Johnson, G. C.: Shallow Overturning Circulations of the Tropical-Subtropical Oceans, in: Earth's Climate, edited by: Wang, C., Xie, S. P., and Carton, J. A., American Geophysical Union, Washington, D.C., 261–304, https://doi.org/10.1029/147GM15, 2004.

Smith, S. D.: Coefficients for sea surface wind stress, heat flux, and wind profiles as a function of wind speed and temperature, J. Geophys. Res., 93, 15467–15472, https://doi.org/10.1029/JC093iC12p15467, 1988.

Smyth, W. D., Durland, T. S., and Moum, J. N.: Energy and heat fluxes due to vertically propagating Yanai waves observed in the equatorial Indian Ocean, J. Geophys. Res.-Oceans, 120, 1–15, https://doi.org/10.1002/2014JC010152, 2015.

Talley, L. D., Pickard, G. L., Emery, W. J., and Swift, J. H.: Descriptive Physical Oceanography: an introduction, 6th Edn., Elsevier, 555 pp., 2011.

Ullman, D. S. and Hebert, D.: Processing of underway CTD data, J. Atmos. Ocean. Tech., 31, 984–998, https://doi.org/10.1175/JTECH-D-13-00200.1, 2014.

Wang, W. and Huang, R.: Wind Energy Input to the Ekman Layer*, J. Phys. Oceanogr., 34, 1267–1275, https://doi.org/10.1175/1520-0485(2004)034<1267:WEITTE>2.0.CO;2, 2004.

Watanabe, M. and Hibiya, T.: Global estimates of the wind-induced energy flux to inertial motions in the surface mixed layer, Geophys. Res. Lett., 29, 1239, https://doi.org/10.1029/2001GL014422, 2002.

Wijffels, S., Firing, E., and Bryden, H.: Direct Observations of the Ekman Balance at 10° N in the Pacific, J. Phys. Oceanogr., 24, 1666–1679, https://doi.org/10.1175/1520-0485(1994)024<1666:DOOTEB>2.0.CO;2, 1994.

Wijffels, S. E., Toole, J., and Bryden, H.: The water masses and circulation at 10° N in the Pacific, Deep-Sea Res. Pt. I, 43, 501–544, https://doi.org/10.1016/0967-0637(96)00006-4, 1996.

Wunsch, C. and Heimbach, P.: Estimated Decadal Changes in the North Atlantic Meridional Overturning Circulation and Heat Flux 1993–2004, J. Phys. Oceanogr., 36, 2012–2024, https://doi.org/10.1175/JPO2957.1, 2006.

Zhang, J. and Rothrock, D.: Modeling Arctic sea ice with an efficient plastic solution, J. Geophys. Res., 105, 3325, https://doi.org/10.1029/1999JC900320, 2000.

Zweng, M. M., Reagan, J. R., Antonov, J. I., Locarnini, R. A., Mishonov, A. V., Boyer, T. P., Garcia, H. E., Baranova, O. K., Johnson, D. R., Seidov, D., and Biddle, M. M.: World Ocean Atlas 2013, Volume 2: Salinity, edited by: Levitus, S., technically edited by: Mishonov, A., NOAA Atlas NESDIS 74, 39 pp., 2013.

14

Validation of an ocean shelf model for the prediction of mixed-layer properties in the Mediterranean Sea west of Sardinia

Reiner Onken

Helmholtz-Zentrum Geesthacht (HZG), Centre for Materials and Coastal Research, Max-Planck-Straße 1, 21502 Geesthacht, Germany

Correspondence to: Reiner Onken (reiner.onken@hzg.de)

Abstract. The Regional Ocean Modeling System (ROMS) has been employed to explore the sensitivity of the forecast skill of mixed-layer properties to initial conditions, boundary conditions, and vertical mixing parameterisations. The initial and lateral boundary conditions were provided by the Mediterranean Forecasting System (MFS) or by the MERCATOR global ocean circulation model via one-way nesting; the initial conditions were additionally updated through the assimilation of observations. Nowcasts and forecasts from the weather forecast models COSMO-ME and COSMO-IT, partly melded with observations, served as surface boundary conditions. The vertical mixing was parameterised by the GLS (generic length scale) scheme (Umlauf and Burchard, 2003) in four different set-ups. All ROMS forecasts were validated against the observations which were taken during the REP14-MED survey to the west of Sardinia. Nesting ROMS in MERCATOR and updating the initial conditions through data assimilation provided the best agreement of the predicted mixed-layer properties with the time series from a moored thermistor chain. Further improvement was obtained by the usage of COSMO-ME atmospheric forcing, which was melded with real observations, and by the application of the k-ω vertical mixing scheme with increased vertical eddy diffusivity. The predicted temporal variability of the mixed-layer temperature was reasonably well correlated with the observed variability, while the modelled variability of the mixed-layer depth exhibited only agreement with the observations near the diurnal frequency peak. For the forecasted horizontal variability, reasonable agreement was found with observations from a ScanFish section, but only for the mesoscale wave number band; the observed sub-mesoscale variability was not reproduced by ROMS.

1 Introduction

In ocean acoustics research, the diagnostics and prediction of selected mixed-layer properties, such as the mixed-layer depth and the mixed-layer temperature, are of primary interest because they have a profound impact on the propagation of sound in the ocean. In this article, a high-resolution ocean circulation numerical model is presented which provides nowcasts and forecasts of these properties. The objectives are (i) to evaluate the sensitivity of the properties to different set-ups of the initial conditions, lateral boundary conditions, atmospheric forcing patterns, vertical grid, and vertical mixing parameterisations and (ii) to find a set-up which reproduces and best predicts the depth and the temperature of the mixed layer and the associated spatio-temporal variabilities obtained from observations.

By definition, temperature and salinity are constant in the mixed layer, and the sound speed increases slightly with depth due to the pressure effect (Dietrich et al., 1975). Therefore, sound rays in the mixed layer are refracted upwards and reflected at the sea surface. Hence, the mixed layer acts as a surface duct (Katsnelson et al., 2012). On the other hand, at a depth greater than the mixed-layer depth and because of the decreasing temperature, the rays are refracted in the other direction, i.e. towards greater depths. Consequently, in terms of passive acoustic monitoring, if a sound source is within the mixed layer, the sound cannot be "heard" at depths greater than the mixed-layer depth. If the sound source is located below the mixed layer, it cannot be heard in the mixed layer. The equivalent is true for the location of objects by active sonar: if the sonar is within the mixed layer, the acoustic signal can hardly reach an object at a greater depth, and vice versa. This is, of course, an idealised model based on ray theory which does not take account of the non-linear and frequency-dependent effects, but it clearly emphasises that knowledge about the depth of the mixed layer is mandatory for the planning and conduction of acoustic experiments.

The sound speed c in seawater is a function of temperature T, salinity S, and pressure p. Hence, small changes dc in the sound speed can be described by the total differential

$$dc = \left.\frac{\partial c}{\partial T}\right|_{S_0, p_0} dT + \left.\frac{\partial c}{\partial S}\right|_{T_0, p_0} dS + \left.\frac{\partial c}{\partial p}\right|_{T_0, S_0} dp, \quad (1)$$

where the subscripts T_0, S_0, and p_0 indicate that T, S, and p, respectively, are held constant during the execution of the partial differential. For the mid-latitudes and close to the sea surface ($T_0 = 15\,°C$, $S_0 = 35$, $p_0 = 0\,dbar$), the partial differentials in Eq. (1) yield $\partial c/\partial T \approx 3.2\,m\,s^{-1}\,°C^{-1}$ and $\partial c/\partial S \approx 1.2\,m\,s^{-1}$, which means that the fractional change in the sound speed with temperature is about 3 times larger than the change with salinity (Chen and Millero, 1977). Moreover, as typical spatio-temporal variations in temperature are $\mathcal{O}(10\,°C)$, but those of salinity are only $\mathcal{O}(1)$ at best, the first two terms in Eq. (1) yield 31.2 and $1.2\,m\,s^{-1}$. Hence, changes in the sound speed are largely controlled by changes in the temperature, and the impact of salinity variations in the mixed layer can confidently be ignored for the calculation of the sound speed. However, one may note that this is only true for the open ocean. In coastal areas, estuaries, and in polar regions, the salinity variations are frequently larger and the concurrent variations in temperature smaller.

Besides the temperature and the depth of the mixed layer, the temporal and horizontal variability of these two quantities require special attention (Pace and Jensen, 2002). The temporal variability at a fixed location is affected by temporal changes in the following:

– air–sea fluxes in momentum, heat, and fresh water;

– sea state conditions and internal waves;

– horizontal advection;

– vertical motion; and

– optical properties of the seawater.

The horizontal variability is due to spatial differences of the same quantities and, in addition, to the presence of mesoscale and sub-mesoscale features like fronts, meanders, eddies, and filaments (e.g. Medwin and Clay, 1998). Both the temporal and the horizontal variability impact the sound speed and the underwater sound propagation.

The objective of this article is to find a model set-up which predicts in the best possible way the mixed-layer properties and their spatio-temporal variabilities. While for the temporal variabilities the main focus of attention is directed at timescales between $\mathcal{O}(1\,h)$ and $\mathcal{O}(10\,days)$, the intention is to resolve the horizontal variabilities on scales of 10 km and

below. This requires a circulation model with a built-in vertical mixing scheme that accurately reproduces the diurnal cycle. A state-of-the-art scheme has recently been published by Ling et al. (2015). It is an enhancement of the turbulence closure model of Noh et al. (2011), which is similar to Mellor and Yamada (1982) but additionally takes into account the effects of wave breaking and Langmuir circulation. Ling et al. (2015) developed new numerical techniques and improved the schemes for the physics in Noh's model, which amongst others intensified the diurnal amplitude of the simulated sea surface temperature. Noh et al. (2016) incorporated Noh's model into a global ocean general circulation model, and they could show that the new mixing scheme helped to correct too-high mixed-layer temperatures and too-shallow mixed-layer depths in the high-latitude ocean. A similar approach was pursued by a series of papers by Bernie et al. (2005, 2007, 2008). In the first publication, a one-dimensional mixed-layer model was developed based on the K-profile parameterisation of Large et al. (1994). The model was forced with observed fluxes from a mooring in the tropical Pacific Ocean, and it qualitatively reproduced the observed diurnal variability in the sea surface temperature over a period of about 4 months. However, most of the time, the modelled temperature was higher than the observed one by up to $1\,°C$. Bernie et al. (2007) implemented the turbulent kinetic energy scheme of Blanke and Delecluse (1993) in an ocean circulation model, and this circulation model was coupled with an atmospheric circulation model (Bernie et al., 2008). The major outcome of the latter publication was that the inclusion of the diurnal cycle leads to a tropic-wide increase in the mean sea surface temperature, and, in addition, the authors could demonstrate that the modelled diurnal cycle was modulated by intraseasonal variations. The vertical mixing in all papers mentioned above was accomplished by turbulence closure models. By contrast, Gentemann et al. (2009) improved the parameterisation of the absorption of solar radiation in the diurnal heating bulk model of Fairall et al. (1996a). This change, combined with a reduction in accumulated heat and momentum, increased the model's responsiveness to changes in the surface heat flux and surface stress. Amongst others, the improved model predicted the vertical temperature profile within the diurnal thermocline, increased warming at low wind speeds, and decreased warming at high wind speeds.

The experimental area addressed in this study is situated in the Mediterranean Sea to the west of Sardinia (see Figs. 3, 4, 5, and 7). Here, the observational data from a June 2014 oceanographic survey are used to drive the aforementioned ocean circulation model and to validate the model results (Onken et al., 2014, 2016).

The reader may note that within this article all dates refer to the year 2014 and all times are in UTC (Coordinated Universal Time) except where otherwise stated.

2   Observational data

The observational data originate from the REP14-MED experiment, which took place in the eastern Sardo-Balearic Sea in the period of 6–25 June 2014. The collection of in situ data was accomplished by the NATO Research Vessel Alliance, the Research Vessel Planet of the German Ministry of Defence, a fleet of underwater gliders, surface drifters, one subsurface float, and six oceanographic moorings. Throughout this article, however, the author will refer only to the data of one mooring, denoted as M1, and to the CTD (conductivity–temperature–depth) data collected by the survey vessels and the gliders. For more details of the observations, see Onken et al. (2016).

2.1   Mooring M1

M1 (Fig. 1) was launched on 8 June at 07:14 at 8°12.98′ E, 39°30.80′ N (Fig. 3) and recovered on 20 June at 13:55. The water depth at the launch position was ≈ 150 m. M1 consisted of the central mooring M1$_{CTR}$ and a sideways-extending appendix M1$_{APP}$ floating at the sea surface. M1$_{CTR}$ was equipped with an upward-looking ADCP (acoustic doppler current profiler) mounted at a nominal depth of 100 m below the sea surface, a CTD probe at 1 m of depth, and a meteorological buoy at the surface. The appendix was connected by a 50 m long rope to M1$_{CTR}$ and extended to about 40 m in the vertical direction. Forty Starmon thermistors (Star-Oddi, Gardabaer, Iceland) were mounted along the vertical cable to record the temperature with high vertical resolution. In addition, four RBR data loggers (RBR, Ottawa, Canada) determined the actual depth of the Starmons. The nominal and actual vertical positions and the recorded parameters of the sensors are summarised in Table 1.

The Starmons recorded time t (10 s) and temperature T in intervals of 10 s. The RBRs additionally recorded pressure p. The depth z was calculated internally by the RBR software. As the Starmons did not have a pressure gauge, their actual vertical position was evaluated thereafter from the depth records of the RBRs because the positions of the Starmons relative to the RBRs was known. This was accomplished in the following way:

– the Starmon records at 3.0 and 7.0 m of depth were rejected because the recorders failed (sensor positions 7 and 15 in Table 1);

– all records before 8 June at 07:18 and after 20 June at 13:30 were clipped and then sub-sampled in 5 min intervals;

– at each instant, a second-order polynomial fit was applied to the actual depth of the RBRs versus their nominal depth; and

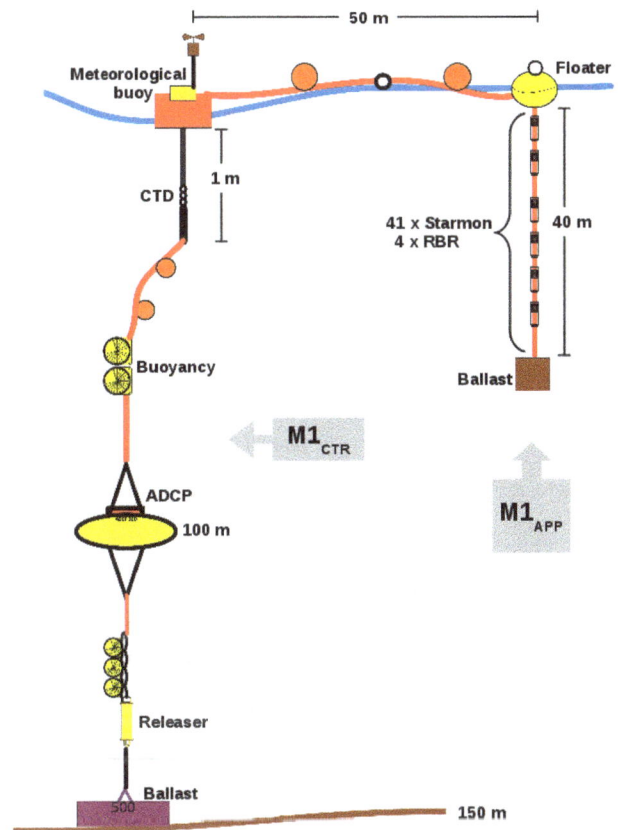

Figure 1. The design drawing of mooring M1. For explanations, see the text.

– the actual depths of the Starmon recorders were calculated with the polynomial using the previously calculated coefficients.

This procedure was advisable in order to correct for potential depth changes of the Starmons due to the actions of waves, internal waves, horizontal advection, and vertical shear. However, it turned out that these corrections were rather small: right at the sea surface at sensor position 1, the actual depth of the Starmon (Table 1) varied between 0 and 1.85 m, and the depth of the sensors at position 41 varied between 39.89 and 41.81 m. Hence, the applied corrections were around ±1 m.

The time series of the recorded temperature and the vertical temperature gradient are shown in Fig. 2, several properties of which may be challenging to reproduce for the circulation model.

– The near-surface temperature varied between about 22 °C and more than 24 °C. At the beginning of the recording period, it was around 23 °C, then it rose slowly and reached the maximum value on 12 and 14 June; afterwards, it decreased. The minimum of around 22 °C was reached on 17 June, and during the final 3 days, the temperature rose again by about 1 °C.

Table 1. The nominal and actual depths of the Starmon and RBR sensors mounted on the appendix of mooring M1. For the meaning of the recorded variables, see the text.

Sensor position	Sensor type	Recorded variables	Nominal depth [m]	Mean actual depth [m]	Remarks
1	Starmon	t, T	0.0	0.81	
2	Starmon	t, T	0.5	1.30	
3	Starmon	t, T	1.0	1.79	
4	Starmon	t, T	1.5	2.28	
5	Starmon	t, T	2.0	2.77	
6	Starmon	t, T	2.5	3.26	
7	Starmon	t, T	3.0	3.76	failed after 9 Jun 19:39
8	Starmon	t, T	3.5	4.25	
9	Starmon	t, T	4.0	4.74	
10	Starmon	t, T	4.5	5.23	
11	Starmon	t, T	5.0	5.72	
12	Starmon	t, T	5.5	6.23	
13	Starmon	t, T	6.0	6.71	
14	Starmon	t, T	6.5	7.21	
15	Starmon	t, T	7.0	7.70	broken
16	Starmon	t, T	7.5	8.19	
17	Starmon	t, T	8.0	8.69	
18	Starmon	t, T	8.5	9.18	
19	Starmon	t, T	9.0	9.68	
20	Starmon	t, T	9.5	10.17	
21	Starmon, RBR	t, T, p	10.0	10.69	
22	Starmon	t, T	11.0	11.66	
23	Starmon	t, T	12.0	12.65	
24	Starmon	t, T	13.0	13.65	
25	Starmon	t, T	14.0	14.64	
26	Starmon	t, T	15.0	15.63	
27	Starmon	t, T	16.0	16.63	
28	Starmon	t, T	17.0	17.63	
29	Starmon	t, T	18.0	18.63	
30	Starmon	t, T	19.0	19.62	
31	Starmon, RBR	t, T, p	20.0	20.63	
32	Starmon	t, T	22.0	22.63	
33	Starmon	t, T	24.0	24.64	
34	Starmon	t, T	26.0	26.65	
35	Starmon	t, T	28.0	28.67	
36	Starmon, RBR	t, T, p	30.0	30.69	
37	Starmon	t, T	32.0	32.71	
38	Starmon	t, T	34.0	34.74	
39	Starmon	t, T	36.0	36.77	
40	Starmon	t, T	38.0	38.81	
41	Starmon, RBR	t, T, p	40.0	40.85	

- The mixed-layer depth may be defined approximately by means of the depth at which the vertical temperature gradient is maximal. Between about 15 and 20 June, there is clear evidence for such a signal: the mixed-layer depth varied between about 4 m on 15 June and about 13 m on 18 and 19 June. However, between 8 and 14 June, the signal is rather indistinct: the nighttime mixed-layer depth ranged from about 2 m on 9 June to about 6 m on 10 June, but during daylight hours the maximum gradient was sometimes found right at the surface. Thus, a mixed layer in the "classical" sense did not exist.

- There are clear signals for temporal variability for both temperature and the depth of the mixed layer. The variability occurred on all scales between about 2 weeks and the Nyquist period of 10 min (twice the sampling interval; see above).

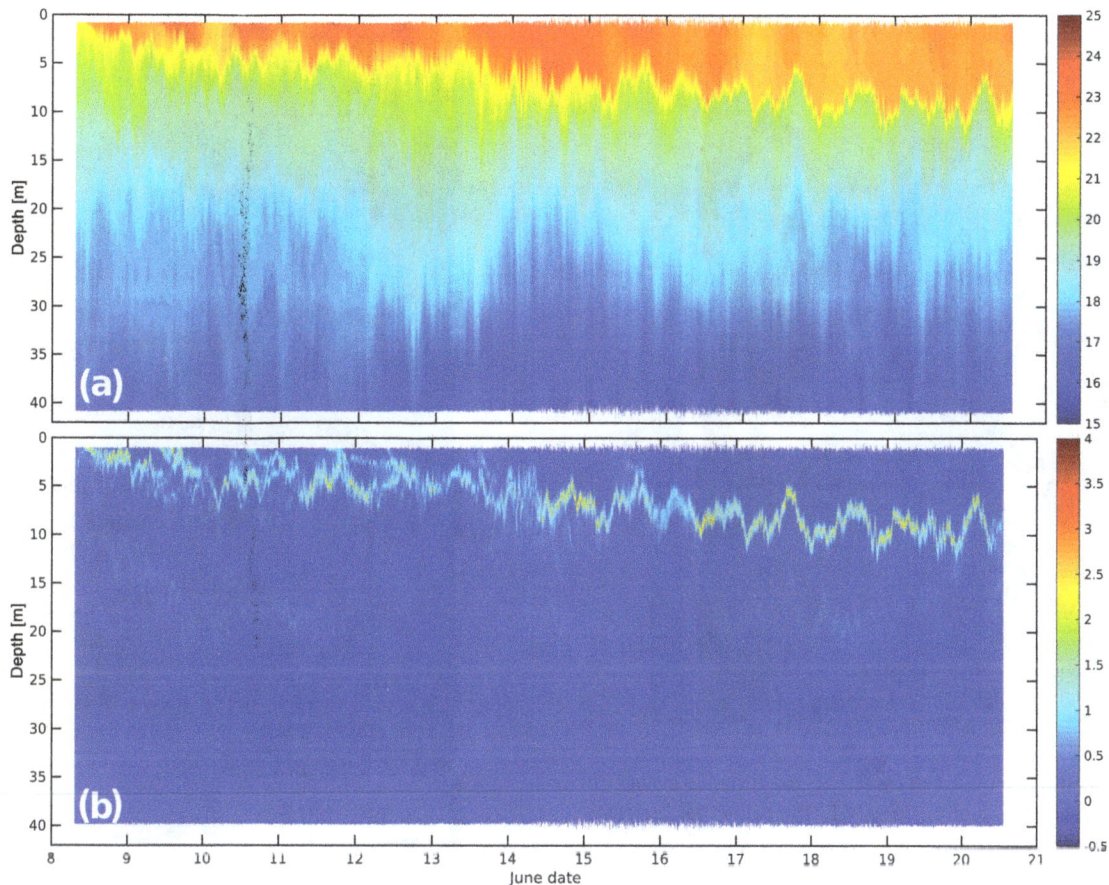

Figure 2. (a) The observed temperature [°C] at mooring M1 and **(b)** the vertical temperature gradient [°C m^{-1}].

2.2 Data collected by lowered CTD, gliders, and towed measuring systems

On both survey vessels, casts with lowered CTD probes were conducted during the entire survey, but only the casts taken during the 7–11 June period were used here (for a more detailed description of the probes and their calibration, see Onken et al., 2016). These casts belonged to the initialisation survey, the purpose of which was to provide realistic temperature and salinity data for the initialisation of the numerical models. In total, 108 casts were taken on a regular horizontal grid with a mesh size of ≈ 10 km (Fig. 3), resolving the internal Rossby radius of deformation, the first mode of which is around 13 km (Grilli and Pinardi, 1998). Eleven gliders (for their payloads, see Onken et al., 2016) were deployed on 8 and 9 June at the positions marked "L" in Fig. 4 and directed to their nominal zonal tracks. The scheduled tracks were arranged parallel to the CTD sections (Fig. 3) but offset by 5 km in the meridional direction. For the validation of the model forecasts, Alliance conducted a survey during 21–24 June with a ScanFish (EIVA, Skanderborg, Denmark). The tracks are shown in Fig. 5.

3 The circulation model

3.1 ROMS

The employed numerical ocean circulation model is ROMS, the Regional Ocean Modeling System. ROMS is a hydrostatic, free-surface, primitive equations ocean model, the algorithms of which are described in detail in Shchepetkin and McWilliams (2003, 2005). In the vertical, the primitive equations are discretised over variable topography using stretched terrain-following coordinates, or so-called *s* coordinates (Song and Haidvogel, 1994). In the version used for this article, spherical coordinates on a staggered Arakawa C grid are applied in the horizontal, and the horizontal mixing of the momentum and the tracers is along isopycnic surfaces. The vertical mixing is parameterised by means of the GLS (generic length scale) scheme (Umlauf and Burchard, 2003) using the stability function of Kantha and Clayson (1994). The air–sea interaction boundary layer in ROMS is based on the bulk parameterisation of Fairall et al. (1996b).

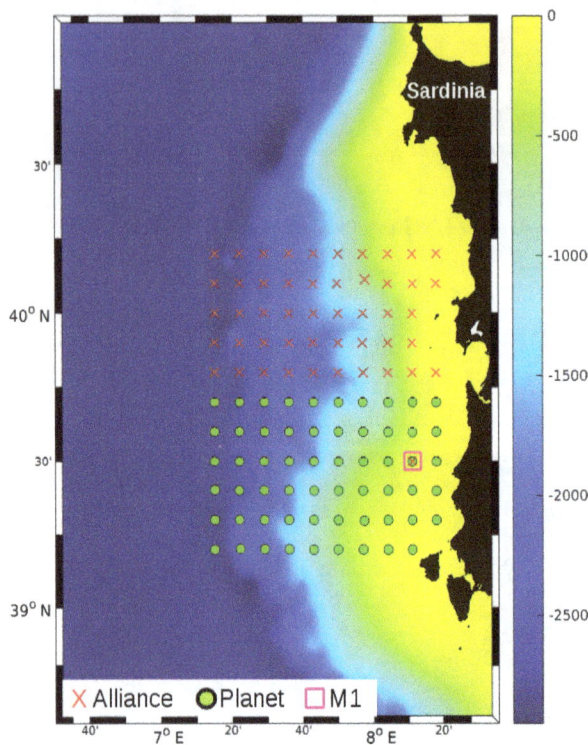

Figure 3. The CTD casts taken by Planet and Alliance during 7–11 June and the position of mooring M1. The colour bar indicates the water depth [m].

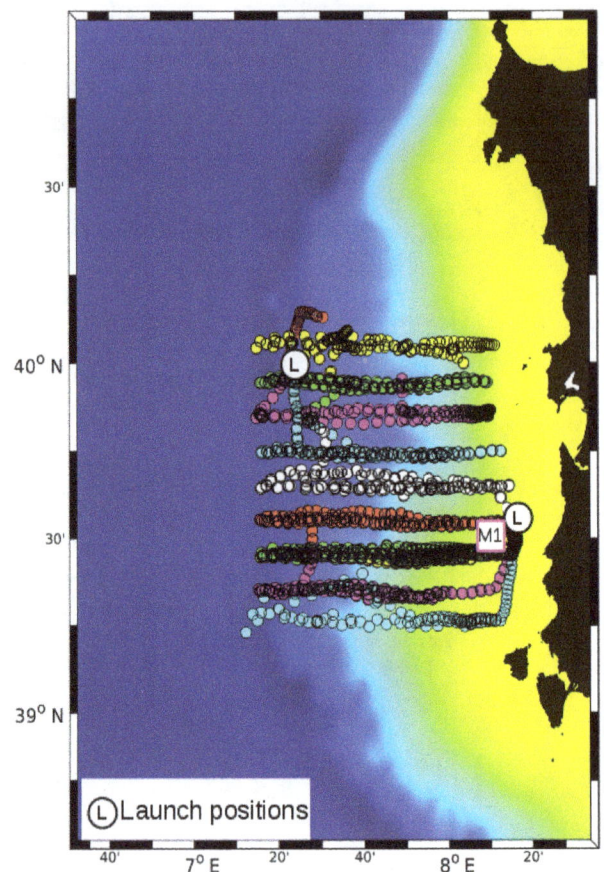

Figure 4. The actual glider tracks during 8–23 June. The small circles along the tracks show the surfacing positions. Each glider is marked by a different colour. The colour code for the bathymetry is the same as in Fig. 3.

3.2   The model domain and discretisation

The model domain is situated to the west of Sardinia and it is identical to the area shown in Fig. 3. The west and east boundaries are at 6°30.5 and 8°35.5′ E, while in the south and north the domain is limited by the 38°36.4 and 40°59.6′ N latitude circles, respectively. In the east–west direction, the domain is separated into 120 grid cells and in the south–north direction into 178 cells, which yields an average grid spacing of $\Delta x \approx \Delta y \approx 1500$ m in the zonal and meridional direction, respectively. A comparison with Fig. 3 reveals that the domain boundaries are kept away from the observations by about 30 arcmin; this was intended in order to mitigate a deterioration of the model solution at the observational sites due to false advection from the open boundaries.

Bathymetry data from the General Bathymetric Chart of the Oceans (GEBCO) with a spatial resolution of 1 arcmin were provided by the British Oceanographic Data Centre (BODC), and coastline data were obtained from NOAA (National Oceanic and Atmospheric Administration). In order to avoid the crowding of the s coordinates in shallow water regions, the bathymetry was clipped at 20 m, which is the minimum allowed water depth. For the smoothing of the bathymetry, a second-order Shapiro filter was applied. After smoothing, the so-called $rx0$ parameter resulted in 0.31, which is about 50 % higher than the maximum value

of 0.2 recommended by Haidvogel et al. (2000). However, $rx0$ is still lower than 0.4 as suggested in the ROMS forum (https://www.myroms.org/forum).

In the vertical direction, the model domain was separated into 70 s layers, the position of which is controlled by three parameters (θ_s, θ_b, h_c) and two functions, V_{tr} and V_{str}. Here, V_{tr} is the transformation equation, V_{str} is the vertical stretching function, θ_s and θ_b are the surface and bottom control parameters, and h_c is the critical depth controlling the stretching (for more details, see https://www.myroms.org/wiki/). For all ROMS runs shown below, the functions and parameters were selected as $V_{tr} = 2$, $V_{str} = 1$, $\theta_s = 5$, and $\theta_b = 0.4$, while h_c was kept a variable.

3.3   Initialisation

ROMS was initialised from nowcasts of the coarser Mediterranean Forecasting System (MFS; Dobrowsky et al., 2009; Tonani et al., 2014) or the MERCATOR global ocean circulation model (Drévillon et al., 2008). In either case, the downscaling from the parent to the child was accomplished first

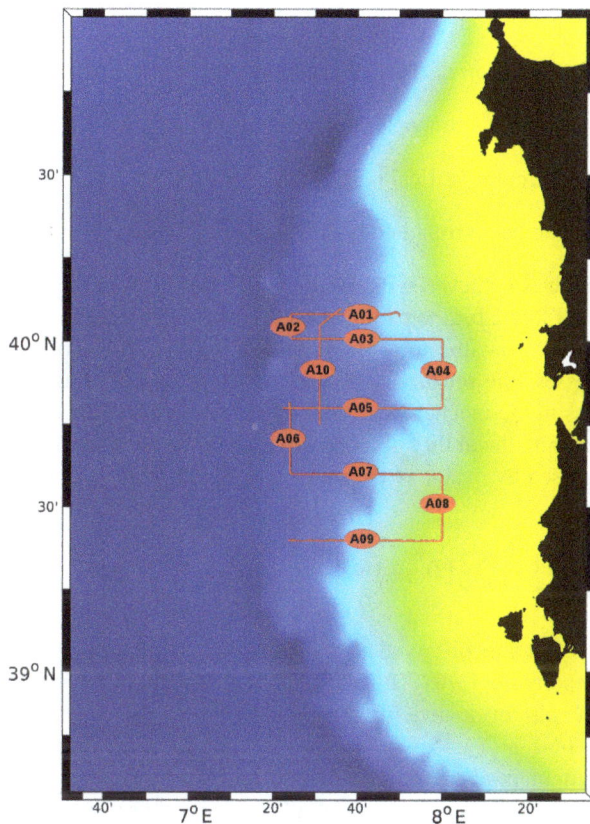

Figure 5. The ScanFish tracks of Alliance during 21–24 June. The colour code for the bathymetry is the same as in Fig. 3.

by the linear horizontal interpolation of the prognostic fields on the ROMS grid. As the maximum horizontal resolution of MFS was close to 7 km ($1/16°$) and that of MERCATOR was 9.25 km ($1/12°$), the scale factors were around 4.7 and 6.2, respectively. Thereafter, all fields were interpolated vertically from the horizontal depth levels to the s coordinates. A special issue was the alignment of the land masks of the parent and the child; if any wet grid cell of the child was covered by a dry grid cell of the parent, a smooth transition of all variables was created by taking the average of the surrounding parent cells. However, as this may lead to a violation of continuity by non-zero horizontal velocities normal to the land mask, all horizontal velocities next to the ROMS land mask were set to zero.

3.4 Lateral boundary conditions and nesting

The ROMS code includes various methods for the treatment of open boundaries. After extensive sensitivity studies, it was found that the following algorithms served best for the posed problem: for the sea surface elevation, the Chapman condition was selected (Chapman, 1985), and for all other quantities (i.e. barotropic and baroclinic momentum, turbulent kinetic energy, temperature, and salinity), the mixed radiation-

nudging conditions after Marchesiello et al. (2001) were applied.

The lateral time-dependent boundary conditions were provided by the parent by means of one-way nesting. However, the information from the parent was not instantaneously superimposed to the ROMS solution; additional nudging was applied to all prognostic variables (except for the sea surface elevation), which allowed these fields to adjust slowly to the parent values at the boundaries within an e-folding timescale of 2 days. In addition, a factor of 5 was used for the nudging timescales, which caused a stronger nudging on the inflow.

3.5 Surface boundary conditions

At the sea surface, the boundary conditions for the air–sea exchange of fresh water, momentum, and heat were evaluated from the outputs of two numerical weather prediction models and from the measurements of the meteorological buoy on top of M1 (see Fig. 1). This was accomplished by means of the wind field at 10 m (2 m for M1) of height, air temperature and relative humidity at 2 m, air pressure at sea level, total cloud cover (not available from M1), net shortwave radiation, and precipitation. The output of the weather prediction models was made available by the Italian weather service CNMCA (Centro Nazionale di Meteorologia e Climatologia Aeronautica) in two different set-ups, COSMO-ME and COSMO-IT. COSMO-ME covers the entire Mediterranean Sea with a horizontal resolution of 7 km and provided 72 h forecasts, while COSMO-IT encompasses Italy and the adjacent waters at the very high resolution of 2.2 km; however, the forecast range was only 24 h. The temporal resolution of both models was 1 h. The time series of all available variables from COSMO-ME, COSMO-IT, and the meteorological buoy are shown in Fig. 6 at the M1 position.

3.6 Data assimilation

In most of the model runs presented below, the temperature and salinity data from the shipborne CTD probes and gliders were assimilated using objective analysis (OA; see Bretherton et al., 1976; Carter and Robinson, 1987; Thomson and Emery, 2014). Namely, ROMS includes a module which enables data assimilation with the 4D-Var method. However, as 4D-Var is based on variational methods, it is rather expensive in terms of computer resources; according to parallel ROMS runs using 4D-Var (A. Funk, personal communication, 2016), the CPU time increases by about a factor of 10 compared to OA. During the integration of ROMS, the engine conducting the data assimilation was invoked every day at midnight, and it was controlled by six parameters:

- W: the width of the time window which determines what data are assimilated. In all ROMS runs below, $W = 48$ h; this setting was found to provide the best forecast skill (Onken, 2017). Hence, all temperature and salinity data of the previous and the following 24 h were selected for assimilation.

- *C*: the isotropic correlation length scale. $C = 15$ km was used throughout, which is approximately the internal Rossby radius of the Western Mediterranean in summer (Grilli and Pinardi, 1998). Isotropic correlation is a strong assumption, especially close to the coast. However, according to the observations from the ADCP measurements (I. Borrione, personal communication, 2016), predominantly meridional currents prevailed only in a 10 km wide strip along the Sardinian coast, while the rest of the 180 km wide model domain was characterised by an eddy field with alternating currents. Here, the usage of a non-isotropic correlation scale would deteriorate the results.

- δT_{obs}, δS_{obs}: the observational errors of temperature and salinity, respectively. $\delta T_{obs} = 0.5\,°\mathrm{C}$ and $\delta S_{obs} = 0.16$ were used throughout. These values were obtained from the variance of all CTD casts in the upper thermocline.

- δT_{clim}, δS_{clim}: the climatology errors. $\delta T_{clim} = 5 \times \delta T_{obs} = 2.5\,°\mathrm{C}$ and $\delta S_{clim} = 5 \times \delta S_{obs} = 0.8$ were applied.

3.7 Integration and output

All ROMS runs presented below were initialised on 1 June at 00:00 and integrated forward for 24 days until 25 June at 00:00. To satisfy the horizontal and the vertical CFL criterion, a baroclinic time step $\Delta t = 108$ s (800 steps per day) was chosen, and the number of barotropic time steps between each baroclinic time step was 40. Harmonic mixing along the isopycnals with an eddy diffusivity coefficient of $5\,\mathrm{m^2\,s^{-1}}$ was used for the horizontal diffusion of the tracers T and S, and a horizontal viscosity coefficient of $1\,\mathrm{m^2\,s^{-1}}$ was selected for the diffusion of momentum. Further on, a quadratic law using a coefficient of 0.003 was applied for the bottom friction, and the pressure gradient term was computed using the standard density Jacobian algorithm of Shchepetkin and Williams (2001).

The three-dimensional volume of all prognostic fields was written to an output file at 6 h intervals. For comparison of the ROMS results with the observed records at mooring M1, the time series of the vertical temperature profiles right at the position of M1 were written to an extra file at the full temporal resolution.

4 Sensitivity of near-surface temperature and mixed-layer depth

The purpose of this section is to investigate the impacts of the following on the temperature between the surface and about 42 m of depth (which was the vertical range of the M1 observations) and the depth of the mixed layer:

- initialising ROMS from different data sets;

- the set-up of the vertical grid;

- different atmospheric forcing patterns;

- different vertical mixing schemes; and

- the background eddy diffusivity.

This was achieved with 5 series of ROMS runs named A–E (see Table 2 for the parameter settings and the results of each model run) with a total of 28 runs. The task of each series was to assess the sensitivity of the ROMS forecast skill to variations in the mechanisms mentioned in the bullets above. For each run, the ability of ROMS to predict the temperature was assessed by means of the root mean square (rms) difference

$$\Delta T = \left[\frac{1}{N} \sum_{1}^{N} (T_{\mathrm{ROMS}} - T_{\mathrm{obs}})^2 \right]^{\frac{1}{2}} \tag{2}$$

between the observed temperature T_{obs} and the predicted temperature T_{ROMS} at each depth level of the observations. ΔT was evaluated for the period of 15 June at 00:00 to 20 June at 13:55 where N observations were available in 5 min intervals (Sect. 2.1). This interval was selected because it enabled the comparison of all runs with those which were forced by data assimilation until 12 June at 00:00. The 3-day lag between the last assimilation on 12 June and the start of the evaluation period on 15 June was granted to ROMS in order to recover from "assimilation shocks" which frequently become noticeable in the form of strong inertial oscillations. The experience from the precursor model runs has shown that such oscillations die off after about three to four inertial periods (18.7 h at 40° N). In order to synchronise the modelled and the observed temperature, T_{ROMS} was linearly interpolated in space and time on the observations. The equivalent method was also applied to the mixed-layer depth, D, which due to the lack of salinity observations at the M1 position was defined as the depth at which the temperature was 1 °C colder than the temperature at the surface for the first time (Lamb, 1984; Wagner, 1996). Hence,

$$\Delta D = \left[\frac{1}{N} \sum_{1}^{N} (D_{\mathrm{ROMS}} - D_{\mathrm{obs}})^2 \right]^{\frac{1}{2}} \tag{3}$$

is the rms difference of the mixed-layer depths.

4.1 Series A: initialising ROMS from different data sets

In this series, $h_c = 10$ m was selected for the critical depth. In the first run, referred to as A1, ROMS was initialised from MFS, while in A2 the initial conditions were provided by MERCATOR. A3 was initialised from MERCATOR as well, but temperature and salinity data from the CTD casts and 10 gliders taken during 7–12 June at 00:00 were additionally

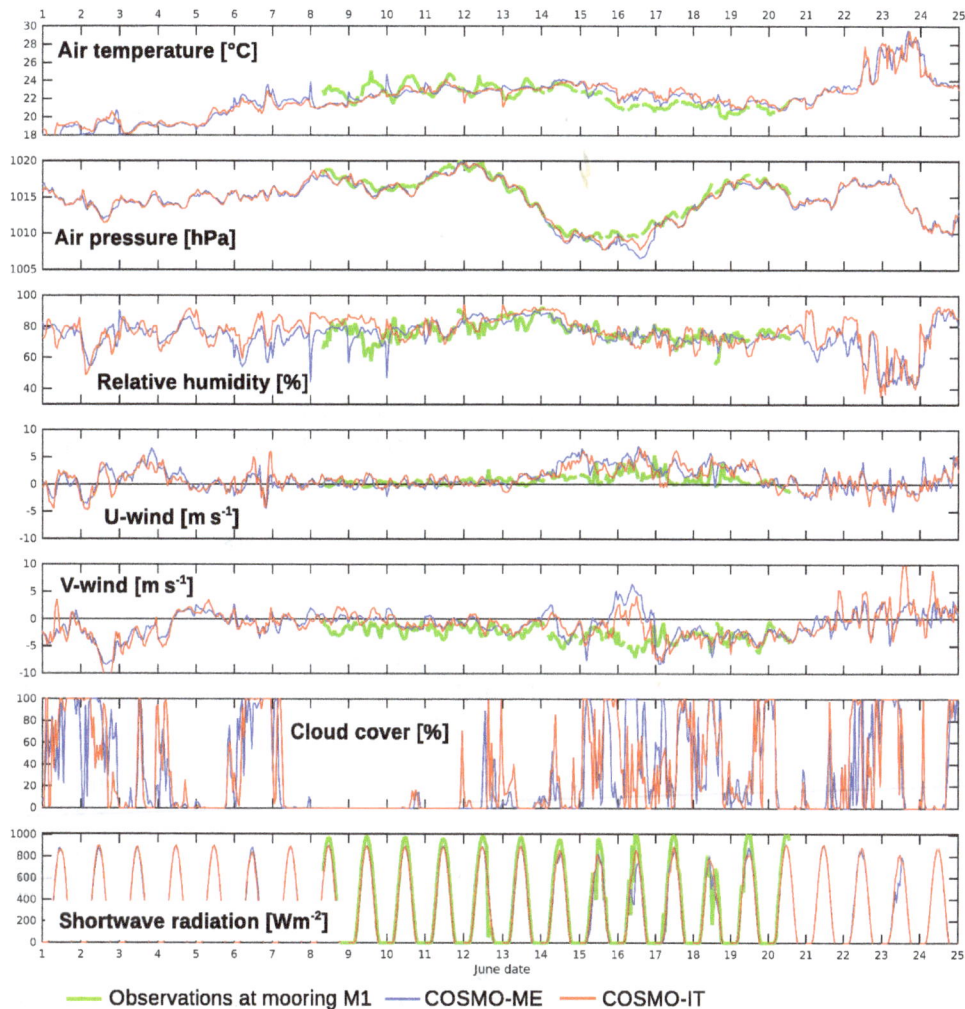

Figure 6. The time series of the measured and predicted atmospheric parameters at the site of mooring M1 from the observations of the meteorological buoy on top of M1, COSMO-ME, and COSMO-IT. U-wind and V-wind denominate the zonal and meridional wind components, respectively. The cloud cover was not recorded at M1. The precipitation is not shown because no precipitation was predicted or measured during the entire period.

assimilated (Sect. 2.2 and Figs. 3, 4, 7). The surface boundary conditions of all runs in the A series were provided by COSMO-ME.

Figure 8a shows the time series of the near-surface temperatures at 0.81 m of depth from runs A1–A3 in comparison with the corresponding observations of the uppermost Starmon sensor in M1 at the same depth level. In A1 and A2, the predicted temperatures agree reasonably well with the observations after 15 June, but before then the temperature exceeds the observations by several degrees. Extreme differences are visible during 12–14 June with differences of close to 3 °C. Figure 6 shows that during this period the predicted and observed wind speeds were close to $0\,\mathrm{m\,s^{-1}}$ and the shortwave radiation flux reached maximum values of more than $800\,\mathrm{W\,m^{-2}}$. Hence, as these quantities are the major drivers of the mixed-layer temperature, it is concluded that the selected parameterisation of the vertical mixing in

ROMS is not adequate for such calm situations. By contrast, as soon as the wind became stronger after 14 June, the maximum difference between the predicted and measured temperature is less than 1 °C. In A3 before 12 June, there is better agreement between the modelled and the observed temperature. However, as can be seen from the sudden drop in the modelled temperature at midnight on 10–12 June, the data assimilation led to an underestimation of the surface temperature. The reasons for this are twofold: first, some of the assimilated profiles started at 2 or even 3 m of depth because the measurements close to the surface were not reliable. In such cases, the uppermost measurements were extended to the surface and led to an underestimation of the near-surface temperature, which was sometimes significant because of the extremely shallow or even non-existent mixed layer. Second, the OA "advected" properties from the neighbouring casts which were not representative for the M1 position. On

Table 2. The parameter settings and results of the ROMS runs in series A–E. The bold text indicates the parameters or boundary forcing patterns which are varied within the respective series. The best run in each series is marked by an asterisk and serves as the control run for the successive series.

Run	h_c [m]	$rx1$	Parent	Mixing scheme	A_{VT} [$m^2\,s^{-1}$]	Atmospheric forcing	Assimilation	ΔT [°C]	$\overline{\Delta T}$ [°C]	ΔD [m]
Series A										
A1	10	21	**MFS**	GLS generic	1×10^{-6}	COSMO-ME	**no**	0.30	1.16	3.47
A2	10	21	**MERCATOR**	GLS generic	1×10^{-6}	COSMO-ME	**no**	0.53	1.12	2.97
A3*	10	21	**MERCATOR**	GLS generic	1×10^{-6}	COSMO-ME	**yes**	0.51	0.90	2.62
Series B										
B1*	**10**	21	MERCATOR	GLS generic	1×10^{-6}	COSMO-ME	yes	0.51	0.90	2.62
B2	**20**	27	MERCATOR	GLS generic	1×10^{-6}	COSMO-ME	yes	0.49	0.89	2.67
B3	**50**	23	MERCATOR	GLS generic	1×10^{-6}	COSMO-ME	yes	0.49	0.91	2.70
B4	**100**	25	MERCATOR	GLS generic	1×10^{-6}	COSMO-ME	yes	0.46	0.89	2.68
B5	**200**	27	MERCATOR	GLS generic	1×10^{-6}	COSMO-ME	yes	0.44	0.89	2.75
Series C										
C1	10	21	MERCATOR	GLS generic	1×10^{-6}	**COSMO-ME**	yes	0.51	0.90	2.62
C2	10	21	MERCATOR	GLS generic	1×10^{-6}	**COSMO-IT**	yes	0.42	0.98	3.45
C3*	10	21	MERCATOR	GLS generic	1×10^{-6}	**M1**	yes	0.80	0.70	3.28
Series D										
D1	10	21	MERCATOR	**GLS generic**	1×10^{-6}	M1	yes	0.80	0.70	3.28
D2	10	21	MERCATOR	**GLSk-kl**	1×10^{-6}	M1	yes	0.50	0.61	2.86
D3	10	21	MERCATOR	**GLSk-ϵ**	1×10^{-6}	M1	yes	0.51	0.60	2.95
D4*	10	21	MERCATOR	**GLSk-ω**	1×10^{-6}	M1	yes	0.41	0.61	2.71
Series E										
E1	10	21	MERCATOR	GLS k-ω	$\mathbf{1 \times 10^{-6}}$	M1	yes	0.41	0.61	2.71
E2	10	21	MERCATOR	GLS k-ω	$\mathbf{5 \times 10^{-6}}$	M1	yes	0.38	0.62	2.74
E3	10	21	MERCATOR	GLS k-ω	$\mathbf{1 \times 10^{-5}}$	M1	yes	0.35	0.59	2.60
E4	10	21	MERCATOR	GLS k-ω	$\mathbf{2 \times 10^{-5}}$	M1	yes	0.31	0.57	2.49
E5	10	21	MERCATOR	GLS k-ω	$\mathbf{3 \times 10^{-5}}$	M1	yes	0.31	0.56	2.36
E6	10	21	MERCATOR	GLS k-ω	$\mathbf{4 \times 10^{-5}}$	M1	yes	0.35	0.57	2.25
E7	10	21	MERCATOR	GLS k-ω	$\mathbf{5 \times 10^{-5}}$	M1	yes	0.44	0.54	2.13
E8	10	21	MERCATOR	GLS k-ω	$\mathbf{6 \times 10^{-5}}$	M1	yes	0.49	0.56	2.11
E9	10	21	MERCATOR	GLS k-ω	$\mathbf{7 \times 10^{-5}}$	M1	yes	0.55	0.58	2.05
E10	10	21	MERCATOR	GLS k-ω	$\mathbf{8 \times 10^{-5}}$	M1	yes	0.66	0.57	2.13
E11	10	21	MERCATOR	GLS k-ω	$\mathbf{9 \times 10^{-5}}$	M1	yes	0.72	0.55	2.08
E12	10	21	MERCATOR	GLS k-ω	$\mathbf{1 \times 10^{-4}}$	M1	yes	0.80	0.57	2.15
E13	10	21	MERCATOR	GLS k-ω	$\mathbf{2 \times 10^{-4}}$	M1	yes	1.37	0.60	2.92

13 June, the modelled temperature again exceeds the observations by almost 2 °C, but the difference is less than in A1 and A2. After 15 June, the A3 temperature is rather close to the temperatures in A1 and A2. As a skill measure for the forecasted near-surface temperature, ΔT was evaluated for all runs and resulted in $\Delta T = 0.30$ °C in A1, $\Delta T = 0.53$ °C in A2, and $\Delta T = 0.51$ °C in A3 (see also the legend box in Fig. 8 and Table 2).

The temporal evolution of the mixed-layer depth is displayed in Fig. 8b. As revealed by the decreasing rms differences ΔD between the modelled and observed mixed-layer depths, the forecast skill increases from A1 to A2 and from A2 to A3. The close agreement between the observed and modelled mixed-layer depth in A3 before 12 June, which was forced by the assimilation, is noteworthy. The mismatch between the model and the observations during 12–15 June is also remarkable as another indication of an inadequate parameterisation of the mixed-layer dynamics at low wind speeds.

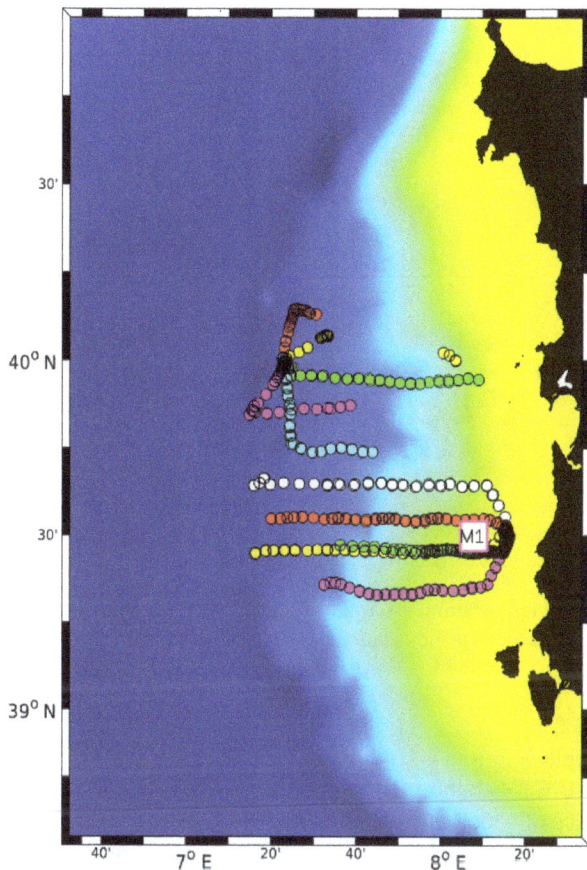

Figure 7. The actual surfacing positions of all assimilated gliders during 7–11 June. Each glider is marked by a different colour. The colour code for the bathymetry is the same as in Fig. 3.

just below, T_{A3} is up to more than $3\,°C$ lower than the observed temperature. This aberrant cold layer can be identified during the whole model run. Apparently, the modelled mixed-layer depth is shallower than the observed one. This is illustrated by the vertical temperature gradient in Fig. 10e. Namely, a comparison with Fig. 10d reveals that the generally descending trend of the maximum gradient is similar, but the depth of the modelled maximum is always less than the observed one. Moreover, the observed variability is significantly higher than the modelled one. While for the entire period there is clear evidence of a strong diurnal variability in the observations (e.g. the deep mixed layer in the early morning and the shallow mixed layer in the afternoon), the modelled variability is much less pronounced. Another feature worth mentioning is that the thermocline is too warm during the assimilation phase before 12 June (Fig. 10c). It has been verified that this was caused by the assimilation of the glider data because this feature is not present in a run where only casts from lowered CTD were assimilated (not shown). As can be seen from Figs. 3 and 7, two CTD casts were taken exactly at the M1 position, while numerous glider casts are close to M1 (note that the meridional offset of the glider tracks with respect to the CTD meridional sections was 5 km). Thus, as the correlation scale of the OA was 15 km, the modelled temperature at M1 was primarily determined by the glider measurements because the large number of glider profiles reduced the statistical weight of the two CTD casts.

4.2   Series B: sensitivity to the set-up of the vertical grid

If the transformation equation, the vertical stretching function, and the total number of layers are held constant, the layer thicknesses of the ROMS vertical grid are controlled by the surface and bottom control parameters, θ_s and θ_B, and the critical depth, h_c. For mixed-layer modelling in shelf areas, it would be desirable to have a high vertical resolution close to the surface, which can be achieved by either increasing θ_s or decreasing h_c. However, as increasing θ_s would make the vertical transformation more non-linear, it was decided to keep $\theta_s = 5$ constant and vary only h_c. In this series, the sensitivity of the ROMS results to five different settings of the critical depth is investigated using $h_c \in \{10, 20, 50, 100, 200\}\,m$. For each of these choices, the impact on the layer thicknesses at the position of mooring M1 is illustrated in Fig. 11. A minimum layer thickness of $0.27\,m$ right at the sea surface is achieved by $h_c = 10\,m$ in run B1, while the thickness of that layer gradually increases in B2–B5. In the latter ($h_c = 200\,m$), the thickness is close to $1.3\,m$. B1, because it is identical to A3, is the control run.

For all runs in this series, the temporal evolution of the mixed-layer properties is displayed in Fig. 12. Although still too high around 14 June, the near-surface temperatures in all runs of this series most resemble the observations during the entire integration period, which is also expressed by the corresponding low values for ΔT; the minimum $\Delta T = 0.44\,°C$

The vertical distribution of the rms temperature differences ΔT of all runs in the A series is shown in Fig. 9. It is demonstrated that at most depth levels, ΔT is lower or equal in A2 compared to A1. The assimilation in A3 led to a further significant decrease between about 4 and $35\,m$ of depth; only above $4\,m$ and below $35\,m$ of depth is ΔT higher in A3. The generally better forecast skill of A3 is also supported by $\overline{\Delta T}$, the vertical mean of ΔT, which is greater than $1\,°C$ in A1 and A2 but only $0.90\,°C$ in A3 (see also Table 2). In summary, nesting ROMS in MERCATOR and assimilating CTD profiles provided the best forecasts for the temperature and the depth of the mixed layer and the thermocline temperature below about $4\,m$ of depth. Therefore, all runs in the B series will be based on A3.

The temporal evolution of the modelled temperature in A3 at the position of mooring M1 is shown in Fig. 10b. In comparison with Fig. 10a, the modelled temperature close to the sea surface is too high on 13 and 14 June, while at depths greater than about $3–10\,m$, T_{A3} appears too low. This is confirmed by Fig. 10c, which exhibits the temperature difference $T_{A3} - T_{M1}$: in approximately the upper $2\,m$ depth range, T_{A3} partly exceeds T_{M1} by about $2\,°C$ on these days, and

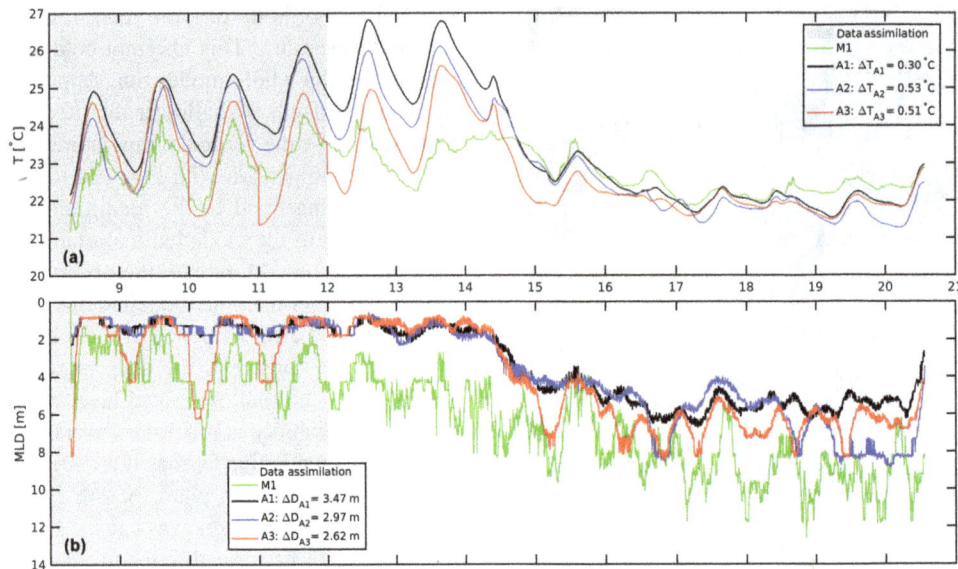

Figure 8. ROMS runs A1, A2, and A3: the time series of the (a) near-surface temperature at 0.81 m of depth, (b) the mixed-layer depth (MLD), and the corresponding observations at mooring M1. The numbers on the abscissae indicate June dates. The period for which the data are assimilated is highlighted with grey shading.

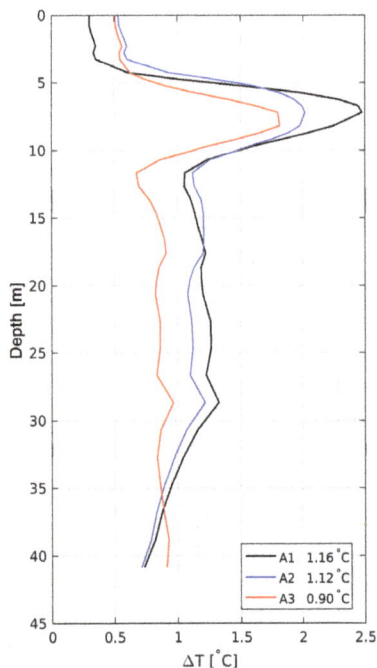

Figure 9. ROMS runs A1, A2, and A3: the rms temperature differences ΔT [°C] between the modelled temperature T_{ROMS} and the observed temperature T_{obs} evaluated at the actual depths of the observations. The vertical mean $\overline{\Delta T}$ is written in the second column of the legend box. ΔT was computed only for the period after 15 June at 00:00.

is obtained from B5, while the highest is found in B1 ($\Delta T = 0.51\,°C$). For the mixed-layer depth, there is no clear evidence of which run might do best. ΔD varies only in a rather narrow range between 2.62 m in B1 and 2.75 m in B5. The vertical distributions of ΔT (Fig. 13) and the vertical averages $\overline{\Delta T}$ are almost identical for all runs. However, right at the surface, ΔT is minimal in B5 as shown in Fig. 12a. As the above results did not reveal a clear tendency of which choice for h_c yielded the best results, it was decided to continue with B1 ($h_c = 10\,m$) as the control run in series C below. This decision was guided by Bernie et al. (2008), who asserted that a minimum vertical resolution of 1 m is mandatory to resolve the diurnal cycle of the sea surface temperature. Another criterion for this decision was the $rx1$ grid parameter (i.e. the Haney condition, after Haney, 1991) being at a minimum in B1 (see Table 2).

4.3 Series C: sensitivity to atmospheric forcing

Series C consists of three model runs, C1, C2, and C3. C1 is identical to B2; in C2, the surface boundary conditions were provided by COSMO-IT instead of COSMO-ME. In C3, the atmospheric forcing was defined by means of the observations of the meteorological buoy on top of mooring M1. Here, the observations were spread uniformly across the entire model domain whenever available. If no observations were available, i.e. before 8 June and after 20 June, the atmospheric fields of COSMO-ME were used. As observations of cloudiness were not available from M1, the corresponding fields from COSMO-ME were used throughout.

According to Fig. 14a, the predicted near-surface temperature from C2 closely resembles that of C1, except for 14–

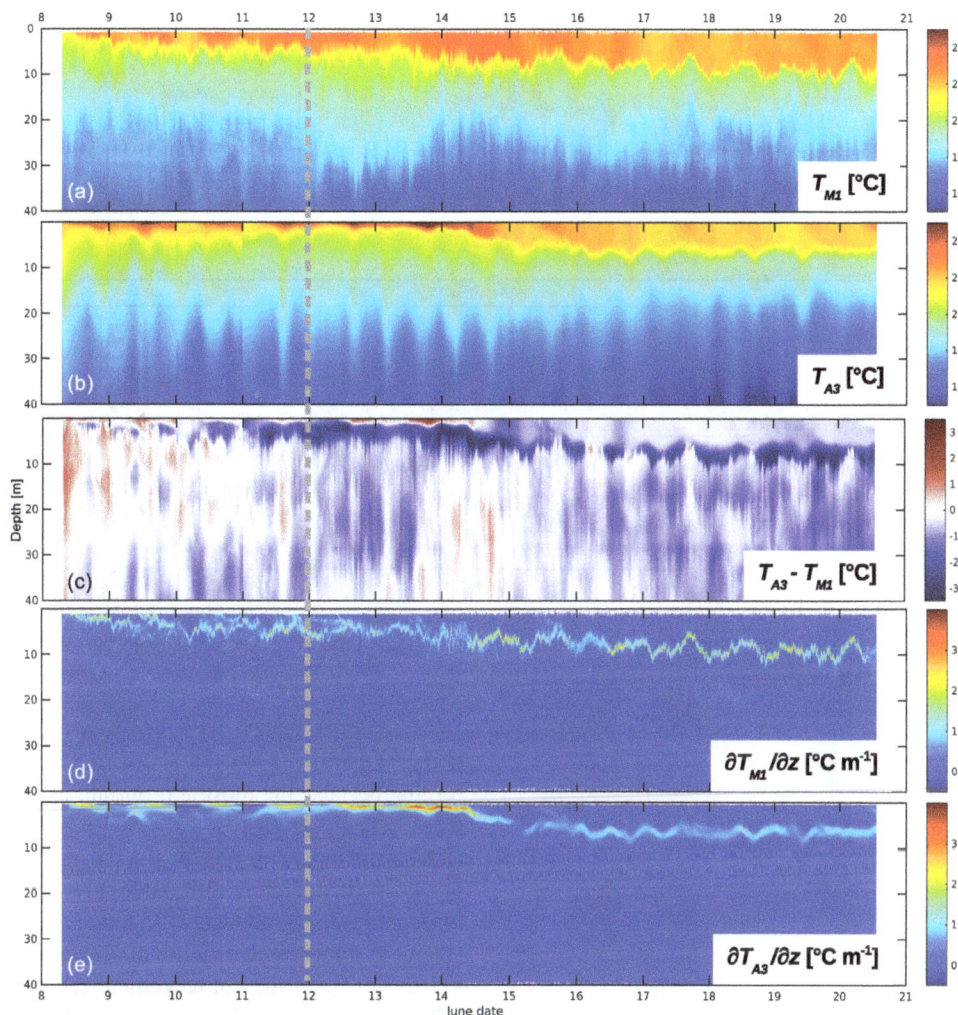

Figure 10. (a) The observed temperature at mooring M1, **(b)** the modelled temperature from ROMS run A3, and **(c)** the difference between the modelled and the observed temperature. The vertical temperature gradient from **(d)** the observations and **(e)** from A3. The instant of the last data assimilation is indicated by the the grey dashed vertical line.

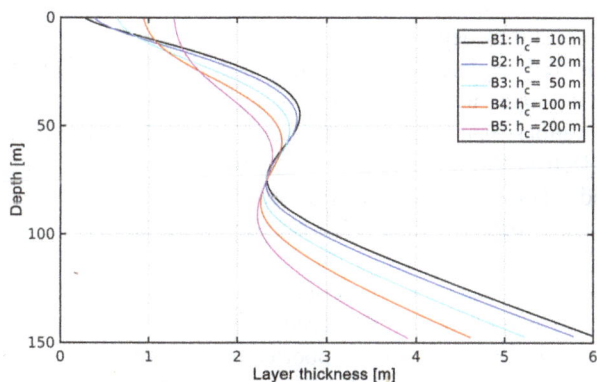

Figure 11. The layer thicknesses at the position of mooring M1 for various assumptions of the critical depth h_c.

17 June when the temperatures in C2 are about $1\,°C$ higher. Apparently, this was driven by the different wind forecasts of the weather prediction models (Fig. 6). Before 14 June, the wind forecasts of both models were almost identical, but for the following 2 days during a period of stronger winds, the forecasts differ from each other. Overall, the near-surface temperature does not appear to be very sensitive to the choice of the weather forecast models. This is also expressed by ΔT, which attains similar values of 0.51 and $0.42\,°C$. The signature of the temperature changes considerably when ROMS was driven by the weather observed at M1; this is already evident during 8–10 June when the modelled temperature in C3 is different from C1 and C2. After 15 June, it is mostly higher than both the observations and the predictions of C1 and C2, which correspondingly leads to a higher ΔT of $0.80\,°C$. With respect to the modelled mixed-layer depth (Fig. 14b) and based on the ΔD criterion, C1

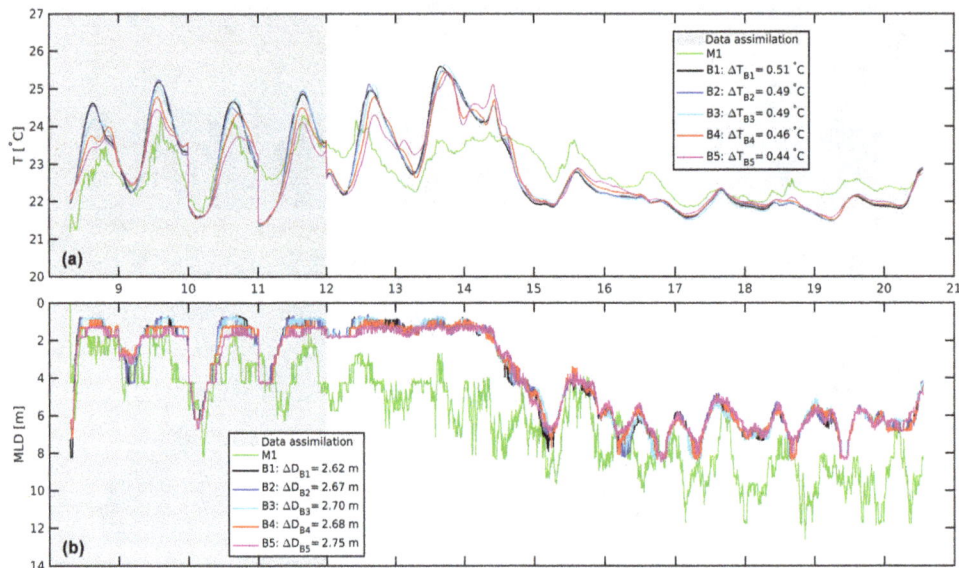

Figure 12. ROMS runs B1–B5: the time series of **(a)** the near-surface temperature at 0.81 m of depth, **(b)** the mixed-layer depth (MLD), and the corresponding observations at mooring M1. The numbers on the abscissae indicate June dates. The period for which the data are assimilated is highlighted with grey shading.

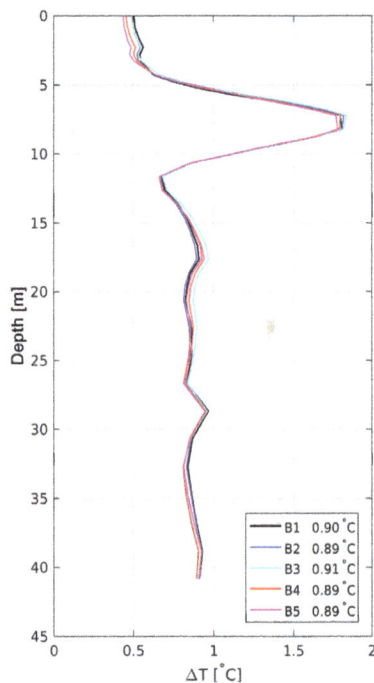

Figure 13. ROMS runs B1–B5: the rms temperature differences ΔT [°C] between the modelled temperature T_{ROMS} and the observed temperature T_{obs} evaluated at the actual depths of the observations. The vertical mean $\overline{\Delta T}$ is written in the second column of the legend box. ΔT was computed only for the period after 15 June at 00:00.

pothesis that the mismatch is not caused by the atmospheric forcing because the most appropriate forcing was applied in C3.

A surprising result was obtained from the vertical structure of the rms temperature difference (Fig. 15). Below about 3 m of depth, ΔT_{C1} is about 0.1 °C lower than ΔT_{C2}, but a considerable improvement in the predicted stratification is provided by C3. In the entire vertical range below about 5 m, ΔT_{C3} is up to 0.4 °C lower than ΔT_{C1}. Only right at the surface is ΔT_{C3} approximately 0.3 °C higher than the corresponding values from C1 and C2, which is obviously due to the above-mentioned mismatch after 15 June. C3 provides the best results for the temperature stratification in the thermocline. As the temperature in this depth range was definitely not affected by the heat exchange at the sea surface ($\approx 90\%$ of the shortwave radiation is absorbed in the uppermost 1 m depth range), its improvement could only be achieved by lateral advection, which is controlled by the wind; apparently, the wind is better represented in the observations than in the weather forecasts. To summarise, the objective skill measure ΔT for the near-surface temperature and ΔD for the mixed-layer depths indicate that C1 provides the best forecast, while ΔT_{C3} is clearly superior to ΔT_{C1} and ΔT_{C2} in the thermocline. The latter confirms the decision to use C3 as a control run in series D because the advective processes are obviously reproduced best.

The decision for C3 is supported by Fig. 16. By visual inspection, the evolution of the predicted temperature pattern in C3 (Fig. 16c) resembles the observations (Fig. 16a) more than in C1 (Fig. 16b). Namely, the near-surface temperature is too high, but the thickness of the warm layer during 16–

is superior to C2 and C3, but the large discrepancies during 12–15 June between the predictions and the observation are still present in all three runs. This corroborates the above hy-

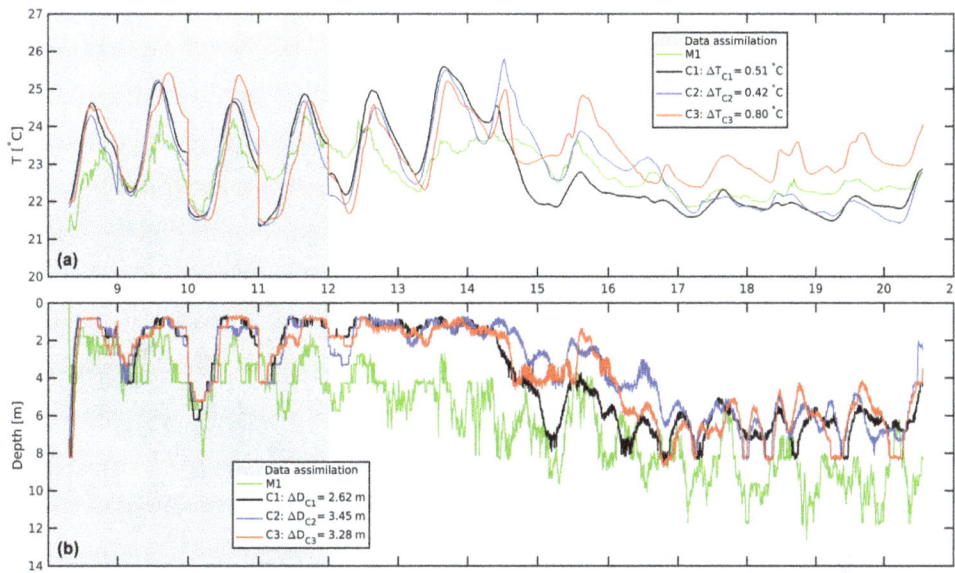

Figure 14. ROMS runs C1, C2, and C3: the time series of (**a**) the near-surface temperature at 0.81 m of depth, (**b**) the mixed-layer depth (MLD), and the corresponding observations at mooring M1. The numbers on the abscissae indicate June dates. The period for which the data are assimilated is highlighted with grey shading.

20 June is roughly the same as in the observations, close to 10 m. Moreover, the depth and the variability of the maximum vertical temperature gradient in C3 resembles the observed pattern to a larger degree (Fig. 16d, e, f), although the vertical temperature gradient is still too weak.

4.4 Series D: sensitivity to the vertical mixing parameterisation

The GLS scheme (Umlauf and Burchard, 2003) provides a generalisation of a class of differential length-scale equations used in turbulence models for oceanic flows. Commonly used models, like the k-kl model of Mellor and Yamada (1982), the k-ϵ model (Rodi, 1987), and the k-ω model (Wilcox, 1988), are recovered as special cases of the generic scheme. Here, k is the turbulent kinetic energy, l is the length scale of the turbulence, ϵ is the dissipation rate, and ω is the specific dissipation rate. In series A–C, the GLS vertical mixing scheme was applied using its generic parameters as formulated by Umlauf and Burchard (2003). In the following, D1 is identical to C3 serving as the control run, the GLS scheme with the k-kl parameterisation is applied in D2, the k-ϵ parameters are applied in D3, and the k-ω parameterisation, which was adjusted to oceanic conditions by Umlauf et al. (2003), is applied in D4.

After 12 June, the near-surface temperature of all runs is correlated with the observations (Fig. 17a), but is mostly still too high. Moreover, the graphs indicate that the temperatures from D2, D3, and D4 are closer to the observed ones, which is also expressed by $\Delta T_{D2} = 0.50\,°C$, $\Delta T_{D3} = 0.51\,°C$, and $\Delta T_{D4} = 0.41\,°C$, while $\Delta T_{D1} = 0.80\,°C$. For the mixed-layer depth (Fig. 17b), the best agreement with the

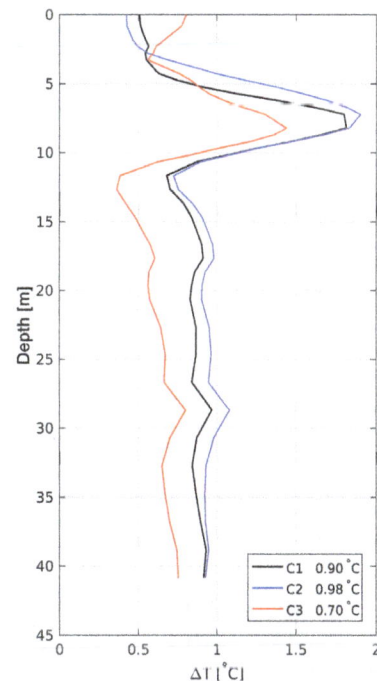

Figure 15. ROMS runs C1, C2, and C3: the rms temperature differences ΔT between the modelled temperature T_{ROMS} and the observed temperature T_{obs} evaluated at the actual depths of the observations. ΔT was computed only for the period after 15 June at 00:00.

observations was obtained from D4 with $\Delta D_{D4} = 2.71\,m$. However, the mixed layer was mostly too shallow in all runs in this series. Hence, based on the ΔT and ΔD criteria, the

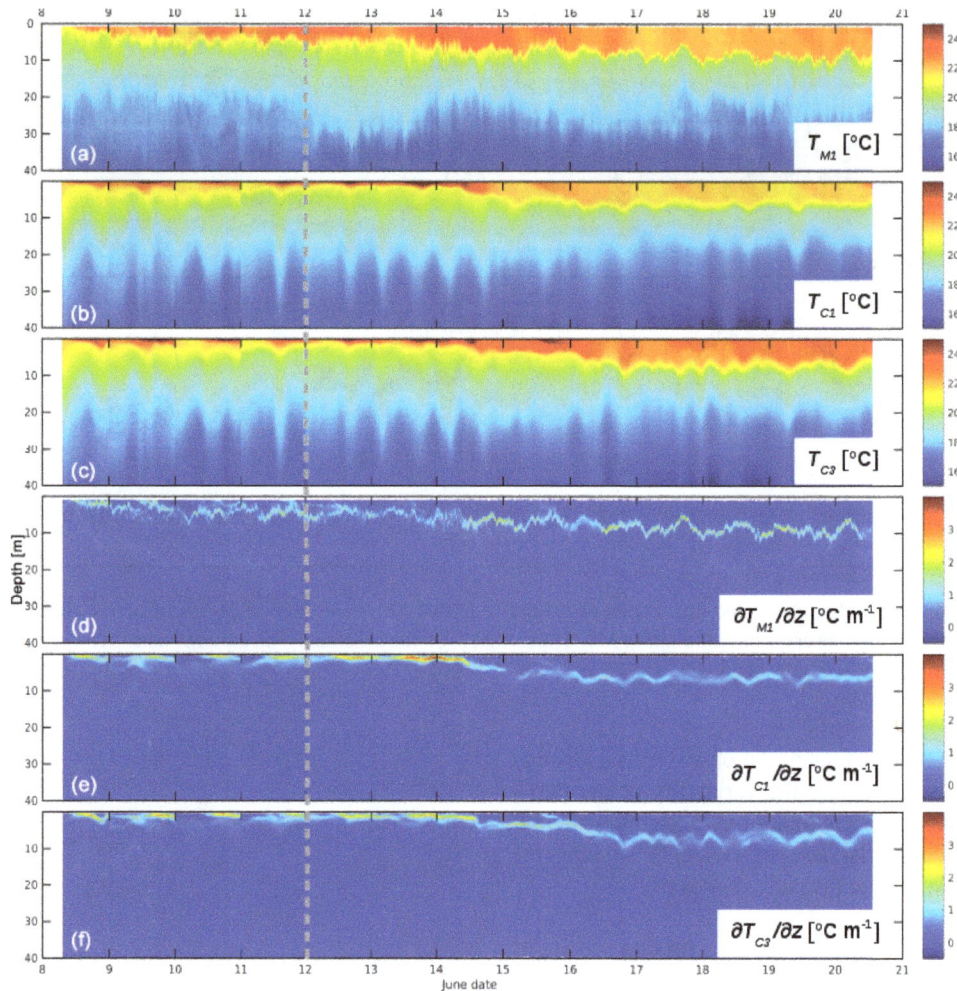

Figure 16. (a) The observed temperature at mooring M1, (b) the modelled temperature from ROMS runs C1 and (c) C3, (d) and the vertical temperature gradient from M1, (e) C1, and (f) C3. The instant of the last data assimilation is indicated by the the grey dashed vertical line.

k-ω mixing scheme in D4 definitely performs the best. This is also supported by the vertical structure of ΔT displayed in Fig. 18. There is clear evidence that the k-kl scheme (D2), the k-ϵ scheme (D3), and the k-ω scheme (D4) do better than the generic GLS (D1). Between the surface and about 5 m of depth, the best result was obtained from D4. Therefore, D4 will serve as the control run in the following E series.

An indicator of why the k-ω parameterisation performed better than the other closure schemes is possibly found in the publication of Reffray et al. (2015). Here, a one-dimensional model implemented in a three-dimensional circulation model was used to investigate physical and numerical turbulent-mixing behaviour. Amongst others, the k-kl, the k-ϵ, and the k-ω scheme were compared to each other. It turned out that the k-ω scheme was the most sensitive to the vertical resolution. In a coarse (about 10 m) resolution model, k-kl and k-ϵ clearly did better than k-ω, while at a high (about 1 m) resolution, all three schemes yielded suitable results. In the D series, the vertical resolution close to the sea surface is

0.27 m (see Fig. 11 and Sect. 4.2 above). Hence, one may speculate that the k-ω formulation becomes superior to the other schemes when the vertical resolution is increased.

4.5 Series E: sensitivity to the background vertical eddy diffusivity

The shortcoming of all the model runs conducted so far was that the mixed layer was too warm and too shallow, and the thermocline was too cold with respect to the observational data. This is also in agreement with the findings of Reffray et al. (2015). Hence, it was conjectured that the parameter-isation of the vertical transport of heat and/or momentum was not adequate. Several attempts were undertaken to fine-tune the D4 results by varying the vertical eddy viscosity co-efficient and the turbulent closure parameters, but the out-comes were sobering; a significant improvement in the fore-cast skills for the mixed-layer properties was not achieved. Hence, in this series, the background vertical eddy diffusiv-

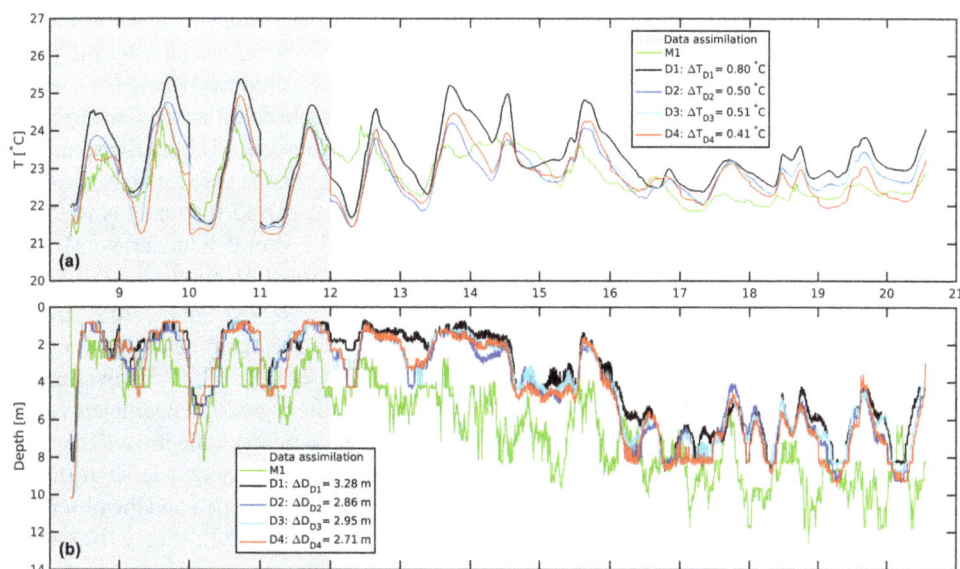

Figure 17. ROMS runs D1–D4: the time series of (**a**) the near-surface temperature at 0.81 m of depth, (**b**) the mixed-layer depth (MLD), and the corresponding observations at mooring M1. The numbers on the abscissae indicate June dates. The period for which the data are assimilated is highlighted with grey shading.

ity A_{VT} was increased gradually from $1 \times 10^{-6}\,\mathrm{m^2\,s^{-1}}$ in E1 (which is the control run identical to D4) to $2 \times 10^{-4}\,\mathrm{m^2\,s^{-1}}$ in E13. The forecast skill of each run was again assessed by means of ΔT at 0.81 m of depth and by ΔD. The dependency of these parameters on A_{VT} is shown in Fig. 19. ΔT exhibits minimum values of 0.31 °C (≈ 0.1 °C lower than in D4) for $A_{VT} \leq 2 \times 10^{-5} \leq 3 \times 10^{-5}\,\mathrm{m^2\,s^{-1}}$ in E4 and E5, which is somewhat higher than $(1.7 \pm 0.2) \times 10^{-5}\,\mathrm{m^2\,s^{-1}}$ obtained from the tracer measurements in the thermocline during the North Atlantic Tracer Release Experiment (Ledwell et al., 1998; Thorpe, 2007). By contrast, the minimum of $\Delta D = 2.05$ m is found in E9 for $A_{VT} = 7 \times 10^{-5}\,\mathrm{m^2\,s^{-1}}$.

Figure 20 shows the near-surface temperature in E4 and the mixed-layer depth in E9 together with the corresponding quantities of the control run E1 and the observations. After 15 June, the increase in A_{VT} from 1×10^{-6} to $2 \times 10^{-5}\,\mathrm{m^2\,s^{-1}}$ shifted the near-surface temperature by about 0.1 °C closer to the observations. Most of the time, the modelled signal is correlated with the observations, although the modelled maximum and minimum temperatures are frequently lagged a few hours behind the observed extreme values. Similar features were also described by Gentemann et al. (2009) when comparing time series of observed sea surface temperatures with those generated by the model of Fairall et al. (1996a). In their improved model (see the Introduction), they demonstrated that the peak warming in the afternoon was shifted earlier. For the mixed-layer depth, the increase in the eddy diffusivity to $7 \times 10^{-5}\,\mathrm{m^2\,s^{-1}}$ caused a significant reduction in ΔD from 2.71 m in E1 to 2.05 m in E9. While in the precursor series the mixed layer was always too shallow, it now agrees remarkably well with the observations, except for

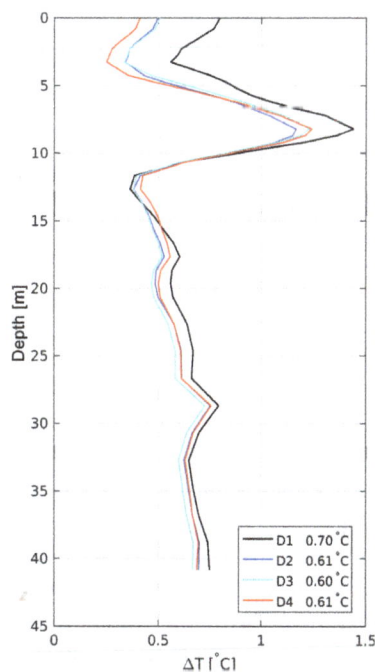

Figure 18. ROMS runs D1–D4: the rms temperature differences ΔT between the modelled temperature T_{ROMS} and the observed temperature T_{obs} evaluated at the actual depths of the observations. ΔT was computed only for the period after 15 June at 00:00.

large discrepancies on 19 and 20 June where the predicted mixed layer is up to 4 m shallower than the observed one. As the M1 wind speed was very low on these days (Fig. 6), other processes leading to a deepening of the mixed layer were

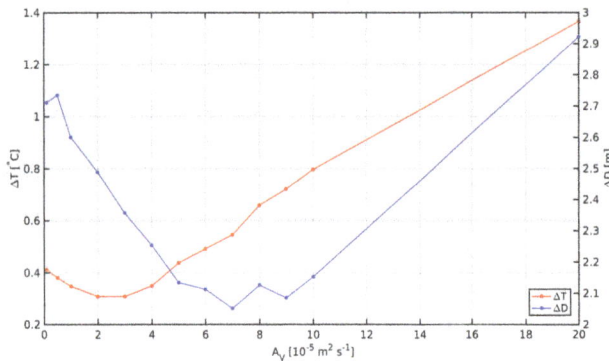

Figure 19. Series E: ΔT and ΔD for ROMS runs E1–E13. Both quantities were computed only for the period after 15 June at 00:00.

probably inadequately parameterised, such as Langmuir circulation and wave breaking (Noh et al., 2011, 2016).

5   Temporal variability

In order to assess the modelled temporal variability of the temperature and the depth of the mixed layer, the normalised spectra of the near-surface temperature amplitude $\hat{T}$ at 0.81 m of depth and of the mixed-layer depth amplitude $\hat{D}$ were computed by the Fourier transform, both from the observations and the ROMS outputs of runs E4 and E9, respectively. To enable a sufficient spectral resolution for the cycle periods of around 1 day, the entire time series between 8 and 20 June was used as input for the Fourier transform. At first glance, the modelled spectrum of the near-surface temperature (Fig. 21a1) resembles the observations in the cycle period range between about 0.1 and 1 days, but significant differences are evident in the bands between about 0.1 and 0.4 days where the modelled amplitude is up to 1 order of magnitude different from the observed one. This mismatch is not surprising, because here the temporal variability is controlled mainly by internal waves that are either not reproduced or are only partially reproduced by the model. By contrast, the range of 0.4–0.8 days (10–19 h) is dominated by tides and inertial motions. Theoretically, at $40°$ N, the inertial peak is at 18.7 h (0.78 days), but a correspondingly small peak is only visible in the modelled spectrum; no such peak is noticeable in the observations. Probably, the modelled peak is a leftover of the assimilation shock on 12 June. Additional peaks are found in both the modelled and the observed spectrum at about 0.4, 0.5, and 0.6–0.7 days (≈ 10, ≈ 12, and $\approx 14–17$ h). While the sources of the first and the latter are unknown, the 12 h peak might be related to a semi-diurnal tidal component. However, as there was no tidal forcing in the ROMS version utilised in this study and the MERCATOR forcing at the lateral boundaries was defined by means of daily averages, the semi-diurnal variability could only be caused by tides in the atmosphere. Both the

modelled and the observed spectrum are dominated by the diurnal variability represented by the peak at 1 day. In the red part of the spectrum between 1 and about 10 days, the modelled and observed amplitudes exhibit some weak correlation, and they are of about the same order of magnitude. This matter is not discussed here because it is potentially impacted by long-period fluctuations in the forcing at the surface and at the lateral boundaries. More detailed information on the correlation $\mathrm{corr}\left(r_{\hat{T}_{\mathrm{ROMS}}}, \hat{T}_{\mathrm{obs}} \right)$ between the modelled and the observed temperature amplitudes is shown in Fig. 21a2. The correlation coefficient $r = 0.74$ together with the p value $p = 3.05 \times 10^{-22}$ proves a high significant correlation, and the regression coefficients $a_0 = 0$ and $a_1 = 1.74$ indicate that, in general, the modelled amplitudes are overestimated. By contrast, there is less but still significant correlation between the modelled and the observed mixed-layer amplitudes $\hat{D}_{\mathrm{ROMS}}$ and $\hat{D}_{\mathrm{obs}}$ (Fig. 21b2), which is indicated by $\mathrm{corr}\left(r_{\hat{D}_{\mathrm{ROMS}}}, \hat{D}_{\mathrm{obs}} \right) = 0.50$, and $p = 4.52 \times 10^{-9}$. This finding is also supported by the spectrum (Fig. 21b1) in which a slight correlation of the amplitudes is only found for the diurnal and semi-diurnal cycles.

6   Horizontal variability

In order to assess the capability of ROMS to reproduce and predict the horizontal variability of mixed-layer properties, the results of run E9 were analysed along the ScanFish tracks A03, A05, A07, A09, and A10 (see Fig. 5) and compared with the data collected by the towed device. E9, using $A_{\mathrm{VT}} = 7 \times 10^{-5} \, \mathrm{m^2 \, s^{-1}}$, was selected for this comparison because both ΔT and ΔD were acceptable. The details of the ScanFish tracks are summarised in Table 3. As ROMS output was only available in 6 h intervals starting at midnight, in each case the output cycle was used which fell within the time window when the tracks were conducted. This assumed synopticity of the ScanFish tracks is justified by the fact that the maximum duration of the tracks was 5 h 28 min for A03.

To make the ScanFish observations and the ROMS products comparable, the ScanFish temperature was interpolated vertically on 1 dbar standard levels, and the ROMS temperature was mapped on the same levels. As the upper inflection point of the ScanFish varied between about 5 and 10 dbar, there was frequently no information on the near-surface temperature available. In such cases, the temperature at the inflection level was extended to the surface. The same method was applied to the ROMS temperature, which was not defined right at the surface but in the centre of the first s layer below the surface. In deep-water regions, this was located at about 3 m of depth.

Figure 22a shows a temperature section from the Scan-Fish measurements along the central track of A05, and the corresponding section from ROMS is displayed in Fig. 22b. The overall features of both sections resemble each other, but

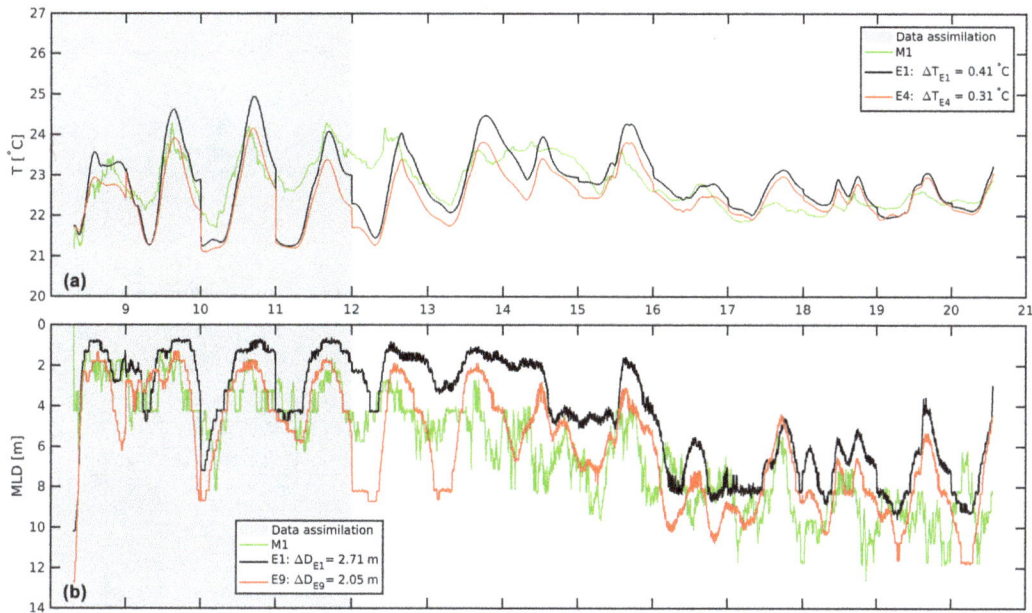

Figure 20. The time series of (**a**) the near-surface temperature at 0.81 m of depth from E1 and E4, (**b**) the mixed-layer depth (MLD) from E1 and E9, and the corresponding observations at mooring M1. The numbers on the abscissae indicate June dates. The period for which the data are assimilated is highlighted with grey shading.

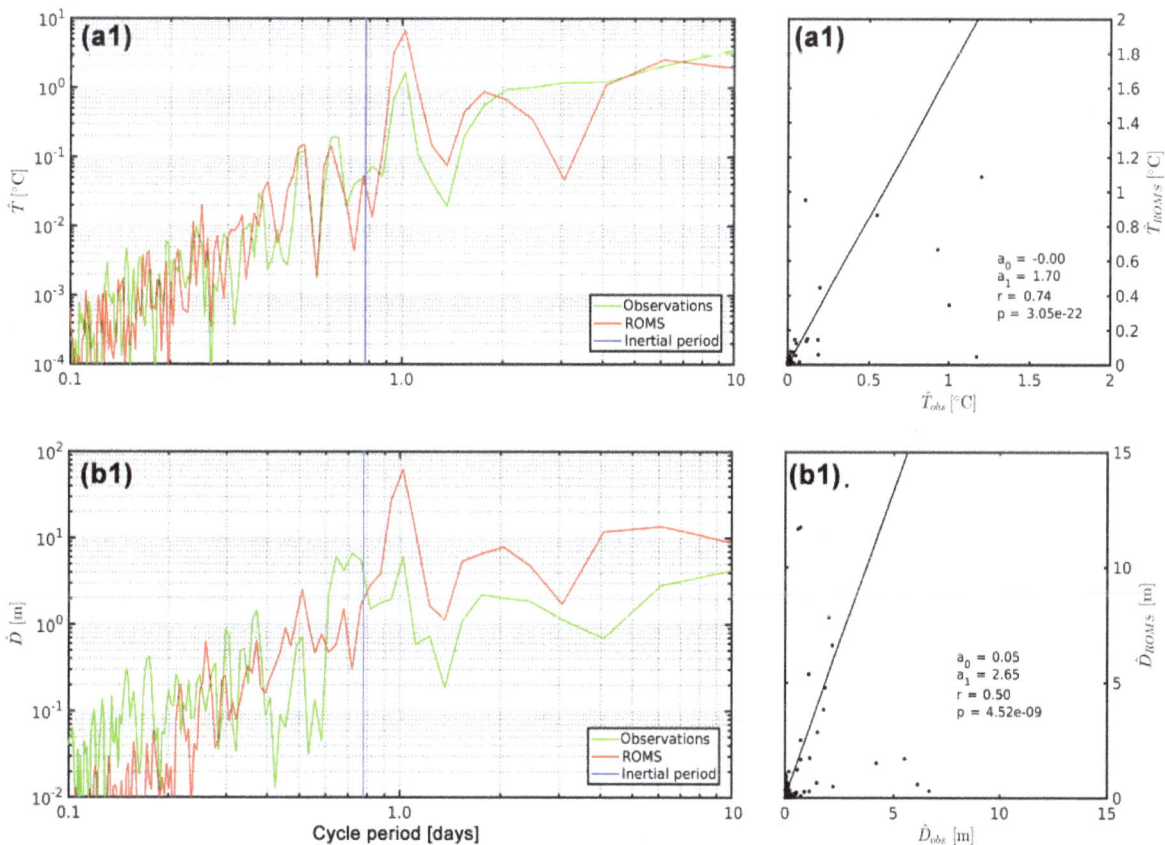

Figure 21. The spectra and correlation parameters of the modelled and observed amplitudes (**a**) $\hat{T}$ of the near-surface temperature at 0.81 m of depth from run E4 and (**b**) $\hat{D}$ of the mixed-layer depth from run E9. The spectra were evaluated for the entire time series during 8–20 June where observations were available.

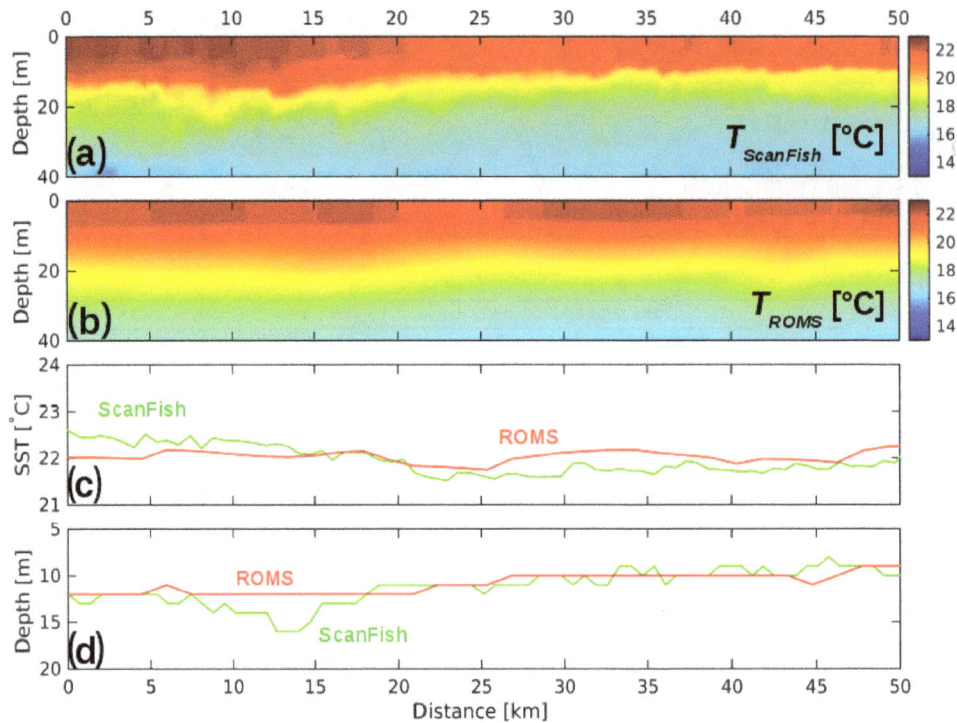

Figure 22. The sections along the A05 ScanFish track (cf. Fig. 5) at 40°48′ N: **(a)** the temperature recorded by the ScanFish, **(b)** the temperature predicted by ROMS, **(c)** the sea surface temperature SST, and **(d)** the mixed-layer depth evaluated from the ScanFish measurements and ROMS. No interpolation was used for the contour plots.

Table 3. The timing and nominal positions of the ScanFish tracks considered in this study (cf. Fig. 5). The ROMS analysis determines the instant of the model output which was used for comparison.

Track	Type	Nominal position	Start time	End time	Duration	ROMS analysis
A01	zonal	40°06′ N	21 Jun 14:03	21 Jun 18:15	4:12	21 Jun 18:00
A03	zonal	40°00′ N	21 Jun 19:10	22 Jun 00:38	5:28	22 Jun 00:00
A05	zonal	39°48′ N	22 Jun 03:00	22 Jun 08:00	5:00	22 Jun 06:00
A07	zonal	39°36′ N	22 Jun 12:57	22 Jun 18:05	5:08	22 Jun 18:00
A09	zonal	39°24′ N	22 Jun 20:17	23 Jun 01:16	4:59	23 Jun 00:00
A10*	meridional	07°31′ E	23 Jun 18:20	23 Jun 22:15	3:55	24 Jun 00:00

* Only the strictly meridional fraction of A10 was utilised.

the small-scale horizontal variability of the ScanFish temperature was not reproduced by ROMS. This is probably due to the smoothing effect of the OA, the combined action of the horizontal eddy diffusivity, and the numerical diffusion. However, as the last assimilation cycle was conducted on 12 June, 10 days prior to the ScanFish observations, one may exclude the possibility that the OA removed the small-scale features. Moreover, the vertical temperature gradient is much weaker in ROMS, which was already noted above. Hence, this is apparently not caused by the increased vertical diffusivity but by the vertical resolution of ROMS. The sea surface temperatures and the mixed-layer depths from the ScanFish and ROMS are displayed in Fig. 22c and d. For the surface temperature, the observed large-scale west–east trend is reproduced by ROMS, but there are differences of up to 0.5 °C in the central portion of the section. The maximum differences between the modelled and the observed mixed-layer depth in the 0–20 km range are close to 5 m at 13 km of distance, while in the eastern half of the section, the modelled and the observed mixed-layer depths approach each other. However, the smaller-scale $\mathcal{O}(1\,\mathrm{km})$ observed variability was not reproduced by ROMS for both the sea surface temperature and the mixed-layer depth.

To investigate why the small-scale variability was not predicted correctly, run E9 was repeated using a smaller horizontal eddy diffusivity coefficient of $1\,\mathrm{m^2\,s^{-1}}$ instead of $5\,\mathrm{m^2\,s^{-1}}$, which was used for all the model runs so far. However, no significant changes were noticeable. Thus, one has to settle for the fact that the present set-up of ROMS is only able to reproduce the horizontal variability of mixed-layer properties on scales which are comparable to the Rossby radius.

7 Conclusions

ROMS has been utilised to diagnose and predict the properties of the ocean mixed layer. The sensitivity of the model results to the choice of the initial and boundary conditions, the set-up of the vertical grid, and the vertical mixing schemes were investigated. The initial and lateral boundary conditions for ROMS were taken from two different parent models through one-way nesting. At the surface, ROMS was forced by two different weather forecasts or by observations. All ROMS nowcasts and forecasts were validated against observations which were taken in June 2014 to the west of Sardinia in the Mediterranean Sea.

To explore the sensitivity of the near-surface temperature and the mixed-layer depth to the choice of the initial conditions, ROMS was alternatively initialised by the Mediterranean Forecasting System (MFS) and the global MERCATOR model. In addition, observed temperature and salinity data were assimilated. For validation, the time series of temperature were compared with the observations from a mooring. Initialising ROMS from MERCATOR instead of MFS provided better agreement between the model and the observations, but significant improvement was obtained from a ROMS run initialised from MERCATOR and updated with assimilated data from CTD casts and gliders. This applied both to the near-surface temperature and the mixed-layer depth as well as to the temperature distribution in the upper thermocline.

To investigate the impact of the surface boundary conditions, atmospheric forcing fields were taken from the weather prediction models COSMO-ME and COSMO-IT and from the observations of a meteorological buoy acting as a point source. With respect to the mixed-layer depth, the best agreement with the observations was obtained from a model run forced with COSMO-ME, while the near-surface temperature exhibited the best match when ROMS was forced by COSMO-IT. However, the stratification in the upper thermocline was best represented when the point source was applied. The obvious reason for this surprising result is that the momentum forcing was overestimated by both COSMO-ME and COSMO-IT.

For the vertical mixing, four different configurations of the GLS scheme of Umlauf and Burchard (2003) were applied, representing the generic version: the k-kl model of Mellor and Yamada (1982), the k-ϵ model (Rodi, 1987), and the k-ω model (Wilcox, 1988). The best performance was obtained from the k-ω model.

Regardless of which initial conditions or surface boundary conditions were applied, the modelled mixed layer was always too shallow and too warm. Therefore, the background vertical eddy diffusivity coefficient, A_{VT}, was varied over more than 1 order of magnitude. The best agreement of the mixed-layer temperature was obtained for $A_{VT} \approx 2 \times 10^{-5}\,\mathrm{m^2\,s^{-1}}$, while $A_{VT} = 7 \times 10^{-5}\,\mathrm{m^2\,s^{-1}}$ provided the best match of the mixed-layer depth with the observations.

A positive and significant correlation was found between the modelled and the observed temporal variability in the mixed-layer temperature. The modelled variability resembled the observed variability predominantly for cycle periods in the spectral ranges between about 0.5 and 1 days. By contrast, less correlation was found between the modelled and the observed variability in the mixed-layer depth. Slight agreement was only found for the diurnal period.

The horizontal variability was validated against measurements from a high-resolution zonal ScanFish section. Both the modelled mixed-layer temperature and the mixed-layer depth closely resembled the observations, but only on the larger scales of $\mathcal{O}(10\,\mathrm{km})$. Hence, the mesoscale variability was rather well reproduced, but the sub-mesoscale variability was not.

Competing interests. The author declares that he has no conflict of interest.

Acknowledgements. The author would like to thank the masters and crews of the NRV Alliance and the RV Planet for their professionalism during the conduction of the experiments at sea. The data from COSMO-ME and COSMO-IT were provided by the Italian weather service Centro Nazionale di Meteorologia e Climatologia Aeronautica, and the MFS and MERCATOR data sets were downloaded from the Copernicus Marine Environment Monitoring Service (http://marine.copernicus.eu). REP14-MED was sponsored by HQ Supreme Allied Command Transformation (Norfolk, VA, USA).

Edited by: J. Chiggiato

References

Bernie, D. J., Woolnough, S. J., and Slingo, J. M.: Modeling diurnal and intraseasonal variability of the ocean mixed layer, J. Climate, 18, 190–1202, 2005.

Bernie, D. J., Guilyardi, E., Madec, G., Slingo, J. M., Woolnough, S. J., and Cole, J.: Impact of resolving the diurnal cycle in an ocean-atmosphere GCM. Part 1: a diurnally forced OGCM, Clim. Dynam., 29, 6, 575–590, doi:10.1007/s00382-007-0249-6, 2007.

Bernie, D. J., Guilyardi, E., Madec, G., Slingo, J. M., Woolnough, S. J., and Cole, J.: Impact of resolving the diurnal cycle in an ocean-atmosphere GCM. Part 2: a diurnally coupled GCM, Clim. Dynam., 31, 909–925, doi:10.1007/s00382-008-0429-z, 2008.

Blanke, B. and Delecluse, P.: Variability of the tropical Atlantic Ocean simulated by a general circulation model with two-different mixed-layer physics, J. Phys. Oceanogr., 23, 1363–1388, 1993.

Bretherton, F. P., Davies, R. E., and Fandry, C. B.: Technique for Objective Analysis and design of oceanographic experiments applied to Mode-73, Deep-Sea Res., 23, 559–582, 1976.

Carter, E. F. and Robinson, A. R.: Analysis models for the estimation of oceanic fields, J. Atmos. Ocean. Tech., 4, 49–74, 1987.

Chapman, D. C.: Numerical treatment of cross-shelf open boundaries in a barotropic coastal ocean model, J. Phys. Oceanogr., 15, 1060–1075, 1985.

Chen, C. T. and Millero, F. J.: Speed of sound in seawater at high pressures, Journal of the Acoustic Society of America, 62, 1129–1135, 1977.

Dietrich, G., Kalle, K., Krauss, W., and Siedler, G.: Allgemeine Meereskunde. 3rd edition, Gebrüder Bornträger, Berlin, 593 pp., 1975.

Dombrowsky E., Bertino, L., Brassington, G. B., Chassignet, E. P., Davidson, F., Hurlburt, H. E., Kamachi, M., Lee, T., Martin, M. J., Meu, S., and Tonani, M.: GODAE Systems in Operation, Oceanography, 22, 83–95, 2009.

Drévillon, M., Bourdallé-Badie, R., Derval, C., Lellouche, J. M., Rémy, E., Tranchant, B., Benkiran, M., Greiner, E., Guinehut, S., Verbrugge, N., Garric, G., Testut, C. E., Laborie, M., Nouel, L., Bahurel, P., Bricaud, C., Crosnier, L., Dombrowsky, E., Durand, E., Ferry, N., Hernandez, F., Le Galloudec, O., Messal, F., and Parent, L.: The GODAE/Mercator-Ocean global ocean forecasting system: results, applications and prospects, Journal of Operational Oceanography, 1, 51–57, doi:10.1080/1755876X.2008.11020095, 2008.

Fairall, C. W., Bradley, E. F., Godfrey, J. S., Wick, A., Edson, J. B., and Young, G. S.: Cool-skin and warm-layer effects on sea surface temperature. J. Geophys. Res., 101, 1295–1308, doi:10.1029/95JC03190, 1996a.

Fairall, C. W., Bradley, E. F., Rogers, D. P., Edson, J. B., and Young, G. S.: Bulk parameterization of air-sea fluxes for Tropical Ocean-Global Atmosphere Coupled-Ocean Atmosphere Response Experiment, J. Geophys. Res., 101, 3747–3764, doi:10.1029/95JC03205, 1996b.

Gentemann, C. L., Minnett, P. J., and Ward, B.: Profiles of ocean surface heating (POSH): A new model ocean diurnal warming, J. Geophys. Res., 114, C07017, doi:10.1029/2008JC004825, 2009.

Grilli, F. and Pinardi, N.: The computation of Rossby radii of deformation for the Mediterranean Sea, MTP news, 6, p. 4, 1998.

Haney, R. L.: On the pressure gradient force over steep topography in sigma coordinate models, J. Phys. Oceanogr., 21, 610–619, 1991.

Haidvogel, D. B., Arango, H. G., Hedstrøm, K., Beckmann, A., Malanotte-Rizzoli, P., and Shchepetkin, A. F.: Model evaluation experiments in the North Atlantic Basin: simulations in nonlinear terrain-following coordinates, Dynam. Atmos. Oceans, 32, 239–281, 2000.

Kantha, L. H. and Clayson, C. A.: An improved mixed-layer model for geophysical applications, J. Geophys. Res., 99, 25235–25266, 1994.

Katsnelson, B., Petnikov, V., and Lynch, J.: Fundamentals of Shallow Water Acoustics, Springer, New York, 540 pp., doi:10.1007/978-1-4419-9777-7, 2012.

Lamb, P. J.: On the mixed-layer climatology of the north and tropical Atlantic, Tellus A, 36, 292–305, 1984.

Large, W., McWilliams, J., and Doney, S.: Oceanic vertical mixing: A review and a model with a nonlocal boundary layer parameterization, Rev. Geophys., 32, 363–403, 1994.

Ledwell, J. R., Watson, A. J., and Law, C. S.: Mixing of a tracer in the pynocline, J. Geophys. Res., 103, 21499–21529, 1998.

Ling, T., Xu, M., Liang, X.-Z., Wang, J. X. L., and Noh, Y.: A multi-level ocean mixed layer model resolving the diurnal cyle: Development and validation, Journal of Advances in Modeling Earth Systems, 7, 1680–1692, doi:10.1002/2015MS000476, 2015.

Marchesiello, P., McWilliams, J. C., and Shchepetkin, A. F.: Open boundary conditions for long-term integration of regional ocean models, Ocean Model., 3, 1–20, 2001.

Mellor, G. L. and Yamada, T.: Development of a turbulence closure model for geophysical fluid problems, Rev. Geophys. Space Ge., 20, 851–875, 1982.

Medwin, H. and Clay, C. S.: Fundamentals of Acoustic Oceanography, Academic Press, San Diego, 712 pp., doi:10.1016/B978-012487570-8/50000-3, 1998.

Noh, Y., Lee, E., Kim, D.-H., Hong, S.-Y., Kim, M.-J., and Ou, M.-L.: Prediction of the diurnal warming of sea surface temperature using an atmosphere-ocean mixed layer coupled model, J. Geophys. Res., 116, C11023, doi:10.1029/2011JC006970, 2011.

Noh, Y., Ok, H., Lee, E., Toyoda, T., and Hirose, N.: Parameterization of Langmuir circulation in the ocean mixed layer model using LES and its application to the OGCM. J. Phys. Oceanogr., 46, 57–78, doi:10.1175/JPO-D-14-0137.1, 2016.

Onken, R.: A Relocatable Ocean Prediction System applied to REP14-MED, Ocean Sci., in preparation, 2017.

Onken, R., Ampolo-Rella, M., Baldasserini, G., Borrione, I., Cecchi, D., Coelho, E., Falchetti, S., Fiekas, H.-V., Funk, A., Jiang, Y.-M., Knoll, M., Lewis, C., Mourre, B., Nielsen, P., Russo, A., and Stoner, R.: REP14-MED Cruise Report, CMRE-CR-2014-06-REP14-MED, CMRE, La Spezia, 76 pp., 2014.

Onken, R., Fiekas, H.-V., Beguery, L., Borrione, I., Funk, A., Hemming, M., Heywood, K. J., Kaiser, J., Knoll, M., Poulain, P.-M., Queste, B., Russo, A., Shitashima, K., Siderius, M., and Thorp-Küsel, E.: High-Resolution Observations in the Western Mediterranean Sea: The REP14-MED Experiment, Ocean Sci. Discuss., doi:10.5194/os-2016-82, in review, 2016.

Pace, N. G. and Jensen, F. B. (Eds.): Impact of Littoral Environmental Variability on Acoustic Predictions and Sonar Performance, Springer, the Netherlands, 620 pp., doi:10.1007/978-94-010-0626-2, 2002.

Reffray, G., Bourdalle-Badie, R., and Calone, C.: Modelling turbulent vertical mixing sensitivity using a 1-D version of NEMO, Geosci. Model Dev., 8, 69–86, doi:10.5194/gmd-8-69-2015, 2015.

Rodi, W.: Examples of calculation methods for flow and mixing in stratified fluids, J. Geophys. Res., 92, 5305–5328, 1987.

Shchepetkin, A. F., and Williams, J. C.: A family of finite-volume methods for computing pressure gradient force in an ocean model with a topography-following vertical coordinate, available at: http://www.atmos.ucla.edu/~alex/ROMS/pgf1A.ps (last access: 10 March 2017), 24 pp., 2001.

Shchepetkin, A. F. and McWilliams, J. C.: A method for computing horizontal pressure gradient force in an oceanic model with a nonaligned vertical coordinate, J. Geophys. Res., 108, 3090, doi:10.1029/2001JC001047, 2003.

Shchepetkin, A. F. and McWilliams, J. C.: The Regional Ocean Modeling System: A split-explicit, free-surface, topography following coordinates ocean model, Ocean Model., 9, 347–404, 2005.

Song, Y. and Haidvogel, D. B.: A semi-implicit ocean circulation model using a generalized topography-following coordinate system, J. Comput. Phys., 115, 228–244, 1994.

Thomson, R. E. and Emery, W. J.: Data Analysis Methods in Physical Oceanography, 3rd Edn., Elsevier, Amsterdam, 728 pp., 2014.

Thorpe, S. A.: An Introduction to Ocean Turbulence, Cambridge University Press, Cambridge, 293 pp., 2007.

Tonani M., Teruzzi, A., Korres, G., Pinardi, N., Crise, A., Adani, M., Oddo, P., Dobricic, S., Fratianni, C., Drudi, M., Salon, S., Grandi, A., Girardi, G., Lyubartsev, V., and Marino, S.: The Mediterranean Monitoring and Forecasting Centre, a component of the MyOcean system, Proceedings of the Sixth International Conference on EuroGOOS 4–6 October 2011, Sopot, Poland, edited by: Dahlin, H., Fleming, N. C., and Petersson, S. E., EuroGOOS Publication no. 30, ISBN 978-91-974828-9-9, 2014.

Umlauf, L. and Burchard, H.: A generic length-scale equation for geophysical turbulence models, J. Mar. Res., 61, 235–265, 2003.

Umlauf, L., Burchard, H., and Hutter, K.: Extending the k-ω turbulence model towards oceanic applications, Ocean Model., 5, 195–218, 2003.

Wagner, R. G.: Decadal scale trends in mechanisms controlling meridional sea surface temperature trends in the tropical Atlantic, J. Geophys. Res., 101, 16683–16694, 1996.

Wilcox, D. C.: Reasessment of the scale-determining equation for advanced turbulence models, AIAA J., 26, 1299–1310, 1988.

Preface: Oceanographic processes on the continental shelf: observations and modeling

Sandro Carniel[1], Judith Wolf[2], Vittorio E. Brando[3,a], and Lakshmi H. Kantha[4]

[1]Institute of Marine Science, National Research Council (ISMAR-CNR), Venice, Italy
[2]National Oceanography Center, Liverpool, UK
[3]Institute of Electromagnetic Sensing of the Environment, National Research Council (IREA-CNR), Milan, Italy
[4]University of Colorado, Boulder, CO 80309, USA
[a]present address: Institute for the Study of Atmosphere and Climate, National Research Council (ISAC-CNR), Rome, Italy

Correspondence to: Sandro Carniel (sandro.carniel@cnr.it)

1 Introduction

Oceanographic processes in the shallow continental shelf and coastal regions have a major impact on human life, since a large fraction of human population lives within 100 km of the shoreline (Halpern et al., 2008). At the same time, the processes occurring in these regions are difficult to analyze and disentangle, because of their intrinsic complexity, the variability of temporal and spatial scales, their multidisciplinary nature, and the influence of offshore boundary conditions (Dickey, 2003; Mitchell et al., 2015).

To improve our knowledge of processes typical of these regions, there is a strong need for an integrated approach, combining numerical coupled systems (of ocean, atmosphere, waves, biology, and sediments) at selected scales (Carniel et al., 2016a), validated with data resulting from either distributed coastal observatories or remote sensing approaches (point-wise data from multivariable buoys, high-frequency radar images, satellite images, drifters, AUVs, gliders, etc.). This scientific challenge has to take into consideration a wide range of processes involving tides, resuspension, stratification, mixing, land boundaries, surrounding land use, river discharges, distributed run off, pollutants from densely populated areas, etc. (e.g., Mitchell et al., 2015 and references there in).

All these aspects are even more relevant nowadays, in a framework of changing climate (Collins et al., 2012). Shallow coastal and transitional areas, wetlands and lagoons, coastal cities, and valuable infrastructures are being threatened by potential impact of climate-change-induced hazards, such as inundation of low-lying areas, exposure to acceler-

ated sea-level rise, and increased rates of coastal erosion. At the same time, these are also the regions where it may be feasible to harvest renewable energy economically, or where state-of-the-art prototypes can be more readily deployed for specific studies.

To improve understanding of shelf processes and to identify key parameters that allow detection and monitoring of likely changes, we invited investigators to contribute original research articles, resulting in the special issue "Oceanographic processes on the continental shelf: observations and modeling".

In Table 1, we summarize how the papers in this special issue have addressed some of the specific aspects that characterize shelf sea process studies as a sort of *fil rouge*: the spatial scale of the processes investigated (regional, meso- and sub-mesoscale, and fine scale); the need to address them using different measurements (in situ, remote sensing, physical or biogeochemical parameters); how and when numerical models can integrate existing data (representing only specific processes like hydrodynamics or waves, or presenting a "coupled" approach); and the length or timescale of the events described (single event, short period, seasonal, yearly, etc.). Readers can therefore identify the most significant characteristics of each paper with respect to these key aspects.

2 Bringing together data and numerical models

Coastal observatories provide sustained information for the thorough understanding of the mechanisms regulating shelf

Table 1. Summary of some specific aspects that characterize the study of shelf processes: the spatial scale of the processes investigated, the timescale of the events described, the need to address them using different measurements, and numerical models typology. Readers can therefore identify the most significant characteristics of each paper with respect to these key aspects.

Papers	Spatial scale				Event/timescale				Measurements			Models				
	Reg	Meso	Sub	Fine/Small	Years	Months	Days	Severe/flood/Bora/storm	Satellite	In situ phys	In situ biogeo	Operational	Coupled phys	Wave	Bio Geo	Statistical/reanalysis method
Brando			X	X			X	X	X			X	X			
Falcieri			X	X			X	X		X						
McKiver			X			X	X	X		X		X	X			
Iuppa					X									X		
Umgiesser			X		X					X			X			
Lanotte			X				X			X		X	X			
Licer			X				X	X		X			X	X		
Olita	X				X	X			X				X			
Grifoll			X		X			X		X			X			
Samaras	X							X								
Barbariol			X		X			X		X						X
Gutierrez		X	X		X			X	X							X
Kraus		X			X						X				X	X
Bonamano			X	X		X			X	X	X		X		X	
Signell		X	X			X	X		X	X	X	X	X			

regions (e.g., Lynch et al, 2014). However, as they are relatively scarce and sparse, they do not often provide sufficient spatial coverage to observe extreme events (Dickey et al., 2003). Brando et al. (2015, this special issue) examine how they can be integrated with high-resolution satellite observations and into coastal numerical model outputs. Namely, sea surface temperature (SST) and turbidity (T) maps derived from Landsat 8 imagery at 30 m resolution were used to characterize river plumes in the northern Adriatic Sea during a significant flood event in November 2014. Circulation patterns and sea surface salinity (SSS) from an operational coupled ocean–wave model supported the interpretation of the plumes' interaction with the receiving waters. A good agreement was found between SSS, T, and SST fields at the sub-mesoscale and mesoscale delineation of the major river plumes, enabling also the description of smaller plume structures, such as the different plumes' reflectance spectra related to the lithological fingerprint of the sediments in the river-catchment basins.

Most of the coastal measurements and data available in coastal regions rely on state-of-the-art measurements such as CTD (conductivity, temperature, and depth) or ADCPs (acoustic current doppler profilers), which nowadays constitute the benchmark for improving our understanding and assessing numerical models. However, relatively uncommon, but very useful, data exploring the small scale are becoming more available (Thorpe, 2005; Carniel et al., 2012). As an example, Falcieri et al. (2016, this special issue) present the very first turbulence observations in the Gulf of Trieste (northern Adriatic Sea), acquired during different water column stratifications. Almost 500 microstructure profiles allowed the demonstration that, during the 2014 winter, the water column in the gulf was not completely mixed, due to the influence of bottom water intruding from the open sea. One type of water intrusion comes from the northern coast of the Adriatic Sea (i.e., cooler, fresher, and more turbid water), which acted as a barrier to wind-driven turbulence. A different water mass, coming from the open sea in front of the Po Delta (i.e., warmer, saltier, less turbid, and with a smaller vertical density gradient) was not able to suppress downward penetration of turbulence from the surface.

Sea-truth data can then be used directly in order to validate modeling tools implemented to describe shelf sea processes in coastal regions (Usui et al., 2015); given the fact that there are several existing typologies of such numerical models, in each case the use of the most appropriate one is required. Bricheno et al. (2014) show the importance of resolving the appropriate spatial scales and using suitable metrics to compare models and data in the nearshore zone. McKiver et al. (2016, this special issue) compare the ability of a finite-difference (SHYFEM, shallow water hydrodynamic finite-element model) and a finite-element model (MITgcm, Massachusetts Institute of Technology general circulation model, Sannino et al., 2014) to simulate coastal processes in the northern Adriatic Sea. The study focused on the northern

Adriatic Sea during a severe event that occurred at the beginning of 2012, and gave the opportunity to understand how these events (related to dense water formation) may affect coastal processes, like upwelling and downwelling, and how they interact with estuarine dynamics. Both models capture the dense water event, though each displays biases in different regions, showing large differences in the reproduction of surface patterns and highlighting the relevance of identifying suitable bulk formulas for the correct simulation of the thermohaline structure of the coastal zone. McKiver et al. (2016, this special issue) highlight that, while a coarser resolution offshore is acceptable for the reproduction of the dense water event (during which the non-hydrostatic processes were found to have little importance), a finer horizontal resolution in the coastal zone is important to reproduce the effect of the complex coastal morphology on the hydrodynamics.

3 Planning the coastal maritime space

Sea regions close to the continental shelf are also those from which it could be feasible to extract renewable energy with the highest efficiency and lowest cost (Cruz, 2008). Iuppa et al. (2015, this special issue) discuss potential sites around the island of Sicily for energy extraction from surface gravity waves, with the aim of selecting possible sites for the implementation of wave energy converters (WECs). A third-generation wave model was adopted to reconstruct the wave data along the coast over a period of 14 years, which allowed the characterization of the most productive areas on the western side of the island and in the Strait of Sicily (i.e., relatively high wave energy and proximity to the coast), which makes them possible sites for the implementation of WEC farms.

Coastal lagoons represent peculiar and fragile situations that can often be in direct contact with coastal and shelf processes. Umgiesser et al. (2016, this special issue) explore the variability of water renewal due to heavy river discharges in the very shallow Curonian Lagoon, connected by a very narrow strait to the Baltic Sea. The lagoon is simulated, using a finite-element hydrodynamic model, to reproduce the circulation patterns for 10 years, focusing on the salinity distribution and the renewal times of the system when forced by river runoff, wind, and Baltic Sea sea-level fluctuations. Results demonstrated how the river discharge within the lagoon was the most important factor triggering the water renewal time.

As stressed above, numerical models are extremely useful for integrating the paucity of marine data available in order to better disentangle different dynamical contributions and provide a synoptic picture of the oceanographic shelf processes (Warner et al., 2010). A careful blending of observations and model data makes it possible to conceive of functional tools to control, for instance, the horizontal spreading of small organisms or substance concentrations, thus being relevant for marine biology and pollutant dispersion as well as oil spill applications. In this special issue, Lanotte et al. (2016) study the role of vertical shear on oceanic horizontal dispersion of passive tracer particles on the continental shelf in the southern Mediterranean, by means of observation and model data. In situ current measurements reveal that vertical gradients of horizontal velocities in the upper mixed layer decorrelate quite fast (~ 1 day), whereas an eddy-permitting ocean model, such as the Mediterranean Forecasting System, tends to overestimate such decorrelation times (possibly due to unresolved scale motions and mesoscale motions that are largely smoothed out at scales close to the grid spacing).

4 The need for a coupled approach

Although the different processes characterizing the shelf regions are intrinsically connected, in order to simplify the numerical approach, historically these different components (e.g., atmosphere, ocean, wave, sediment, biology etc.) have been modeled separately (Mihanovic et al., 2013). However, mostly thanks to increases in the understanding of mutual feedbacks and advances in computer power, this reductionist approach can be now overcome. Licer et al. (2016, this special issue) describe work dealing with a one-way and two-way coupled ocean–atmosphere system during an intense Bora event in the northern Adriatic. Comparing modeled atmosphere–ocean fluxes and sea temperatures from both model setups to platform and CTD measurements from three locations in the northern Adriatic, Licer et al. (2016, this special issue) found that, using two-way coupling, ocean temperatures exhibit a root mean square error (RMSE) 4 times lower than those from a one-way coupled system. Sensible heat fluxes were also improved in the coupled approach, at all stations, while coupled and uncoupled circulations in the northern Adriatic (being predominantly wind-driven) did not show significant mesoscale differences.

There are, of course, several other interesting aspects that should be encompassed when dealing with coupled numerical models (Carniel et al., 2016b). In this special issue, Olita et al. (2015) study the impact of current speeds on the parameterization of surface fluxes and their feedback on regional-scale ocean dynamics. The computations of heat and momentum fluxes in uncoupled models generally happen through standard (Fairall et al., 2003) bulk formulas, where the wind speeds do not take account of their relative effects with respect to the ocean currents. From the results obtained from twin numerical experiments around the island of Sardinia (western Mediterranean), Olita et al. (2015, this special issue) demonstrated that, even at local scales and in temperate regions, it would be preferable to take into account such a contribution in flux computations. The modification of the original code, substantially cost-free in terms of numerical computation, improves the model response in terms of surface fluxes (SST validated) and it also likely improves the

dynamics, as suggested by qualitative comparison with satellite data.

Complementing numerical model results, Grifoll et al. (2016, this special issue) used a set of observations to investigate the inner-shelf response due to the storm passage in the inner-shelf of the NW Mediterranean Sea. The two-peak storm induced an interesting evolution in the momentum balance terms: the appearance of fluctuations with both super-inertial (12–16 h) and sub-inertial (1–2 days) periods. In contrast to the first peak of the storm, during the second one the temporal sequence of increased acceleration reoccurred, but with the along-shelf flow largely influenced by the sub-inertial (likely topographic) waves. The work encompassed water-current observation analysis and the application of theoretical models to describe the shelf wave propagation and the shelf response to the wind.

Although risks associated with climate change may indeed change the frequency and nature of storms in the Mediterranean Sea (Lionello et al., 2012), shelf regions are also prone to other risks, such as coastal inundation related to tsunami generation and propagation. Samaras et al. (2015, this special issue) presented an advanced tsunami-generation, propagation and coastal inundation 2-D (horizontal) model based on the higher-order Boussinesq equations, applied to simulate representative earthquake-induced tsunami scenarios in the eastern Mediterranean. Two areas of interest were selected after evaluating tsunamigenic zones and possible sources in the region: one at the southwest of the island of Crete in Greece and one at the east of the island of Sicily in Italy. Model results are presented in the form of extreme water elevation maps, sequences of snapshots of water elevation during the propagation of the tsunamis, and inundation maps of the studied low-lying coastal areas. This work marks one of the first successful applications of a fully non-linear model for the 2-D horizontal simulation of tsunami-induced coastal inundation; acquired results are indicative of the model's capabilities, also showing how areas in the eastern Mediterranean would be affected by potential larger events.

5 Detecting a changing sea

Characterizing the meteo-oceanographic climate in coastal regions is a fundamental step in being able to distinguish between natural and human-related fluctuations and to detect extreme events (Rockel et al., 2007). When analyzing long-term series, a number of statistical approaches can be evaluated. In this special issue, Barbariol et al. (2016) presented wave extreme characterization for the wave climate at the "Acqua Alta" oceanographic tower (northern Adriatic Sea, Italy), during the period 1979–2008, using self-organizing maps (SOMs, Liu et al., 2006). An application of the proposed two-step approach demonstrated that a proper representation of the extreme wave climate leads to enhanced quantification of, for instance, the alongshore component of the wave energy flux in shallow water. Focusing also on the peaks of the storms, Barbariol et al. showed how practical oceanographic and engineering applications can benefit from the novel SOM processing strategies developed. Besides improving the statistical analysis of long-term wave series, in recent years increasing attention has been devoted to wave reanalysis as a powerful source of information for wave climate research and engineering applications. However, the problem remains that, in coastal areas or shallow water, waves are poorly described due to a lack of spatial resolution, and wave downscaling procedures are needed; there is also a need for higher-resolution wind fields (e.g., Rockel et al., 2007; O'Neil et al., 2017).

Gutierrez et al. (2016, this special issue) demonstrated the feasibility of the use of wind fields detected with synthetic aperture radar (SAR) for the wave climate downscaling of the northern Adriatic Sea, by using a hybrid methodology and global wave and wind reanalysis as forcing. The wave fields produced were compared to wave fields produced with SAR winds that represent the two dominant wind regimes in the area: the Bora (east-northeast direction) and Sirocco (southeast direction). Although differences existed between SAR and modeled wind fields, a good correlation was found for the downscaled waves forced with different wind products. Overall this work showed how Earth observation products, such as SAR wind fields, can be successfully taken up into oceanographic modeling, producing similar downscaled wave fields when compared to waves forced with reanalysis wind.

The relevance of long-term data acquired at sea, also from the biological perspective, was confirmed by Kraus et al. (2016, this special issue), who explored the factors favoring phytoplankton blooms in the northern Adriatic Sea analyzing an oceanographic data set derived from monthly oceanographic cruises covering the 1990–2004 period. Kraus et al. (2016) found that while in winter and early spring the phytoplankton abundances depended on circulation fields, in summer and autumn they were related to Po River discharge rates up to 15 days earlier and on concomitant circulation fields. On the other hand, late spring phytoplankton abundances increased 1–3 days after high Po River discharge rates regardless of the circulation fields. These findings create the basis for the construction of an empirical ecological model of the northern Adriatic, which can ultimately be used in the sustainable economy of the region, as well as for validation of a numerical ecological model which is currently being developed for the region.

6   Towards integrated ocean observing systems

Last but not least, in order to converge towards an integrated ocean observing system capable of providing useful information and contributing to effective management and plan-

ning activities, all data collected in our shelf regions should be brought in contact and integrated with existing numerical models (Williams et al., 2011). In this special issue volume, Bonamano et al. (2016) presented a multiplatform observing network in the coastal marine area of Civitavecchia (Latium, Italy), integrated with numerical models, to analyze coastal processes at high spatial and temporal resolution. The in situ data acquired at long-term fixed stations and during dedicated surveys are integrated with satellite observations (e.g., temperature, chlorophyll a, and TSM), and then used to feed and validate numerical models to describe the dynamics of pollutant dispersion under different conditions. Such integrated ocean observatory systems turn out to be very useful during the activity of marine planning and management (e.g., bathing water quality assessment, evaluation of the effects of the dredged activities on *Posidonia* meadows) and are a practical tool to improve the conflict resolution between anthropic and conservation uses in coastal sensitive areas. They should become more and more commonly used involving transnational actors in order to reach an integrated system capable of connecting national efforts.

In recent years it appeared more and more clear how, in order to be really effective, the increasing amount of collected data (either from single point or remotely) and model output currently available need to be quickly accessed and distributed among the scientific community (see Bergamasco et al., 2012). Signell and Camossi (2016, this special issue) present a solution that allows even small research groups to provide meteorological and ocean model data through standardized web services and tools. A simple, local brokering approach was presented that lets modelers continue producing custom data, but virtually aggregates and standardizes the data using NetCDF Markup Language. The THREDDS Data Server is used for data delivery, pycsw for data search, NCTOOLBOX (MATLAB®) and Iris (Python) for data access, and Ocean Geospatial Consortium Web Map Service for data preview. Such an approach dramatically improves the effectiveness of data distribution and sharing in research communities with limited IT resources, (i) making it simple for providers to enable web service access to existing output files; (ii) using technology that is free, and that is easy to deploy and configure; and (iii) providing tools to communicate with web services that work in existing research environments.

We therefore hope that *Ocean Science* readers will then find much of the material in this special issue of interest, paradigmatic of processes that can be analyzed in other geographical contexts with respect to those presented, and a point of reference for cutting-edge ideas in theory, numerical models, and observations.

Acknowledgements. Sandro Carniel thanks the RITMARE National Flagship project, Phase I and Phase II. Judith Wolf acknowledges support from the UK Natural Environment Research Council. Vittorio E. Brando was supported by the RITMARE Flagship project and the European Union (FP7-427 People Co-funding of Regional, National and International Programmes, GA no. 600407). Lakshmi H. Kantha thanks CNR-ISMAR for providing the opportunity to interact with European oceanographers. All authors gratefully acknowledge the support of *Ocean Science* Executive Editors and Editorial assistants, and the useful suggestions received from John M. Huthnance.

References

Barbariol, F., Falcieri, F. M., Scotton, C., Benetazzo, A., Carniel, S., and Sclavo, M.: Wave extreme characterization using self-organizing maps, Ocean Sci., 12, 403–415, https://doi.org/10.5194/os-12-403-2016, 2016.

Bergamasco, A., Benetazzo, A., Carniel, S., Falcieri, F., Minuzzo, T., Signell, R. P., and Sclavo, M.: From interoperability to knowledge discovery using large model datasets in the marine environment: the THREDDS Data Server example, Adv. Oceanogr. Limnol., 3, 41–50, https://doi.org/10.1080/19475721.2012.669637, 2012.

Bonamano, S., Piermattei, V., Madonia, A., Paladini de Mendoza, F., Pierattini, A., Martellucci, R., Stefanì, C., Zappalà, G., Caruso, G., and Marcelli, M.: The Civitavecchia Coastal Environment Monitoring System (C CEMS): a new tool to analyze the conflicts between coastal pressures and sensitivity areas, Ocean Sci., 12, 87–100, https://doi.org/10.5194/os-12-87-2016, 2016.

Brando, V. E., Braga, F., Zaggia, L., Giardino, C., Bresciani, M., Matta, E., Bellafiore, D., Ferrarin, C., Maicu, F., Benetazzo, A., Bonaldo, D., Falcieri, F. M., Coluccelli, A., Russo, A., and Carniel, S.: High-resolution satellite turbidity and sea surface temperature observations of river plume interactions during a significant flood event, Ocean Sci., 11, 909–920, https://doi.org/10.5194/os-11-909-2015, 2015.

Bricheno, L. M., Wolf, J., and Brown, J.: Impacts of high resolution model downscaling in coastal regions, Cont. Shelf Res., 87, 7–16, 2014.

Carniel, S., Kantha, L. H., Book, J.W., Sclavo, M., and Prandke, H.: Turbulence variability in the upper layers of the Southern Adriatic Sea under a variety of atmospheric forcing conditions, Cont. Shelf Res., 44, 39–56, 2012.

Carniel, S., Bonaldo, D., Benetazzo, A., Bergamasco, A., Boldrin, A., Falcieri, F. M., Sclavo, M., Trincardi, F., Falcieri, F. M., Langone, L., and Sclavo, M.: Off-Shelf Fluxes across the Southern Adriatic Margin: Factors Controlling Dense-Water-Driven Transport Phenomena, Mar. Geol., 375, 44–63, https://doi.org/10.1016/j.margeo.2015.08.016, 2016a.

Carniel, S., Benetazzo, A., Bonaldo, D., Falcieri, F. M., Miglietta, M. M., Ricchi, A., and Sclavo, M.: Scratching beneath the surface while coupling atmosphere, ocean and waves: Analysis of a dense water formation event, Ocean Modell., 101, 101–112, https://doi.org/10.1016/j.ocemod.2016.03.007, 2016b.

Collins, M., Chandler, R. E., Cox, P. M., Huthnance, J. M., Rougier, J., and Stephenson, D. B.: Quantifying future climate change, Nature Climate Change, 2, 304–409, 2012.

Cruz, J.: Ocean wave energy: current status and future perspectives, Green Energy and Technology, Springer, 2008.

Dickey, T.: Emerging ocean observations for interdisciplinary data assimilation systems, J. Marine Syst., 40–41, 5–48, https://doi.org/10.1016/S0924-7963(03)00011-3, 2003.

Fairall, C. W., Bradley, E. F., Hare, J. E., Grachev, A. A., and Edson, J. B.: Bulk parameterization of air–sea fluxes: Updates and verification for the COARE algorithm, J. Climate, 16, 571–591, 2003.

Falcieri, F. M., Kantha, L., Benetazzo, A., Bergamasco, A., Bonaldo, D., Barbariol, F., Malacic, V., Sclavo, M., and Carniel, S.: Turbulence observations in the Gulf of Trieste under moderate wind forcing and different water column stratification, Ocean Sci., 12, 433–449, https://doi.org/10.5194/os-12-433-2016, 2016.

Grifoll, M., Aretxabaleta, A. L., Pelegrí, J. L., and Espino, M.: Temporal evolution of the momentum balance terms and frictional adjustment observed over the inner shelf during a storm, Ocean Sci., 12, 137–151, https://doi.org/10.5194/os-12-137-2016, 2016.

Gutiérrez, O. Q., Filipponi, F., Taramelli, A., Valentini, E., Camus, P., and Méndez, F. J.: On the feasibility of the use of wind SAR to downscale waves on shallow water, Ocean Sci., 12, 39–49, https://doi.org/10.5194/os-12-39-2016, 2016.

Halpern, B. S., Walbridge, S., Selkow, K. A., Kappel, C. V., Micheli, F., D'Agrosa, C., Bruno, J. F., Casey, K. F., Ebert, C., Fox, H. F., Fujita, R., Heinemann, D., Lenihan, H. S., Madin, E. M., Perry, M. T., Selig, E. R., Spalding, M., Steneck, R., and Watson, R.: A global map of human impact on marine ecosystems, Science, 319, 948–952, https://doi.org/10.1126/science.1149345, 2008.

Iuppa, C., Cavallaro, L., Vicinanza, D., and Foti, E.: Investigation of suitable sites for wave energy converters around Sicily (Italy), Ocean Sci., 11, 543–557, https://doi.org/10.5194/os-11-543-2015, 2015.

Kraus, R., Supic, N., and Precali, R.: Factors favouring phytoplankton blooms in the northern Adriatic: towards the northern Adriatic empirical ecological model, Ocean Sci., 12, 19–37, https://doi.org/10.5194/os-12-19-2016, 2016.

Lionello, P., Galati, M. B., and Elvini, E.: Extreme storm surge and wind wave climate scenario simulations at the Venetian littoral, Phys. Chem. Earth, Parts A/B/C, 40–41, 86–92, 2012.

Lanotte, A. S., Corrado, R., Palatella, L., Pizzigalli, C., Schipa, I., and Santoleri, R.: Effects of vertical shear in modelling horizontal oceanic dispersion, Ocean Sci., 12, 207–216, https://doi.org/10.5194/os-12-207-2016, 2016.

Licer, M., Smerkol, P., Fettich, A., Ravdas, M., Papapostolou, A., Mantziafou, A., Strajnar, B., Cedilnik, J., Jeromel, M., Jerman, J., Petan, S., Malacic, V., and Sofianos, S.: Modeling the ocean and atmosphere during an extreme bora event in northern Adriatic using one-way and two-way atmosphere-ocean coupling, Ocean Sci., 12, 71–86, https://doi.org/10.5194/os-12-71-2016, 2016.

Liu, Y., Weisberg, R. H., and He, R.: Sea surface temperature patterns on the West Florida Shelf using growing hierarchical self-organizing maps, J. Atmos. Ocean. Tech., 23, 325–338, https://doi.org/10.1175/JTECH1848.1, 2006.

Lynch, T. P., Morello, E. B., Evans, K., Richardson, A. J., Steinberg, C. R., Roughan, M., Thompson, P., Middleton, J. F., Feng, M., Sherrington, R. B., Brando, V. E., Tilbrook, B., Ridgway, K., Allen, S., Doherty, P., Hill, K., and Moltmann, T. C.: IMOS National Reference Stations: a continental scaled physical, chemical and biological coastal observing system, Plos ONE, 9, e113652, https://doi.org/10.1371/journal.pone.0113652, 2014.

McKiver, W. J., Sannino, G., Braga, F., and Bellafiore, D.: Investigation of model capability in capturing vertical hydrodynamic coastal processes: a case study in the north Adriatic Sea, Ocean Sci., 12, 51–69, https://doi.org/10.5194/os-12-51-2016, 2016.

Mihanovic, H., Vilibic, I., Carniel, S., Tudor, M., Russo, A., Bergamasco, A., Bubic, N., Ljubešic, Z., Vilicic, D., Boldrin, A., Malacic, V., Celio, M., Comici, C., and Raicich, F.: Exceptional dense water formation on the Adriatic shelf in the winter of 2012, Ocean Sci., 9, 561–572, https://doi.org/10.5194/os-9-561-2013, 2013.

Mitchell, S. B., Jennerjahn, T. C., Vizzini, S., and Zhang, W.: Changes to processes in estuaries and coastal waters due to intense multiple pressures–An introduction and synthesis, Estuarine, Coast. Shelf Sci., 156, 1–6, https://doi.org/10.1016/j.ecss.2014.12.027, 2015.

Olita, A., Iermano, I., Fazioli, L., Ribotti, A., Tedesco, C., Pessini, F., and Sorgente, R.: Impact of currents on surface flux computations and their feedback on dynamics at regional scales, Ocean Sci., 11, 657–666, https://doi.org/10.5194/os-11-657-2015, 2015.

O'Neill, A. C., Erikson, L. H., and Barnard, P. L.: Downscaling wind and wave fields for 21st century coastal flood hazard projections in a region of complex terrain, Earth Space Sci., 4, 314–334, https://doi.org/10.1002/2016EA000193, 2017.

Rockel, B. and Woth, K.: Extremes of near-surface wind speed over Europe and their future changes as estimated from an ensemble of RCM simulations, Climatic Change, 81, 267–280, 2007.

Samaras, A. G., Karambas, Th. V., and Archetti, R.: Simulation of tsunami generation, propagation and coastal inundation in the Eastern Mediterranean, Ocean Sci., 11, 643–655, https://doi.org/10.5194/os-11-643-2015, 2015.

Sannino, G., Sanchez Garrido, J. C., Liberti, L., and Pratt, L.: Exchange flow through the Strait of Gibraltar as simulated by a coordinate hydrostatic model and a z-coordinate nonhydrostatic model, in: The Mediterranean Sea: Temporal Variability and Spatial Patterns, John Wiley & Sons Inc., Oxford, UK, 25–50, 2014.

Signell, R. P. and Camossi, E.: Technical note: Harmonising metocean model data via standard web services within small research groups, Ocean Sci., 12, 633–645, https://doi.org/10.5194/os-12-633-2016, 2016.

Thorpe, S. A.: The Turbulent Ocean, Cambridge University Press, Cambridge, UK, 439 pp., 2005

Umgiesser, G., Zemlys, P., Erturk, A., Razinkova-Baziukas, A., Mežine, J., and Ferrarin, C.: Seasonal renewal time variability in the Curonian Lagoon caused by atmospheric and hydrographical forcing, Ocean Sci., 12, 391–402, https://doi.org/10.5194/os-12-391-2016, 2016.

Usui, N., Fujii, Y., Sakamoto, K., and Kamachi, M.: Development of a Four-Dimensional Variational Assimilation System for Coastal Data Assimilation around Japan, Mon. Weather Rev., 143, 3874–3892, https://doi.org/10.1175/MWR-D-14-00326.1, 2015.

Warner, J. C., Armstrong, B., He, R., and Zambon, J. B.: Development of a coupled ocean–atmosphere–wave–sediment transport (COAWST) modeling system, Ocean Model., 35, 230–244, https://doi.org/10.1016/j.ocemod.2010.07.010, 2010.

Williams P. D., Cullen, M. J. P. and, Huthnance, J. M.: How mathematical models can aid our understanding of climate, EOS Transactions of the American Geophysical Union, 92, p. 482, 2011.

PERMISSIONS

LIST OF CONTRIBUTORS

Rosemary Morrow, Alice Carret, Florence Birol and Fernando Nino
LEGOS, IRD, CNRS, Université de Toulouse, Toulouse, 31400, France

Guillaume Valladeau
CLS Ramonville, St.-Agne, 31520, France

Francois Boy
CNES, Toulouse, 31400, France

Celine Bachelier
IRD, Brest, 29280, France

Bruno Zakardjian
Université de Toulon, CNRS, IRD, Mediterranean Institute of Oceanography (MIO), UM 110, 83957 La Garde, France
Aix Marseille Université, CNRS, IRD, Mediterranean Institute of Oceanography (MIO), UM 110, 13288 Marseille, France

Ana Carrasco
Norwegian Meteorological Institute, Henrik Mohns plass 1, 0313 Oslo, Norway

Jean-Raymond Bidlot
European Centre for Medium-Range Weather Forecasts, Shinfield Park, Reading, RG2 9AX, UK

Kai Håkon Christensen
Norwegian Meteorological Institute, Henrik Mohns plass 1, 0313 Oslo, Norway
Department of Geosciences, University of Oslo, Sem Sælands vei 1, 0316, Oslo, Norway

Øyvind Breivik
Norwegian Meteorological Institute, Henrik Mohns plass 1, 0313 Oslo, Norway
Geophysical Institute, University of Bergen, Allégaten 70, 5007, Bergen, Norway

KathrinWahle, Joanna Staneva, Wolfgang Koch, Ha T. M. Ho-Hagemann, and Emil V. Stanev
Institute of Coastal Research, Helmholtz-Zentrum Geesthacht, Geesthacht, Germany

Luciana Fenoglio-Marc
Institute of Geodesy and Geoinformation, University of Bonn, Bonn, Germany

Céline Heuzé
Department of Marine Sciences, University of Gothenburg, Box 115, 405 30 Göteborg, Sweden

Xiao-Dong Shang, and Gui-Ying Chen
State Key Laboratory of Tropical Oceanography, South China Sea Institute of Oceanology, Chinese Academy of Sciences, Guangzhou 510301, China

Chang-Rong Liang
State Key Laboratory of Tropical Oceanography, South China Sea Institute of Oceanology, Chinese Academy of Sciences, Guangzhou 510301, China
University of Chinese Academy of Sciences, Beijing 100049, China

Jiliang Xuan
State Key Laboratory of Satellite Ocean Environment Dynamics, Second Institute of Oceanography, State Oceanic Administration, Hangzhou, China

Daji Huang and Feng Zhou
State Key Laboratory of Satellite Ocean Environment Dynamics, Second Institute of Oceanography, State Oceanic Administration, Hangzhou, China
Ocean College, Zhejiang University, Zhoushan, China

Ruibin Ding
Ocean College, Zhejiang University, Zhoushan, China
State Key Laboratory of Satellite Ocean Environment Dynamics, Second Institute of Oceanography, State Oceanic Administration, Hangzhou, China

Thomas Pohlmann, Jian Su, Bernhard Mayer
Institute of Oceanography, University of Hamburg, Hamburg, Germany

M. M. Amrutha and V. Sanil Kumar
Ocean Engineering Division, CSIR-National Institute of Oceanography (Council of Scientific and Industrial Research), Dona Paula, Goa 403 004, India

Yair De-Leon and Nathan Paldor
Fredy and Nadine Herrmann Institute of Earth Sciences, The Hebrew University of Jerusalem, Edmond J. Safra Campus, Givat Ram, Jerusalem, 9190401, Israel

Guilherme Franz, Lígia Pinto, and Ramiro Neves
MARETEC, Instituto Superior Técnico, Universidade de Lisboa, Av. Rovisco Pais, 1049-001, Lisboa, Portugal

Matthias T. Delpey
Centre Rivages Pro Tech, SUEZ, 2 allée Théodore Monod, Bidart, France

David Brito
ACTION MODULERS, Estrada Principal, no. 29, Paz, 2640-583 Mafra, Portugal

Paulo Leitão
HIDROMOD, Rua Rui Teles Palhinha, no. 4, Leião, 2740-278 Porto Salvo, Portugal

M. Anjali Nair and V. Sanil Kumar
Ocean Engineering Division, Council of Scientific&Industrial Research-National Institute of Oceanography, Dona Paula, 403 004 Goa, India

Joao Marcos Azevedo Correia de Souza
Centro de Investigación Cientifica y de Educación Superior de Ensenada, Baja California (CICESE), Carretera Ensenada-Tijuana No. 3918, Zona Playitas, C.P. 22860, Ensenada, B.C., Mexico
Department of Oceanography, University of Hawaii, 1000 Pope Rd., MSB, Honolulu, 96822 HI, USA

Brian Powell
Department of Oceanography, University of Hawaii, 1000 Pope Rd., MSB, Honolulu, 96822 HI, USA

Shengli Chen, Daoyi Chen, and Jiuxing Xing
Shenzhen Key Laboratory for Coastal Ocean Dynamic and Environment, Graduate School at Shenzhen, Tsinghua University, Shenzhen 518055, China

Yao Fu and Johannes Karstensen
GEOMAR Helmholtz Centre for Ocean Research Kiel, Kiel, Germany

Peter Brandt
GEOMAR Helmholtz Centre for Ocean Research Kiel, Kiel, Germany
Christian-Albrechts-Universität zu Kiel, Kiel, Germany

Reiner Onken
Helmholtz-Zentrum Geesthacht (HZG), Centre for Materials and Coastal Research, Max-Planck-Straße 1, 21502 Geesthacht, Germany

Sandro Carniel
Institute of Marine Science, National Research Council (ISMAR-CNR), Venice, Italy

Judith Wolf
National Oceanography Center, Liverpool, UK

Vittorio E. Brando
Institute of Electromagnetic Sensing of the Environment, National Research Council (IREA-CNR), Milan, Italy
Institute for the Study of Atmosphere and Climate, National Research Council (ISAC-CNR), Rome, Italy

Lakshmi H. Kantha
University of Colorado, Boulder, CO 80309, USA

Index